Lecture Notes in Computer Science   8422

Commenced Publication in 1973
Founding and Former Series Editors:
Gerhard Goos, Juris Hartmanis, and Jan van Leeuwen

Sourav S. Bhowmick   Curtis E. Dyreson
Christian S. Jensen   Mong Li Lee
Agus Muliantara   Bernhard Thalheim (Eds.)

Database Systems
for Advanced Applications

19th International Conference, DASFAA 2014
Bali, Indonesia, April 21-24, 2014
Proceedings, Part II

 Springer

Volume Editors

Sourav S. Bhowmick
Nanyang Technological University, Singapore
E-mail: assourav@ntu.edu.sg

Curtis E. Dyreson
Utah State University, Logan, UT, USA
E-mail: curtis.dyreson@usu.edu

Christian S. Jensen
Aalborg University, Denmark
E-mail: csj@cs.aau.dk

Mong Li Lee
National University of Singapore, Singapore
E-mail: leeml@comp.nus.edu.sg

Agus Muliantara
Udayana University, Badung, Indonesia
E-mail: muliantara@cs.unud.ac.id

Bernhard Thalheim
Christian-Albrechts-Universität zu Kiel, Germany
E-mail: thalheim@is.informatik.uni-kiel.de

ISSN 0302-9743   e-ISSN 1611-3349
ISBN 978-3-319-05812-2   e-ISBN 978-3-319-05813-9
DOI 10.1007/978-3-319-05813-9
Springer Cham Heidelberg New York Dordrecht London

Library of Congress Control Number: 2014934170

LNCS Sublibrary: SL 3 – Information Systems and Application, incl. Internet/Web and HCI

Typesetting: Camera-ready by author, data conversion by Scientific Publishing Services, Chennai, India

Printed on acid-free paper

Springer is part of Springer Science+Business Media (www.springer.com)

Preface

It is our great pleasure to present to you the proceedings of the 19th International Conference on Database Systems for Advanced Applications, DASFAA 2014, which was held in Bali, Indonesia. DASFAA is a well-established international conference series that provides a forum for technical presentations and discussions among researchers, developers, and users from academia, business, and industry in the general areas of database systems, web information systems, and their applications.

The call for papers attracted 257 research paper submissions with authors from 29 countries. After a comprehensive review process, where each paper received at least three reviews, the Program Committee accepted 62 of these, yielding a 24% acceptance rate. The reviewers were as geographically diverse as the authors, working in industry and academia in 27 countries. Measures aimed at ensuring the integrity of the review process were put in place. Both the authors and the reviewers were asked to identify potential conflicts of interest, and papers for which a conflict was discovered during the review process were rejected. In addition, care was taken to ensure diversity in the assignment of reviewers to papers. This year's technical program featured two new aspects: an audience voting scheme for selecting the best paper, and poster presentation of all accepted papers.

The conference program includes the presentations of four industrial papers selected from thirteen submissions by the Industrial Program Committee chaired by Yoshiharu Ishikawa (Nagoya University, Japan) and Ming Hua (Facebook Inc., USA), and it includes six demo presentations selected from twelve submissions by the Demo Program Committee chaired by Feida Zhu (Singapore Management University, Singapore) and Ada Fu (Chinese University of Hong Kong, China).

The proceedings also includes an extended abstract of the invited keynote lecture by the internationally known researcher David Maier (Portland State University, USA). The tutorial chairs, Byron Choi (Hong Kong Baptist University, China) and Sanjay Madria (Missouri University of Science and Technology, USA), organized three exciting tutorials: "Similarity-based analytics for trajectory data: theory, algorithms and applications" by Kai Zheng (University of Queensland, Australia), "Graph Mining Approaches: From Main memory to Map/reduce" by Sharma Chakravarthy (The University of Texas at Arlington, USA), and "Crowdsourced Algorithms in Data Management" by Dongwon Lee (Penn State University, USA). The panel chairs, Seung-won Hwang (Pohang University of Science and Technology, South Korea) and Xiaofang Zhou (University of Queensland, Australia), organized a stimulating panel on database systems for new hardware platforms chaired by Aoying Zhou (East China Normal University, China). This rich and attractive conference program of DASFAA 2014 is accompanied by two volumes of Springer's *Lecture Notes in Computer Science* series.

Beyond the main conference, Shuigeng Zhou (Fudan University, China), Wook-Shin Han (Pohang University of Science and Technology, South Korea), and Ngurah Agus Sanjaya, (Universitas Udayana, Indonesia), who chaired the Workshop Committee, accepted five exciting workshops: the Second International DAS-FAA Workshop on Big Data Management and Analytics, BDMA; the Third International Workshop on Data Management for Emerging Network Infrastructure, DaMEN; the Third International Workshop on Spatial Information Modeling, Management and Mining, SIM^3; the Second International Workshop on Social Media Mining, Retrieval and Recommendation Technologies, SMR; and the DASFAA Workshop on Uncertain and Crowdsourced Data, UnCrowd. The workshop papers are included in a separate proceedings volume also published by Springer in its *Lecture Notes in Computer Science* series.

The conference would not have been possible without the support and hard work of many colleagues. We would like to express our gratitude to the honorary conference chairs, Tok Wang Ling (National University of Singapore, Singapore) and Zainal Hasibuan (University of Indonesia, Indonesia), for their valuable advice on many aspects of organizing the conference. Our special thanks also go to the DASFAA Steering Committee for its leadership and encouragement. We are also grateful to the following individuals for their contributions to making the conference a success:

- General Chairs - Stéphane Bressan (National University of Singapore, Singapore) and Mirna Adriani (University of Indonesia, Indonesia)
- Publicity Chairs - Toshiyuki Amagasa (University of Tsukuba, Japan), Feifei Li (University of Utah, USA), and Ruli Manurung (University of Indonesia, Indonesia)
- Local Chairs - Made Agus Setiawan and I. Made Widiartha (Universitas Udayana, Indonesia)
- Web Chair - Thomas Kister (National University of Singapore, Singapore)
- Registration Chair - Indra Budi (University of Indonesia, Indonesia)
- Best Paper Committee Chairs - Weiyi Meng (Binghamton University, USA), Divy Agrawal (University of California at Santa Barbara, USA), and Jayant Haritsa (Indian Institute of Science, India)
- Finance Chairs - Mong Li Lee (National University of Singapore, Singapore) and Muhammad Hilman (University of Indonesia, Indonesia)
- Publication Chairs - Bernhard Thalheim (Christian-Albrechts-University, Germany), Mong Li Lee (National University of Singapore, Singapore) and Agus Muliantara (Universitas Udayana, Indonesia)
- Steering Committee Liaison - Rao Kotagiri (University of Melbourne, Australia)

Our heartfelt thanks go to the Program Committee members and external reviewers. We know that they are all highly skilled scientists with many demands on their time, and we greatly appreciate their efforts devoted to the timely and careful reviewing of all submitted manuscripts. We also thank all authors for submitting their papers to the conference. Finally, we thank all other individuals who helped make the conference program attractive and the conference successful.

April 2014   Sourav S. Bhowmick
                                                      Curtis E. Dyreson
                                                      Christian S. Jensen

Organization

Honorary Conference Co-Chairs

Tok Wang Ling   National University of Singapore, Singapore
Zainal Hasibuan   University of Indonesia, Indonesia

Conference General Co-Chairs

Stéphane Bressan   National University of Singapore, Singapore
Mirna Adriani   University of Indonesia, Indonesia

Program Committee Co-Chairs

Sourav S. Bhowmick   Nanyang Technological University, Singapore
Curtis E. Dyreson   Utah State University, USA
Christian S. Jensen   Aalborg University, Denmark

Workshop Co-Chairs

Shuigeng Zhou   Fudan University, China
Wook-Shin Han   POSTECH, South Korea
Ngurah Agus Sanjaya   University Udayana, Indonesia

Tutorial Co-Chairs

Byron Choi   Hong Kong Baptist University, Hong Kong
Sanjay Madria   Missouri University of Science & Technology, USA

Panel Co-Chairs

Seung-won Hwang   POSTECH, South Korea
Xiaofang Zhou   University of Queensland, Australia

Demo Co-Chairs

Feida Zhu   Singapore Management University, Singapore
Ada Fu   Chinese University of Hong Kong, Hong Kong

Industrial Co-Chairs

Yoshiharu Ishikawa   Nagoya University, Japan
Ming Hua   Facebook Inc., USA

Best Paper Committee Co-Chairs

Weiyi Meng   Binghamton University, USA
Divy Agrawal   University of California Santa Barbara, USA
Jayant Haritsa   IISc, India

Steering Committee Liaison

R. Kotagiri   University of Melbourne, Australia

Publicity Co-Chairs

Toshiyuki Amagasa   University of Tsukuba, Japan
Feifei Li   University of Utah, USA
Ruli Manurung   University of Indonesia, Indonesia

Publication Co-Chairs

Bernhard Thalheim   Christian-Albrechts-University, Kiel, Germany
Mong Li Lee   National University of Singapore, Singapore
Agus Muliantara   University Udayana, Indonesia

Finance Co-Chairs

Mong Li Lee   National University of Singapore, Singapore
Muhammad Hilman   University of Indonesia, Indonesia

Registration Chairs

Indra Budi   University of Indonesia, Indonesia

Local Co-Chairs

Made Agus Setiawan   University Udayana, Indonesia
I. Made Widiartha   University Udayana, Indonesia

Web Chair

Thomas Kister   National University of Singapore, Singapore

Research Track Program Committee

Nikolaus Augsten	University of Salzburg, Austria
Srikanta Bedathur	IIIT Delhi, India
Ladjel Bellatreche	Poitiers University, France
Boualem Benatallah	University of New South Wales, Australia
Bishwaranjan Bhattacharjee	IBM Research Lab, USA
Cui Bin	Peking University, China
Athman Bouguettaya	CSIRO, Australia
Seluk Candan	Arizona State University, USA
Marco A. Casanova	Pontifícia Universidade Católica do Rio de Janeiro, Brazil
Sharma Chakravarthy	University of Texas Arlington, USA
Chee Yong Chan	National University of Singapore, Singapore
Jae Woo Chang	Chonbuk National University, South Korea
Sanjay Chawla	University of Sydney, Australia
Lei Chen	Hong Kong University of Science & Technology, Hong Kong
James Cheng	Chinese University of Hong Kong, Hong Kong
Reynold Cheng	University of Hong Kong, Hong Kong
Gao Cong	Nanyang Technological University, Singapore
Sudipto Das	Microsoft Research, USA
Khuzaima Daudjee	University of Waterloo, Canada
Prasad Deshpande	IBM India, India
Gill Dobbie	University of Auckland, New Zealand
Eduard C. Dragut	Purdue University, USA
Cristina Dutra de Aguiar Ciferri	Universidade de São Paulo, Brazil
Sameh Elnikety	Microsoft, USA
Johann Gamper	Free University of Bozen-Bolzano, Italy
Shahram Ghandeharizadeh	University of Southern California, USA
Gabriel Ghinita	University of Massachusetts Boston, USA
Le Gruenwald	University of Oklahoma, USA
Chenjuan Guo	Aarhus University, Denmark
Li Guoliang	Tsinghua University, China
Ralf Hartmut Gting	University of Hagen, Germany
Takahiro Hara	Osaka University, Japan
Haibo Hu	Hong Kong Baptist University, Hong Kong
Mizuho Iwaihara	Waseda University, Japan
Adam Jatowt	Kyoto University, Japan
Panos Kalnis	King Abdullah University of Science & Technology, Saudi Arabia
Kamal Karlapalem	IIIT Hyderabad, India
Panos Karras	Rutgers University, USA
Norio Katayama	National Institute of Informatics, Japan
Sangwook Kim	Hanyang University, South Korea

Hiroyuki Kitagawa   University of Tsukuba, Japan
Jae-Gil Lee   KAIST, South Korea
Sang-Goo Lee   Seoul National University, South Korea
Wang-Chien Lee   University of Pennsylvania, USA
Hou U. Leong   University of Macau, China
Ulf Leser   Humboldt University Berlin, Germany
Hui Li   Xidian University, China
Lipyeow Lim   University of Hawai, USA
Xuemin Lin   University of New South Wales, Australia
Sebastian Link   University of Auckland, New Zealand
Bin Liu   NEC Lab, USA
Changbin Liu   AT & T, USA
Boon Thau Loo   University of Pennsylvania, USA
Jiaheng Lu   Renmin University, China
Qiong Luo   Hong Kong University of Science & Technology,
                           Hong Kong
Matteo Magnani   Uppsala University, Sweden
Nikos Mamoulis   University of Hong Kong, Hong Kong
Sharad Mehrotra   University of California Irvine, USA
Marco Mesiti   University of Milan, Italy
Prasenjit Mitra   Penn State University, USA
Yasuhiko Morimoto   Hiroshima University, Japan
Miyuki Nakano   University of Tokyo, Japan
Wolfgang Nejdl   University of Hannover, Germany
Wilfred Ng   Hong Kong University of Science & Technology,
                           Hong Kong
Makoto Onizuka   NTT Cyber Space Laboratories, Japan
Stavros Papadoupoulos   Hong Kong University of Science & Technology,
                           Hong Kong
Stefano Paraboschi   Università degli Studi di Bergamo, Italy
Sanghyun Park   Yonsei University, South Korea
Dhaval Patel   IIT Rourkee, India
Torben Bach Pedersen   Aalborg University, Denmark
Jian Pei   Simon Fraser University, Canada
Jeff Phillips   University of Utah, USA
Evaggelia Pitoura   University of Ioannina, Greece
Pascal Poncelet   Université Montpellier 2, France
Maya Ramanath   IIT New Delhi, India
Uwe Röhm   University of Sydney, Australia
Sherif Sakr   University of New South Wales, Australia
Kai-Uwe Sattler   Ilmenau University of Technology, Germany
Markus Scheider   University of Florida, USA
Thomas Seidl   Aachen University, Germany
Atsuhiro Takasu   National Institute of Informatics, Japan
Kian-Lee Tan   National University of Singapore, Singapore

Nan Tang	Qatar Computing Research Institute, Qatar
Dimitri Theodoratos	New Jersey Institute of Technology, USA
Wolf Tilo-Balke	University of Hannover, Germany
Hanghang Tong	CUNY, USA
Kristian Torp	Aalborg University, Denmark
Vincent Tseng	National Cheng Kung University, Taiwan
Vasilis Vassalos	Athens University of Economics and Business, Greece
Stratis Viglas	University of Edinburgh, UK
Wei Wang	University of New South Wales, Australia
Raymond Wong	Hong Kong University of Science & Technology, Hong Kong
Huayu Wu	Institute for Infocomm Research, Singapore
Yinghui Wu	University of California at Santa Barbara, USA
Xiaokui Xiao	Nanyang Technological University, Singapore
Jianliang Xu	Hong Kong Baptist University, Hong Kong
Bin Yang	Aarhus University, Denmark
Man-Lung Yiu	Hong Kong Polytechnic University, Hong Kong
Haruo Yokota	Tokyo Institute of Technology, Japan
Xike Xie	Aalborg University, Denmark
Jeffrey Xu Yu	Chinese University of Hong Kong, Hong Kong
Aoying Zhou	East China Normal University, China
Wenchao Zhou	Georgetown University, USA
Roger Zimmermann	National University of Singapore, Singapore

Industrial Track Program Committee

Alfredo Cuzzocrea	ICAR-CNR and Unversity of Calabria, Italy
Yi Han	National University of Defense Technology, China
Kaname Harumoto	Osaka University, Japan
Jun Miyazaki	Tokyo Institute of Technology, Japan
Yang-Sae Moon	Kangwon National University, South Korea
Chiemi Watanabe	University of Tsukuba, Japan
Kyoung-Gu Woo	Samsung Advanced Institute of Technology, South Korea
Chuan Xiao	Nagoya University, Japan
Ying Yan	Microsoft Research, Asia, China
Bin Yao	Shanghai Jiaotong University, China

Demonstration Program Committee

| Palakorn Achananuparp | Singapore Management University, Singapore |
| Jing Gao | University at Buffalo, USA |

Yunjun Gao   Zhejiang University, China
Manish Gupta   Microsoft Bing Research, India
Hady Lauw   Singapore Management University, Singapore
Victor Lee   John Carroll University, USA
Zhenhui Li   Penn State University, USA
Siyuan Liu   Carnige Mellon University, USA
Weining Qian   East China Normal University, China
Victor Sheng   University of Central Arkansas, USA
Aixin Sun   Nanyang Technological University, Singapore
Yizhou Sun   Northeastern University, USA
Jianshu Weng   Accenture Analytics Innovation Center,
                                 Singapore
Tim Weninger   University of Notre Dame, USA
Yinghui Wu   University of California at Santa Barbara, USA
Peixiang Zhao   Florida State University, USA

External Reviewers

Ibrahim Abdelaziz   Soumyava Das
Ehab Abdelhamid   Ananya Dass
Yeonchan Ahn   Jiang Di
Cem Aksoy   Aggeliki Dimitriou
Amin Allam   Lars Döhling
Yoshitaka Arahori   Philip Driessen
Nikolaos Armenatzoglou   Ines Faerber
Sumita Barahmand   Zoé Faget
Christian Beecks   Qiong Fang
Brigitte Boden   Xing Feng
Selma Bouarar   Sergey Fries
Ahcene Boukorca   Chuancong Gao
Sebastian Breß   Ming Gao
Yilun Cai   Azadeh Ghari-Neat
Yuanzhe Cai   Gihyun Gong
Jose Calvo-Villagran   Koki Hamada
Mustafa Canim   Marwan Hassani
Brice Chardin   Sven Helmer
Wei Chen   Silu Huang
Sean Chester   Fuad Jamour
Ricardo Rodrigues Ciferri   Min-Hee Jang
Xu Cui   Stéphane Jean

Minhao Jiang
Salil Joshi
Akshar Kaul
Georgios Kellaris
Selma Khouri
Jaemyung Kim
Henning Koehler
Hardy Kremer
Longbin Lai
Thuy Ngoc Le
Sang-Chul Lee
Hui Li
John Liagouris
Wenxin Liang
Xumin Liu
Cheng Long
Yi Lu
Yu Ma
Zaki Malik
Xiangbo Mao
Joseph Mate
Jun Miyazaki
Basilisa Mvungi
Adrian Nicoara
Sungchan Park
Youngki Park
Paolo Perlasca
Peng Peng
Jianbin Qin
Lizhen Qu
Astrid Rheinländer
Avishek Saha
Shuo Shang
Jieming Shi
Juwei Shi
Masumi Shirakawa
Md. Anisuzzaman Siddique

Thiago Luís Lopes Siqueira
Guanting Tang
Yu Tang
Aditya Telang
Seran Uysal
Stefano Valtolina
Jan Vosecky
Sebastian Wandelt
Hao Wang
Shenlu Wang
Xiang Wang
Xiaoyang Wang
Yousuke Watanabe
Huanhuan Wu
Jianmin Wu
Xiaoying Wu
Fan Xia
Chen Xu
Yanyan Xu
Zhiqiang Xu
Mingqiang Xue
Da Yan
Shiyu Yang
Yu Yang
Zhen Ye
Jongheum Yeon
Adams Wei Yu
Kui Yu
Qi Yu
Chengyuan Zhang
Zhao Zhang
Zhou Zhao
Jingbo Zhou
Xiangmin Zhou
Linhong Zhu
Anca Zimmer
Andreas Zuefle

Table of Contents – Part II

Data Mining

Spatio-temporal Data Management

Graph Data Management

Security, Privacy and Trust

Web and Social Data Management

Keyword Search

Data Stream Management

Data Quality

Industrial Papers

Demo Papers

Tutorials

Table of Contents – Part I

Data Mining

Probabilistic and Uncertain Data Management

Web and Social Data Management

Ensemble Pruning: A Submodular Function Maximization Perspective

Chaofeng Sha[1], Keqiang Wang[2], Xiaoling Wang[2], and Aoying Zhou[2]

[1] School of Computer Science
Shanghai Key Laboratory of Intelligent Information Processing
Fudan University, Shanghai 200433, China
[2] Shanghai Key Laboratory of Trustworthy Computing,
East China Normal University, Shanghai 200062, China
cfsha@fudan.edu.cn

Abstract. Ensemble pruning looks for a subset of classifiers from a group of trained classifiers to make a better prediction performance for the test set. Recently, ensemble pruning techniques have attracted significant attention in the machine learning and the data mining community. Unlike previous heuristic approaches, in this paper we formalize the ensemble pruning problem as a function maximization problem to strike an optimal balance between quality of classifiers and diversity within the subset. Firstly, a quality and pairwise diversity combined framework is proposed and the function is proved to be submodular. Furthermore, we propose a submodular and monotonic function which is the composition of both quality and entropy diversity. Based on the theoretical analysis, although this maximization problem is still NP-hard, the greedy search algorithm with approximation guarantee of factor $1 - \frac{1}{e}$ is employed to get a near-optimal solution. Through the extensive experiments on 36 real datasets, our empirical studies demonstrate that our proposed approaches are capable of achieving superior performance and better efficiency.

Keywords: ensemble pruning, pairwise diversity, entropy, submodularity, greedy algorithm.

1 Introduction

Given the ensemble of trained individual learners (classifiers), rather than combining all of them, ensemble pruning tries to select a subset of individual learns (sub-ensemble) to comprise the ensemble [23].Usually an ensemble is significantly more accurate than a single classifier [23]. Representative ensemble methods include AdaBoost [6], Bagging [2], and random decision trees [5], to name a few. In spite of the significant contributions made by those work, the ensembles generated by existing techniques such as Bagging are sometimes unnecessarily large, which can lead to large memory cost, computational costs, and occasional decreases in effectiveness [21,23].

S.S. Bhowmick et al. (Eds.): DASFAA 2014, Part II, LNCS 8422, pp. 1–15, 2014.

Former work on ensemble pruning can be categorized into clustering based approaches, ordering based methods, and optimization based solutions [21]. In the last category, the ensemble pruning is viewed as a combinatorial optimization problem with the goal to find a sub-ensemble to optimize a predefined criterion [21].

Different from previous optimization approaches, such as those based on semi-definite programming, we formalize the ensemble pruning as a submodular function maximization problem to obtain a set of k classifiers from the collection of trained classifiers that can well generalize to the test set. The submodular function can be used to characterize diminishing return property that the marginal gain of adding an element to a smaller subset is higher than that of adding it to its supersets. Consider a set function $F : 2^E \to R$, which maps subsets $S \subseteq E$ of a finite ground set E to real numbers. F is called submodular if, $\forall S \subseteq T \subseteq E$, $\forall e \in E \setminus T$, $F(S \cup \{e\}) - F(S) \geq F(T \cup \{e\}) - F(T)$. Submodular function maximization technique has been used for problems such as influence maximization [9], feature selection [11], active learning [8], and dictionary selection [10]. **This paper develops the ensemble pruning as an submodular function maximization problem.**

This subset selection problem is a combinatorial optimization problem and thus finding the exact optimal solution is NP-hard [19]. So this maximization problem turns to be a challenging problem, even for some simple submodular functions, such as mutual information [11]. Fortunately, the greedy algorithm can be used to find a $1 - \frac{1}{e}$ approximation to the optimal solution when the submodular function is normalized and monotone [17]. To measure the diversity among classifiers from different aspects, the pairwise diversity and entropy diversity are employed. We also conduct the submodularity analysis and prove that our object function with entropy diversity is submodular and monotonic. Based on these theoretical analysises, we are able to get a good approximate solution efficiently by best-first greedy search algorithm.

The main contributions can be summarized as follows.

1. We formalize the ensemble pruning as a submodular function maximization problem.
2. We propose two different ensemble pruning framework: (1) quality and pairwise diversity combined approach, (2) quality and entropy combined method. At the same time, we conduct the theoretical submodularity analysis for the maximization problem with these objective functions.
3. Based on the theoretical analysis, we design the efficient greedy algorithm to optimize our objectives.
4. We conduct extensive experiments on real datasets, which include performance comparisons with state-of-the-arts and time cost. The experimental results demonstrate the efficiency and effectiveness of our proposed solutions.

The rest of the paper is organized as follows. In Section 2, the related work is reviewed. Section 3 is devoted to the statement of the problem, the formulation of the framework and the proposed algorithm. Then we show the ensemble pruning process in detail including search strategy and parameter selection in Section 4. Experimental evaluation using real data are shown in Section 5. We conclude with future work in Section 6.

2 Related Work

In this section, we review some previous work, which can be categorized into clustering-based approaches, ordering-based methods and optimization-based solutions.

[15] is the first work to prune AdaBoost on this topic. Instead of pruning ensembles generated by sequential methods, [25] and [3] respectively studied on pruning ensembles generated by parallel methods such as Bagging [2] and parallel heterogeneous ensembles consisting of different types of individual classifiers, and it was shown that better performance can be obtained at smaller ensemble sizes [23].

[16] and [14] are two representative ordering-based ensemble methods. Based on the assumption that near-optimal subensemble of increasing size can be constructed incrementally by incorporating at each step the classifier that is expected to produce the maximum reduction in the generalization error, the ordering-based approach proposed in [16] to ensemble pruning is to modify the original random aggregation ordering in the ensemble. Ensemble Pruning via Individual Contribution Ordering (EPIC) is introduced in [14], which orders individual classifiers in an ensemble in terms of their importance to subensemble construction.

From optimization perspective, [13] proposes a regularized selective ensemble algorithm that chooses the weights of the base classifiers through minimizing a regularized risk function, which can be formulated as a quadratic program with an ℓ^1-norm constraint on the weight vector. In [22], the ensemble pruning problem formulated as a strict mathematical programming problem and the semi-definite programming relaxation techniques is used to obtain a good approximate solution. This work is followed by [21] which also formulates ensemble pruning as an optimization problem. Both individual accuracy and the pairwise diversity of the subensemble are also combined into their pruning criterion. A relaxation of the original integer programming is transformed into a constrained eigen-optimization problem, which can be solved efficiently with an iterative algorithm with global convergence guarantee [21].

Recently, [12] presents the first PAC-style analysis on the effect of diversity in voting. Guided by this result, a greedy ensemble pruning method called DREP is proposed in [12] to explicitly exploit the diversity regularization. However, they focus on the binary classification. A general framework for the greedy ensemble selection algorithm by abstracting the main 27 aspects of existing methods is proposed in [18]. These aspects are the direction of search, the evaluation dataset, the evaluation measure and the size of the final ensemble. In [24], the authors employ the mutual information $I(X_S; Y)$ as the objective function which composes of mutual information between the class label and classifiers and the (conditional) multi-information between classifiers. However, as pointed in [11], the mutual information is not submodular.

Different from the existing work, we present our work through a submodular function maximization perspective and we design the pairwise diversity regularization into our objective which is proved to be submudular and monotonic. Based on these theoretical results, we employ the greedy algorithm to optimize it to get the results with a $1 - \frac{1}{e}$ approximation to the optimal solution.

3   Ensemble Pruning as Submodular Function Maximization

In this section, we review the formal setting of ensemble pruning problem firstly. Then we propose two unified framework to solve the ensemble pruning problem. The first one combines the quality and pairwise diversity of classifiers which is submodular while not monotonic. Then a submodular and monotonic one combining quality and entropy diversity is developed.

3.1   Ensemble Pruning Definition

Let $D = \{(x_i, y_i), i = 1, 2, \cdots, N\}$ be a labeled training set where each example consists of a feature vector x_i and a class label $y_i \in \{1, \cdots, C\}$. $H = \{h_t, t = 1, 2, \cdots, T\}$ is the set of classifiers or hypotheses of an ensemble, where each classifier h_t maps an instance x to a class label y, $h_t(x) = y$. In the following sections, in some sense of abuse of notation, we use $h_i \in H$ or $i \in H$. $S \subseteq H$, is the current subensemble. After pruning, each example x is classified using the majority rule $y = \arg\max_c |\{i \in S : h_i(x) = c\}|$. Let $q(h)$ denote the quality of the classifier h (i.e.we use the accuracy to measure the quality), and $sim(h_i, h_j)$ denote the similarity between two classifiers h_i and h_j.

Ensemble pruning is to obtain a set of k classifiers from the collection of trained classifiers that can best predict the test set. There are three questions need to consider: (1) how to measure the quality $q(h_i)$ of each classifier; (2) how to measure the diversity between classifiers; (3) how to balance between quality and diversity, and find out the top-k classifiers as the representative ones efficiently?

In order to get the mathematical formulation of the ensemble pruning problem, we give one approximation of error or accuracy of the overall ensemble through the combination of quality of individual classifier and diversity or similarity between them. Minimizing (maximizing) this approximate ensemble error (accuracy) function is the objective of the optimization formulation.

3.2   Ensemble Pruning with Pairwise Diversity

Firstly we discuss the following ensemble pruning framework through submodular function maximization perspective:

$$\max_{S \subseteq H:|S|=k} f_{within}(S) := \sum_{i \in S} q(h_i) - \lambda \sum_{i,j \in S, i \neq j} sim(h_i, h_j). \tag{1}$$

The intuition is to choose a subset of classifiers with high quality and with low similarity between the classifiers within the subensemble. The regularization parameter λ is used to balance the quality and diversity which can be set through cross-validation. This setting is justified in [12]. Below we show that $f_{within}(\cdot)$ is submodular.

Theorem 1. *The function defined in (1) is submodular.*

Proof. For any $S \subseteq T \subseteq H$ and $i \notin T$ we have

$$
\begin{aligned}
& f_{within}(S \cup \{i\}) - f_{within}(S) \\
&= \sum_{j \in S \cup \{i\}} q(h_j) - \lambda \sum_{j,k \in S \cup \{i\}} sim(h_j, h_k) - \sum_{j \in S} q(h_j) + \lambda \sum_{j,k \in S} sim(h_j, h_k) \\
&= q(h_i) - \lambda \sum_{j \in S} sim(h_i, h_j) \geq q(h_i) - \lambda \sum_{j \in T} sim(h_i, h_j) \\
&= f_{within}(T \cup \{i\}) - f_{within}(T).
\end{aligned}
$$

However, the function would not be monotone thanks to the fact that the marginal gain $q(h_i) - \lambda \sum_{k \in S} sim(h_i, h_k)$ would not be nonnegative. Therefore the greedy algorithm's constant-factor approximation guarantee of [17] does not apply in this case.

When implementing this ensemble pruning framework, we employ accuracy as the quality of individual classifier. For the pairwise diversity between classifiers, we incorporate two measures: (1) the first one, $\widetilde{G}_{ij} = \frac{1}{2} \left(\frac{G_{ij}}{G_{ii}} + \frac{G_{ij}}{G_{jj}} \right)$, is defined in [22], where G_{ii} is the total number of errors made by classifier i and G_{ij} is the number of common errors of classifier pair i and j; (2) the second one is transformed from a $0/1$ loss based disagreement measure, which was proposed by Ho [7], to characterize the pair-wise diversity for ensemble members [14]. Given two classifiers h_i and h_j, let $N^{(01)}$ denote the number of data points incorrectly predicted by h_i but correctly predicted by h_j, and $N^{(10)}$ is the opposite of $N^{(01)}$. The diversity of h_i and h_j, denoted by $Div_{i,j} = \frac{N^{(01)} + N^{(10)}}{N}$, is the ratio between the sum of the number of data points correctly predicted by one of the classifiers only and the total number of data points. This measure has some resemblance to $\widetilde{G}_{i,j}$. We transform it to a similarity measure as $sim(h_i, h_j) = 1 - Div_{i,j}$.

3.3   Ensemble Pruning with Entropy

The above framework with pairwise diversity measures is submodular while not monotonic. In this subsection, we propose another ensemble pruning method with entropy as the diversity measure, which is submodular and monotonic.

The joint entropy [4] between subset of classifiers $H_S = \{h_1, \cdots, h_{|S|}\}$ is denoted as

$$
Ent(H_S) = - \sum_{z \in \{0,1\}^{|S|}} p(z) \log p(z) \tag{2}
$$

where the random variable z_i is the indicator variable i.e. $z_i = 1\{h_i(x) = y\}$ for example (x, y).

The entropy is often used to measure the diversity or uncertainty between random variables. When $|S| = 2$, it measure the joint entropy between classifiers such as h_1 and h_2, where $p(0,0)$ denotes the probability that both classifiers make an error on a random example. Now we arrive at the following objective function composed of the quality of the classifiers and entropy between them:

$$\max_{S \subseteq H: |S|=k} f_{ent}(S) := \sum_{i \in S} q(h_i) + \lambda Ent(H_S). \tag{3}$$

Theorem 2. *The function defined in (3) is monotonic and submodular.*

Proof. For any $S \subseteq T \subseteq H$ and $i \notin T$ we have

$$f_{ent}(S \cup \{i\}) - f_{ent}(S)$$
$$= \sum_{j \in S \cup \{i\}} q(h_j) + \lambda Ent(H_{S \cup \{i\}}) - \sum_{j \in S} q(h_j) - \lambda Ent(H_S)$$
$$= q(h_i) + \lambda Ent(h_i | H_S) \geq q(h_i) + \lambda Ent(h_i | H_T)$$
$$= f_{ent}(T \cup \{i\}) - f_{ent}(T).$$

Therefore we know that $f_{ent}(\cdot)$ is monotonically increasing thanks to the non-negativity of both terms, $q(h_i)$ and $Ent(h_i | H_S)$. And the inequality is due to the fact that $Ent(Z|X, Y) \leq Ent(Z|X)$ [4]. The submodularity of $f_{ent}(\cdot)$ is followed.

According to Theorem 2, it is feasible to employ the greedy algorithm to solve this NP-hard problem with a $1 - \frac{1}{e}$ approximation to the optimal solution. In the following section, we design a best-search greedy algorithm to this ensemble pruning problem.

4 Search Strategy

After proposing two ensemble pruning framework as submodular function maximization, we describe the search strategy to optimize the objectives in this section. As shown that the problem is in general NP-hard, therefore we are only able to solve them in polynomial time through approximate algorithm.

The whole ensemble pruning process is outlined in Figure 1. In the experiments, we use a bagging ensemble of l decision trees and have a training set to train and a validation subset to prune. We vary the parameter $\lambda \in [0, 1]$ with decimal steps. Using the pruning set, we can compute the accuracy of every classifier and $sim(h_i, h_j)$ for all $h_i, h_j \in H$. Then we order the classifiers for every λ value by using Best-First Search strategy and then we can compute the best λ for different size$(1, 2, ..., l)$ of S. We use greedy algorithm to obtain the ordered classifiers list L_λ for every λ (Line 1-8). Then we exhaust every λ to get best S_k and λ_k for every size of S (Line 9-13). The whole search process is outlined in Algorithm 1.

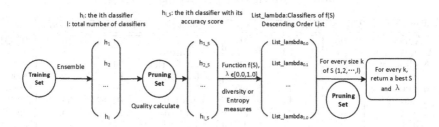

Fig. 1. Ensemble Pruning Process

Algorithm 1. SubmEP approach

Input : Set of classifiers H, f
Output: $S_1, S_2, \cdots, S_l \subseteq H$ and $\lambda_{|S|}$ for $|S| = 1, 2, ..., l$

1 **for** $\lambda = 0.0$ *to* 1.0 **do**
2 $\quad$ $L_\lambda \leftarrow \{\arg\max_{h_i \in H} q(h_i)\}$;
3 $\quad\quad\quad$ /*L_λ is a order list of classifiers*/
4 $\quad$ **while** $|L_\lambda| < l$ **do**
5 $\quad\quad$ $i \leftarrow \arg\max_{i' \in H \setminus L_\lambda} f(L_\lambda \cup \{i'\}) - f(L_\lambda)$
6 $\quad\quad$ $L_\lambda \leftarrow L_\lambda \cup \{i\}$
7 $\quad$ **end**
8 **end**
9 **for** $k = 1$ *to* l **do**
10 $\quad$ $S_k \leftarrow$ select $L_{\lambda,k}$ that $\max_\lambda$ accuracy of $L_{\lambda,k}$ for pruning set
11 $\quad\quad\quad$ /*$L_{\lambda,k}$ is the top k classifiers of L_λ*/
12 $\quad$ $\lambda_k \leftarrow \arg\max_\lambda$ accuracy of $L_{\lambda,k}$ for pruning set
13 **end**
14 **return** S

---

5 Experiments

We conduct extensive experiments on real datasets to study the performance of the proposed frameworks and search algorithms.

5.1 Experimental Settings

We use 36 real datasets selected from the UCI Machine Learning Repository [1]. The detailed statistics of the these datasets are given in Table 1.

We compare the proposed method (named as SubmEP hereafter) against the three baseline approaches:

1. Bagging[2]. Bagging is used as the original ensemble in previous work;
2. OrderedBagging [16]. OrderedBagging is based on bagging in descending order of accuracy;
3. Eigcons[21]. We implement this method according to [25].

Table 1. A breif description of the data sets

Data Set	Classes	Dimensions	Size	Data Set	Classes	Dimensions	Size
Anneal	6	38	898	Letter	26	16	20000
Arrhythmia	16	279	452	Lymph	4	18	148
Audiology	24	69	226	Mfeat-fourier	10	76	2000
Autos	7	25	205	Mfeat-pixel	10	240	2000
Balance Scale	3	4	625	Nursery	5	8	12960
Balloons	2	4	76	Optdigits	10	64	5620
Breast-w	2	9	699	Pendigits	10	16	10992
Bridges	6	11	108	Primary-tumor	22	17	339
Car	4	6	1728	Segment	7	19	2310
Cmc	3	9	1473	Sonar	2	60	208
Credit-g	2	20	1000	Soybean	19	35	683
Dermatology	6	34	366	Spambase	2	57	4601
Flag	6	27	194	Splice	3	61	3190
Glass	7	9	194	Tae	3	5	151
Heart-h	5	13	294	Tic-tac-toe	2	9	958
Hypothyroid	4	29	3772	Vehicle	4	18	846
Ionosphere	2	34	351	Vowel	11	13	990
Kr-vs-kp	2	36	3196	Wine	3	13	178

Two pairwise diversity measures and entropy measure are used in the proposed SubmEP method, and there are :

1. SubmEP_$\widetilde{G}$. The common error based pairwise diversity measure discussed in Section 3.2 is used in our proposed SubmEP function;
2. SubmEP_div. Another pairwise diversity measure proposed in Section 3.2 is used in our SubmEP;
3. SubmEP_ent. The entropy-based measure discussed in Section 3.3 is used in the proposed SubmEP.

In the experiments, the base classifier we use is J48, which is a Java implementation of C4.5 decision tree method in Weka [20]. All the approaches including the baselines are implemented in Java and run on a PC with Intel Duo-Core 2.66GHz 2 CPUs, one thread and 8GB memory.

The experimental setting follows [14]. Each data set was randomly divided into three subsets with equal sizes. There are six permutations of the three subsets. Experiments on each data set consisted of six sets of sub-experiments. Each set of sub-experiments used one of the subsets as the training set, one of the subsets as the testing set, and the other one as the pruning set, corresponding to the order of one of the six permutations. And each set of subexperiments consisted of 50 independent trials. Therefore a total of $500 \times \binom{3}{2} = 300$ trials of experiments are conducted on each data set. In each trial, a bagging ensemble of 100 decision trees is trained then pruned, the subensemble is then evaluated on testing set.

There is one parameter λ in our algorithm to be determined. For each λ, we explore results obtained through varying the parameter in the range [0,1] with decimal steps and get the best λ by verifying the error rate on pruning set.

5.2   Performance Evaluation

(1). Efficiency Comparison

In the first set of experiments, we study the efficiency of the proposed SubmEP approach with different pairwise diversity measures and entropy measure. Two kinds of experiments are conducted: (1) running time in the same data set with different number of classifiers; (2) different data sets with the same number of classifiers. In both experiments, the pruning ratio is set 20% and the average running time is reported here. For different optimization methods and different diversity measures, the running time with Eigcons or SubmEP method have different measurements. The formulas of running time are calculated as follows:

1. For Eigcons method,
$$T_{eigcons} = t_{eigen\_opt}.$$

We only consider the time of eigen-optimization algorithm as in [21].

2. For our SubmEP_div approach,
$$T_{subm\_pw} = \sum_{\lambda \in \{0.0, 0.1, \cdots, 1.0\}} t_\lambda / l + t_{select\_best\_\lambda},$$

where $t_{select\_best\_\lambda}$ is the running time of line 9-12, $\sum_{\lambda \in \{0.0, 0.1, \cdots, 1.0\}} t_\lambda$ is the running time of line 1-7 in Algorithm 1 respectively and l is the number of classifiers. Here, because we can compute $\sum_{i,j \in S, i \neq j} sim(h_i, h_j)$ incrementally, we can finish the computation of diversity from 1 to l classifiers one time. So for the pruning ratio is set 20%, we only need the diversity of $l/5$ classifiers and have t_λ / l running time averagely for every λ. But it is not possible to use the same method to compute entropy diversity. So the running time for SubmEP_ent is as follows.

3. For our SubmEP_ent approach,
$$T_{subm\_ent} = \sum_{\lambda \in \{0.0, 0.1, \cdots, 1.0\}} t_{\lambda, k} + t_{select\_best\_\lambda},$$

where $\sum_{\lambda \in \{0.0, 0.1, \cdots, 1.0\}} t_{\lambda, k}$ is the running time of line 1-7 in algorithm 1 when we set $l = k$ and k is the size of subset.

And this procedure is repeated 50 times then the average running time is reported. Table 2 and Table 3 show the running time with different situations. For

Table 2. Running Time(s) with Different Number of Classifiers

Number of classifiers	50	100	150
Eigcons	0.206	1.029	2.837
SubmEP_$\widetilde{G}$	0.015	0.035	0.061
SubmEP_div	0.018	0.034	0.053
SubmEP_ent	0.404	1.50	3.28

Table 3. Running Time(s) with Different Data Sets

Data set	Sonar	Splice	Letter
Eigcons	1.029	1.705	3.99
SubmEP_$\widetilde{G}$	0.035	0.256	1.523
SubmEP_div	0.034	0.246	1.640
SubmEP_ent	1.50	23.69	146.2

the first situation, we use only one "sonar" data set. From Table 2, it can be seen that the running time of Eigcons grows more fast with more classifiers. While the others except SubmEP_ent have a slower growth trend and the running time is much less than Eigcons method. For SubmEP_ent, the running time is worse than others. The reason is that the estimation of $Ent(H_S)$ is time consuming. In searching procedure, we calculate entropy of different subsets $k * l$ time, while we only need to calculate pairwise $sim(h_i, h_j)$ once.

The running time on three data sets with different sizes are shown in Table 3. Although the running time of SubmEP method with pairwise diversity measures is also less than the Eigcons method, the running time will be longer than the later when the size of data set is large (for example, the size maybe 10^5 or 10^6). It means that the running time of SubmEP method with pairwise diversity measures is growing faster than Eigcons method when the size of dataset become larger. And the reason is that when we estimate the parameter λ, it spends a lot of time on voting the class label. So in the future, we should use some methods to reduce the time of voting. Noting that the running time of SubmEP method with entropy measure is becoming much longer with larger datasets, such as "Splice" or "letter".

(2). Effectiveness with Different Similarity Measures

Next we study the effectiveness of our proposed methods. We use 36 data sets from the UCI machine learning for evaluations whose statistics is described in Table 1. We conduct experiments to evaluate the effectiveness of SubmEP method with different diversity measures on these 36 UCI data sets.

The effectiveness comparison is shown in Figure 2. It can be seen that the effectiveness of our method with $\widetilde{G}$ and $1 - Div$ measure is better than with entropy diversity. When the number of classifiers is very small, the SubmEP_ent method has a good results. However the SubmEP_ent performs worse when the number of classifiers grows. The sizes of data sets in table 1 is less than 20,000, which leads to error in the estimation of entropy. The value of $Ent(H_S)$ is almost equal to $Ent(H_{S \cup \{i\}})$ when the size of S is larger than 14^1, because the average of every situation[2] is less than one and maybe many ones don't appear at all. And this situation leads to that uncertainty of classifiers' results is not large. It is to say that the entropy diversity of classifiers is not obvious when the number of classifiers is large. We can find that the computation cost of the entropy is growing fast as the size of data set becomes very large. So the SumbEP with

---

[1] $\log 20000 \approx 14.29$.

[2] If the size is k, there are 2^k situations for joint entropy.

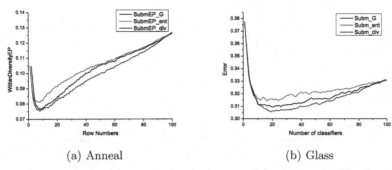

(a) Anneal   (b) Glass

Fig. 2. Comparison of prediction errors on data sets "Anneal" and "Glass" with pairwise diversity and entropy measures

Table 4. Classification error percentages of bagging, Eigcons and SubmEP_$\widetilde{G}$ based on two pairwise diversity measures or entropy measure

Data Set	Bagging(full)	Eigcons	SubmEP_$\widetilde{G}$	SubmEP_div	SubmEP_ent
Anneal	12.70±3.11	8.86±2.46	8.14±2.32	8.47±2.30	9.06± 2.24
Arrhythmia	28.20±3.44	27.72±3.20	27.37±3.14	27.57±3.07	27.64 ± 3.25
Audiology	30.53±5.74	27.51±5.72	26.54±5.40	26.76±5.44	27.79 ± 5.51
Autos	36.98±7.05	32.58±6.46	32.35±6.55	32.48±6.45	33.05±6.68
Balance Scale	17.05±2.70	16.00±2.37	15.58±2.40	15.72±2.37	16.21± 2.56
Balloons	32.47±8.71	29.77±8.84	28.89±9.01	29.32±8.75	31.53 ± 8.98
Breast-w	6.31±2.18	5.55±1.59	5.24±1.48	5.22±1.47	5.10 ± 1.45
Bridges	41.53±8.52	39.01±7.91	38.59±7.52	38.72±7.81	38.75±7.99
Car	12.57±1.67	11.27±1.61	11.03±1.57	11.15±1.57	11.56 ± 1.59
Cmc	47.83±2.06	48.02±1.99	47.73±1.96	47.76±2.05	47.55 ± 2.01
Credit-g	26.48±2.33	26.32±2.25	26.08±2.10	26.14±2.09	26.13±2.16
Dermatology	7.77±3.81	5.12±2.62	4.79±2.51	5.02±2.52	5.17 ± 2.63
Flag	36.19±6.83	32.81±5.99	32.44±6.03	32.45±5.86	33.68±6.21
Glass	33.07±5.66	30.92±5.04	30.66±5.08	33.30±6.36	31.63±5.06
Heart-h	19.63±3.62	19.14±3.45	19.22±3.58	19.51±3.61	19.51 ± 3.58
Hypothyroid	0.72±0.29	0.63±0.28	0.60±0.24	0.58±0.24	0.60 ± 0.24
Ionosphere	9.06±2.79	7.80±2.50	7.65±2.40	7.78±2.39	8.26 ± 2.51
Kr-vs-kp	1.38±0.57	1.06±0.41	0.97±0.38	0.98±0.39	1.03±0.41
Letter	10.04±0.61	10.13±0.54	10.09±0.54	10.10±0.53	10.24±0.54
Lymph	22.85±5.91	20.80±5.55	20.52±5.60	20.78±5.51	21.86±5.97
Mfeat-fourier	21.56±1.67	21.43±1.57	21.34±1.57	21.41±1.59	21.65±1.60
Mfeat-pixel	18.74±2.84	15.21±2.66	15.13±2.64	15.11±2.64	15.82±2.69
Nursery	4.86±0.42	4.53±0.42	4.46±0.41	4.48±0.40	4.56±0.39
Optdigits	5.07±0.80	4.60±0.62	4.58±0.61	4.63±0.64	4.74±0.64
Pendigits	2.69±0.35	2.44±0.31	2.42±0.30	2.47±0.30	2.51±0.31
Primary-tumor	61.08±4.31	60.46±4.11	60.14±4.11	60.17±4.17	60.23±3.90
Segment	4.67±0.92	4.09±0.86	4.03±0.86	4.08±0.85	4.24±0.85
Sonar	25.63±4.95	24.42±5.02	24.07±4.98	24.25±4.80	25.67±4.94
Soybean	13.34±2.87	10.74±2.54	10.50±2.48	10.74±2.50	11.25±2.56
Spambase	7.01±0.69	6.73±0.67	6.65±0.69	6.71±0.72	6.79±0.66
Splice	7.56±0.85	7.00±0.81	6.85±0.81	6.87±0.81	7.02±0.83
Tae	56.45±7.65	53.76±7.00	53.33±7.01	53.25±6.95	53.76±7.17
Tic-tac-toe	16.20±2.82	13.51±2.50	13.36±2.49	13.72±2.45	14.45±2.48
Vehicle	27.61±2.62	27.01±2.50	27.04±2.54	26.96±2.46	27.24±2.56
Vowel	24.26±3.18	22.88±2.84	22.82±2.93	22.72±3.08	23.64±2.95
Wine	8.83±4.64	5.41±3.30	5.36±3.14	5.41±3.23 8	6.67±3.99

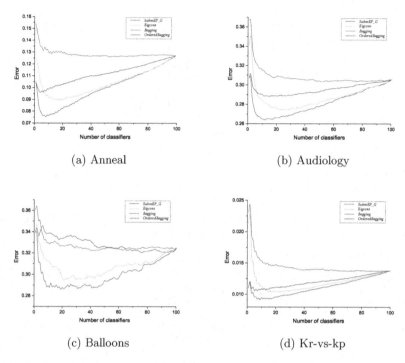

(a) Anneal

(b) Audiology

(c) Balloons

(d) Kr-vs-kp

Fig. 3. Comparison of prediction errors on data sets "Anneal","Autos", "Balloons" and "Kr-vs-kp"

entropy measure is not desirable unless we can find a approximate way to solve the above problems.

(3). Effectiveness Comparison with Eigcons Methods

In this part, we test SubmEP with three measures: $\widetilde{G}$,$1 - Div$ and Entropy measures. In these experiments, we set the size of the subensembles to 20%, due to the observation on the results in Figure 3 and 4. Table 4 shows the mean and standard deviation of prediction error on 300 trials.

In most cases both Eigcons method and our methods outperform Bagging as observed in previous ensemble pruning work. It can also been observed that the results of the SubmEP with different pairwise diversity measures do not differ too much. On most data sets, our proposed submEP_$\widetilde{G}$ outperforms all others.

As in [21], we vary the sizes of subensembles from 1 to the size of the full ensemble, 100, and then plot the error curves of compared algorithms with the increase of the number of decision trees included in the subensemble in Figure 3 and 4. Figure 3 shows that SubmEP_$\widetilde{G}$ outperforms other ensemble pruning methods. Our methods have lower prediction error when the size of subensemble is small. This is obvious in data sets "Anneal", "Audiology" and "Kr-vs-kp". It can be seen that these data sets have a common phenomenon that when the size of subensembles is 1, the performance has been good even by the classifier with

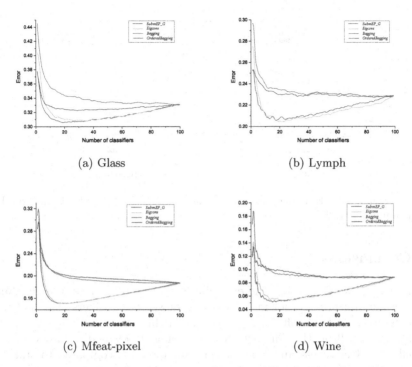

(a) Glass

(b) Lymph

(c) Mfeat-pixel

(d) Wine

Fig. 4. Comparison of prediction errors on data sets "Glass", "Lymph", "Mfeat-pixel" and "Wine"

highest accuracy on pruning set. However our methods also show an outstanding result when the size of subensemble increases, at least is not worse than Eigcons method. The results in Figure 4 show that the performance by Eigcons and SubmEP_$\widetilde{G}$ is almost the same on those data sets.

From the figures, we can also see that our methods have the best performance when the pruning ratio is set between 8% and 20%. However, the Eigcons method does the best between 15% and 35%. So in a word, our methods can get a better performance with less classifiers. And this is the reason that we choose the pruning ratio 20% for the Table 4 and the efficiency experiments.

(4). Additional Tests: Effectiveness of SubmEP with fixed λ

In the above experiments, our method has a drawback that it is time consuming to exhaust λ when the size of dataset is large. To examine the robustness of our methods, we also prune the ensembles from 1 to the size of the full ensemble by only using the λ when the pruning ratio is set 20%. We choose the dataset "Anneal" and "Audiology". And the results are showed in Figure 5. It can be seen that the SubmEP_G_fixed method has a similar performance with SubmEP_$\widetilde{G}$ method. The running time can be much less than Eigcons method in the ensemble pruning from 1 to the size of the full ensemble even if the size of dataset is very large. The results demonstrate the effectiveness of our proposed methods.

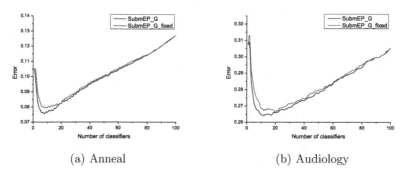

(a) Anneal   (b) Audiology

Fig. 5. Comparison of prediction errors on data sets "Anneal" and "Audiology" between tuning λ and fixed λ

6   Conclusions

In this paper, we formalized the ensemble pruning problem as a submodular function maximization problem to search a subset of classifiers that had nice accuracy-diversity trade-off. Using greedy algorithm, we found a good approximate solution efficiently. The extensive experiments on real datasets demonstrated the effectiveness and efficiency of our approach compared to state-of-the-art methods for ensemble pruning.

To our best knowledge, this work is the first one to solve ensemble pruning problem via submodular function maximization. We will investigate more optimization objective function for ensemble pruning.

Acknowledgments. This work was supported by the 973 project (No. 2010CB328106), NSFC grant (No. 61033007 and 61170085), Program for New Century Excellent Talents in University (No. NCET-10-0388) and Shanghai Knowledge Service Platform Project(No. ZF1213).

References

1. Asuncion, A., Newman, D.: Uci machine learning repository (2007),
   http://www.ics.uci.edu/~mlearn/MLRepository.html
2. Breiman, L.: Bagging predictors. Machine Learning 24(2), 123–140 (1996)
3. Caruana, R., Niculescu-Mizil, A., Crew, G., Ksikes, A.: Ensemble selection from libraries of models. In: ICML, pp. 18–25 (2004)
4. Cover, T.M., Thomas, J.A.: Elements of Information Theory, 2nd edn. Wiley-Interscience (2006)
5. Fan, W., Wang, H., Yu, P.S., Ma, S.: Is random model better? on its accuracy and efficiency. In: ICDM, pp. 51–58 (2003)
6. Freund, Y., Schapire, R.E.: A decision-theoretic generalization of on-line learning and an application to boosting. JCSS 55(1), 119–139 (1997)

7. Ho, T.K.: The random subspace method for constructing decision forests. IEEE Transactions on Pattern Analysis and Machine Intelligence 20(8), 832–844 (1998)
8. Hoi, S.C.H., Jin, R., Zhu, J., Lyu, M.R.: Batch mode active learning and its application to medical image classification. In: ICML, pp. 417–424 (2006)
9. Kempe, D., Kleinberg, J., Tardos, E.: Maximizing the spread of influence through a social network. In: KDD (2003)
10. Krause, A., Cevher, V.: Submodular dictionary selection for sparse representation. In: ICML (2010)
11. Krause, A., Guestrin, C.: Near-optimal nonmyopic value of information in graphical models. In: Uncertainty in Artificial Intelligence, pp. 324–331 (2005)
12. Li, N., Yu, Y., Zhou, Z.-H.: Diversity regularized ensemble pruning. In: KDD, pp. 330–345 (2012)
13. Li, N., Zhou, Z.-H.: Selective ensemble under regularization framework. In: Benediktsson, J.A., Kittler, J., Roli, F. (eds.) MCS 2009. LNCS, vol. 5519, pp. 293–303. Springer, Heidelberg (2009)
14. Lu, Z., Wu, X., Zhu, X., Bongard, J.: Ensemble pruning via individual contribution ordering. In: KDD (2010)
15. Margineantu, D., Dietterich, T.: Pruning adaptive boosting. In: ICML, pp. 211–218 (1997)
16. Martinez-Munoz, G., Suarez, A.: Pruning in ordered bagging ensembles. In: ICML (2006)
17. Nemhauser, G., Wolsey, L.A., Fisher, M.: An analysis of approximations for maximizing submodular set functions - i. Mathematical Programming 14(1), 265–294 (1978)
18. Partalas, I., Tsoumakas, G., Vlahavas, I.: A study on greedy algorithms for ensemble pruning. Technical Report TR-LPIS-360-12, LPIS, Dept. of Informatics, Aristotle University of Thessaloniki, Greece (2012)
19. Tamon, C., Xiang, J.: On the boosting pruning problem. In: Lopez de Mantaras, R., Plaza, E. (eds.) ECML 2000. LNCS (LNAI), vol. 1810, pp. 404–412. Springer, Heidelberg (2000)
20. Witten, I., Frank, E.: Data Mining: Practical machine learning tools and techniques. Morgan Kaufmann (2005)
21. Xu, L., Li, B., Chen, E.: Ensemble pruning via constrained eigen-optimization. In: ICDM (2012)
22. Zhang, Y., Burer, S., Street, W.N.: Ensemble pruning via semi-definite programming. Journal of Machine Learning Research 7 (2006)
23. Zhou, Z.-H.: Ensemble Methods: Foundations and Algorithms. Chapman and Hall CRC (2012)
24. Zhou, Z.-H., Li, N.: Multi-information ensemble diversity. In: El Gayar, N., Kittler, J., Roli, F. (eds.) MCS 2010. LNCS, vol. 5997, pp. 134–144. Springer, Heidelberg (2010)
25. Zhou, Z.-H., Wu, J., Tang, W.: Ensembling neural networks: Many could be better than all. Artificial Intelligence 137 (2002)

Identify and Trace Criminal Suspects in the Crowd Aided by Fast Trajectories Retrieval

Jianming Lv, Haibiao Lin, Can Yang,
Zhiwen Yu, Yinghong Chen, and Miaoyi Deng

South China University of Technology, Guangzhou, 510006, China
{jmlv,cscyang,zhwyu}@scut.edu.cn,lin.hb@mail.scut.edu.cn,
chenyinghong.com@qq.com,136221375@qq.com

Abstract. Aided by the wide deployment of surveillance cameras in cities nowadays, capturing the video of criminal suspects is much easier than before. However, it is usually hard to identify the suspects only according to the content of surveillance video due to the low resolution rate, insufficient brightness or occlusion. To address this problem, we consider the information of when and where a suspect is captured by the surveillance cameras and achieve a spatio-temporal sequence ζ_i. Then we search the records of mobile network to locate the mobile phones which have compatible trajectories with ζ_i. In this way, as long as the suspect is carrying a mobile phone when he is captured by surveillance cameras, we can identify his phone and trace him by locating the phone. In order to perform fast retrieval of trajectories, we propose a threaded tree structure to index the trajectories, and adopt a heuristics based query optimization algorithm to prune unnecessary data access. Extensive experiments based on real mobile phone trajectory data show that a suspect's phone can be uniquely identified with high probability while he is captured by more than four cameras distributed in different cells of the mobile network. Furthermore, the experiments also indicate that our proposed algorithms can efficiently perform the search within 1 second in the trajectory dataset containing 104 million records.

Keywords: trajectory, search, efficiency, suspect identification.

1 Introduction

Surveillance cameras have been widely used to monitor public places and help the police in criminal investigation to recognize and trace suspects in the crowd. The quality of the surveillance video seriously affects the success rate of recognition. Unfortunately, most of time it is not easy to identify a person only according to the video content due to the low resolution rate of the image, insufficient brightness or occlusion.

Different from the traditional research concentrated in pattern recognition of images, we take the time and location information of a surveillance video into account to identify a person according to his trajectory. Suppose a criminal suspect P_k is captured by a camera Υ_i located in the place Loc_i at the time T_i,

S.S. Bhowmick et al. (Eds.): DASFAA 2014, Part II, LNCS 8422, pp. 16–30, 2014.

we denote this surveillance event as a spatio-temporal tuple (T_i, Loc_i). If this suspect is captured by multiple cameras $\{\varUpsilonΥ_i|i > 1\}$, we can collect all these events as a spatio-temporal sequence $\zeta_k = \{(T_i, Loc_i)|i > 1\}$, which is exactly the suspect's motion trajectory. At the same time, if the suspect is carrying a mobile phone, his trajectory is also recorded by the mobile service providers. We can search in the mobile phone trajectory database for the phones which have a trajectory compatible with ζ_k. Ideally, if there is only one matched mobile phone returned, we can identify the suspect's phone and trace him through locating this phone in the mobile network.

There exist two fundamental problems in above trajectory retrieval mechanism:

- Search efficiency problem: How to efficiently search for the mobile phone trajectories compatible with a given spatio-temporal sequence in a large scale trajectory dataset?
- Identification precision problem: What is the precision of identifying a suspect? Given a spatio-temporal sequence, is the number of retrieved compatible trajectories small enough to facilitate identification of the suspect's phone?

To solve the efficiency problem, we propose a threaded tree structure to index the phone trajectories, which can reduce the time of querying correlated spatio-temporal tuples. Further more,we develop a heuristics based query optimization algorithm to prune unnecessary data access. Experiments show that the time to search for a compatible trajectory in a dataset containing 104 million records is within one second.

To answer the question about the identification precision, we study which property has an impact on the precision. We theoretically prove that the number of retrieved compatible trajectories is a decreasing function of the overlapping degree of human motion trajectories, and a increasing function of the length of the searched spatio-temporal sequence. Experiments on real mobile phone trajectory dataset show that a suspect's phone can be uniquely identified with high probability while he is captured by more than four cameras distributed in different cells of the mobile network.

Recently, trajectory retrieval has attracted a lot of interesting research [9–14] in this field. This paper differs with those proposed search algorithms in the following aspects:

- The spatio-temporal points of the trajectories recorded by the mobile network are coarse-grained. The position in each point is presented as an ID of a cell, which is a land area to be supplied with radio service from one base station. The size of a cell is about hundreds of meters in the dense city zone. Furthermore, the temporal information of each point is not a exact timestamp, but a time interval when a mobile phone appears in the cell.
- Different from the similarity based search, the task in this paper is to retrieve the compatible trajectories which exactly match a given spatio-temporal sequence.

The main contributions of this paper are summarized as follows:

- To the best of our knowledge, we are the first to pioneer a systematic approach to locate and trace suspects by integrating mobile phone trajectory retrieval with surveillance spatio-temporal events detection.
- We propose a kind of threaded tree structure to store and index the phone trajectories to improve the efficiency of searching compatible trajectories.
- We also present a heuristic based query optimization algorithm to further filter unnecessary data access while executing the search.
- We illustrate the effectiveness and efficiency of this trajectory retrieval system based on both theoretical proof and comprehensive experiments on real trajectory dataset.

The remainder of this paper is organized as follows. Section 2 reviews the related work. Section 3 offers the clear definition of the trajectory retrieval problem. Section 4 present the our proposed methods. Section 5 evaluates the performance of this system by implement the prototype and conducting experiments in real dataset. We conclude the work in Section 6.

2   Related Work

The most recent report about identifying individual person in the crowd based on the motion trajectories is proposed by Montjoye et al. [2]. It showed that human mobility traces are so highly unique, that four randomly selected spatio-temporal points are enough to uniquely identify 95% of the individuals. This analysis result theoretically supports the effectiveness of our research. Beyond the basic data analysis in [2], we are the first to propose a novel systematic scheme to apply the identification technique in criminal tracing and develop efficient retrieval algorithms to solve the problems.

A lot of trajectory retrieval algorithms have been proposed recently. The basis of the search problem is to define the distance between two trajectories. Several typical similarity functions for different applications are presented including Euclidean Distance [3], Dynamic Time Warping (DTW) [5], Longest Common Subsequence(LCSS) [6], Edit Distance with Real Penalty (ERP) [7],Edit Distance on Real Sequences (EDR) [8], and other similarity measurement [4].

Tree-based structures are widely deployed to index trajectories to enhance retrieval efficiency. The indexing strategy can be roughly classified into two categories according to [15]. The first class is the structures based on data partitioning, such as R*-tree [13], TPR-tree [16] and its variant the TPR*-tree [17]. The second category of indexes are constructed through space partitioning, such as the B+-tree based indexes [15, 18] and the grid based proposals [19, 20].

Based on above indexing structures, different trajectory search algorithms [9–14] are presented to satisfying different constraint conditions in real applications. Chen et al.[9] study the problem of searching top k best-connected trajectories connecting the designated locations geographically. A new similarity function is proposed to measure how well a trajectory connects the query locations while considering the spatial distance and order constraint of locations. Sherkat et al. [10] propose a summary based structure to approximate high dimensional trajectories for efficient similarity based retrieval. Chen et. al [11] introduced a novel distance measurement of trajectories called EDR to eliminate unexpected noises, which bring serious side effect while performing similarity search. Some pruning techniques were also developed to be combined with EDR to improve the retrieval efficiency. Vlachos et. al [12] proposed an index structure that supports multiple distance measures. They organized MBRs extracted from data sequences in an R-tree and pruned irrelevant data sequences based on the intersection of the query MBRs with those in the index. Lee et al [13] partition data sequences into sub-sequences, which are approximated as MBRs (minimum bounding rectangles) and organized in an R*-tree index structure. The query is also presented as MBR to search the intersected MBRs in the indexing tree. Cai et al. [14] used the coeffients of chebyshef polynomials as features to approximate timing series and indexed the coeffent points in the multi-dimensional indexing trees. The lower bound property of approximate distance based on the coefficients is used to prune the search results while guaranteeing no false positives.

3   Problem Definition

3.1   Overview

The design goal of the system is to identify and trace a criminal suspect by using the trajectory information recorded in the mobile network.

Fig. 1 illustrates the architecture of the system. For a person P_i captured by surveillance cameras while he is moving, a spatio-temporal sequence about his motion trajectory can be obtained as shown in Fig. 1(a). Specifically, the time when he is captured by the three cameras is $T_1 = 2 : 30'10''$, $T_2 = 2 : 30'30''$ and $T_3 = 2 : 31'20''$ in sequence, and the locations of three cameras are Loc_1, Loc_2 and Loc_3.

At the same time, the cellular mobile network records the trajectories of all mobile phones in the service area. Fig. 1(b) illustrates that the mobile network divided the land area into cells, each of which is supplied with radio service from one base station. The size of a cell is about hundreds of meters in dense city zones. While a mobile phone is moving in a service area, its trajectory is recorded, which is formed as a sequence of the tuple $(P_k, TF_k, TE_k, C_k, A_k)$ as shown in the Table 1. Each tuple indicates the phone P_k is in the cell C_k from the time TF_k to TE_k. If the person P_i is carrying a mobile phone while walking, his phone trajectory must be recorded. We can identify his phone by searching for the phone trajectory which satisfying the following conditions: at the time T_j, the phone appears at the cell zone containing Loc_j $(1 \leq j \leq 3)$. Ideally, if there is only a few matched trajectories returned, the police can identify and trace the suspects through looking up the communication records of the corresponding phones and locating the phones in the mobile network. The example in Table 1 shows the best case that only P_2's trajectory is matched and it is exactly the target person's phone.

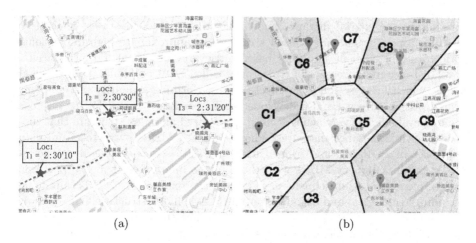

(a)   (b)

Fig. 1. An example to identify a suspect's phone with a particular motion trajectory. a) A moving suspect is captured by three surveillance cameras distributed in three different areas. b) The distribution of the cells in the mobile network. There are 9 cells ($C_i(1 \le i \le 9)$) in the area. The records of mobile phones are illustrated in Table 1.

3.2   Problem Formulation

Below we give the formal definition of the problem solved in this paper. For clarity, the main notations are summarized in Table 2.

Definition 1 (*Surveillance Trajectory*). When a person is captured by a surveillance camera at the time T_i and at the location Loc_i, this event is defined as a *spatio-temporal point* (*STP* for short) $\varpi_i =< T_i, Loc_i >$. While he is captured by multiple cameras, all the corresponding STPs are linked into a *surveillance trajectory* as $\zeta_i = \{\varpi_{i1}, \varpi_{i2}, ..., \varpi_{in}\}$ in the order of time.

Definition 2 (*Phone Trajectory*). The cellular mobile network divides the land area into cells, each of which is serviced by one base station. While a phone is moving, its trajectory is recorded as the a sequence of spatio-temporal segment: $\xi_i = \{\sigma_{i1}, \sigma_{i2}, ..., \sigma_{in}\}$. Here σ_{ij} is a *spatio-temporal segment*(*STS* for short) which is formulated as $\sigma_i =< U_i, TF_i, TE_i, C_i, A_i >$. It indicates a segment of time $[TF_i, TE_i]$ when the phone U_i is staying in a same cell C_i. A_i is some attached description information recorded in the mobile network.

Definition 3 (*Overlapping Degree of STSs.*) The overlapping degree of two STSs is defined as follows:

$$\Phi(\sigma_i, \sigma_j) = \begin{cases} 0 & (if \ \sigma_i.C \ne \sigma_j.C) \\ |[\sigma_i.TF, \sigma_i.TE] \cap [\sigma_j.TF, \sigma_j.TE]| & (if \ \sigma_i.C = \sigma_j.C) \end{cases} \quad (1)$$

Here $\sigma_x.TF$, $\sigma_x.TE$, and $\sigma_x.C$ ($x = i, j$)mean the start time, end time and the cell ID of σ_x. The operation $\cap(.,.)$ means the intersection of two time segments. The another operation $|.|$ means the length of a time segment.The equation indicates the length of the overlapping time of two persons staying in a same cell.

Table 1. Records of Phone Trajectories

P_k	TF_k	TF_k	C_k
P_1	2:20'05"	2:40'10"	C_1
P_1	2:40'11"	3:00'20"	C_2
P_2	2:25'08"	2:30'20"	C_2
P_2	2:30'21"	2:31'05"	C_5
P_2	2:31'06"	2:39'17"	C_9
P_3	2:26'09"	3:00'19"	C_5
P_4	2:27'10"	2:30'18"	C_2
P_4	2:30'19"	2:35'21"	C_3
P_4	2:35'21"	2:38'30"	C_4
P_5	2:30'11"	5:04'10"	C_9

Table 2. Summary of Notations

Notation	Description
$\varpi_i = <T_i, Loc_i>$	a *spatio-temporal point* (STP for short) about a person captured by a surveillance camera at the location Loc_i and at the time T_i
$\zeta_i = \{\varpi_{i1}, \varpi_{i2}, ..., \varpi_{in}\}$	a surveillance trajectory of a person captured by cameras
$\sigma_i = <U_i, TF_i, TE_i, C_i, A_i>$	a *spatio-temporal segment*(STS for short) of a phone which means the phone U_i is at the cell C_i from the time TF_i to TE_i. A_i is some attached description information.
$\xi_i = \{\sigma_{i1}, \sigma_{i2}, ..., \sigma_{in}\}$	a phone trajectory formed as a sequence of STSs.
$|\xi_i|$	the length of the phone trajectory ξ_i
$\Phi(\sigma_i, \sigma_j)$	the overlapping degree of two STSs.
$\Psi(\xi_i, \xi_j)$	the overlapping degree of two phone trajectories
$\xi_i \triangleright \zeta_j$	The phone trajectory ξ_i is compatible with the surveillance trajectoryζ_j

Definition 4(*Overlapping Degree of phone trajectories*). The overlapping degree of two phone trajectories is defined as follows:

$$\Psi(\xi_i, \xi_j) = \frac{\sum_{\sigma_x \in \xi_i, \sigma_y \in \xi_j} \Phi(\sigma_x, \sigma_y)}{\sum_{\sigma_x \in \xi_i} (\sigma_x.TE - \sigma_x.TF)} \tag{2}$$

$\Psi(\xi_i, \xi_j)$ defined above indicates the ratio of the time interval of the trajectory ξ_i overlapped with another one ξ_j.

Definition 5 (*Compatible Trajectory*). A phone trajectory ξ_i is said to be compatible with another surveillance trajectoryζ_j (denoted as $\xi_i \triangleright \zeta_j$), if and only if the following condition is satisfied. For each STP $\varpi_j = <T_j, Loc_j>$ belonging to ζ_j, there exists a STS $\sigma_i = <U_i, TF_i, TE_i, C_i, A_i>$ belonging to ξ_i that has the following properties: $TF_i \le T_j \le TE_i$ and the location Loc_j is in the cell C_i.

When a person moves and passes by several surveillance cameras, a surveillance trajectory ζ_j can be obtained. If he carries a mobile phone while walking, a phone trajectory ξ_i is also recorded in the mobile service provider and it must be compatible with ζ_j. If we can find the phone trajectories compatible with ζ_j in the massive trajectories recorded by the mobile service provider, it will be quite helpful for us to determine the target phone and further trace the person. Thus, the problem raised in this paper can be concentrated on how to search for the phone trajectories compatible with a given surveillance trajectory ζ_j.

3.3   Problem Analysis

Input with a surveillance trajectory, the number of returned compatible phone trajectories decides the precision to identify a suspect. In the following Theorem 1, we try to discuss the factors affecting the retrieval result.

Theorem 1. Given any surveillance trajectory ζ_x of any person P_x composed of n spatio-temporal points $\{< T_k, Loc_k > | 1 \leq k \leq n\}$, the expectation of the number of phone trajectories compatible with ζ_x is equal to $N * E(\Psi^n(a,b))$. Here N is the total number of the phone trajectories, $\Psi(a,b)$ is the overlapping degree of any pair of phone trajectory a and b, and the operation $E(.)$ means the expectation of value in the brace.

Proof. Suppose that the phone trajectory of P_x is ξ_x, and he is captured by the cameras at n randomly selected time while he is moving.

For any phone trajectory ξ_y of another person P_y, the overlapping degree between ξ_x and ξ_y is $\Psi(\xi_x, \xi_y)$. That also means, at the time T_k $(1 \leq k \leq n)$, the probability for the persons P_x and P_y staying in a same cell is $\Psi(\xi_x, \xi_y)$. Thus they stay in same cells at all of the n time points $\{T_k | 1 \leq k \leq n\}$ with the probability $\Psi^n(\xi_x, \xi_y)$. Based on Definition 5, we can deduce further that ξ_y is compatible with ζ_x with a probability $\Psi^n(\xi_x, \xi_y)$.

While considering all of the phone trajectories, we can calculate the expectation of the number of compatible trajectories M as follows:

$$
\begin{aligned}
M &= E\left(\sum_{\xi_y \in \Omega} \Psi^n(\xi_x, \xi_y) \right) \\
&= \sum_{\xi_y \in \Omega} E(\Psi^n(\xi_x, \xi_y)) \\
&= N * E(\Psi^n(a,b)) \quad\quad\quad (3)
\end{aligned}
$$

Here Ω means the collection of all phone trajectories. a and b are any pair of phone trajectories. ∎

Theorem 1 shows that the number of retrieved compatible trajectories is a increase function of the overlapping degree between trajectories, which indicates the probability to discriminate the trajectories of different persons. On the other hand, the more cameras capturing the person, the less compatible trajectories can be returned to identify the person precisely.

4 Trajectory Retrieval Methods

4.1 Basic Search Method

In order to support efficient retrieval of the phone trajectories, we can build the B+ tree index on the STSs of trajectories as illustrated in Fig. 2(a). For each STS $\sigma_i = < U_i, TF_i, TE_i, C_i, A_i >$ of the phone P_i, we can build compound key by linking C_i, TF_i and TE_i in order. Fig.2(a)shows an example about the data arranged in the tree. Based on this simple indexing structure, the algorithm to search phone trajectories compatible with a given surveillance trajectory is shown in Algorithm 1.

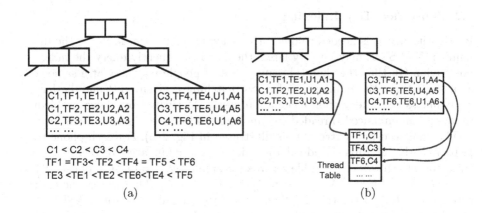

Fig. 2. (a) The tree structure (b)The threaded tree structure

---

Algorithm 1. Search phone trajectories compatible with a given surveillance trajectory

---

Input:
$\zeta_i = \{\varpi_{i1}, \varpi_{i2}, ..., \varpi_{in}\}$ - A surveillance trajectory of the person P_i. Specifically, $\varpi_{ij} = < T_{ij}, Loc_{ij} >$ is a STP about P_i captured by a surveillance camera at the location Loc_{ij} and at the time T_{ij}
n - The length of ζ_i.
Output: S - A set of phones with trajectories compatible with the ζ_i.
Method:
1 **for** $j = 1$ to n
2   **do** $T = \varpi_{ij}.T$ //the time of ϖ_{ij}
3        $C = cell(\varpi_{ij}.Loc)$ //the cell covering $\varpi_{ij}.Loc$
4        $G_i \leftarrow \{\sigma_k | \sigma_k.TF \leq T \leq \sigma_k.TE, \sigma_k.C = C\}$ // search for STSs in the index tree
5        $S_i \leftarrow \{\sigma_k.U | \sigma_k \in G_i\}$ //phone IDs of the STSs in G_i.
6 $S = \bigcap_{i=1}^{n} S_i$

In step 4 of Algorithm 1, for each STP ϖ_{ij} in the surveillance trajectory, we search the index tree for STSs overlapping with ϖ_{ij}. In step 5, the phone IDs of the retrieved STSs are gathered into a collection S_i. Finally, the phones with compatible trajectories are retrieved as the intersection of the sets $\{S_i | 1 \leq i \leq n\}$.

Suppose the total number of STSs in the index tree is N and the average number of STSs matching a given STP is m. The time spent to search the B+ tree index for STSs matching one STP in step 4 is $\Theta(\log N + m)$. Considering the n iterations are run in the algorithm, the total time for searching is $\Theta(n \log N + nm)$. Moreover, step 6 calculates intersection of n sets with average size m, so the time is $\Theta(nm \log m)$. Thus, the total time spent in this algorithm is $\Theta(n \log N + nm \log m)$.

4.2   Threaded Tree Indexing

In Algorithm 1, it is needed to query the index tree for n times. While the total number N of STSs may be very large, the cost to frequently query the index is not trivial. On the other hand, because a lot of people may stay at a same cell at a same time, the number m of STSs matching a given STP may be large. That makes the intersection calculation costly. Motivated by above problems, we design an enhanced threaded tree indexing to improve the search efficiency.

The enhanced index structure is illustrated in Fig. 2(b). For each phone trajectory, a thread table is added to maintain the summaries of STSs on the trajectory. Each row of the table records the summary $\varepsilon_i = (T_i', C_i')$ of each STS $\sigma_i = <U_i, TF_i, TE_i, C_i, A_i>$, where $T_i' = TF_i$, $C_i' = C_i$. That means the phone enters the cell C_i' at the time T_i'. The rows of the thread table are ranked by T_i'. We also add one pointer on each STS recording the offset of its summary in the thread table.

Given a surveillance trajectory $\zeta_i = \{\varpi_{i1}, \varpi_{i2}, ..., \varpi_{in}\}$, we can search the index tree for STSs matching the first STP ϖ_{i1}, and then we can use the thread tables of the retrieved STSs to efficiently prune the results unmatched with the left STPs of ζ_i. Only one-pass search of the tree index is required and no intersection of large collections is necessary any longer. The detailed procedure is illustrated in Algorithm 2.

In step 3 of Algorithm 2, we search the index tree for the STSs overlapping with ϖ_{i1}, which is the first STP on the surveillance trajectory ζ_i. All the phones corresponding to the retrieved STSs form the candidate result S in step 4. For each matched STS σ_k of the phone u, we check whether u's trajectory has STSs overlapping with the left STPs of ζ_i in step 7 to 12.

Specifically, in step 10, we locate u's STS summary having overlapping time with ϖ_{ij}, which is one STP of ζ_i. The detailed search procedure can firstly follow the pointer on the STS σ_k to locate its summary $\varepsilon_k = (T_k', C_k')$ in u's thread table. Starting from ε_k, if $\varpi_{ij}.T$ is bigger than T_k', we search for the summary forward in the thread table. Otherwise, we search backward. After we find the proper summary ε_x, we verify whether ε_x is at the same location as ϖ_{ij} in step 11. If their locations are different, that means u's phone trajectory is not compatible with ζ_i and we can directly filter it out of the result as step 12.

Algorithm 2. Search phone trajectories compatible with a given surveillance trajectory through threaded tree index.

Input: $\zeta_i = \{\varpi_{i1}, \varpi_{i2}, ..., \varpi_{in}\}$ - A surveillance trajectory of a person P_i. The same as Algorithm 1.
Output: S - A set of phones whose trajectories are compatible with the ζ_i.
Method:
1 $T = \varpi_{i1}.T$ //the time of ϖ_{i1}, the first STP of ζ_i
2 $C = cell(\varpi_{i1}.Loc)$ //the cell of ϖ_{i1}
3 $G_0 \leftarrow \{\sigma_k | \sigma_k.TF \leq T \leq \sigma_k.TE, \sigma_k.C = C\}$ // search the index tree
4 $S \Leftarrow \{\sigma_k.U | \sigma_k \in G_0\}$ //the candidate result
5 **for** each $\sigma_k \in G_0$ **do**
6      $u = \sigma_k.U$ //the phone ID of σ_k
7   **for** $j = 2$ to n **do**
8          $T_j \leftarrow \varpi_{ij}.T$ //the time of ϖ_{ij}, the No. j STP of ζ_i
9          $C_j \leftarrow cell(\varpi_{ij}.Loc)$ //the cell of ϖ_{ij}
10          $\varepsilon_x \leftarrow u$'s summary satisfying $\varepsilon_x.T' \leq T_j \leq \varepsilon_{x+1}.T'$ //search thread table
11   **if** $\varepsilon_x.C' \neq C_j$ **then**
12   remove u from S

Compared with Algorithm 1, there is only one-pass search of B+ tree in step 3 of Algorithm 2. The thread tables make the browse of each phone trajectory efficient. Furthermore, the calculation of intersection of large sets is not needed any longer. Similar with the analysis of Algorithm 1, the time complexity of Algorithm 2 can be proved to be $\Theta(\log N + nm)$. Detailed experimental comparison of the performance will be presented in the following section 5.

4.3 Heuristics Based Query Optimization

In Algorithm 2, the number of STSs returned by the first tree-index based search determines the number of iterations in step 5. This motivate us to re-examine the search procedure to reduce the size of the search result in step 3.

In Algorithm 2, we search in the tree index for STSs overlapping with the first STP ϖ_{i1} of ζ_i in step 1 to 4 , and then filter the result by matching the left STPs in step 5 to 12. In fact, it is not necessary to start the search from the first STP ϖ_{i1}. We can modify step 1 to 4 to search for the STSs overlapping with any STP $\varpi_{ij} \in \zeta_i$, which may lead smallest search result in step 3, and then filter the result by matching left STPs.

However, we cannot judge exactly which STP on ζ_i can produce smallest search result before performing the search. To solve this problem, we present a heuristics based method to estimate the search cost. For any STP $\varpi_{ij} \in \zeta_i$, the number of STSs overlapping with ϖ_{ij} is the number of the phones in the cell C_j containing $\varpi_{ij}.Loc$ at the time $\varpi_{ij}.T$. We estimate this number based on the historical statistics about the average number of phones in the cell C_j. This can be achieved from the records in the mobile network and updated every hour. The smaller cells servicing less phones may have less chances to find phones with

STSs overlapping with ϖ_{ij}, so we can select the STP corresponding to these cells to achieve smaller search result.

The detailed enhanced search algorithm is illustrated in Algorithm 3. The major differences made by Algorithm 3 lie in step 1 to 3, where a STP corresponding to a smallest cell is selected to perform search first. In step 8 to 13, the left STPs are used to filter the search result. The experimental result to show the effectiveness of Algorithm 3 will be presented in the following section 5.

Algorithm 3. Search phone trajectories compatible with a given surveillance trajectory through heuristics based method.

Input: $\zeta_i = \{\varpi_{i1}, \varpi_{i2}, ..., \varpi_{in}\}$ - A surveillance trajectory of a person P_i. The same as Algorithm 1.
$H(C_i)$ - The average number of phones in the cell C_i.
Output: S - A set of phones whose trajectories are compatible with the ζ_i.
Method:
1   $\varpi_{im} = \underset{\varpi_x \in \zeta_i}{\arg\min} H(cell(\varpi_x.Loc))$ //select the STP on the cell with minimum phones.
2   $T = \varpi_{im}.T$ //the time of ϖ_{im}
3   $C = cell(\varpi_{im}.Loc)$ //the cell of ϖ_{im}
4   $G_0 \leftarrow \{\sigma_k | \sigma_k.TF \leq T \leq \sigma_k.TE, \sigma_k.C = C\}$ // search the index tree
5   $S \leftarrow \{\sigma_k.U | \sigma_k \in G_0\}$ //the candidate result
6   **for** each $\sigma_k \in G_0$ **do**
7       $u = \sigma_k.U$ //the phone ID of σ_k
8   **for** each $\varpi_{ij} \in \zeta_i(j \neq m)$ **do**
9           $T_j \leftarrow \varpi_{ij}.T$ //the time of ϖ_{ij} which is the No. j STP of ζ_i
10          $C_j \leftarrow cell(\varpi_{ij}.Loc)$ //the cell of ϖ_{ij}
11          $\varepsilon_x \leftarrow u$'s summary satisfying $\varepsilon_x.T' \leq T_j \leq \varepsilon_{x+1}.T'$ //search thread table
12   **if** $\varepsilon_x.C \neq C_j$ **then**
13   remove u from S

---

5   Evaluation

In this section, we conduct experiments on a real dataset which consists of 5.9 million mobile phone trajectories distributed in the urban area containing 24,370 cells in Guangzhou, China. The data is collected from 0:00 to 8:00 at Feb 2,2013 by a mobile service provider and totally 104 million STSs are recorded. The algorithms are implemented on a windows platform with Intel Core 2 CPU(2.93 GHz) and 2.GB memory.

The main metrics we adopt for measuring the performance are the *Number of Compatible Trajectory* that reflects how many retrieved phone trajectories compatible with a given surveillance trajectory, and the *Query Time* that indicates how fast a query can be returned.

In order to simulate a surveillance trajectory with a length of n, we randomly selects a phone trajectory and randomly pick out n STSs from it. For each

STS $\sigma_i(i \leq 5)$, we randomly construct a STP ϖ_i statisfying $(\sigma_i.TF \leq \varpi_i.T \leq \sigma_i.TE, cell(\varpi_i.Loc) = \sigma_i.C)$. Then we compose the surveillance trajectory as $\zeta_x = \{\varpi_i | 1 \leq i \leq n\}$.

5.1   Effectiveness

The effectiveness of the suspect identification mechanism proposed in this paper is correlated to the number of returned phone trajectories compatible with an given surveillance trajectory. The theorem 1 in the section 3.3 shows that the number is proportional to the overlapping degree between trajectories, which is a kind of natural property of human society. Fig. 3 illustrates the distribution of the overlapping degree in the dataset. It shows that more than 99.9% pairs of trajectories are totally not overlapped, and the average of the overlapping degree is very low. This is good news to distinguish different persons based their phone trajectories.

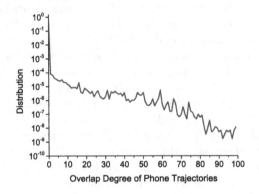

Fig. 3. The distribution of the overlapping degree of trajectories

We divide the dataset into 8 subsets, each of which contains one hour of records, and test the search performance on each subset. Fig. 4(a) shows the search result. It indicates that the *Number of Comptible Trajectory* decreases while increasing the length of the surveillance trajectory. This verifies Theorem 1 in a visualized manner. More important, Fig. 4(a) shows that a surveillance trajectory with 3 STPs may leads to a search result with less than 10 compatible phone trajectories, while more than 4 STPs can lead to only one or two search results. That means, if a person is captured by 3 cameras, we can achieve a suspect set including his phone with a size less than 10. Further, if he is captured by more than 4 cameras in different cells, we can identify his phone uniquely with high probability.

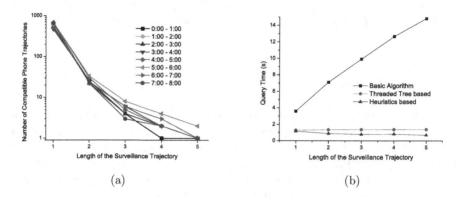

Fig. 4. Search Performance. (a) The number of compatible phone trajectories vs the length of the surveillance trajectory. (b)Time spent to perform trajectory search.

5.2   Efficiency

In this section, we test the timing efficiency of the search algorithms proposed in this paper. Fig. 4(b) shows how the *Query Time* in different algorithms changes while increasing the length of queried surveillance trajectory. 'Basic Algorithm' here means Algorithm 1 based on B+ tree proposed in the section 4.1. 'Threaded Tree based' is corresponding to Algorithm 2 based on the threaded tree structure presented in the section 4.2. Moreover, 'Heuristics based' indicates Algorithm 3 equipped with the Heuristics based optimization presented in the section 4.3.

Fig. 4(b) shows that the 'Threaded Tree based' algorithm outperform the 'Basic Algorithm' obviously. This result reveals that the thread table adopted in the algorithm can effectively reduce the *Query Time* by pruning unnecessary search on the index tree. Moreover, compared with the 'Threaded Tree based' algorithm, the 'Heuristics based' algorithm can further reduce about half of the Query Time. This validates the effectiveness of the heuristics based query optimization. On the other hand, Fig. 4(b) also indicates that the 'Heuristics based' algorithm can finish a query efficiently within one second and the time is not influenced by increasing the length of the surveillance trajectory to be matched.

5.3   Prototype

Beyond the above simulations, we also develop a prototype to implement the search mechanism proposed in this paper. As illustrated in the Fig. 5, the prototype is a web based application run on the Apache Tomcat web server. The map information is provided by invoking the google map APIs. Specifically, Fig. 5(a) shows the locations of the base stations in the map. Fig. 5(b) illustrates in the map the retrieved phone trajectories compatible with an input surveillance trajectory.

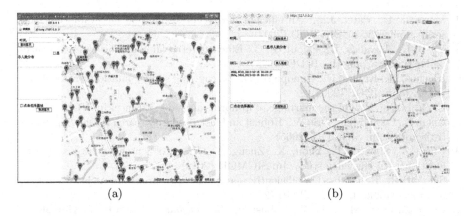

(a)   (b)

Fig. 5. The user interfaces of the web based prototype. a)The distribution of base stations. b) The result of trajectory retrieval.

6   Conclusions

In this paper, we propose a novel systematic approach to locate and trace a person by integrating mobile phone trajectory with surveillance spatio-temporal events. We combine a threaded tree structure with a heuristics based query optimization algorithm to improve the search performance. Experiments show that, if a person is captured by three cameras, a candidate set with a size less than 10 including the target person can be retrieved. Further, if he is captured by more than four cameras, he can be identified uniquely with high probability. The search procedure in a dataset including 104 million records can be finished within one second efficiently. We also develop a web-based prototype to validate the proposed algorithms and provide friendly visual interfaces.

Acknowledgement. The work was supported by the grants from National Natural Science Foundation of China (61300221), the Fundamental Research Funds for the Central Universities (2014ZZ0038), the Comprehensive Strategic Cooperation Project of Guangdong Province and Chinese Academy of Sciences (2012B090400016) and the Technology Planning Project of Guangdong Province (2012A011100005).

References

1. Smith, T., Waterman, M.: Identification of Common Molecular Subsequences. Journal of Molecular Biology 147(1), 195–197 (1981)
2. Montjoye, Y., Hidalgo, C., Verleysen, M., Blondel, V.: Unique in the Crowd: The privacy bounds of human mobility. Scientific Reports, 651–659 (2013)
3. Agrawal, R., Faloutsos, C., Swami, A.: Efficient similarity search in sequence databases. In: Lomet, D.B. (ed.) FODO 1993. LNCS, vol. 730, pp. 69–84. Springer, Heidelberg (1993)

4. Morse, M.D., Patel, J.M.: An efficient and accurate method for evaluating time series similarity. In: SIGMOD (2007)
5. Yi, B.-K., Jagadish, H., Faloutsos, C.: Efficient retrieval of similar time sequences under time warping. In: ICDE (1998)
6. Vlachos, M., Kollios, G., Gunopulos, D.: Discovering similar multidimensional trajectories. In: ICDE (2002)
7. Chen, L., Ng, R.: On the marriage of lp-norms and edit distance. In: VLDB (2004)
8. Chen, L., Özsu, M.T., Oria, V.: Robust and fast similarity search for moving object trajectories. In: SIGMOD (2005)
9. Chen, Z., Shen, H.T., Zhou, X., Zheng, Y., Xie, X.: Searching trajectories by locations: An efficiency study. In: SIGMOD (2010)
10. Sherkat, R., Rafiei, D.: On efficiently searching trajectories and archival data for historical similarities. In: VLDB (2008)
11. Chen, L., Ozsu, M., Oria, V.: Robust and fast similarity search for moving object trajectories. In: SIGMOD (2005)
12. Vlachos, M., Hadjieleftheriou, M., Gunopulos, D., Keogh, E.: Indexing multidimensional time-series with support for multiple distance measures. In: SIGKDD (2003)
13. Lee, S., Chun, S., Kim, D., Lee, J., Chung, C.: Similarity search for multidimensional data sequences. In: ICDE (2000)
14. Cai, Y., Ng, R.: Indexing spatio-temporal trajectories with Chebyshev polynomials. In: SIGMOD (2004)
15. Chen, S., Ooi, B., Tan, K., Nascimento, M.: STB-tree: A self-tunable spatio-temporal b+tree index for moving objects. In: SIGMOD (2008)
16. Saltenis, S., Jensen, C., Leutenegger, S.T., Lopez, M.A.: Indexing the Positions of Continuously Moving Objects. In: SIGMOD (2000)
17. Tao, Y., Papadias, D., Sun, J.: The TPR*-Tree: An Optimized Spatio-Temporal Access Method for Predictive Queries. In: VLDB (2003)
18. Yiu, M.L., Tao, Y., Mamoulis, N.: The Bdual-Tree: Indexing Moving Objects by Space Filling Curves in the Dual Space. VLDB J. 17(3), 379–400 (2008)
19. Mouratidis, K., Papadias, D., Hadjieleftheriou, M.: Conceptual Partitioning: An Efficient Method for Continuous Nearest Neighbor Monitoring. In: SIGMOD (2005)
20. Xiong, X., Mokbel, M., Aref, W.: SEA-CNN: Scalable Processing of Continuous K-Nearest Neighbor Queries in Spatio-temporal Databases. In: ICDE 2005 (2005)

Multi-Output Regression with Tag Correlation Analysis for Effective Image Tagging

Hongyun Cai[1,3], Zi Huang[1], Xiaofeng Zhu[2], Qing Zhang[3], and Xuefei Li[1]

[1] The University of Queensland, QLD 4072 Australia
{h.cai2,x.li14}@uq.edu.au,
huang@itee.uq.edu.au
[2] Guangxi Normal University, Guilin, 541004 China
seanzhuxf@gmail.com
[3] Australia E-Health Research Center, QLD 4006 Australia
qing.zhang@csiro.au

Abstract. Automatic image tagging is one of the most important research topics in multimedia. How to achieve accurate image tagging to bridge the semantic gap between images' content and users' semantic understanding has been widely studied in the last decade. One common approach is to convert image tagging to a multi-task learning problem. However, most existing methods ignore tag correlations in the learning process. In this paper, we show the importance of tag correlations in conducting multi-task learning. We formulate image tagging as a multi-output regression problem accounting for tag correlations, which are captured by the covariance matrix of the regression coefficients and the noise across all tags respectively. The combination of multi-output regression with tag correlation analysis takes advantage of the latent dependencies among tags to overcome limitations of existing work. Extensive experiments have been conducted on two benchmark datasets, and the results confirm that our approach outperforms the state-of-the-art methods.

Keywords: Image tagging, multi-output regression, correlation analysis.

1 Introduction

With the prevalence of image capture devices and the rapid advancement of computer vision techniques, images have been playing an increasingly important role in all kinds of social websites in recent decades. One prominent research task is automatic image tagging. By means of bridging semantic gap [19] between visual representations and people's interpretations of the same image, well labelled images benefit a lot of multimedia applications, such as image retrieval, indexing and visual event detection. However, among around 100 billion images existing on the Internet, only a very limited percentage of them are annotated [20]. Consequently, developing an efficient and effective automatic image tagging model is in high demand.

Automatic image tagging is often converted into a multi-label classification problem, where the image visual features are considered as the input and the

S.S. Bhowmick et al. (Eds.): DASFAA 2014, Part II, LNCS 8422, pp. 31–46, 2014.

corresponding tags are the output. A straightforward way to perform multi-label classification is to decompose the original problem into a number of individual binary classification problems [4], each of which is considered as one task to predict one tag. In contrast, multi-task learning (MTL) [3] is another way for image tagging, which takes the correlations among tasks into consideration and learns multiple tasks jointly by analyzing data from all tasks at the same time. One instance of MTL is multi-output regression [2], which has been widely used in image tagging. However, existing methods have the following two main limitations. Firstly, to the best of our knowledge, only a few literatures (e.g., [1]) consider the high dimensionality issue in the tag feature space, although the number of dimensions in tag space (i.e., dictionary size) is always very large in practice. In [1], a predefined tag ontology is used to construct a graph structure hierarchy to reduce the dimensionality of the tag feature space. However, such prior knowledge on tag structure is often not available in real applications. Secondly, some tags are usually correlated with each other, which has not been well studied and exploited in existing image tagging methods. In [23], the authors consider four aspects: low-rank, content consistency, tag correlation and error sparsity when predicating tags for images. However, their method is tag refinement approach which requires a couple of known tags as input while we aim to propose an image tagging algorithm recommending tags based on pure images with the consideration of tag correlations. Intuitively, intrinsic tag correlations are expected to benefit classifier learning by bringing in more information.

Inspired by [17], we devise a novel image tagging approach, named Multi-output regression with Tag correlation analysis (MorTca), in this paper. [17] is the first multi-output regression model which takes both unknown output and task structures into consideration. However, it is proposed for low-dimensional data, hence the optimization process is very inefficient and space consuming for high-dimensional data like images and tags. The proposed MorTca analyzes and utilizes the tag correlations in a multi-output regression model to achieve effective image tagging. Dimensionality reduction is performed on visual and tag features respectively to reduce the overall computational cost of the learning process. Furthermore, a more efficient optimization algorithm is designed to solve the problem for high-dimensional data. As illustrated in Figure 1, MorTca can be divided into an offline classifier training process and an online image tagging process. At the beginning of the training process, a regularized least squares regression model and the classical Principal Component Analysis (PCA) [9] are applied to generate a reduced tag and visual feature subspace respectively. After that, a multi-output regression model is employed to learn a joint tagging classifier on the reduced visual and tag features. With tag correlation analysis, the multi-output regression model takes into account not only the relevance structure of the regression coefficients of multiple tasks but also the noise covariance learnt from data itself. In brief, we refer these two underlying structures as task correlation and noise correlation respectively, which will be discussed in Section 3. The estimation of these two covariance matrices and the regression coefficients is performed simultaneously in the model training process as

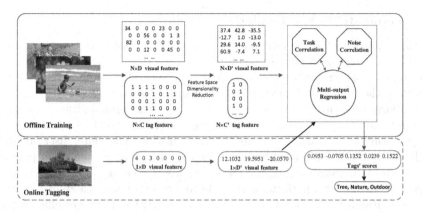

Fig. 1. The Framework of MorTca. Two main components in the offline training process, dimensionality reduction and a multi-output regression model with task and noise correlations. During online tagging, after PCA, the reduced visual feature is mapped into corresponding tags based on the learnt multi-output model.

introduced in Section 4. In the online process, given a test image, its visual feature is firstly mapped into the visual feature subspace by PCA, on which the learnt classifier from the offline training process is executed to derive the tags for the test image.

The main contributions of this paper go as follows.

- We devise an effective multi-output regression model for image tagging with the analysis of tag correlations learnt from the training data and we propose a very efficient optimization solution for the objective function which significantly improves the algorithm in [17] in both memory usage and executing time.
- We propose a new tag feature space dimensionality reduction method by feature selection to further speed up the training process.
- We have conducted comprehensive experiments on two benchmark image datasets to prove the superiority of the proposed tagging approach over a bunch of state-of-the-art methods.

The remainder of this paper is organized as follows. In Section 2, we review the related work on image tagging and multi-output regression. The proposed MorTca model is illustrated in Section 3, followed by the optimizations for the objective functions in Section 4. The evaluation results are presented in Section 5. Finally, we complete this paper in Section 6.

2   Related Work

Our proposed MorTca approach tackles the image tagging problem with a new multi-output regression model. In this section, we elaborate the related work on image tagging and multi-output regression respectively.

2.1 Automatic Image Tagging

The multimedia research community has witnessed a prolific growth of automatic image tagging techniques recently. The majority of these methods can be categorized as *generative* or *discriminative*. The generative method recommends tags from the joint distribution of image contents and labels [14]. However, it needs a large quantity of training data to learn the joint probabilities of semantic concepts and image visual features and cannot guarantee the optimal predictive performance due to data likelihood maximization. Therefore, discriminative methods are preferred in the literature. TagProp has been proven to outperform most up-to-date image tagging methods in terms of accuracy [8]. However, it acquires this exceptional performance at the cost of sacrificing scalability on large data sets. Random Forest for Image Annotation (RFIA) improves efficiency by adding semantic factors though compromising its accuracy [7]. Nevertheless the above two approaches use 15 visual descriptors to represent every image, causing both time and space concerns. In our approach, using only one visual descriptor, our performance is very competitive compared against most recent approaches.

2.2 Multiple Output Regression

Multiple output regression has been widely used in a variety of domains such as stock prices prediction, pollution prediction, etc. It was first noticed by Breiman and Friedman [2] that through utilizing correlations between outputs the regression accuracy can be improved. Subsequently many research efforts have been devoted on mining different underlying correlations in outputs. Generally these methods aim at mining two types of correlations: the task correlation and the noise correlation. The majority focus on finding only one type of correlation, either task correlation [10,21] or noise correlation [18]. Considering that only one type of correlations will inevitably lead to an incomplete model, in [11], the authors propose an approach including both types of correlations, albeit the noise correlation is predefined. It is easy to see that this method is rarely practicable in real data due to the lack of prior knowledge. Instead, a multi-output regression with output and task structures approach is introduced in [17], which simultaneously mines two types of correlations and consequently achieved better regression performance. However its extremely high time complexity limits its scalability on high-dimensional data. This motives our feature space dimensionality reduction that will be discussed in Section 3.3 and our efficient optimization solution in Section 4.2.

3 Multi-Output Regression with Tag Correlation Analysis (MorTca)

In this section, we present the proposed MorTca, which mainly consists of two steps: feature space dimensionality reduction and multi-output regression with tag correlation analysis. Before that, we first provide the notations used in the paper and formulate the problem.

3.1   Notations

Given a set of training data consisting of N images, each image I_i is associated with a D-dimensional visual feature, denoted as $x_i = (x_{i1}, x_{i2}, \cdots, x_{iD}) \in \mathbb{R}^D$ and a C-dimensional tag feature, denoted as $y_i = (y_{i1}, y_{i2}, \cdots, y_{iC}) \in \{0,1\}^C$, where $y_{ij} = 1$ if the j-th tag is assigned to the i-th image and $y_{ij} = 0$ otherwise. Hence $X = [x_1, x_2, \cdots, x_N]^T$ is an $N \times D$ visual feature matrix and $Y = [y_1, y_2, \cdots, y_N]^T$ is the corresponding $N \times C$ tag feature matrix.

3.2   Problem Formulation

As a standard image tagging task, given the visual feature matrix X and the tag feature matrix Y over all the training images, we aim to discover the correlations between X and Y, based on which automatic tagging can be achieved for the unseen images. A multi-output regression model is learnt to predict the tags for images, which is usually written as follows:

$$Y = XB + E \tag{1}$$

where $B = [B_1, \cdots, B_C]$ is a $D \times C$ regression coefficient matrix, each element B_j denotes the regression coefficient of the j-th tag. $E = [\varepsilon_1, \cdots, \varepsilon_N]^T$ is an $N \times C$ matrix, where $\varepsilon_i = (\varepsilon_{i1}, \cdots, \varepsilon_{iC}) \in \mathbb{R}^C$ denotes the errors on each tag prediction introduced by the i-th sample.

The estimation of B can be obtained by minimizing the loss function:

$$\underset{B_1, \cdots, B_C}{\arg\min} \sum_{j=1}^{C} (\text{tr}((Y_j - XB_j)(Y_j - XB_j)^T) + \lambda \times tr(B_j B_j{}^T)) \tag{2}$$

where $tr(\cdot)$ denotes matrix trace and $tr(B_j B_j{}^T)$ is the regularizer on the regression coefficient matrix B. Usually ℓ_2 norm is picked as the regularizer if we assume the independent, zero-mean Gaussian priors on the regression coefficients. $Y_j \in \mathbb{R}^N$ and $B_j \in \mathbb{R}^D$ are the j-th column of Y and B respectively, which form the j-th task of the total C individual linear regression tasks.

3.3   Feature Space Dimensionality Reduction

To enhance the scalability and efficiency of the multi-output regression model, the dimensionalities of both visual feature space and tag feature space should be reduced. This is because in image tagging, the data from both sides are usually high-dimensional. Thus, we first discuss the issue of dimensionality reduction in two feature spaces, before we illustrate the multi-output regression model with tag correlation analysis for image tagging in the next subsection.

We follow the classic approach to reduce the high dimensionality of visual features by PCA, which utilizes a set of principal components in a subspace to represent the original data. However, the situation is different for tag feature space. In our work, we aim to find a subspace from the original tag feature space by remaining a set of representative dimensions, each of which represents an individual tag. The motivation is to remove those insignificant, noisy or redundant tags. To this end, instead of reducing the dimensionality by projecting the data into a transformed subspace linearly or nonlinearly, we propose to design a feature selection method to optimally select a subset of tags from the original tag feature space. A regularized least squares regression model is employed to compute a matrix $A \in \mathbb{R}^{C \times C}$ for feature selection from the original C-dimensional tag feature space $Y \in \mathbb{R}^{N \times C}$, i.e.,

$$\arg\min_{A} \quad tr((Y - YA)(Y - YA)^T) + \gamma \|A\|_{1,2} \tag{3}$$

where

$$\|A\|_{1,2} \stackrel{def}{=} \sum_{i=1}^{C} \sqrt{\sum_{j=1}^{C} A_{i,j}^2}$$

The first term $tr((Y - YA)(Y - YA)^T)$ is the standard least squares empirical risk in linear regression models. To avoid the identity matrix I as the ineffective solution, we apply $\|A\|_{1,2}$ as a regularizer, which is the ℓ_1 norm of the ℓ_2 norm across the rows of A. This regularizer enforces the joint group sparsity on the individual rows. The non-zero rows in the optimized A demonstrate the existence of the corresponding columns in Y, which form the reduced subspace. The regularization parameter γ controls the sparsity of the representative dimension selection. The bigger γ is, the less dimensions will be selected to construct the subspace. More discussions are provided in Section 5.

3.4   Tag Correlation Analysis

In image tagging, there usually exist semantic correlations among tags. For example, "sea" and "beach" are two closely related tags, which are often assigned to the same images. Compared with learning C regression models for all C tags independently, it is expected to achieve superior effectiveness when taking into account the implicit relationships among tags, which are reflected on the following two underlying structures.

The first one is the relevance structure of the regression coefficients. As shown in Equation (2), image tagging can be converted into a multi-task learning problem, where predicting each individual tag is considered as a single task. The regression coefficient vector B_j for the j-th task represents the relationship between the sample image visual features and the j-th tag. The relevance structure of the regression coefficients reveals the relationships among the classification tasks. The profit of leveraging the information contained in parallel tasks has been proved [3]. The relevance structure of the regression coefficients is represented by their covariance matrix.

The second one is the underlying structure among noise, which are also known as errors. Traditionally, it is always assumed that $\varepsilon_1, \cdots, \varepsilon_N$ in Equation (1) are independent and identically distributed random variables. In our tagging model, we assume that noise is correlated. As proven in [18], jointly estimating all regression coefficient vectors of B, accounting for the correlated noise, outperforms estimating each of them separately. The underlying structure among noise is captured by the covariance matrix of the noise across all tags.

Unlike most of the existing work, which do not consider both correlations, the proposed MorTca aims to simultaneously learn both structures of regression coefficients and the noise with the estimation of regression coefficients. A group of regression coefficient vectors $B_1, \cdots, B_C$ are obtained for all tags after the offline learning process. Given the fact that tag correlations are not always available as prior knowledge, we propose to use multiple regularizers in the objective function to discover the above two correlations from the data itself.

Two covariance matrices $\Sigma \in \mathbb{R}^{C \times C}$ and $\Omega \in \mathbb{R}^{C \times C}$ are derived for capturing the task correlation and noise correlation respectively. According to the posterior distribution of the regression coefficient matrix, our objective function is proposed based on the negative log-posterior of regression coefficients, which is shown as following:

$$
\begin{aligned}
\underset{B,\Omega^{-1},\Sigma^{-1}}{\operatorname{argmin}} \ &\operatorname{tr}((Y - XB)\Omega^{-1}(Y - XB)^T) - N \log\left|\Omega^{-1}\right| + \lambda_1 tr(BB^T) \\
&+\lambda_2 tr(B\Sigma^{-1}B^T) - D \log\left|\Sigma^{-1}\right| + \lambda_3 tr(\Omega^{-1}) + \lambda_4 tr(\Sigma^{-1})
\end{aligned}
\tag{4}
$$

where $|.|$ denotes the determinant of a matrix.

The inverse covariance matrix Ω^{-1} couples the correlated noise across tags, and similarly, Σ^{-1} obtained relationships among the multiple tasks' regression coefficients. Apparently, both Ω^{-1} and Σ^{-1} are learnt from the training data rather than pre-defined prior knowledge. The last two terms $tr(\Omega^{-1})$ and $tr(\Sigma^{-1})$ are the regularizers, which impose the matrix variate Gaussian priors on both $\Omega^{-1/2}$ and $\Sigma^{-1/2}$ to solve the overfitting problem that occurs in the solution of Equation (4) when the tag feature dimensionality is of the same order as the visual feature dimensionality.

Algorithm 1. MorTca

    Input: $X \in \mathbb{R}^{N \times D}$, $Y \in \mathbb{R}^{N \times C}$, γ, λ_1, λ_2, λ_3, λ_4 and a test image $x \in \mathbb{R}^D$.
    Output: Generated tags $y \in \mathbb{R}^C$ for x.
1 Apply PCA on X to obtain $\bar{X}$;
2 Update Y to obtain $\bar{Y}$ according to Section 4.1 ;
3 Learn the multi-output regression model $\bar{X} \to \bar{Y}$ by Algorithm 2;
4 Generate y for x based on the learnt model after mapping x to $\bar{x}$;
5 **return** y;

---

4 Optimizations

The major steps in MorTca are outlined in Algorithm 1. Basically, it consists of two major steps, which are dimensionality reduction (lines 1-2) and multi-output regression model learning (line 3). Given a test image, represented by its visual feature x, it is first mapped into the reduced visual feature subspace, followed by which the multi-output regression model is employed to generate the tags of the test image (line 4). In this section, we will discuss the detailed optimization processes for objective functions in Equation (3) and (4).

4.1 Estimation of A

To get the solution of Equation (3), we set the derivative of the objective function with respect to A as zero, i.e.,

$$
\begin{aligned}
&\frac{\partial}{\partial A}(tr((Y - YA)(Y - YA)^T) + \gamma\|A\|_{1,2}) = 0 \\
&\Rightarrow Y^T(YA - Y) + \gamma LA = 0 \\
&\Rightarrow (Y^TY + \gamma L)A = Y^TY
\end{aligned}
\tag{5}
$$

where L is a diagonal matrix with its i-th diagonal element calculated as:

$$
L_{i,i} = \frac{1}{2\|A_i\|_2}
\tag{6}
$$

where A_i denotes the i-th row of matrix A. According to Equation (5), given a fixed L, A can be calculated as:

$$
A = (Y^TY + \gamma L)^{-1}Y^TY
\tag{7}
$$

Since A and L depend on each other, an iterative algorithm is applied to optimize Equation (3) by alternatively computing one of the two variables while fixing the other one, i.e., iteratively updating Equation (7) and Equation (6) alternatively until convergence.

4.2 Estimation of B, Ω^{-1} and Σ^{-1}

The objective function (Equation (4)) of the multi-output regression model is not jointly convex in all variables but individually convex in each variable while others are fixed. Hence the optimization problem is divided into three sub-problems, in each of which, one of the three variables is optimized. The complete optimization process is depicted step by step in Algorithm 2. The estimation of each variable is expatiated one by one as following.

With Ω^{-1} and Σ^{-1} fixed, the estimation of B is obtained by setting the derivative of objective function in Equation (4) to zero. Consequently, we get:

$$
\begin{aligned}
&\frac{\partial}{\partial B}(tr(Y\Omega^{-1}Y^T - XB\Omega^{-1}Y^T - Y\Omega^{-1}B^TX^T \\
&+XB\Omega^{-1}B^TX^T) + \lambda_1 tr(BB^T) + \lambda_2 tr(B\Sigma^{-1}B^T)) = 0 \\
&\Rightarrow 2X^TXB\Omega^{-1} + 2\lambda_1 B + 2\lambda_2 B\Sigma^{-1} = 2X^TY\Omega^{-1} \\
&\Rightarrow X^TXB + \lambda_1 B\Omega + \lambda_2 B\Sigma^{-1}\Omega = X^TY
\end{aligned}
\tag{8}
$$

Since $\lambda_1 \Omega + \lambda_2 \Sigma^{-1} \Omega$ is systemic and positive-definite, the Cholesky factorization is performed on it to produce lower triangular matrix P:

$$\lambda_1 \Omega + \lambda_2 \Sigma^{-1} \Omega = PP^T \tag{9}$$

Then Equation (8) can be expressed as

$$X^T X B + BPP^T = X^T Y \tag{10}$$

Let $X = U_1 \Sigma_1 V_1^T$ and $P = U_2 \Sigma_2 V_2^T$ be the SVD of X and P respectively, where $U_1 \in R^{N \times N}$, $\Sigma_1 \in R^{N \times D}$, $V_1 \in R^{D \times D}$, $U_2 \in R^{C \times C}$, $\Sigma_2 \in R^{C \times C}$ and $V_2 \in R^{C \times C}$. Equation (10) can be expressed as

$$V_1 \Sigma_1^T U_1^T U_1 \Sigma_1 V_1^T B + BU_2 \Sigma_2 V_2^T V_2 \Sigma_2^T U_2^T = X^T Y$$
$$\Rightarrow \Sigma_1^T \Sigma_1 V_1^T BU_2 + V_1^T BU_2 \Sigma_2 \Sigma_2^T = V_1^T X^T YU_2 \tag{11}$$

Set

$$\tilde{B} = V_1^T BU_2,$$

$$S = V_1^T X^T YU_2,$$

$$\Sigma_1^T \Sigma_1 = diag(\sigma_1{}^1, \cdots, \sigma_1{}^D) \in R^{D \times D},$$

$$\Sigma_2 \Sigma_2^T = diag(\sigma_2{}^1, \cdots, \sigma_2{}^C) \in R^{C \times C},$$

Based on Equation (11), $\tilde{B}$ is calculated as:

$$\tilde{B}_{ij} = \frac{S_{ij}}{\sigma_1{}^i + \sigma_2{}^j} \tag{12}$$

Finally, B is obtained as:

$$B = V_1 \tilde{B} U_2^T \tag{13}$$

Notably, in [17], the Kronecker product is adopted for solving the value of B. With fixed Ω^{-1} and Σ^{-1}, the Kronecker product will generate a $DC \times DC$ matrix which has high space complexity if D and C are large. We utilize the Cholesky factorization and singular value decomposition to solve the problem and get a much more efficient solution for the proposed objective function. The optimization time of the original multi-output regression model in [17] and that of our proposed MorTca are listed in Table 1, which demonstrate the outstanding efficiency gained by our method.

With Σ^{-1} and B fixed, the estimation of Ω^{-1} is obtained by setting the derivative of objective function in Equation (4) to zero:

$$\frac{\partial}{\partial \Omega^{-1}} (tr((Y - XB)\Omega^{-1}(Y - XB)^T) - N \log |\Omega^{-1}| + \lambda_3 tr(\Omega^{-1})) = 0$$
$$\Rightarrow (Y - XB)^T (Y - XB) - N\Omega + \lambda_3 I_C = 0 \tag{14}$$
$$\Rightarrow \Omega^{-1} = \left(\frac{(Y - XB)^T (Y - XB) + \lambda_3 I_C}{N} \right)^{-1}$$

where $I_C \in \mathbb{R}^{C \times C}$ is an identity matrix and M^{-1} denotes the inverse matrix of the matrix M.

Similarly, with Ω^{-1} and B fixed, the estimation of Σ^{-1} is obtained by setting the derivative of objective function in Equation (4) to zero:

$$\frac{\partial}{\partial \Sigma^{-1}} \left(\lambda_2 tr(B\Sigma^{-1}B^T) - D \log \left| \Sigma^{-1} \right| + \lambda_4 tr(\Sigma^{-1}) \right) = 0$$

$$\Rightarrow \lambda_2 B^T B - D\Sigma + \lambda_4 I_C = 0 \qquad (15)$$

$$\Rightarrow \Sigma^{-1} = \left(\frac{\lambda_2 B^T B + \lambda_4 I_C}{D} \right)^{-1}$$

Algorithm 2. Weight Coefficients Estimation

Input: $X \in \mathbb{R}^{N \times D}, Y \in \mathbb{R}^{N \times C}$, λ_1, λ_2, λ_3 and λ_4;
Output: The weight matrix $B \in \mathbb{R}^{D \times C}$;
1 Initialize Ω and Σ as two $C \times C$ identity matrices and $t = 0$;
2 **repeat**
3 | Compute SVD as $X^{(t+1)} = U_1 \Sigma_1 V_1^T$ and $P^{(t+1)} = U_2 \Sigma_2 V_2^T$;
4 | Compute $\tilde{B}^{(t+1)}$ by Equation (12);
5 | Update $B^{(t+1)}$ by Equation (13);
6 | Update $\Sigma^{-1(t+1)}$ by Equation (15);
7 | Update $\Omega^{-1(t+1)}$ by Equation (14);
8 | $t = t + 1$;
9 **until** *Convergence*;
10 **return** B;

---

5   Experiments

In this section, we demonstrate the superiority of our model over a set of state-of-the-art methods with extensive experiments on two benchmark datasets.

5.1   Datasets

We conduct the experiments on two publicly available benchmark datasets: Core5k [16] and NUS-WIDE [6]. Corel5k is a widely adopted dataset for keywords based image retrieval and image annotation, which consists of 4999 images with 260 tags. The average number of tags per image is 3.4. A fixed set of 4500 images are used for training and the rest are used for testing. As for the visual feature, we adopt the 1000-dimensional BoW representation provided by [8], which is constructed on multiscale grid-based densely extracted SIFT feature. NUS-WIDE is a Web image dataset created by Lab for Media Search in National University of Singapore for Web image annotation and retrieval. It includes 269,648 Flickr images associated with 5,018 unique tags and 81 concepts.

There are 5.8 words for each image averagely. Similar to the previous work [13], we select those images having at least 5 tags to construct a smaller dataset named NUS-WIDE-SUB, which consists of 23,818 images. We randomly split the whole dataset into two equally sized parts, where one part is for training and the other one is for testing. The 500-dimensional BoW representation based on SIFT descriptions is used for this dataset.

For each dataset, we firstly use PCA to reduce the visual feature's dimensionality by preserving 90% of the carried energy, resulting in the reduced 338- and 241-dimensional visual feature subspaces for Corel5k and NUS-WIDE-SUB respectively. In the meanwhile, we use the proposed feature selection to reduce the dimensionality of the tag feature space.

We recommend the top 1 to 5 tags for each image in Corel5k and the top 4 to 10 tags for the images in NUS-WIDE-SUB respectively. Regarding the evaluation criteria, we use precision and recall as two key performance indicators. The mean precision and recall over all the testing images are calculated to measure the final performance.

5.2 Parameters' Tuning

We first tune the five parameters used in our MorTca model. The first one is γ in Equation (3). The selection of γ determines the dimensionality of the reduced tag feature space. As observed from Table 1, which shows the results corresponding to different dimensionalities of the reduced tag feature space caused by different γ values, feature selection effectively saves the elapsed time of the optimization process as the dimensionality decreases without affecting the precision and recall noticeably. We set $\gamma = 100$ for its best precision and recall, which corresponds to the reduced 188-dimensional tag feature. An extremely small γ value (e.g., 0.01) suggests that no reduction is performed on the tag feature space. Notably, due to the proposed optimization algorithm, the proposed MorTca (column 4) reduces the optimization time greatly compared to the original multi-output regression model proposed in [17] (column 3), which demonstrates the efficiency of the proposed optimization process.

The rest four parameters are λ_1, λ_2, λ_3 and λ_4 in multi-output regression model (Equation (4)). The effect of λ_1 and λ_2 on dataset Corel5k is illustrated in Figure 2(a) and 2(b) respectively. As can be seen, both precision and recall are not sensitive to the changes of λ_1 and λ_2. When λ_1 and λ_2 change from 10

Table 1. Effect of γ on Time, Precision and Recall

γ	Dimensionality	Original MOR Time(s)	MorTca Time(s)	Precision	Recall
0.01	260	15395	6.989	0.3154	0.4449
100	188	1661	4.158	0.3158	0.4456
140	174	1304	3.954	0.3150	0.4446
170	158	1098	3.588	0.3146	0.4441
210	134	700	3.283	0.3110	0.4396

to 100,000, the tagging precision fluctuates slightly between 0.304 to 0.316 and recall varies from 0.427 to 0.456. Moreover, when these two parameters are less than 10, the precision and recall remain stable. So we fix both λ_1 and λ_2 as $1e5$, and use two-fold cross-validation with respect to precision and recall for the selection of λ_3 and λ_4. The results are in Figure 2(c) and 2(d), indicating that both λ_3 and λ_4 have significant impacts on the tagging performance. According to the results, we set both λ_3 and λ_4 to be $1e3$. We use the same process to tune parameters for NUS-WIDE-SUB, and get $1e5$ for λ_1 and λ_2, and $1e4$ for λ_3 and λ_4 in the following experiments.

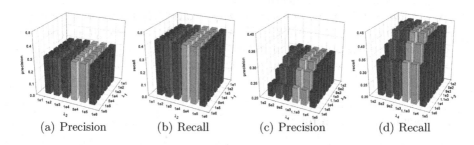

(a) Precision   (b) Recall   (c) Precision   (d) Recall

Fig. 2. Parameters' Tuning on Corel5k

5.3   Compared Algorithms

We compare our MorTca method with seven existing state-of-the-art tagging algorithms, including:

- **ML-LGC** [22] (Multi-Label Local and Global Consistency) decomposes the multi-label image annotation problem into a set of independent binary classification problems, which are solved by a label propagation procedure.
- **CNMF** [15] (Constrained Non-negative Matrix Factorization) formalizes the annotation problem as a constrained Non-negative Matrix Factorization (NMF) problem and optimizes the consistency between image visual (input patterns) and semantic (class memberships) similarities to label images.
- **SMSE** [5] (Semi-supervised Multi-label learning via Sylvester Equation) constructs two graphs on the instance level and the category level respectively and builds a regularization framework to combine two regularization terms for the two graphs to annotate images.
- **TagProp** [8] constructs a weighted nearest-neighbour model for annotation and combines a couple of similarity metrics by integrating metric learning which leads to a wide coverage of various image contents' aspects.
- **M-E Graph** [13] (Multi-Edge Graph) encodes each image as a region bag and constructs a multi-edge graph for tag propagation.
- **TagSearcher** [12] performs a tag-related random search process over the graphical model made up of range-constrained visual neighbours. It takes advantage of both visual and textual correlations for tag score prediction.

- **RFIA** [7] (Random Forest for Image Annotation) generates random trees based on tags, and designs semantic nearest neighbour and semantic similarity measure for its tag ranking algorithm.

A Naïve MorTca, which refers to our MorTca without considering the regression coefficients structure and noise structure, is also compared as a baseline.

5.4   Results and Analysis

Table 2 shows the performance comparisons with the above 10 algorithms over two datasets. From the results in Table 2, we can make the following observations:

- Among all image tagging methods, our proposed MorTca achieves the best performance in all metrics for both datasets except for precision on Corel5K. Although TagProp generates 3% higher precision, its recall is 6.7% lower than MorTca (i.e., 0.42 vs. 0.45).
- The unsatisfactory performance of Naïve MorTca proves the effectiveness of learning tag correlations in our model. Without considering the regression coefficients covariance and noise covariance, the precision declines 37.5% and 31% on Corel5k and NUS-WIDE-SUB respectively. Furthermore, there are bigger declines in the recall, i.e., 47% and 34% on two datasets respectively.

More detailed results of MorTca are illustrated in Figure 3. The precision and recall at various K values, i.e., the number of recommended tags, (K=1...8 for Core5k and K=4...10 for NUS-WIDE-SUB) are reported in Figure 3(a) and 3(b). The corresponding precision-recall curves are demonstrated in Figure 3(c). When the number of recommended tags is close to the average number of tags per image in the training set, we can get relatively good performance for both precision and recall. This shows that our experimental results are consistent with the real tag distribution among images.

Considering Naïve MorTca as a baseline method, we present some real tagging examples to show the advantages of considering tag correlations in MorTca. Take

Table 2. Performance Comparison of Different Methods on Benchmark Datasets

	Corel5k		NUS-WIDE-SUB	
method	Precision	Recall	Precision	Recall
ML-LGC [22]	0.22	0.24	0.28	0.29
CNMF [15]	0.24	0.27	0.29	0.31
SMSE [5]	0.23	0.28	0.32	0.32
TagProp [8]	**0.33**	0.42	-	-
M-E Graph [13]	0.25	0.31	0.35	0.37
TagSearcher [12]	0.32	0.35	-	-
RFIA [7]	0.29	0.41	-	-
Naïve MorTca	0.20	0.24	0.29	0.25
MorTca	0.32	**0.45**	**0.42**	**0.38**

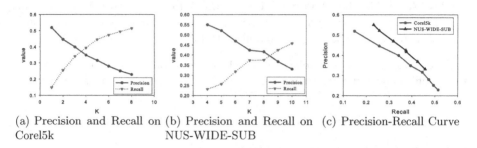

(a) Precision and Recall on Corel5k (b) Precision and Recall on NUS-WIDE-SUB (c) Precision-Recall Curve

Fig. 3. Precision and Recall on Corel5k and NUS-WIDE-SUB

Figure 4 (a) as an example. "building" and "window" are two high frequency tags in the training data set. They are recommended by Naïve MorTca. However, the proposed MorTca successfully remove these two tags and promote an important tag "beach" to a higher rank (i.e., from the 6th place to the 3rd place), because "beach" is highly correlated with some other recommended tags, such as "ocean" and "water".

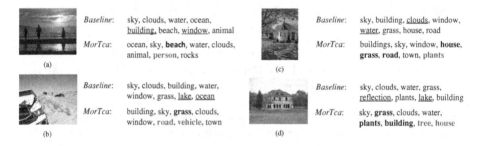

Fig. 4. Real Tagging Examples from NUS-WIDE Dataset. For each example image, we provide two lists of tags, recommended by Naïve MorTca and the proposed MorTca respectively. The underlined words are the irrelevant tags produced by Naïve MorTca according to the ground-truth. The words in bold are the tags, which are promoted by MorTca, compared with their ranking in the Naïve MorTca list. The words in red are correct tags, which cannot be found by Naïve MorTca.

6 Conclusions and Future Work

In this paper, we have presented a novel image tagging approach named MorTca. The approach converts the challenging annotation problem to a multi-output regression learning problem with task correlation and noise correlation analysis. Moreover, a feature selection method is designed to reduce the high dimensionality of tag feature space. We demonstrate the competitiveness of our approach to other state-of-the-art image tagging methods with extensive experimental results on two benchmark datasets. For the future work, we plan to directly integrate the dimensionality reduction step into the multi-output regression step to further improve the performance.

References

1. Bi, W., Kwok, J.T.: Multilabel classification on tree- and dag-structured hierarchies. In: ICML, pp. 17–24 (2011)
2. Breiman, L., Friedman, J.H.: Predicting multivariate responses in multiple linear regression. J. R. Stat. Soc. Ser. B Stat. Methodol. 59(1), 3–4 (2002)
3. Caruana, R.: Multitask learning. Machine Learning 28(1), 41–75 (1997)
4. Chang, E.Y., Goh, K., Sychay, G., Wu, G.: Cbsa: Content-based soft annotation for multimodal image retrieval using bayes point machines. IEEE Trans. Circuits Syst. Video Techn. 13(1), 26–38 (2003)
5. Chen, G., Song, Y., Wang, F., Zhang, C.: Semi-supervised multi-label learning by solving a sylvester equation. In: SDM, pp. 410–419 (2008)
6. Chua, T.-S., Tang, J., Hong, R., Li, H., Luo, Z., Zheng, Y.: Nus-wide: A real-world web image database from national university of singapore. In: CIVR (2009)
7. Fu, H., Zhang, Q., Qiu, G.: Random forest for image annotation. In: Fitzgibbon, A., Lazebnik, S., Perona, P., Sato, Y., Schmid, C. (eds.) ECCV 2012, Part VI. LNCS, vol. 7577, pp. 86–99. Springer, Heidelberg (2012)
8. Guillaumin, M., Mensink, T., Verbeek, J.J., Schmid, C.: Tagprop: Discriminative metric learning in nearest neighbor models for image auto-annotation. In: ICCV, pp. 309–316 (2009)
9. Jolliffe, I.T.: Principal Component Analysis, 2nd edn. Springer (October 2002)
10. Kim, S., Sohn, K.-A., Xing, E.P.: A multivariate regression approach to association analysis of a quantitative trait network. Bioinformatics 25(12) (2009)
11. Kim, S., Xing, E.P.: Tree-guided group lasso for multi-response regression with structured sparsity, with an application to eqtl mapping. Ann. Appl. Stat. 6(3), 1095–1117 (2012)
12. Lin, Z., Ding, G., Hu, M., Wang, J., Sun, J.: Automatic image annotation using tag-related random search over visual neighbors. In: CIKM, pp. 1784–1788 (2012)
13. Liu, D., Yan, S., Rui, Y., Zhang, H.-J.: Unified tag analysis with multi-edge graph. In: ACM Multimedia, pp. 25–34 (2010)
14. Liu, J., Wang, B., Li, M., Li, Z., Ma, W.-Y., Lu, H., Ma, S.: Dual cross-media relevance model for image annotation. In: ACM Multimedia, pp. 605–614 (2007)
15. Liu, Y., Jin, R., Yang, L.: Semi-supervised multi-label learning by constrained non-negative matrix factorization. In: AAAI, pp. 421–426 (2006)
16. Müller, H., Marchand-Maillet, S., Pun, T.: The truth about corel - evaluation in image retrieval. In: Lew, M., Sebe, N., Eakins, J.P. (eds.) CIVR 2002. LNCS, vol. 2383, pp. 38–49. Springer, Heidelberg (2002)
17. Rai, P., Kumar, A., Daumé III, H.: Simultaneously leveraging output and task structures for multiple-output regression. In: NIPS, pp. 3194–3202 (2012)
18. Rothman, A.J., Levina, E., Zhu, J.: Sparse multivariate regression with covariance estimation. J. Comput. Graph. Statist. 19(4), 947–962 (2010)
19. Smeulders, A.W.M., Worring, M., Santini, S., Gupta, A., Jain, R.: Content-based image retrieval at the end of the early years. IEEE Trans. Pattern Anal. Mach. Intell. 22(12), 1349–1380 (2000)

20. Wu, F., Yuan, Y., Rui, Y., Yan, S., Zhuang, Y.: Annotating web images using nova: Non-convex group sparsity. In: ACM Multimedia, pp. 509–518 (2012)
21. Zhang, Y., Yeung, D.-Y.: A convex formulation for learning task relationships in multi-task learning. In: UAI, pp. 733–442 (2010)
22. Zhou, D., Bousquet, O., Lal, T.N., Weston, J., Schölkopf, B.: Learning with local and global consistency. In: NIPS (2003)
23. Zhu, G., Yan, S., Ma, Y.: Image tag refinement towards low-rank, content-tag prior and error sparsity. In: ACM Multimedia, pp. 461–470 (2010)

The Ranking Based Constrained Document Clustering Method and Its Application to Social Event Detection

Taufik Sutanto and Richi Nayak

Queensland University of Technology,
Brisbane 4000, Australia
{taufikedy.sutanto,r.nayak}@qut.edu.au

Abstract. With the growing size and variety of social media files on the web, it's becoming critical to efficiently organize them into clusters for further processing. This paper presents a novel scalable constrained document clustering method that harnesses the power of search engines capable of dealing with large text data. Instead of calculating distance between the documents and all of the clusters' centroids, a neighborhood of best cluster candidates is chosen using a document ranking scheme. To make the method faster and less memory dependable, the in-memory and in-database processing are combined in a semi-incremental manner. This method has been extensively tested in the social event detection application. Empirical analysis shows that the proposed method is efficient both in computation and memory usage while producing notable accuracy.

Keywords: constrained clustering, ranking, social event detection.

1 Introduction

With media creation and sharing technology becoming ubiquitous, the current abundance of media files on the web is giving birth to many challenges and opportunities. It has become natural to mine social events related information for knowledge discovery and information retrieval [1–3]. Unsupervised document clustering is a popular method for grouping document collections based on the common characteristics that they share; however, this method suffers because of the sparsity and high dimensionality of the data [4]. Accuracy of this method can be improved by incorporating background knowledge [5], particularly in the domain of social media where some prior information is available. Nevertheless, the computational issue dealing with *big data* remains an interesting research challenge for clustering methods.

On the other hand, full-text search engines such as *Lucene*[1], *Solr*[2], and *Sphinx*[3] have shown their proficiency in handling large document collections

---

[1] http://lucene.apache.org
[2] http://lucene.apache.org/solr
[3] http://sphinxsearch.com

S.S. Bhowmick et al. (Eds.): DASFAA 2014, Part II, LNCS 8422, pp. 47–60, 2014.
© Springer International Publishing Switzerland 2014

for information retrieval. For example, the open source full-text search server *Sphinx* is capable of indexing more than 25 billion documents from over 9TB of data (6-12 MB of raw text per second per single CPU core) and seeking answers for queries in milliseconds [6]. Search engine technology, more specifically the ranking concept, has the potential to be applied to the area of large scale document clustering. However, document clustering and ranking in search engines are two different problems; embedding ranking in the clustering framework may well solve the scalability problem inherent with clustering methods.

Inspired by the concept of document ranking employed in search engines, we propose a novel efficient method, Constrained Incremental Clustering via Ranking (CICR), to improve scalability in semi-supervised document clustering. During the clustering process, for assigning a document into the appropriate cluster, a neighborhood of best cluster candidates is chosen using a document ranking scheme instead of calculating distance between the document and all of the clusters' centroids (or the rest of the documents). The method is semi-supervised because initial cluster centroids are generated using the label information available in some documents. Note that the proposed method can work in an unsupervised manner too by guessing initial centroids instead of being guided by some labeled data. Instead of putting all of the documents' information in memory as in batch processing, CICR incrementally process sequence of documents from the database to the algorithm. The proposed method can adjust the number of clusters as necessary during the clustering process according to data characteristics.

We tested the efficacy of our proposed method in the real-world social event detection (SED) task [2] as well as analyzing its performance using synthetic data. The SED problem is interesting not only because of the size of the dataset, but also because it has large and unfixed total number of clusters and it contains multi-domain attributes (a mixture of text and non-text data). In this paper, we explain the distinct properties of our proposed method that makes it suitable for clustering large data. Empirical analysis shows that the proposed method is computational and memory efficient without significant loss of clustering solution accuracy.

The rest of the paper is organized as follows. In section 2 related work on clustering and ranking is presented, and the distinction between our proposed method and existing works is highlighted. Section 3 details the proposed method including the ranking schemes used. Experimental setup and empirical analysis on the latest real-world problems from social media are reported in section 4. Finally, section 5 concludes this paper.

2   Related Work

There has been a long history of research that relates to document clustering and *Information Retrieval* (IR) . The bridge between the two is the legacy assumption that relevant documents should be grouped together in the same cluster [7]. Some studies have shown that clustering improves the relevancy of document retrieval [8, 9].

Meanwhile in the data mining community, a considerably newer trend of clustering, semi-supervised clustering has emerged. The semi-supervised clustering methods use additional prior knowledge for grouping the data, and have been reported to improve the clustering quality compared to the unsupervised clustering methods [5]. Based on how the prior knowledge is used, a semi-supervised clustering algorithm can be classified as (1) constrained clustering algorithm by using the known label information [10]; (2) specific clustering algorithm by adapting the similarity measures [11]; or (3) a mix of the two [12]. Our proposed method falls into the category of constrained clustering due to its use of limited labeled data in generating the initial centroids. We use the labeled data as the instance level constraint must-link, to decide which documents must belong to the same cluster.

The constrained KMeans clustering method (CKMeans) [5] has shown its effectiveness in clustering large, sparse and high dimensional data. However, the KMeans family of methods is found computationally challenged when the number of the cluster is large, due to the need of calculating the distance between the documents and all cluster centroids. Furthermore, this method is unable to adjust the varied number of clusters during the process, the K needs to be set in the beginning.

The proposed CICR method utilizes the concept of ranking from IR to develop a scalable incremental semi-supervised clustering method. To the best of our knowledge, it is the first time that a ranking scheme is used in constrained document clustering. Furthermore the concept of ranking is utilized in the clustering algorithm in order to make it more scalable without compromising accuracy. The closest work that we can find are Luo et al. [13] and Davidson et al. [14]. Luo et al. use neighbors to initialize a cluster in unsupervised clustering problem, and propose a new similarity function using the combination of cosine and link functions [13]. While, Davidson et al. use constraints incrementally in a non-distance learning clustering algorithm [14], in contrast CICR uses the neighborhood concept to select a small subset of best cluster candidates to gain better computational efficiency without having to introduce a new similarity function. Proposing an efficient method to solve clustering problems with large and unfixed total number of clusters is important as it is the common characteristic and requirement of current real-world databases.

3   Constrained Incremental Clustering via Ranking

In this section we elaborate the proposed method CICR. We will start by stating the problem formulation. Let the dataset be the collection of N documents: where $D = D_{train} \cup D_{test}$ and $D_{train} \cap D_{test} = \emptyset$. D_{train} denotes the set of labeled documents (called as *training data*) and D_{test} denotes a set of unlabeled document (called as *test data*). Each document is a set of terms:

$$d_i = \{t_1^i, t_2^i, \ldots, t_{n_i}^i\}, \quad i = 1, 2, 3, \ldots, N, \tag{1}$$

where n_i represents the length (number of terms) of document d_i. A set of clusters C is notated as $C = \{c_1, c_2, \ldots, c_K\}$. Initially, we set K equal to the number of clusters present in the training data (labeled documents). Since the documents in the test data is allowed to form new clusters, the value of K might change to K', where $K' > K$. Each cluster is the set of relevant documents, in which its atomic element is the union of terms from the documents that are contained within. It is denoted as:

$$c_i = \{d_1, d_2, \ldots, d_{p_i}\} = \bigcup_{j=1}^{p_i} d_j = \bigcup_{j=1}^{p_i} \{t_1^j, t_2^j, \ldots, t_{n_j}^j\}, \quad (2)$$

where p_i corresponds to the number of documents assigned to cluster c_i and n_j is the cardinality of set of terms $d_1 \cap d_2 \cap \ldots \cap d_{p_i}$. The term weight in the cluster is the average weight of the term appearing in all documents within the cluster.

We use the document length normalized tf-idf (term frequency-inverse document frequency) term weighting [15] to represent a document vector. For each term $t \in d_i$, $i = 1, 2, , N$, the weight is calculated as

$$w(t, d_i) = w_t^{d_i} = \left(\frac{log(|\{t \in d_i\}|) + 1}{\sum_{t \in d_i} log(|t \in d_i|) + 1}\right)\left(\frac{|u_t^i|}{1 + 0.0115|u_t^i|}\right)log\left(\frac{N - td}{td}\right), \quad (3)$$

where $|\{t \in d_i\}|$ is the number of terms t in d_i (term frequency), $|u_t^i|$ is the number of unique terms in document d_i and $td = \{d \in D : t \in d\}$ is the set of all documents in D that contains the term t. We further normalize all term weights to unit vector as in [4] such that $w_t^{d_i} = \frac{w_t^{d_i}}{||w^{d_i}||}$ for all terms in the document.

3.1   Ranking Scheme

In IR, a ranking function is used to determine the relevance level of a document according to a query. The CICR method proposes the use of ranking to determine the subset of relevant clusters neighborhood for a document. Four ranking schemes, tf-idf, BM25, BM25 with proximity ($BM25p$), and Sphinx specific ranking scheme $SPH04$ [6] have been utilized in CICR and experimented to find out the most appropriate scheme for CICR. The ranking formula used is briefly described below.

The BM25 implementation is slightly different from the original BM25 implementation [7]. In the original implementation, if $Q = \{q_1, q_2, \ldots, q_r\}$ is the given query of r terms then the original BM25 score of a document d given a query Q is

$$f_{BM25}(Q, d) = \sum_{i=1}^{r} idf(q_i)\frac{|\{q_i \in d_i\}|(k + 1)}{|\{q_i \in d_i\}| + k(1 - b + b\frac{d}{D})} \quad (4)$$

where $idf(q_i) = log(\frac{N - |qd| + 0.5}{|qd| + 0.5})$ and $|\overline{D}|$ is the average document length in D. The constant b is usually set to 0.75 and $k \in [1.2, 2.0]$.

In Sphinx BM25 implementation, document length is ignored. In other words the constant b is set to zero and $k = 1.2$. To be precise the following is how the

BM25 score is calculated in *Sphinx*

$$f_{BM25}(Q,d) = \sum_{i=1}^{r} |dq|(\frac{log(\frac{N-|dq|+1}{|dq|})}{log(1+N)})(|dq| + 1.2)^{-1}. \qquad (5)$$

There are several other differences of *BM25* implementation in *Sphinx* (including normalization), further details can be found in [6].

The *BM25* proximity rank is defined as

$$f_{BM25p}(Q,d) = 1000d_w + \lceil 999 f_{BM25}(Q,d) \rceil, \qquad (6)$$

where d_w is the document phrase weight calculated as the longest common subsequence (LCS) length between the query and the document. This value is the length of the sub-phrase found in the same order in the document.

The *Sphinx* specific ranking score *SPH04* is based on proximity *BM25* and calculated as

$$f_{SPH04}(Q,d) = 1000f_w + \lceil 999 f_{BM25}(Q,d) \rceil, \qquad (7)$$

where f_w is $\sum_{\forall fields} u_w(4 * LCS + efm(Q,f))$. u_w is user field weight (1 by default) and $efm(Q,f) = 3$ if there is an exact sub phrase match and equal to 2 if the first keyword matches.

3.2   Finding a Neighborhood of Clusters

We now explain how the neighborhood of clusters is formed. For simplicity *BM25* is used as an illustration, however, the same reasoning is applicable for other ranking schemes. Following the well-established research in IR, we also believe in the conjecture that documents that are relevant to one another should be grouped into the same cluster [7].

Let the set of unique terms in the metadata of the events' images be the query terms that will represent the document. Note that we consider the metadata in SED dataset as short-medium documents, in longer documents, terms selection or document summarization might be used to shorten the query. The relevance of clusters to a document is decided by the ranking function between the document and the clusters. Formally, given an arbitrary document d_i, $f_{BM25}^m(d_i, c)$ is calculated for all $c \in C$. The m most relevant clusters ranked with $f_{BM25}^m(d_i, c)$ are then used as a small subset of best candidates for cluster assignment of d_i (instead of all $c \in C$) (as illustrated in Figure 1). This process is repeated for all $d_i \in D$. We postulate that, for a given document d_i and two distinct clusters c_a and c_b, if $f_{BM25}(d_i, c_a) \geq f_{BM25}(d_i, c_b)$ then $sim^{cos}(d_i, c_a) \geq sim^{cos}(d_i, c_b)$ in majority of cases, where $sim^{cos}(d, c)$ denotes cosine similarity distance between a document d and a cluster c.

$$sim^{cos}(d_i, c_a) = \frac{[w_{t \in d_i}^i][w_{t \in c_j}^j]'}{||w^i||.||w^j||} = [w_{t \in d_i}^i][w_{t \in c_j}^j]'. \qquad (8)$$

The denominator $||w^i||.||w^c|| = 1$ because we normalized the term weight to unit length.

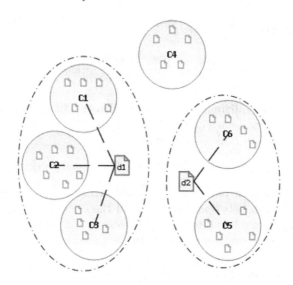

Fig. 1. CICR creates neighborhood of best cluster candidates

3.3   The Proposed Algorithm

In this section we will show how the ranking is incorporated to propose an efficient clustering algorithm. We start by restating the standard KMeans objective function, which is to locally minimize the sum of squared distances between documents d_i and cluster centroids μ_{c_i}:

$$\underset{\{\mu_{c_1},\mu_{c_2},\dots,\mu_{c_K}\}}{\text{minimize}} \quad \sum_{c_i \in C} \sum_{d_i \in D} \|d_i - \mu_{c_i}\|, \tag{9}$$

In CICR there are some distinctions: the number of cluster is not fixed, documents in training data are constrained to the label (cluster) assignment given, and the objective function is a function of text, space, and time data. Let $\delta(d_i, c_j)$ be an assignment of document d_i to cluster c_j, then the objective function of CICR under the text information only, can be represented as a constrained optimization:

$$\underset{\{\mu_{c_1},\mu_{c_2},\dots,\mu_{c_{K'}}\}}{\text{minimize}} \quad \sum_{c_i \in C'} \sum_{d_i \in D_{test}} \|d_i - \mu_{c_i}\| \tag{10}$$
$$\text{subject to} \qquad \delta(d_i, c_j), \qquad\qquad\qquad d_i \in D_{train}, c_j \in C$$

where $K' > K$ is the number of cluster that will be found in the clustering process using a pre-determined threshold value γ and $C' = \{c_1, c_2, \dots, c_{K'}\}$.

The detailed algorithm of CICR is given in Algorithm 1. The input is preprocessed documents from the training and test dataset as explained in section 4.1. The labeled training data is used to generate initial cluster centres. Documents from test data is incrementally assigned to these clusters or form new clusters if the distance of a document to available clusters is less than a threshold value γ. In the assignment phase, a small portion of best cluster candidates is chosen for each document in test data. A specific similarity function is used to decide the final cluster assignment. Document assignment to clusters is recorded to make the cluster centroid re-calculation more efficient.

---

input : Set of documents D, initial clusters $C = \{c_1, c_2, \ldots, c_K\}$,
             neighborhood size m, cluster threshold γ.
output: K' disjoint partitions of D, where $K' \geq K$ and the CICR objective
            function is optimized.

Using all the labeled data in D_{train}, initialize clusters $C = \{c_1, c_2, \ldots, c_K\}$
and set an array to store changed clusters $G = \emptyset$;
repeat
    for *each $d_i \in D_{test}$* **do**
        calculate set of clusters neighborhood $B = \{f^m_{rank}(d_i, c)\}$;
        for *each $c \in B$* **do**
            calculate $c* = max_c\{sim(d_i, c), c \in B\}$;
            if $sim(d_i, c*) > \gamma$ **then**
                Assign document d_i to cluster $c*$;
                $G = G \cup \{c*\}$;
            else
                Form a new cluster with d_i as its first member;
                $C = C \cup \{\text{the new cluster}\}$
            end
        end
    end
    Recalculate centroid based on G;
until *convergence*;

Algorithm 1. Constrained Incremental Clustering via Ranking

Assuming that the most outer loop requires R iterations to converge, the complexity yield by most existing KMeans based clustering methods such as CKMeans is $O(N \times K \times R)$. In CICR under the assumption that a query can be done in a (near) constant time, then complexity is reduced to $O(N \times m \times R)$ where $m << K$. In big data processing, it is common that the number of clusters (K) is large, hence even though it is in the same order, the improvement on performance is significant. An experimental study on synthetic data in the section 4.4 supports this claim. Moreover, it is much less than the document computing methods that need pair-wise documents processing resulting in the complexity of $O(N^2)$.

3.4 CICR Implementation: Making It Scalable

In this section we elaborate how we implemented CICR to make it scalable for large data processing. In order to optimally balance performance and memory usage, we combine the in-memory and in-database processing in a semi-incremental manner. To suppress memory usage, the algorithm received sequences of documents incrementally instead of all at once as in batch processing. The system architecture of CICR is explained in Figure 2.

We utilize *Sphinx* search engine [6] to index the set of texts in all clusters using the RT in-memory delta index. We would like to mention that once the in-memory index reaches its threshold size, the index is flushed to disk index, and the new memory index is created. The flushed disk index is still accessible for document retrieval and still gives descent performance. Detailed studies on the performance of Sphinx queries for various size of dataset can be found in [16].

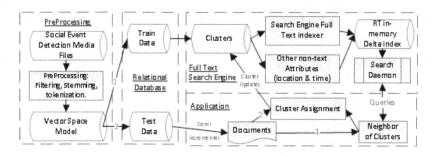

Fig. 2. The system architecture of CICR

4   Empirical Analysis: Application to Social Event Detection

The proposed algorithm is tested for its efficacy on the recently emerged real-world problem of Social Event Detection (*SED*). Social events are defined as activities that are planned, attended and reported by people [2]. A social event can be represented with event-related metadata (e.g., title, location, time, venue, description and performers), example tags or other social information, images, or a combination of the above. For example, an event or a set of events could be described as *"commonwealth game that took place in Delhi in October 2010"* with appropriate tags India, Games, Competition, etc.

The *SED* task is detecting media items related to the interesting happenings within a pool of data (represented by images accompanied with tags, time and location metadata, social information and many more) [2]. This task is apt for semi-supervised clustering as some of the events are labeled whereas some events are unlabeled. The objective of SED task is to identify which group an unlabeled event belongs to by using the limited information on labeled data.

4.1   Problem Statement and Pre-processing

More than 430,000 images from the social media *Flickr* were used in the *SED* task [2]. Metadata from these images were used to organize the media according to some labels given in the training data. Distinct initial cluster numbers in the training data was found to be more than 14,000. A sample of *SED* images is given in Figure 3. The *SED* data consists of several attributes that record information of location (*latitude* and *longitude*), time (*date taken* and *date upload*), and text (*username, description, tag,* and *title*). This task requires semi-supervised clustering, that is to cluster all images in the test data (around 30%) based on initial labels given in the training data (70%) using the metadata information. Further detail of the task and data is given in [2].

The standard text pre-processing were applied. All non-text characters (symbols) were replaced by a single white space and all text data was converted to lowercase. English stop words removal and stemming were applied to the text data. For each image, all text attributes such as *title, tag, username,* and *description* were combined into a text field and were treated as short document that represent an event. The time information attributes (*date upload* and *date taken*) were transformed into day interval between the two dates.

Fig. 3. Example of social event pictures from the *MediaEval* 2013 *SED* data

4.2 Similarity Measure

In order to apply CICR to the *SED* task, a linear combination of similarity measures between multiple domain (text, time, and space) were used. Cosine similarity was used for the text information, day interval for the time information, and geo-distance for the space information. Spatial distances were calculated using the *Haversine* formula [17] from the *latitude* and *longitude* attributes of documents and the mean of *latitude* and *longitude* of clusters. The distance was then normalized to unit value by infinity norm. Formally, let $sim^{space}(d, c)$

defined as a space distance between a document d and a cluster c, as follows:

$$sim^{space}(d,c) = 1 - \frac{H(d,c)}{||H(d,C^*)||_\infty}, \qquad (11)$$

where C^* is set of clusters from the query ranking and $H(d,c)$ is the Haversine formula (great circle distance formula)

$$H(d,c) = R \ cos^{-1}(sin(lat_d)sin(lat_c)+cos(lat_d)cos(lat_c)cos(lon_c-lon_d)), \quad (12)$$

where R is the earth radius (approximately 6378.10 km) and $lat_d, lon_d, lat_c,$ and lon_c are the *latitude* and *longitude* attribute values of the document and the cluster. The time similarity value between a document and a cluster is calculated as a simple difference between the time of an event (document) and the mean time of events in a cluster.

$$sim^{time}(d,c) = 1 - \frac{|t_d - t_c|}{||t||_\infty}, \qquad (13)$$

where t_d and t_c are the time difference between *date upload* and *date taken* attributes in the document and cluster respectively. The denominator is the maximum value of absolute differences between *date taken* and *date upload*.

The total similarity measure between a document d and a cluster c is then given by:

$$sim(d,c) = \alpha sim^{cos}(d,c) + \beta sim^{time}(d,c) + \beta sim^{space}(d,c), \qquad (14)$$

where $0 \le \alpha \le 1$ and $\beta = \frac{1-\alpha}{2}$. When time or location information is missing in the data, parameter α and β were modified such that it will sum to one. We tested the effectiveness of our similarity measure by comparing the combined measure to the clustering result based on text information only (as reported in table 1).

4.3   Accuracy Results

Several clustering solutions were generated based on different ranking methods and similarity measures. Clustering parameters were fine-tuned based on training data. For clustering the test data, threshold value was set to $\gamma = 0.3$, the size of the cluster neighborhood $m = 5$, and similarity weight was set to $\alpha = 0.9$. By setting higher similarity weight, we favor our algorithm to the text information due to three fold reasons: (1). A significant portion of values are missing in the location ($> 50\%$) and time ($> 13\%$) attributes. (2) The location information must be precise to be included in calculation to be reliably used as an event indicator. This was not the case in our dataset. (3) The time information alone is not reliable enough to recognize events.

The first four experiments use *tf-idf*, *BM25*, *BM25p* and *SPH04* ranking formula and the multi-domain similarity measure. The last run in the experiment

Table 1. CICR results using different ranking schemes

	tf-idf	BM25	BM25p	SPH04	SPH04t
F1-Score	0.804	0.811	0.802	**0.812**	0.784
NMI		0.949| 0.953	0.951	**0.954**	0.943

(*SPH04t*) was conducted to test the effectiveness of our similarity measure by measuring only text information and using the *SPH04* ranking formula.

Table 1 shows that the multi-domain similarity measure effectively improves the clustering quality. It ascertains that the other meta information has also some effect on grouping the image data. It can also be noted that the selection of ranking scheme has a minor effect on changing the quality of clustering results. This is partly due to the use of bag-of-words model in CICR, some important phrases that might be used by the ranking scheme to find more relevant clusters is not yet optimally exploited. Nevertheless, the *SPH04* ranking scheme, however, performs the best amongst all schemes.

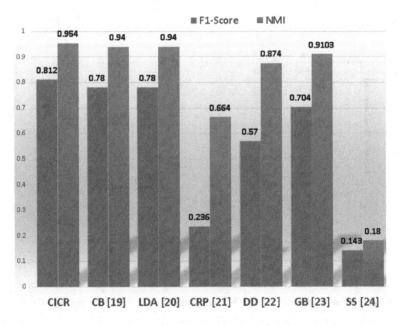

Fig. 4. Summary of results from participants in *SED MediaEval 2013*

We benchmarked CICR with other methods that have used the *SED* task data. Results are shown in Figure 4. *CB* is a constraint based method that uses support vector classifier and query refinement [18]. Topic modeling method using *LDA* was done in [19] and another topic modeling by using Chinese Restaurant Process (*CRP*) similarity measure is used in [20]. *DD* is a data driven method using spatio-temporal information and cluster inter-correlations [21]. *GB* uses the graph based clustering [22] and *SS* is clustering through Semantic Similarity [23]. The superiority of CICR indicates the efficacy of its approach in using query ranking in constrained document clustering.

4.4   Performance and Memory Usage

To illustrate the significant performance gain with CICR, a numerical study was done using a synthetic data. One million documents with 1000 text attributes was generated and several clustering solutions was derived with different number of clusters K and different neighborhoods size m. As shown in Figure 5, CICR performance was relatively unchanged with the increasing number of clusters, whereas, the running time of CKMeans grew exponentially with the number of clusters. Results also show that CICR is not affected by the neighborhood size due to its support of in-memory and in-database processing.

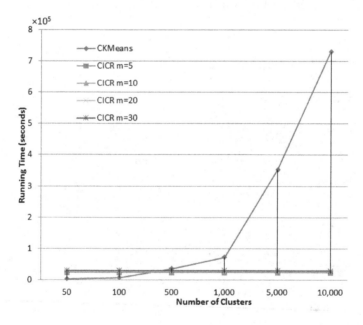

Fig. 5. The performance of CKMeans and CICR on the synthetic data

A simple investigation on the memory usage of CICR and its comparison with *LDA* was done. Using 50,000 records and about 100,000 text attributes, we generated 3,000 topics using the *Matlab modelling LDA toolbox* [24]. Matlab required more than 30GB of computer memory to generate 3,000 topics automatically. In contrast, the combination of in-memory and in-database process of CICR only required less than 1GB of memory for one million records with about 200,000 text attributes.

5   Conclusion

This paper explored the concept of query ranking in clustering large size document collections. We presented a novel constrained document clustering via ranking method that incrementally assigns the documents to clusters. The query ranking is used to create neighborhoods of cluster candidates to speed up computations while matching a document to clusters for forming clusters. Experiments on the real-world problem of social event detection and on synthetic data indicate that the proposed method is not only scalable, but also gives a notable accuracy and requires less memory.

References

1. Reuter, T., Cimiano, P.: Event-based classification of social media streams. In: Proceedings of the 2nd ACM International Conference on Multimedia Retrieval, ICMR 2012, pp. 22:1–22:8. ACM, New York (2012)
2. Reuter, T., Papadopoulos, S., Petkos, G., Mezaris, V., Kompatsiaris, Y., Cimiano, P., de Vries, C., Geva, S.: Social event detection at mediaeval 2013: Challenges, datasets, and evaluation. In: Proceedings of the MediaEval 2013 Multimedia Benchmark Workshop Barcelona, Spain, October 18-19, vol. 1043, CEUR-WS.org (2013)
3. Petkos, G., Papadopoulos, S., Kompatsiaris, Y.: Social event detection using multimodal clustering and integrating supervisory signals. In: Proceedings of the 2nd ACM International Conference on Multimedia Retrieval, ICMR 2012, pp. 23:1–23:8. ACM, New York (2012)
4. Dhillon, I.S., Fan, J., Guan, Y.: Efficient clustering of very large document collections. In: Grossman, R., Kamath, C., Kumar, V., Namburu, R.R. (eds.) Data Mining for Scientific and Engineering Applications, pp. 357–381. Kluwer Academic Publishers (2001) (Invited book chapter)
5. Basu, S., Banerjee, A., Mooney, R.J.: Semi-supervised clustering by seeding. In: Proceedings of the Nineteenth International Conference on Machine Learning, ICML 2002, pp. 27–34. Morgan Kaufmann Publishers Inc., San Francisco (2002)
6. Aksyonoff, A.: Introduction to Search with Sphinx: From installation to relevance tuning. O'Reilly (2011)
7. Jardine, N., van Rijsbergen, C.J.: The use of hierarchic clustering in information retrieval. Information Storage and Retrieval 7(5), 217–240 (1971)
8. Lin, Y., Li, W., Chen, K., Liu, Y.: Model formulation: A document clustering and ranking system for exploring medline citations. Journal of the American Medical Informatics Association 14(5), 651–661 (2007)
9. Cai, X., Li, W.: Ranking through clustering: An integrated approach to multi-document summarization. IEEE Transactions on Audio, Speech, and Language Processing 21(7), 1424–1433 (2013)
10. Basu, S., Davidson, I., Wagstaff, K.: Constrained Clustering: Advances in Algorithms, Theory, and Applications, 1st edn. Chapman & Hall/CRC (2008)
11. Bilenko, M., Basu, S., Mooney, R.J.: Integrating constraints and metric learning in semi-supervised clustering. In: Proceedings of the Twenty-First International Conference on Machine Learning, ICML 2004, pp. 11–18. ACM, New York (2004)
12. Basu, S., Bilenko, M., Mooney, R.J.: A probabilistic framework for semi-supervised clustering. In: Proceedings of the Tenth ACM SIGKDD International Conference on Knowledge Discovery and Data Mining, KDD 2004, pp. 59–68. ACM, New York (2004)

13. Luo, C., Li, Y., Chung, S.M.: Text document clustering based on neighbors. Data and Knowledge Engineering 68(11), 1271–1288 (2009)
14. Davidson, I., Ravi, S.S., Ester, M.: Efficient incremental constrained clustering. In: Proceedings of the 13th ACM SIGKDD International Conference on Knowledge Discovery and Data Mining, KDD 2007, pp. 240–249. ACM, New York (2007)
15. Singhal, A., Buckley, C., Mitra, M.: Pivoted document length normalization. In: Proceedings of the 19th Annual International ACM SIGIR Conference on Research and Development in Information Retrieval, SIGIR 1996, pp. 21–29. ACM, New York (1996)
16. Schutz, J.: Sphinx search engine comparative benchmarks (2011) (Online; accessed January 6, 2014)
17. Sinnott, R.W.: Sky and telescope. Virtues of the Haversine 68(2), 159 (1984)
18. Brenner, M., Izquierdo, E.: Mediaeval 2013: Social event detection, retrieval and classification in collaborative photo collections. In: Working Notes Proceedings of the MediaEval 2013 Multimedia Benchmark Workshop Barcelona, Spain, October 18-19, vol. 1043, CEUR-WS.org (2013)
19. Zeppelzauer, M., Zaharieva, M., del Fabro, M.: Unsupervised clustering of social events. In: Working Notes Proceedings of the MediaEval 2013 Multimedia Benchmark Workshop Barcelona, Spain, October 18-19, 2013. Volume 1043. CEUR-WS.org (2013)
20. Papaoikonomou, A., Tserpes, K., Kardara, M., Varvarigou, T.A.: A similarity-based chinese restaurant process for social event detection. In: Working Notes Proceedings of the MediaEval 2013 Multimedia Benchmark Workshop Barcelona, Spain, October 18-19, vol. 1043. CEUR-WS.org (2013)
21. Rafailidis, D., Semertzidis, T., Lazaridis, M., Strintzis, M.G., Daras, P.: A data-driven approach for social event detection. In: Working Notes Proceedings of the MediaEval 2013 Multimedia Benchmark Workshop Barcelona, Spain, October 18-19, vol. 1043. CEUR-WS.org (2013)
22. Schinas, M., Mantziou, E., Papadopoulos, S., Petkos, G., Kompatsiaris, Y.: Certh @ mediaeval 2013 social event detection task. In: Working Notes Proceedings of the MediaEval 2013 Multimedia Benchmark Workshop Barcelona, Spain, October 18-19, vol. 1043. CEUR-WS.org (2013)
23. Gupta, I., Gautam, K., Chandramouli, K.: Vit@mediaeval 2013 social event detection task: Semantic structuring of complementary information for clustering events. In: Working Notes Proceedings of the MediaEval 2013 Multimedia Benchmark Workshop Barcelona, Spain, October 18-19, vol. 1043. CEUR-WS.org (2013)
24. Steyvers, M., Griffiths, T.: Probabilistic topic models. In: Latent Semantic Analysis: A Road to Meaning. Laurence Erlbaum (2007)

A Skylining Approach to Optimize Influence and Cost in Location Selection

Juwei Shi[1], Hua Lu[2], Jiaheng Lu[1], and Chengxuan Liao[1]

[1] School of Information and DEKE, MOE in Renmin University of China
[2] Department of Computer Science, Aalborg University, Denmark
juwei.shi@gmail.com, luhua@cs.aau.dk,
jiahenglu@ruc.edu.cn, liaochengxuan@gmail.com

Abstract. Location-selection problem underlines many spatial decision-making applications. In this paper, we study an interesting location-selection problem which can find many applications such as banking outlet and hotel locations selections. In particular, given a number of spatial objects and a set of location candidates, we select some locations which maximize the influence but minimize the cost. The influence of a location is defined by the number of spatial objects within a given distance; and the cost of a location is indicated by the minimum payment for such location, which is measured by quality vectors. We show that a straightforward extension of a skyline approach is inefficient, as it needs to compute the influence and cost for all the location candidates relying on many expensive range queries. To overcome this weakness, we extend the Branch and Bound Skyline (BBS) method with a novel spatial join algorithm. We derive influence and cost bounds to prune irrelevant R-tree entries and to early confirm part of the final answers. Theoretical analysis and extensive experiments demonstrate the efficiency and scalability of our proposed algorithms.

1 Introduction

Location selection has been an emerging problem with many commercial applications. For example, telecom service providers store huge volumes of location data to provide data monetization applications to the third party, such as banking outlet and hotel locations selections. In many scenarios, additional attributes besides the location are available in a spatial object. For example, a hotel has a spatial position as well as quality attributes such as star, service, etc. These attributes can improve the price-performance of selected locations. Unfortunately, traditionally location selections only take the spatial distance into account [7, 22, 25], which ignore non-spatial attributes.

In this paper, we select locations in terms of both spatial distances and quality vectors. In particular, we select locations to maximize their *influences* but minimize their *costs*. Given a distance threshold δ, a location l's *influence* is measured by the number of existing spatial objects within the distance δ from l. It indicates how many objects the location can potentially influence. As shown

S.S. Bhowmick et al. (Eds.): DASFAA 2014, Part II, LNCS 8422, pp. 61–76, 2014.

in Figure 1(a), there are four location candidates (l_1, l_2, l_3 and l_4) and a bunch of existing spatial objects. Here l_2 impacts the most number of objects since its δ-neighborhood contains four existing spatial objects.

Given a distance threshold δ, a location l's *cost* is measured by the payment for obtaining minimal quality vectors [1] to *dominate* all the existing objects within the distance δ from l. The concept of *dominance* [3] is proposed to compare two quality vectors. One quality vector v_i dominates another vector v_j if v_i is no worse than v_j on all attributes and better than v_j on at least one attribute. As shown in Figure 1(a), the minimal quality vector to dominate all the objects in l_2's δ-neighborhood is $\langle 5, 10, 5 \rangle$. Furthermore, we assume there is a monotonic function which maps a quality vector to a numerical cost. For instance, the cost of l_2 can be defined as $f(l_2) = f(\langle 5, 10, 5 \rangle) = \frac{1}{2} \cdot (5 + 10 + 5) = 10$.

In this paper, we adopt the skyline query [3] to define our location selection problem. Given a set of objects, the skyline operator returns a subset of objects such that the object in the subset is not dominated by any other objects. In particular, we select locations whose influence and cost are not dominated by any other locations. The skyline points of locations in Figure 1(a) are shown in Figure 1(b). Suppose that the influence and cost are $\langle 3, 10 \rangle$ for l_1, $\langle 4, 10 \rangle$ for l_2, $\langle 2, 5 \rangle$ for l_3, and $\langle 1, 6 \rangle$ for l_4. Then $\langle l_2, l_3 \rangle$ is the result of the skyline location selection.

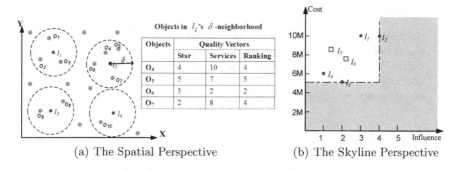

(a) The Spatial Perspective   (b) The Skyline Perspective

Fig. 1. The Example in the Spatial and Skyline Perspective

A straightforward solution to address the skyline location selection problem is that we compute the influence and the cost for each of the location candidates, and then use existing skyline query algorithms to generate the skyline points. For large data sets this is infeasible since it relies on expensive range queries to compute the influence and the cost for all the locations.

We develop an efficient algorithm to answer a skyline location selection query. First, we build two R-trees on the location set L and the object set O, denoted

---

[1] Without loss of generality, we assume that larger values are preferred in the dominance comparison throughout the paper.

as R_L and R_O, respectively. At high-level, the algorithm descends the two R-trees in the branch and bound manner, progressively joining R_L entries with R_O entries to compute the two bounds of the influence and the cost for each entry e_L in R_L; then based on the generated skyline points, the algorithm decides whether to prune an entry e_L, or to access its children until all leaf entries of R_L are accessed. During the two R-trees traversal, we use a min-heap to control the accessing order of R_L entries, which can ensure that an irrelevant R-tree node is not visited before its dominance skyline point is generated.

The contributions of this paper are summarized as follows.

- We define a type of optimal location selection problem that takes into account not only spatial distance but also non-spatial quality vectors associated to locations. Our problem definition returns optimal locations that can potentially influent the largest number of objects in proximity at the lowest costs. Our approach employs the skyline dominance concept that is popular for multi-criteria optimization, and therefore it requires no specific user intervention in selecting optimal locations.
- We propose a novel location selection algorithm. Tight influence and cost bounds are derived to prune irrelevant R-tree entries and to early confirm part of the final answers.
- We provide theoretical analysis on the IO cost of our algorithm based on a spatial join cost model.
- We conduct extensive experiments to evaluate our algorithm under various settings.

The remainder of this paper is organized as follows. Section 2 gives the definitions and the problem statement. Section 3 proposes our algorithm for the problem. Section 4 provides the IO cost analysis of our algorithm. Section 5 evaluates our proposed algorithm experimentally. Section 6 reviews related works. Finally, section 7 concludes the paper and discusses the future work.

2 Problem Definition

In this section, we formally define the location selection problem. A location is a point $\lambda = \langle x, y \rangle$ where x and y are coordinate values in a 2-dimensional space. We assume that there are c quality dimensions and $\mathcal{D} = D_1 \times D_2 \times \ldots \times D_c$ is the quality space. A quality vector is a c-dimensional point $p = (d_1, d_2, \ldots, d_c)$, where $d_i \in D_i (1 \leq i \leq c)$. Then, a spatial object o is composed of a location λ and a quality vector ψ associated with that location, i.e., $o = \langle \lambda, \psi \rangle$. We use $o.loc$ and $o.p$ to denote a location object o's location and quality vector respectively. For a spatial object set O, we use $\pi_P(O)$ to denote all the quality vectors associated with spatial objects in O, i.e. $\pi_P(O) = \{o.p \mid o \in O\}$. We define the key components of our problem as follows.

δ-Neighborhood: Given a location loc, a spatial object set O, and a distance threshold δ, loc's δ-Neighborhood, termed as $N_\delta(loc, O)$, is the subset of objects in O that are within the distance δ from loc. Formally, $N_\delta(loc, O) = \{o \mid o \in O \wedge \|o.loc, o\| \leq \delta\}$. The cardinality of a location's δ-Neighborhood indicates the location's potential influence.

Quality Dominance: Given two quality vectors p and p', p dominates p' if p is no worse than p' on all attributes and p is better than p' on at least one attribute. We use $p \prec p'$ to denote that p dominates p'. Given a quality vector p, there can be multiple quality vectors that dominate p. Each of them is called p's quality dominator. Given a quality vector set P, we use $p \prec P$ to indicate that p is P's dominator.

Cost Functions: Given a quality vector $q \in \mathcal{D}$, the cost function $f^g_{cost}(q)$ returns a cost value of real type, i.e. $f^g_{cost} : \mathcal{D} \to \mathcal{R}$. For example, a cost function can be defined as the weighted sum of all quality attribute values, i.e., $f^c_{cost}(q) = \sum_{i=1}^{c} w_i \cdot q.d_i$. Here, w_i is zero if quality dimension D_i has no impact on the dominance cost; $w_i > 0$ if the dominance cost is proportional to the values on quality dimension D_i; otherwise $w_i < 0$. Note that it is natural to define cost functions that are monotonic with respect to dominance. We say a cost function f_{cost} is *monotonic* if and only if $f_{cost}(q) \geq f_{cost}(q')$ for any two quality vectors that satisfy $q \prec q'$. This is consistent with the intuition that better quality is achieved at a higher cost.

Minimum Quality Dominance Cost: Among all the quality dominators, we are interested in the one with the minimum cost. Given a quality space $\mathcal{D}$, a set of quality vectors $P \subseteq \mathcal{D}$, and a cost function f_{cost}, we use $\mathbb{D}(P)$ to denote P's minimum cost quality dominator such that

1. $\mathbb{D}(P) \prec P$;
2. $\forall p' \in \mathcal{D}$ and $p' \prec P$, $f_{cost}(p') \geq f_{cost}(\mathbb{D}(P))$.

Then, we define $\mathbb{C}(P) = f_{cost}(\mathbb{D}(P))$ to denote P's *Minimum Quality Dominance Cost*.

Note that $\mathbb{D}(P)$ is a quality vector and $\mathbb{C}(P)$ is a scalar value. As we assume large values are preferred in the dominance comparison, $\mathbb{D}(P).d_i = \min\{(p.d_i) \mid p \prec P\}$.

Location Dominance: Given a spatial object set O, a distance threshold δ, and a cost function f_{cost}, a location loc_1 *location dominates* another location loc_2, termed as $loc_1 \prec loc_2$, if and only if

1. $|N_\delta(loc_1, O)| \geq |N_\delta(loc_2, O)|$;
2. $\mathbb{C}(\pi_P(N_\delta(loc_1, O))) \leq \mathbb{C}(\pi_P(N_\delta(loc_2, O)))$;
3. $|N_\delta(loc_1, O)| > |N_\delta(loc_2, O)|$ or $\mathbb{C}(\pi_P(N_\delta(loc_1, O))) < \mathbb{C}(\pi_P(N_\delta(loc_2, O)))$.

With all definitions formulated above, we give the problem statement of optimal location selection as follows.

Problem 1. Given a location set L, a spatial object set O, a distance threshold δ, and a cost function f_{cost}, an optimal location selection returns from L a subset of locations that are not location dominated by any others. Formally, the Optimal Location Selection (OLS) problem is defined as:

$$OLS(L, O, \delta) = \{l \mid l \in L, \nexists l' \in L \wedge l' \prec l\} \qquad (1)$$

3   Algorithms for Optimal Location Selection

In this section, we present algorithms for defined optimal location selection. We start from the naive loop algorithm, and then describe the spatial join based algorithm.

Algorithm 1. Loop(Spatial object set O's R-tree R_O, location set L, distance threshold δ)

1: $S \leftarrow \emptyset$
2: **for** each location $l \in L$ **do**
3:     $N_\delta(l, O) \leftarrow$ Range_Query(l, δ, R_O)
4:     $p \leftarrow (0, \ldots 0)$ $\triangleright$ c-dimensional point
5:     **for** each object $o \in N_\delta(l, O)$ **do**
6:         $p[i] \leftarrow \max(p[i], o.p[i])$
7:     $cost \leftarrow f_{cost}(p)$ $\triangleright$ $\mathbb{C}(\pi_P(N_\delta(l, O)))$
8:     $influence \leftarrow |N_\delta(l, O)|$
9:     $candidate \leftarrow (l, influence, cost)$
10:    dominanceCheck$(S, candidate)$
11: **return** S

Algorithm 2. dominanceCheck(Current skyline S, a candidate $candidate$)

1: $flag \leftarrow$ false
2: **for** each tuple $tp \in S$ **do**
3:     **if** $tp \prec candidate$ **then**
4:         $flag \leftarrow$ true; **break**
5:     **else if** $candidate \prec tp$ **then**
6:         remove tp from S
7: **if** $flag =$ false **then**
8:     add $candidate$ to S

3.1   The Loop Algorithm

We develop a loop algorithm shown in Algorithm 1 as the baseline algorithm. The idea is that: for each location $l \in L$, we issue a range query on the object set O to get l's δ-neighborhood $N_\delta(l, O)$. All objects in the δ-neighborhood are checked to obtain the minimum cost quality dominator. Then, the influence and cost of a location candidate are computed. Finally, the candidate is checked against all generated optimal locations in terms of location dominance. This dominance check is shown in Algorithm 2.

3.2 The Join Algorithm

To reduce the computation overhead of influence and cost, we propose a spatial join based algorithm. The basic idea is that we make use of the R-tree [9] based spatial join [4] to find the δ-neighborhood for location candidates.

Suppose that spatial attributes of the location set L and the object set O are indexed by the R-trees R_L and R_O respectively. The locations in L are joined with the objects in O based on the two R-trees: an entry e_L from R_L are joined with a set of overlapped entries from R_O. These relevant entries are defined as e_L's *join list*. We use $e_L.JL$ to denote e_L's join list. Intuitively, we avoid to find the joint list from the whole object data set, and make use of the spatial join to obtain the joint list only from relevant object R-tree entries. Thus the IO cost of operations on the object set O is significantly reduced.

To further reduce the overhead of influence and cost computations for irrelevant location candidates, we derive the influence bound and the cost bound for all locations in a given location R-tree entry e_L. The two bounds are used in the join algorithm to prune irrelevant R_L and R_O nodes.

Influence Bound. We introduce the δ-Minkow-ski region [2] to derive the influence bound. Given a distance threshold δ, a location entry e_L's δ-Minkow-ski region, denoted by $\Xi(e_L, \delta)$, is the set of all locations whose minimum distance from e_L is within threshold δ. Formally, we define

$$\Xi(e_L, \delta) = \{t \in \mathbb{R}^2 \mid dist_{min}(t, e_L) \leq \delta\} \tag{2}$$

We are interested in those objects from O that fall into Region $\Xi(e_L, \delta)$. Accordingly, we define e_L's δ-Minkowski region with respect to O as follows.

$$\Xi_O(e_L, \delta) = \{o \in O \mid o.loc \in \Xi(e_L, \delta)\} \tag{3}$$

Given a spatial object set O and a distance threshold δ, we define the *influence bound* of e_L, termed as $BI_{O,\delta}(e_L)$, to be the number of objects in $\Xi_O(e_L, \delta)$.

$$BI_{O,\delta}(e_L) = |\Xi_O(e_L, \delta)| \tag{4}$$

If e_L' is a descent entry of e_L, we have $\Xi_O(e_L', \delta) \subseteq \Xi_O(e_L, \delta)$. Therefore, we have the following lemma that guarantees the correctness of the influence bound.

Lemma 1. *Given a spatial object set O, a distance threshold δ, and a location entry e_L, any location l's influence cannot be larger than e_L's influence bound, i.e. $|N_\delta(l, O)| \leq BI_{O,\delta}(e_L)$.*

Proof. The lemma is proved by the fact that $\forall l \in e_L$, $N_\delta(l, O) \subseteq \Xi_O(e_L, \delta)$.

Cost Bound. Given a spatial object set O and a distance threshold δ, we term the *cost bound* of e_L as $BC_{O,\delta}(e_L)$. Intuitively, $BC_{O,\delta}(e_L)$ can be the cost to dominate the most disadvantaged quality vector among all that are associated to locations in e_L. Since we assume large values are preferred in the dominance comparison, the most disadvantaged quality vector is the one that has the minimum value on all attributes. Then, that (virtual) quality vector is defined as $mdq(\Xi_O(e_L, \delta))$ where $mdq.d_i = \min\{p.d_i \mid p \in \pi_P(\Xi_O(e_L, \delta))\}$.

Given a quality cost function f_{cost}, we define the cost bound $BC_{O,\delta}(e_L)$ as the cost to dominate this most disadvantaged quality vector.

$$BC_{O,\delta}(e_L) = f_{cost}(mdq(\Xi_O(e_L, \delta))) \tag{5}$$

The correctness of this cost bound is guaranteed by the following lemma.

Lemma 2. *Given a spatial object set O, a distance threshold δ, and an location entry e_L, e_L's cost bound is larger than or equal to that of its any descent entry e'_L, i.e. $BC_{O,\delta}(e'_L) \geq BC_{O,\delta}(e_L)$.*

Proof. Suppose the most disadvantaged quality vectors in $\Xi_O(e_L, \delta)$ and $\Xi_O(e'_L, \delta)$ are mdq and mdq' respectively. We have $mdq.d_i = \min\{p.d_i \mid p \in \pi_P(\Xi_O(e_L, \delta))\}$ and $mdq'.d_i = \min\{p.d_i \mid p \in \pi_P(\Xi_O(e'_L, \delta))\}$. Since $e'_L \subseteq e_L$, we have $\Xi_O(e'_L, \delta) \subseteq \Xi_O(e_L, \delta)$. Therefore, we have $mdq'.d_i \geq mdq.d_i$. Due to the monotonicity of the quality cost function f_{cost}, we have $f_{cost}(mdq') \geq f_{cost}(mdq)$, i.e., $BC_{O,\delta}(e'_L) \geq BC_{O,\delta}(e_L)$.

As a remark, the cost of a location in the entry e_L satisfies $\mathbb{C}(\pi_P(N_\delta(l, O))) = f_{cost}(\mathbb{D}(\pi_P(N_\delta(l, O)))) \geq BC_{O,\delta}(e_L)$.

The Join Algorithm. We propose the join algorithm in Algorithm 3 and Algorithm 4. To make use of the influence bound, we index the object set O with an aggregate R-tree R_O in which each non-leaf node entry e has an additional filed $e.count$. Here $e.count$ is the total number of all spatial objects in e. Similarly, to make use of the cost bound, each non-leaf node entry e in R_O has another additional filed $e.\psi$, which is a quality vector defined as follows.

$$e.\psi.d_i = \min\{o.p.d_i \mid o \in e\} \tag{6}$$

Thus, we extend each non-leaf node entry e in object R-tree R_O with two extra fields $e.count$ and $e.\psi$. Since the calculation of the two bounds is in the course of the spatial join, no additional IO costs on R_O and R_L are introduced. Accessing all location entries in R_L is prioritized by a min-heap. A location entry e_L is pushed to the heap with a key which equals to $BC_{O,\delta}(e_L) - \sum_{e \in e_L.JL} e.count$, i.e. the difference between the influence and cost bound. When the value of two keys are the same, we randomly select one entry as the lower value key. The min-heap ensures that irrelevant R-tree nodes will not be visited before its dominance skyline point is generated.

As a remark, our algorithm follows the spirit of the well-established Branch-and-Bound Skyline (BBS) algorithm [15] that prioritizes R-tree node access to ensure that skyline points are always generated before their dominating R-tree nodes are visited. The difference is that we integrate the branch-and-bound to the spatial join algorithm such that the θ-neighborhood of a location can be efficiently found.

Algorithm 3. Join_Together(Spatial object set O's combined R-tree R_O, location set L's R-tree R_L, distance threshold δ)

1: $S \leftarrow \emptyset$
2: initialize a min-heap H
3: $e_{root} \leftarrow R_L.root$; $e_{root}.JL \leftarrow \{R_O.root\}$
4: enheap($H, \langle e_{root}, e_{root}.JL, 0, 0, 0 \rangle$)
5: **while** H is not empty **do**
6: 　　$\langle e_L, e_L.JL, count, cost, v \rangle \leftarrow$ deheap(H)
7: 　　**if** $\exists tp \in S$ s.t. $tp \prec (*, count, cost)$ **then**
8: 　　　　**continue**
9: 　　**if** e_L is a leaf entry **then**
10: 　　　　$l \leftarrow$ the location pointed by e_L
11: 　　　　$influence \leftarrow count$
12: 　　　　$candidate \leftarrow (l, influence, cost)$
13: 　　　　add $candidate$ to S
14: 　　**else**
15: 　　　　read the child node CN_L pointed to by e_L
16: 　　　　**for** each entry e_i in CN_L **do**
17: 　　　　　　$count \leftarrow 0$; $e_i.JL \leftarrow \emptyset$
18: 　　　　　　**for** each e_j in $e_L.JL$ **do**
19: 　　　　　　　　**if** $\Xi(e_i, \delta)$ contains e_j **then**
20: 　　　　　　　　　　add e_j to $e_i.JL$; $count \leftarrow count + e_j.count$
21: 　　　　　　　　**else**
22: 　　　　　　　　　　read the child node CN_O pointed to by e_j
23: 　　　　　　　　　　Minkowski($e_i, \delta, CN_O, e_i.JL, count$)
24: 　　　　　　**for** each entry $e \in e_i.JL$ **do**
25: 　　　　　　　　$p[i] \leftarrow \min(p[i], e.\psi[i])$
26: 　　　　　　$cost \leftarrow f_{cost}(p)$
27: 　　　　　　**if** $\exists tp \in S$ s.t. $tp \prec (*, count, cost)$ **then**
28: 　　　　　　　　**continue**
29: 　　　　　　**else**
30: 　　　　　　　　enheap($H, \langle e_i, e_i.JL, count, cost, cost - count \rangle$)
31: **return** S

Algorithm 4. Minkowski(R-tree R_L's entry e_L, distance threshold δ, aggregate R-tree R_O's node CN_O, aggregate R-tree R_O's entry list JL, count v)

1: **for** each child entry e in CN_O **do**
2: 　　**if** $\Xi(e_L, \delta)$ contains e **then**
3: 　　　　add e to JL; $v \leftarrow v + e.count$
4: 　　**else**
5: 　　　　read the child node CN_P pointed to by e
6: 　　　　Minkowski(e_L, δ, CN_P, JL, v)

4　Analysis

In this section, we first prove the correctness of the proposed algorithm. Then, we provide IO cost analysis for our algorithm.

4.1   Correctness of the Algorithm

The proof of the correctness is similar to that proposed in [15]. We use B_i and B_c to denote the influence and cost bound respectively. The difference is that our algorithm visits entries of the location R-tree R_L in ascending order based on the distance between $\langle B_i, B_c \rangle$ and $\langle \infty, 0 \rangle$ on the influence-cost formed coordinate plane. It is straightforward to prove that our algorithm never prunes a location entry of R_L which contains skyline points.

4.2   IO Cost Analysis

To quantify the IO cost of the proposed algorithm, we extend the concept of Skyline Search Region (SSR) proposed by [15]. In this paper, the SSR is the area defined by the skyline points and the two axes of influence and cost. For example, the SSR area is shaded in Figure 1(b). Our algorithm must access all the nodes whose $\langle B_i, B_c \rangle$ falls into the SSR. In other word, if a node does not contain any skyline points but its $\langle B_i, B_c \rangle$ falls into SSR, it will also be visited if it has not been pruned.

Lemma 3. *If the influence and cost bound of an object entry e does not intersect the SSR, then there is a skyline point p whose distance to $\langle \infty, 0 \rangle$ is smaller than the distance between e and $\langle \infty, 0 \rangle$.*

Proof. Since the influence and cost bounds of the object R-tree entry dominate that of all its child node, p dominates all the leaf nodes covered by e.

Theorem 1. *An entry of the location R-tree will be pruned if its influence and cost bounds $\langle B_i, B_c \rangle$ fall into the SSR.*

Proof. We prove it based on Lemma 3 and the fact that we visit R_L in the order of $B_i - B_c$. Based on the min-heap structure, if there is a skyline point that dominates the entry bounded by $\langle B_i, B_c \rangle$, the skyline point will be visited earlier than that entry. Thus the entry will be pruned when it is popped up from the heap.

Next, we derive IO cost of the Join_Together algorithm based on the cost model proposed in [29]. Let $P_L(i)$ be the probability that a level i node's $\langle B_i, B_c \rangle$ is contained by the SSR. The number of node accesses at the ith level of the location R-tree R_L equals:

$$NA_L(i) = \frac{N_L}{f_L^{i+1}} \cdot P_L(i) \tag{7}$$

where N_L is the cardinality of the location candidate data set L and f_L is the node fan-out of R_L. Let $P_L(\alpha, \beta, i)$ be the probability that $\langle B_i, B_c \rangle$ of a level i node of R_L is contained by the rectangle with the corner points $\langle \infty, 0 \rangle$ and $\langle \alpha, \beta \rangle$. The density of the influence and cost equals:

$$D_L(\alpha, \beta, i) = \frac{\partial^2 P(\alpha, \beta, i)}{\partial \alpha \partial \beta} \tag{8}$$

Then we have

$$P_L(i) = \int\int_{\langle x,y \rangle \in SSR} D_L(x,y,i)dxdy \qquad (9)$$

where x and y is the influence and cost bounds of R_L entries respectively. Thus we obtain the IO cost of location R-tree:

$$NA_L = \sum_{i=1}^{h_L-1} NA_L(i) \qquad (10)$$

where h_L is the height of the location R-tree R_L.

Lemma 4. *To get the join list of an entry in R_L, we only expand the join list of its parent.*

Proof. The lemma is proved by the fact that the join list of a entry is computed based on its parent's join list from the heap.

Let $P_O(j)$ be the probability that a level j node from R_O intersects with unpruned entries from R_L, we have

$$NA_O(j) = \frac{N_O}{f_O^{j+1}} \cdot P_O(j) \cdot f_L \qquad (11)$$

where N_O is the cardinality of the object data set O and f_O is the node fan-out of R_O, and we have

$$P_O(i) = \int\int_{\langle x,y \rangle \in L_{unpruned}} D_L(x,y,i)dxdy \qquad (12)$$

where $\langle x,y \rangle \in L_{unpruned}$ denotes the location of a level j node intersects with an unpruned R_L node. Thus we obtain the IO cost of the object R-tree R_O:

$$NA_O = \sum_{j=1}^{h_O-1} NA_O(j) \qquad (13)$$

where h_O is the height of the location R-tree R_O.

Finally, the number of node accesses of both R_L and R_O equals

$$NA = NA_L + NA_O \qquad (14)$$

5   Experimental Studies

5.1   Settings

We use both real and synthetic data sets in our experiments. The real world US hotel (USH) data set consists of 30,918 hotel records with the schema (longitude, latitude, review, stars, price). For all hotel records, their locations (longitude and latitude) are normalized to the domain $[0, 10000] \times [0, 10000]$, and their quality attributes are normalized to the domain $[0, 1]^3$. We perform value conversions on quality attributes to make smaller values preferable. We randomly extract 918 hotels from USH and use their locations as the location set L. The remaining 3000 hotels form the object set O.

We also generate an object set with three independent quality attributes, and another object set with three anti-correlated quality attributes. All quality attribute values are normalized to the range $[0, 1]$. Both object sets contain 100,000 objects whose locations are randomly assigned within the space $[0, 10000] \times [0, 10000]$. As larger quality attribute values are preferred in our setting, we employ a cost function $f_{cost}^c(q) = \sum_{i=1}^c q.d_i$ in all experiments.

We set the page size to 4 KB when building the R-trees. All trees have node capacities between 83 and 169. All algorithms are implemented in Java and run on a Windows platform with Intel Core 2 CPU (2.54GHz) and 2.0 GB memory.

5.2   Performance of Location Selection Algorithms

We report an experimental evaluation of skyline location selection algorithms, namely Loop (Algorithm 1) and Join_Together (Algorithm 3). To study the effect of each bound separately, we add Join_Influence and Join_Cost which use either the influence or the cost bound only.

In the Join_Influence algorithm, each non-leaf Object R-tree entry e has an extra filed $e.count$ that is the total number of all spatial objects in e. Accessing all location entries in R_L is prioritized by a max-heap with a key which equals to the influence bound $\sum_{e \in e_L.JL} e.count$. Similarly, in the Join_Cost algorithm, an extra filed $e.\psi$ is added to the object R-tree entry e. Here $e.\psi$ is defined as $e.\psi.d_i = \min\{o.p.d_i \mid o \in e\}$. e_L is pushed to a min-heap with a key which equals to the cost bound $BC_{O,\delta}(e_L)$.

The Impact of the Number of Query Locations: In order to study the impact of the number of query locations, we vary the number of query locations in L. All locations in each L are generated at random with the spatial domain $[0, 10000] \times [0, 10000]$. We set the distance threshold δ to 800.

Figure 2 and Figure 3 show the results of the loop and the join algorithms, respectively. The join algorithms are significantly more efficient than the loop algorithm under each setting in the experiments. Because the join algorithms leverage the R-tree based spatial join to prune many irrelevant nodes. Figure 2 and Figure 3 indicate that the response time of the skyline location selection increases as the number of locations increases for all the algorithms. Figure 3 indicates that the Join_Together performs better than Join_Influence and Join_Cost. This is because that Join_Together makes use of both bounds to prune more irrelevant nodes.

The Impact of the Distance Threshold δ: Next, we evaluate the impact of distance threshold δ for all the join algorithms. We use 3000 locations on both real and synthetic data sets. We vary δ using 80, 400, 800, 2000 and 5000.

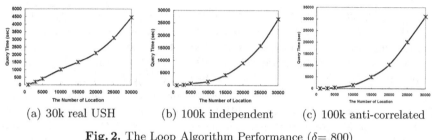

(a) 30k real USH   (b) 100k independent   (c) 100k anti-correlated

Fig. 2. The Loop Algorithm Performance ($\delta = 800$)

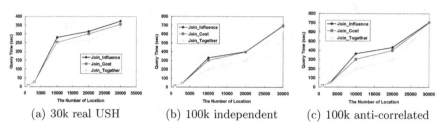

(a) 30k real USH   (b) 100k independent   (c) 100k anti-correlated

Fig. 3. The Join Algorithm Performance ($\delta = 800$)

The results shown in Figure 4 indicate that the response time of skyline location selection increases when the distance threshold δ is increased. Because a larger distance indicates that more spatial objects will be involved in the spatial join and subsequent checks.

The Impact of the Query Location Coverage Area: Finally, we evaluate the impact of the query location coverage area, i.e., the region of all query locations in L. We set the number of query locations to 3000, and the distance threshold δ to 800. We first use a set of small query location coverage areas that varies from 0.4% to 8.0% of the entire space of interest. The result is shown in Figure 5(a). The Join_Cost outperforms the other two algorithms when all query locations are distributed in a very small part of the entire space. It indicates that the cost bound is more effective than the influence bound when all query locations are very close. When locations are close, their Minkowski regions tend to overlap intensively, which weakens the influence bound based pruning that counts on the number of objects in Minkowski regions.

We also use a set of large query location coverage areas that vary from 8% to 50% of the entire space. The result is shown in Figure 5(b). We see that the Join_Together outperforms the other two algorithms. When the query locations cover a larger area, there is less overlap among their Minkowski regions. Then, the influence bound become more effective. Therefore, the combination of both bounds performs the best among all algorithms.

Summary: The experimental results show that our proposed spatial join algorithms outperforms the baseline method in the skyline location selection. The

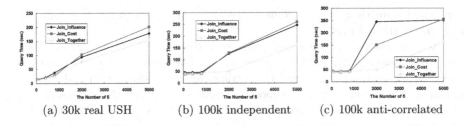

(a) 30k real USH   (b) 100k independent   (c) 100k anti-correlated

Fig. 4. The Effect of δ (3000 query locations)

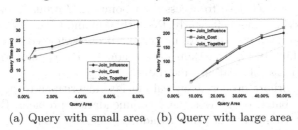

(a) Query with small area (b) Query with large area

Fig. 5. The Query Area, 3000 locations, $\delta = 800$

combined optimization with the influence and the cost bounds achieves the best performance in most cases in our experiments.

6   Related Work

Spatial Location Selection. A nearest neighbor (NN) query returns the locations that are closest to a given location. A NN query can be efficiently processed via an R-tree on the location data set, in either a depth-first search [17] or a best-first search [10]. In contrast, the optimal location selection query in this paper considers not only the spatial distances but also quality attributes.

So far in the literature, various constraints have been proposed to extend the nearest neighbor concept to select semantically optimal locations or objects. Du et al. [7] proposed the optimal-location query which returns a location with maximum influence. Xia et al. [22] defined a different top-t most influential spatial sites query, which returns t sites with the largest influences. Within the same context, Zhang et al. [25] proposed the min-dist optimal-location query. However, these proposals do not consider quality attributes of spatial objects.

Skyline Queries. Borzonyi et al. [3] defined the skyline query as a database operator, and gave two skyline algorithms: *Block Nested Loop* (BNL) and *Divide-and-Conquer* (D&C). Chomicki et al. [5] proposed a variant of BNL called the *Sort-Filter-Skyline* (SFS) algorithm. Godfrey et al. [8] provided a comprehensive analysis of these non-index-based algorithms and propose a hybrid method with improvements. Bartolini et al. [1] proposed a presorting based algorithm that is able to stop dominance tests early. Zhang et al. [26] proposed a dynamic indexing tree for skyline points (not for the data set), which helps reduce CPU costs in sort-based algorithms [1, 5, 8]. None of the above skyline algorithms require any indexing of the data set.

Alternative skyline algorithms require specific indexes. Tan et al. [20] proposed two progressive algorithms: *Bitmap* and *Index*. The former represents points by means of bit vectors, while the latter utilizes data transformation and B^+-tree indexing. Kossmann et al. [6] proposed a *Nearest Neighbor* (NN) method that identifies skyline points by recursively invoking R*-tree based depth-first NN search over different data portions. Papadias et al. [15] proposed a *Branch-and-Bound Skyline* (BBS) method that employs an R-tree on the data set. Lee et al. [12] proposed ZB-tree to access data points in Z-order in order to compute/update skylines more efficiently. Recently, Liu and Chan [14] improved the ZB-tree with a nested encoding to further speed up skyline computation. However, these skyline query algorithms do not address the computation overhead of influence and cost of location candidates.

In [27], the authors proposed several efficient algorithms to process skyline view queries in batch to address the recommendation problem. Hu et al. [28] proposed a deterministic algorithm to address the I/O issue of skyline query. However, none of these works can be directly applied to solve the optimal location selection problem proposed in this paper. The Cost Bound has the same principle as the pseudo documents in IR-tree [16], but we use a spatial join to answer a skyline location selection query.

7   Conclusion and Future Work

In this paper, we defined a skyline location selection problem to maximize the influence and minimize the cost. We proposed a spatial join algorithm that can prune irrelevant R-tree nodes. We also conducted theoretical analysis about the IO cost of the algorithm. The extensive experiments demonstrated the efficiency and scalability of the proposed algorithm. As for the future works, we are extending the skyline location selection algorithm to achieve the low-cost scalability in a distributed data management framework.

Acknowledgment. This paper is partially supported by 863 National High-tech Research Plan of China (No. 2012AA011001), NSF China (No: 61170011), NSSF China (No: 12&ZD220) and Research Funds of Renmin University of China (No.11XNJ003).

References

1. Bartolini, I., Ciaccia, P., Patella, M.: Efficient sort-based skyline evaluation. ACM Trans. Database Syst. (TODS) 33(4) (2008)
2. Böhm, C.: A cost model for query processing in high dimensional data spaces. ACM Trans. Database Syst. (TODS) 25(2), 129–178 (2000)
3. Börzsönyi, S., Kossmann, D., Stocker, K.: The skyline operator. In: Proc. ICDE, pp. 421–430 (2001)

 4. Brinkhoff, T., Kriegel, H.-P., Seeger, B.: Efficient processing of spatial joins using r-trees. In: Proc. SIGMOD, pp. 237–246 (1993)
 5. Chomicki, J., Godfrey, P., Gryz, J., Liang, D.: Skyline with presorting. In: Proc. ICDE, pp. 717–719 (2003)
 6. Kossmann, D., Ramsak, F., Rost, S.: Shooting stars in the sky: An online algorithm for skyline queries. In: Proc. VLDB, pp. 275–286 (2002)
 7. Du, Y., Zhang, D., Xia, T.: The optimal-location query. In: Medeiros, C.B., Egenhofer, M., Bertino, E. (eds.) SSTD 2005. LNCS, vol. 3633, pp. 163–180. Springer, Heidelberg (2005)
 8. Godfrey, P., Shipley, R., Gryz, J.: Maximal vector computation in large data sets. In: Proc. VLDB, pp. 229–240 (2005)
 9. Guttman, A.: R-trees: A dynamic index structure for spatial searching. In: Proc. SIGMOD, pp. 47–57 (1984)
10. Hjaltason, G., Samet, H.: Distance browsing in spatial databases. ACM Trans. on Database Syst. (TODS) 24(2), 265–318 (1999)
11. Korn, F., Muthukrishnan, S.: Influence sets based on reverse nearest neighbor queries. In: ACM SIGMOD Record, vol. 29, pp. 201–212 (2000)
12. Lee, K.C.K., Zheng, B., Li, H., Lee, W.-C.: Approaching the skyline in z order. In: Proc. VLDB, pp. 279–290 (2007)
13. Li, C., Tung, A.K.H., Jin, W., Ester, M.: On dominating your neighborhood profitably. In: Proc. VLDB, pp. 818–829 (2007)
14. Liu, B., Chan, C.-Y.: Zinc: Efficient indexing for skyline computation. PVLDB 4(3), 197–207 (2010)
15. Papadias, D., Tao, Y., Fu, G., Seeger, B.: An optimal and progressive algorithm for skyline queries. In: Proc. SIGMOD, pp. 467–478 (2003)
16. Cong, G., Jensen, C.S., Wu, D.: Efficient retrieval of the top-k most relevant spatial web objects. PVLDB 2(1), 337–348 (2009)
17. Roussopoulos, N., Kelley, S., Vincent, F.: Nearest neighbor queries. In: ACM SIGMOD Record, vol. 24, pp. 71–79 (1995)
18. Stanoi, I., Agrawal, D., Abbadi, A.: Reverse nearest neighbor queries for dynamic databases. In: ACM SIGMOD Workshop on Research Issues in Data Mining and Knowledge Discovery, pp. 44–53 (2000)
19. Stanoi, I., Riedewald, M., Agrawal, D., El Abbadi, A.: Discovery of influence sets in frequently updated databases. In: Proc. VLDB, pp. 99–108 (2001)
20. Tan, K.L., Eng, P.K., Ooi, B.C.: Efficient progressive skyline computation. In: Proc. VLDB, pp. 301–310 (2001)
21. Tao, Y., Papadias, D., Lian, X.: Reverse knn search in arbitrary dimensionality. In: Proc. VLDB, pp. 744–755 (2004)
22. Xia, T., Zhang, D., Kanoulas, E., Du, Y.: On computing top-t most influential spatial sites. In: Proc. VLDB, pp. 946–957 (2005)
23. Xiao, X., Yao, B., Li, F.: Optimal location queries in road network databases. In: Proc. ICDE, pp. 804–815 (2011)
24. Yang, C., Lin, K.: An index structure for efficient reverse nearest neighbor queries. In: Proc. ICDE, pp. 485–492 (2001)
25. Zhang, D., Du, Y., Xia, T., Tao, Y.: Progressive computation of the min-dist optimal-location query. In: Proc. VLDB, pp. 643–654 (2006)
26. Zhang, S., Mamoulis, N., Cheung, D.W.: Scalable skyline computation using object-based space partitioning. In: Proc. SIGMOD, pp. 483–494 (2009)

27. Chen, J., Huang, J., Jiang, B., Pei, J., Yin, J.: Recommendations for two-way selections using skyline view queries. In: Knowledge and Information Systems, pp. 397–424 (2013)
28. Hu, X., Sheng, C., Tao, Y., Yang, Y., Zhou, S.: Output-sensitive Skyline Algorithms in External Memory. In: Proc. SODA, pp. 887–900 (2013)
29. Theodoridis, Y., Stefanakis, E.: T.K. Sellis Efficient Cost Models for Spatial Queries Using R-Trees. IEEE Trans. Knowl. Data Eng (TKDE) 12(1), 19–32 (2000)

Geo-Social Skyline Queries

Tobias Emrich[1], Maximilian Franzke[1], Nikos Mamoulis[2], Matthias Renz[1],
and Andreas Züfle[1]

[1] Ludwig-Maximilians-Universität München
{emrich,franzke,renz,zuefle}@dbs.ifi.lmu.de
[2] The Hong Kong University
nikos@cs.hku.hk

Abstract. By leveraging the capabilities of modern GPS-equipped mobile devices providing social-networking services, the interest in developing advanced services that combine location-based services with social networking services is growing drastically. Based on geo-social networks that couple personal location information with personal social context information, such services are facilitated by geo-social queries that extract useful information combining social relationships and current locations of the users. In this paper, we tackle the problem of geo-social skyline queries, a problem that has not been addressed so far. Given a set of persons $\mathcal{D}$ connected in a social network SN with information about their current location, a geo-social skyline query reports for a given user $U \in \mathcal{D}$ and a given location P (not necessarily the location of the user) the pareto-optimal set of persons who are close to P and closely connected to U in SN. We measure the social connectivity between users using the widely adoted, but very expensive Random Walk with Restart method (RWR) to obtain the social distance between users in the social network. We propose an efficient solution by showing how the RWR-distance can be bounded efficiently and effectively in order to identify true hits and true drops early. Our experimental evaluation shows that our presented pruning techniques allow to vastly reduce the number of objects for which a more exact social distance has to be computed, by using our proposed bounds only.

1 Introduction

In real life, we are connected to people. Some of these connections may be stronger than others. For example, for some individual the strength of a social connection may monotonically decrease from their partner, family, friends, colleagues, and neighbours to strangers. Social connections (or their lack of) define social networks that extend further than just a person's acquaintances: There are friends of friends, stepmothers and contractors that stand in an indirect relation to a person; eventually reaching every person in the network. Such social networks are used, consciously or unconsciously, by everybody to find amiable people for a plethora of reasons: To find people to join a common event such as a concert or to have a drink together or to find help, such as a handyman or an expert in a specific domain. Yet, the person with the strongest social connection to may not be the proper choice due to non-social aspects. This person may not be able to help with a specific problem due to lack of expertise, or the person may simply be too far away to join. For example, in a scenario where your car has

S.S. Bhowmick et al. (Eds.): DASFAA 2014, Part II, LNCS 8422, pp. 77–91, 2014.
© Springer International Publishing Switzerland 2014

broken down, you are likely to accept the help of a stranger. When you are travelling, for example visiting a conference in Bali, the people you are strongly connected to are likely to be too far away to join you for a drink.

Also, another interesting problem arises when travelling and you need a place to stay for the night. Assume you do not want to book a hotel but rather sleep at someone's home (aka "couchsurfing"). Of course it's most convenient if you have a strong social connection to someone close to your destination, but the farther you travel from home, the less likely this becomes, as most of your acquaintances are usually spatially close to you [22]. Trying to find the person best suited to accommodate you, you face the following trade-off: Rather stay with someone you have less social connections to but can provide shelter close to your destination or you accept longer transfer times and choose to stay with someone more familiar. Since this trade-off depends on personal preferences, a skyline query is suitable: By performing a skyline query, a user obtains a list of people, with each person's attributes being a pareto-optimum between social distance to the user and spatial distance to their destination. Driven by such applications, there is a new trend of novel services enabled by geo-social networks coupling social network functionality with location-based services. A geo-social network is a graph where nodes represent users with information about their current location and edges correspond to friendship relations between the users [3]. User locations are typically provided by modern GPS-equipped mobile devices enabling check-in functionality, i.e. the user is able to publish his current location by "checking in" at some place, like a restaurant or a shop. Example applications based on geo-social networks are Foursquare and novel editions of Facebook and Twitter that adopted the check-in functionality recently.

Extracting useful information out of geo-social networks by means of geo-social queries taking both the social relationships and the (current) location of users into account is a new and challenging problem, first appraoches have been introduced recently [3]. In this paper we tackle the geo-social skyline query problem. This is the first approach for this problem. Given a set of persons $\mathcal{D}$ connected in a social network SN with information about their current location a geo-social skyline query reports for a given user $U \in \mathcal{D}$ and a given location P (not necessarily the location of the user) the pareto-optimal set of persons who are close to P and closely connected to U in SN. In particular we present and study initial approaches to compute the geo-social skyline efficiently which is challenging as we apply the very expensive Random Walk with Restart distance to measure the social connectivity between users in the social network. The basic idea of our approach is that for skyline-queries it is not necessary to calculate exact social distances to all users, which is very expensive. In a nutshell, we efficiently determine lower and upper distance bounds used to identify the skyline. The bounds are iteratively refined on demand allowing early termination of the refinement process.

2   Problem Definition

In the following we will define the problem of geo-social skyline query tackled in this paper. Furthermore, we discuss methods for measuring the social similarity in social networks which we apply to compute the geo-social skyline.

2.1   Geo-Social Skyline Query

The problem of answering a geo-social skyline query is formally defined as follows.

Definition 1. *Geo-Social Skyline Query (GSSQ)*
Let $\mathcal{D}$ be a geo-social database, consisting of a set of users U, where each user $u \in U$ is associated with a geo-location $u.loc \in \mathcal{R}^2$. Let $S = (<U>, <L>)$ be a social network consisting of a set $<U>$ of vertices corresponding to users, and a set $<L>$ of weighted links between users. Furthermore, let $dist_{geo} : U \times U \to \mathcal{R}$ be a geo-spatial distance measure, and let $dist_{soc} : U \times U \to \mathcal{R}$ be a social distance measure. A geo-social skyline query (GSSQ) returns, for a given geo-location $q_{geo} \in \mathcal{R}^2$ and a given user $q_{soc} \in U$ the set of users $u \in GSSQ(q_{geo}, q_{soc}) \subseteq U$, s.t.

$$u \in GSSQ(q_{geo}, q_{soc}) \Leftrightarrow$$
$$\neg\exists u' \in U : dist_{soc}(q_{soc}, u') < dist_{soc}(q_{soc}, u) \wedge dist_{geo}(q_{geo}, u') < dist_{geo}(q_{geo}, u)$$

Informally, a *GSSQ* returns, for a given geo-location q_{geo} and a given user q_{soc}, the set of users such that for each user $u \in GSSQ(q_{geo}, q_{soc})$ there exists no other user $u' \in U$ such that user u has both a larger spatial distance to q_{geo} and a larger social distance to q_{soc} than user u'.

By design, we exclude the query node from the result set: In the foreseeable applications including oneself in the result gives no additional information gain, but may actually prune and therefore exclude other nodes from the result. This design decision is without loss of generality; depending on application details the query node itself may be allowed to be a valid result as well.

Yet, we haven't specified the distance measures involved in the *GSSQ*. In the following, we discuss the notion of similarity in the two domains in involved in our query problem, the geo-spatial domain and the social domain and introduce the two similarity distance measures, the geo-spatial distance and social distance, respectively.

An example of a *GSS*-query is shown in Figure 1. The social network, the geo-location of each user and the social distance of each user to Q (numbers in black boxes) are shown in Figure 1(a). The resulting skyline-space is illustrated in Figure 1(b), with users G, E, F and D being the result of the query. The semantics of the result in this example is also very interesting. Ranked by distance we get the following result: user G is more or less a stranger but has the closest distance, user E is a friend-of-a-friend, user F is a close friend and user D is the best friend which however has a very large distance. The result thus gives the user Q free choice in the trade-off between importance of social and spatial distance.

2.2   Geo-Spatial Similarity

For the geo-spatial distance, instead of using the common Euclidean distance, we decided to use the geodetic distance which is the shortest distance between two points on Earth along the Earth's surface (simplified as a sphere). Given that locations are specified by longitude and latitude coordinates, this distance measure is more adequate, in particular for long distance measures, than the Euclidean distance and is depending on

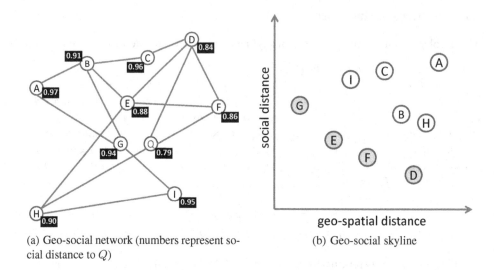

(a) Geo-social network (numbers represent social distance to Q)

(b) Geo-social skyline

Fig. 1. The geo-social skyline of a geo-social network

the Earth's radius r (approx. 6371km). The geo-spatial distance is defined as:

$$dist_{\text{geo}}(u_1^{\text{lat}}, u_1^{\text{long}}, u_2^{\text{lat}}, u_2^{\text{long}}) = r \cdot arccos(sin(u_1^{\text{lat}}) \cdot sin(u_2^{\text{lat}}) + cos(u_1^{\text{long}} - u_2^{\text{long}}))$$

2.3 Social Similarity

There are two main factors that contribute to someone's importance in a social graph: There are on the one hand nodes that have a lot of connections (sometimes referred to as "hubs" or "influencers") - on Twitter this may be Justin Bieber or Barack Obama. On the other hand you have nodes that are close to the query node - these people may be considered important because they are in close relation to the query. Please note that the first set of nodes is query-independent: Justin Bieber will always have the same amount of followers regardless of what node you are looking at. So both measures of importance for the query have to be taken into account. By how much is defined by a personalisation factor α. α may be determined by empirical studies and may be application-dependant or even query-dependant and chosen by the user, so we are considering α as variable and do not suggest or assume any specific value for it, besides it having to be from the range of $[0, 1]$.

Random Walk with Restart. To measure similarity between nodes, a widely adopted model is Random Walk with Restart (RWR). It correlates with how close a node A is to a query node Q by considering not only direct edges or shortest distance (network distance), but also taking into account the amount of paths that exist between Q and A. With RWR, a virtual random walker starts at Q and then chooses any outgoing link by random. The walker will continue to do this, but with every iteration there is a probability of α of jumping back to the query node (restart). The similarity between

A and Q then is the probability of the random walker reaching A when starting at Q. Thus, the walker will visit nodes "close" to Q more likely than others, giving them a higher similarity. With the walker not being dependant on a single route between A and Q, modification of the nodes and links in the graph in general affects the similarity of all the other nodes through the addition or removal of possible paths the walker may choose from. Therefore, precomputing social similarities will become difficult and unpractical: In popular social networks links and nodes are added constantly, making precomputed similarities superfluous.

Bookmark Coloring Algorithm. The Bookmark Coloring Algorithm (BCA) introduced by [4] is mathematically equivalent to RWR, but is more tangible and has other advantages. In a nutshell, it starts by injecting an amount of color into Q. Every node has the same color retainment coefficient α; i.e. that for every amount of color c the node receives, it gets to keep $\alpha \cdot c$, while the rest is forwarded equally spread across all outgoing links. This is the same α as for RWR. While BCA as well as RWR both rely on potentially infinite iterations of a power method to get exact results, one can terminate early to get quite exact approximations of them. But BCA follows more of a breadth-first-approach, while RWR may be compared to a depth-first one. This results in BCA giving all the socially close nodes a first visit much faster than RWR. We will exploit this feature to improve the search for socially close nodes over distant ones. It terminates when a stopping criterion is met; usually if the distributed color falls below a certain threshold or a total minimum sum of color is distributed). The algorithm gives back a vector, containing for every node its similarity to Q.

To derive the required social distance attribute for the skyline, we simply subtract the similarity, which lies within the bounds of $[0, 1]$, from 1. This fulfils the requirement that a node A with a similarity higher than that of node B (and is therefore considered better than B) gets a smaller distance than B (so that it is still better than B).

3   Related Work

Similarity Measures in Social-Networks: While colloquially speaking of "socially close" people, it is necessary to define a method of measuring this proximity. For this purpose, [15] proposed to use Random Walks with Restart ([24]) to measure social proximity. This metric, which is commonly adapted by the research community ([3,9,25]) to measure social similarity, considers all walks between two users, rather than using shortest path distance only ([18]), or using direct friendship relations of a user only [3].

Geo-Social Networks: [3] formulates a framework for geo-social query processing that builds queries based upon atomic operations. These queries (like *Nearest Friends*, *Range Friends*) focus on a specific user and his direct friends. [1] focusses on proximity detection of friends while preserving privacy through the fact that the location of a user is only known to the user himself and his friends, thus also lacking transitivity, where friends-of-friends are considered as well. [3] gives a comprehensive overview of the state-of-the-art in geo-social networks and geo-social queries.

Empirical Analysis: There are studies that examine the data of geo-social networks empirically: [7] detects that one's friends have a high probability of living close to each other, which emphasizes the importance of our introduced skyline queries when going further away from home. Similar findings have been made for the Foursquare geo-social network [22].

Skyline Queries: The skyline operator was introduced in [5]. Additionally, the authors propose block-nested-loop processing and an extended divide-and-conquer approach to process results for their new method. Since then, skyline processing has attracted considerable attention in the database community. [23] proposes two progressive methods to improve the original solutions. The first technique employs Bitmaps and is directed towards data sets being described with low cardinality domains. In other words, each optimization criterion is described by a small set of discrete attribute values. Other solutions for this scenario are proposed in [19]. The second technique proposed in [23] is known as index method and divides the data set into d sorted lists for d optimization criteria. [16] introduces the nearest-neighbor approach which is based on an R-Tree [11]. This approach starts with finding the nearest neighbor of the query point which has to be part of the skyline. Thus, objects being dominated by the nearest neighbor can be pruned. Afterwards, the algorithm recursively processes the remaining sections of the data space and proceeds in a similar way. A problem of this approach is that these remaining sections might overlap and thus, the result has to be kept consistent. To improve this approach, [20] proposes a branch-and-bound approach (BBS) which is guaranteed to visit each page of the underlying R-Tree at most once. There exist several techniques for post-processing the result of skyline queries for the case that the number of skyline points becomes too large to be manually explored (i.e. [21,17]). Though we do not focus on reducing the number of results of our algorithm we can utilize the techniques of the above works in a post-processing step. There has already been some work considering the application of the skyline operator in a setting including road networks (i.e. [8,13,14]). The main difference to our setting is that the network distance in terms of the length of the shortest path is taken into consideration. Rather, in our work we use the RWR distance to compute the score among the individuals in the network graph which we will show poses new problems and challenges.

4 Social Distance Approximation

The exact evaluation of the social distance between two users using the RWR distance is computationally very expensive. Thus we rely on the following lower and upper bounds for the social similarity in order to boost the efficiency of our approach by avoiding having to calculate the social similarity to a precision not necessary for skyline queries, while still maintaining correctness of the results.

4.1 Bounds Derived from Network Distance

Starting from the BCA algorithm, consider the maximum amount of color a node can get: To give a node A as much color as possible, the color has to flow directly from Q to A on the shortest path (that is the path with the least hops). With every hop, an α-portion of the available color is "lost", because it gets assigned to the intermediate node. If the flow is not on the shortest path, at least another hop is added where another α-portion is lost and cannot contribute to A. Assume the shortest path from Q to A contains l edges. Then there are $l - 1$ intermediate nodes plus one start node (Q), each of which gets its α-portion. The maximum amount of color that then can reach A in the best case is $(1 - \alpha)^l$. This is an upper bound and can be computed for every node before even starting the BCA. The bound gets lower and "better", if either A has a larger distance to Q or α is large (high personalization).

In practice, calculating the network distance at query time (online phase) is not suitable. Dijkstra's algorithm for example gets expensive for a large network with many nodes. Therefore, we suggest to introduce a preprocessing step that supports approximating the network distance optimistically. We use graph embedding [10], where the distance from any node to a small set of reference nodes RN is precomputed and then stored in a lookup-table. At query-time, we can then easily derive a conservative as well as an optimistic approximation for the network distance. Let o be the optimistic approximation (that is $o \leq$ actual network distance l). Then

$$(1 - \alpha)^o \geq (1 - \alpha)^l \geq BCA(Q)_A$$

where $BCA(Q)_A$ is the actual amount of color node A gets assigned. Therefore, the approximation derived from precomputed graph embedding provides an upper bound as well.

4.2 Bounds Derived from BCA

In general, it is possible that a node does not receive any color at all, so its generic lower bound is 0. But if the node has already received some color, there is no way of it ever losing this amount of assigned color - so once assigned color can be interpreted as an ever-improving lower bound. Building upon this, another upper bound can be derived from this lower bound: The maximum amount of color this node can get is the color it already has plus the total remaining unassigned color in the BCA-queue.

5 Algorithm

In the following, we propose three algorithms to compute the Geo-Social Skyline: First we provide a straightforward solution which is simple but not practicable for large datasets. Then, we propose a baseline solution which utilizes the bounds proposed in Section 4 and an advanced solution which uses additional pruning criteria to further improve its performance.

5.1 Naive Algorithm

Assuming no index structure, neither for the social nor for the spatial dataset, a naive algorithm for computing the Geo-Social Skyline is shown in Algorithm 1. This approach computes spatial distances and social distances of all users in Line 4 and Line 5 respectively. Computation of social distance requires to call the complete Bookmark Coloring

---

Algorithm 1. Naive Geo-Social Skyline Computation

Require: $SN, q_{geo}, q_{soc}, \alpha$

1. $bca = PerformCompleteBCA(SN, q_{soc}, \alpha)$
2. **for** each $n \in SN$ **do**
3.   **if** $n_{soc} \neq q_{soc}$ **then**
4. $n.\text{spatialDistance} = distance(n_{geo}, q_{geo})$
5. $n.\text{socialDistance} = 1 - bca_{n.id}$
6.   **end if**
7. **end for**
8. compute *skyline* using a scan based skyline algorithm
9. **return** *skyline*

Algorithm in Line 1, which is the main bottleneck of this approach. Given spatial and social distances, we represent each user by a two-dimensional vector and apply traditional approaches to compute two-dimensional skylines in Line 8. This naive algorithm however is very ineffiecient and not practicable in a setting with large datasets. The main problem here is that the runtime complexity of BCA has shown to be $O(n^3)$. Thus the following algorithm makes use of the social distance bounds proposed in Section 4 in order to avoid the complete run of BCA.

5.2   Baseline Algorithm

The baseline algorithm assumes a simple index structure such as a sorted list, a min-heap or a B-Tree to access users in ascending order to their spatial distance to the query location q_{geo}. This algorithm starts by computing spatial distance and initializing distance bounds in Lines 3-10. In Line 8, the lower bound is initialized by network distance bounds described in Section 4.1. The upper bound is initiliazed with the trivial upper bound of 1. Then, nodes are accessed in increasing order of the distance to q_{geo} in Line 12. When a new node c is accessed, a check is performed in Line 14 to see if c's lower bound social distance is already higher than the upper bound social of the last object that has been returned as a skyline result. In this case, c can be pruned, as we can guarantee that the previous result node *previous* has a lower social distance than c and due to accessing nodes ordered by their spatial distance, we can guarantee that *previous* must also have a lower spatial distance. Another check is performed in Line 16, for the case where c can be returned as a true hit, by assessing that c must have a lower social distance than *previous*. If neither of these two checks allows to make any decision, Line 19 calls Algorithm 3 to perform an additional iteration of the BCA algorithm to further refine all current social bounds, until c can either be pruned or returned as a true hit. The main idea of this algorithm is to minimize expensive iterations of the BCA. This is achieved by first considering the spatial dimension. Furthermore, BCA iterations are required only in cases where absolutely necessary. For this reason, the social distance of results may still be an approximation. Nevertheless, this algorithm can guarantee to return the correct skyline.

Note that in the worst case, where every node is contained in the skyline, this approach yields no performance gain compared to the trivial solution. A major disadvantage of this approach is its space consumption and the linear scan over the database

---

Algorithm 2. Geo-Social Skyline Baseline

Require: $q_{\text{geo}}, q_{\text{soc}}, \alpha, SN, RN$

1. $assigned = [], incoming = [], total = 0, results = \emptyset$
2. $queue = \emptyset$ // min-heap sorted ascending by $dist_{\text{geo}}$
3. **for** $n \in SN$ **do**
4. $incoming[n.\text{id}] = 0, assigned[n.\text{id}] = 0$
5.   **if** $n \neq q_{\text{soc}}$ **then**
6. $n.\text{spatialDistance} = dist_{\text{geo}}(n.\text{point}, q_{\text{geo}})$
7. $l = max_{R_i \in RN}(|networkDist(q_{\text{soc}}, R_i) - networkDist(n, R_i)|)$
                // those network distances come from graph embedding
8. $n.\text{socialDist.lower} = 1 - (1 - \alpha)^l, n.\text{socialDist.upper} = 1, queue.add(n)$
9.   **end if**
10. **end for**
11. $incoming[q_{\text{soc}}.\text{id}] = 1, previous = queue.popMin(), results.add(previous)$
12. **while** $queue.\text{size} > 0$ **do**
13. $candidate = queue.popMin()$
14.   **if** $previous.\text{socialDist.upper} < candidate.\text{socialDist.lower}$ **then**
15. $prune(candidate)$
16.   **else if** $candidate.\text{socialDist.upper} < previous.\text{socialDist.lower}$ **then**
17. $results.add(candidate), previous = candidate$
18.   **else**
19. $IncrementalBCA(\alpha, \epsilon, incoming, assigned, total)$ // see Algorithm 3
20. $queue.add(candidate)$ // re-insert
21.   **end if**
22. **end while**
23. **return** $results$

---

to materialize every single node, for which the spatial distances are calculated and the sorting takes place. This means that basically the entire graph has to be loaded into memory for processing. The following algorithm alleviates this problem.

5.3 Improved Algorithm

Our improved algorithm still iterates over nodes from closest to farthest spatially, but avoids having to store the whole graph in memory. Therefore, we use a spatial index structure such as an R-Tree, which supports efficient incremental nearest neighbour algorithms [12] to access nodes sorted by increases distance to the query location q_{geo}. This approach allows to avoid loading spatial locations of nodes into the memory if we can terminate the algorithm early by identifying a time when we can guarantee that all skyline points have been found.

For this purpose, the improved algorithm uses the bounds presented in Section 4.2 by considering the maximum amount of color these nodes can get in future iterations of the BCA. These bounds essentially allow to assess an upper bound of completely unseen, i.e., not yet accessed nodes. Given these bounds, and exploiting that unseen nodes must have a higher spatial distance than all accessed nodes due to accessing nodes in ascending order of their spatial distance, we terminate the algorithm early in Line 3 if the following conditions are met:

Algorithm 3. IncrementalBCA

Require: α, ϵ, *incoming, assigned, total*
1. $color = \max(incoming[])$, $k = incoming.indexOf(color)$
2. $incoming[k] = 0$, $assigned[k] = assigned[k] + \alpha \cdot color$
3. $total = total + \alpha \cdot color$
4. **for** each $l \in k.getLinks()$ **do**
5. $incoming[l] = incoming[l] + (1 - \alpha) \cdot color/k.getLinks().size$
6. **end for**
7. $k.suggestLB(total - assigned[k])$, $k.suggestUB(1 - assigned[k])$
8. **if** $k.\text{UB} - k.\text{LB} < \epsilon$ **then**
9. $k.\text{LB} = k.\text{UB} = (k.\text{LB} + k.\text{UB})/2$
10. **end if**

- The set of candidates contains no entries anymore (i.e. all candidates either became results or were pruned) and
- all unseen nodes can be pruned, thus there is no possibility an unseen node can be contained in the skyline.

6   Experiments

For our experiments, we evaluated our solutions on a geo-social network taken from the Gowalla dataset[1]. It consists of a social network having 196,591 nodes and 950,327 undirected edges, leaving every node on average with approximately ten links. Furthermore, the dataset contains 6,442,890 check-ins of users. Each user is assigned their latest (most recent) check-in location as geo-spatial location. Users having no check-in at all are matched to a special location that has a geo-distance of infinity to any other place. For all geo-social skyline queries performed in this experimental evaluation, if not mentioned otherwise, the spatial query components q_{geo} are obtained by uniformly sampling *(latitude, longitute)* pairs in the range $([-90, 90], [-180, 180])$. The social query components q_{soc} are obtained by uniformly sampling nodes of the social network.

When compared to trending social networks like Facebook or Foursquare, the example dataset is rather small when compared in network size. Unfortunately, data of larger networks is not publicly available. We decided not to run experiments on synthetic data, because the artificial generation of spatial and social data may introduce a bias into the experiments, when correlations between spatial and social distances are different in real and synthetic data. Although there exist indiviual data generators for social-only networks (e.g. [2]) and spatial networks ([6]), the naive combination of both generators is not feasible here.

In Figure 2 we compare all proposed algorithms with a varying value of α. The improved algorithm allows further optional domination checks in lines 12 and 13, which were not done for the *prune unseen* data row. Only the check in Line 12 is done in *prune candidate with hit*, while both checks are done in the *prune candidates with candidates* rows. For the latter, the frequency of the check in Line 13 is varied for every 1,000 respectively 10,000 iterations.

---

[1] http://snap.stanford.edu/data/loc-gowalla.html

Algorithm 4. Geo-Social Skyline Improved

Require: $q_{geo}, q_{soc}, \alpha, SN, RN$
1. $assigned = [], incoming = [], total = 0, results = \emptyset, prunedUnseen = \textbf{false}$
2. $previous = getNearestNeighbour(q_{geo}), results.add(previous), i = 2, incoming[q_{soc}.id] = 1$
3. **while** $i < |SN.\text{Nodes}|$ **and not** $prunedUnseen$ **do**
4. $candidate = getN^{th}NearestNeighbour(q_{geo}, i)$ // get i^{th} nearest neighbour of q_{geo}
5. $i++$
6.   **if not** $candidate.isPruned$ **then**
7.   **while** $[candidate.\text{LB}, candidate.\text{UB}] \cap [previous.\text{LB}, previous.\text{UB}] \neq \emptyset$ **do**
8. $IncrementalBCA(\alpha, \epsilon, incoming, assigned, total)$ // see Algorithm 3
9.   **end while**
10.   **if not** $previous.dominates(candidate)$ **then**
11. $results.add(candidate)$
12.   // optional: check if candidate dominates other candidates (remove them)
13.   // optional: check if elements in candidates dominate each other (remove them)
14. $previous = candidate, prunedUnseen = (candidate.\text{UB} < total)$
15.   **else**
16. $prune(candidate)$
17.   **end if**
18.   **end if**
19. **end while**
20. **return** $results$

The experiments show that both optional checks in Algorithm 4 do not result in better runtimes with our datasets, even when only performed at greater intervals (that is not after every single iteration of BCA). On the other hand, checking whether all unseen nodes can be pruned results in an actual performance increase compared with the baseline algorithm. As a result of this experiment, all following experiments will use the *prune unseen* setting, as this algorithm shows the best runtime performance for almost any $\alpha \in [0, 1]$.

6.1 Skyline Results

For varying α, we experience a rapid drop in runtime for a larger α as seen in Figure 3(a). This can be attributed to the fact that a larger value of α allows the BCA to terminate faster, as in each iteration more color is distributed over the social network, thus yielding tighter bounds in each iteration. This result implies that, since α represents the personalization factor of the query, highly personalized queries can be performed much faster than unpersonalized ones. This is evident, since unpersonalized queries represent an overview of the entire network, thus requiring a deeper and more complete scan of the graph. Choosing an α-value of zero will cause the BCA not terminating, because no color gets assigned. An α-value approaching zero causes the BCA to approach the stationary distribution of the social network, which is completely independent of the social query node q_{soc}. A value of α approaching one will give all the colour to the q_{soc}, thus returing q_{soc} himself as his only match. Furthermore, it can be observed in Figure 3(b) that despite a fairly large social network, the number of results in the Geo-Social Skylines become quite small. We ran several queries for different α-values and in our

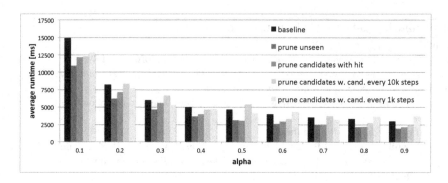

Fig. 2. Runtime Evaluation

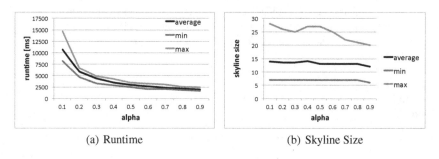

(a) Runtime (b) Skyline Size

Fig. 3. Evaluation of α

experiments the number of elements in the skyline remained low - we never encountered a skyline containing more than approximately 30 elements. This shows that the spatial-social skyline returns useful result sets for real-world applications by reducing a large network of approximately 200k nodes down to a feasible number of elements.

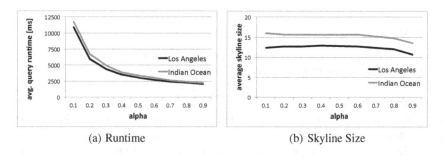

(a) Runtime (b) Skyline Size

Fig. 4. Evaluation of the population density of q_{geo}

6.2   Varying the Spatial Query Point

Considering the motivational example of querying the graph when going on vacation, it becomes interesting to query areas considered "dense" and "sparse" populated places.

In our setting, we considered regions to be dense where a lot of social nodes are located at. In the Gowalla data set, the city of Los Angeles is such a dense place, as the user base of this geo-social network was mainly U.S.-based. In contrast a place in the southern Indian Ocean is chosen as a sparse region. This region is located approximately on the opposite of the U.S. on the Earth, so this region maximizes the spatial distance of the majority of the user base. In the following experiment, we performed pairs of geo-social skyline queries, such that each pair had an identical social query user q_{soc}, but the spatial location q_{geo} differes by being either in Los Angeles and in the Indian Ocean. While we only observed a slight performance gain when querying dense locations (cf. Figure 4(a)), it is interesting to see in Figure 4(b) that querying a sparse place results in a larger skyline. The reason is that users in the U.S. have a higher average number of social links in this data set than users outside of the U.S., since most active users of this geo-social network origin from the U.S. It is more likely for a random user to have a socially close person coming from the vicinity of Los Angeles, therefore having both a small social and a small geographic distance from a query issued in Los Angeles, and thus pruning most of the database.

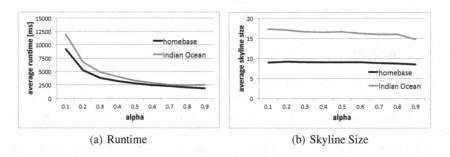

(a) Runtime   (b) Skyline Size

Fig. 5. Evaluation of the distance between the user and q_{geo}

In another set of experiments we evaluate the effect of the spatial distance between a user that performs a geo-social skyline query and the spatial query location q_{geo}. Therefore, we performed pairs of geo-social skyline queries, such that each pair had an identical social query user q_{soc}, but the spatial location q_{geo} differed by either being identical to the user's location, and by being located in the Indian Ocean. The result depicted in Figure 5(b) shows that the skyline for the home-based query only contains roughly half as many elements as the distant one also yielding a lower runtime (cf. Figure 5(a)). The reason is that in geo-social networks, the spatial distance between users is known to be positively correlated with their social distance [22]. Thus, it is likely to have a socially close person in your immediate vicinity - and in the case where a user's location equals the query point q_{geo}, this person dominates most other individuals. At the same time, the other extreme case where q_{geo} is located at the other end of the world, leads to a negative correlation between social distance and distance to q_{geo} thus yielding a larger number of skyline results.

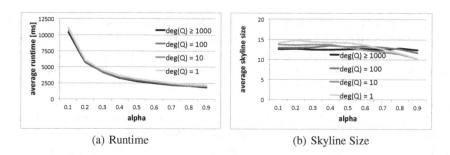

(a) Runtime   (b) Skyline Size

Fig. 6. Evaluation of the number of friends of q_{soc}

6.3 Varying the Number of Friends of q_{soc}

Next, we evaluated in impact of the degree of q_{soc}, i.e., the number of direct friends of the query user, as we expect that social stars may produce skylines shaped differently than users having a few friends only. The experimental results depicted in Figure 6 show that, while making almost no difference in terms of run-time, we can observe, for small values of α, a slight trend towards larger skylines for nodes having a low degree. For large values of α, this effect reverses.

7 Conclusions

In this work, we have defined the problem of answering *Geo-Social Skyline Queries*, a useful new type of query that allows to find, for a specified query user and a specified spatial query location, a set of other users that are both socially close to the query user and spatially close to the query location. To answer such queries, we followed the state-of-the-art approach of measuring social distance using Random-Walk-with-Restart-distance (RWR-distance), which is more meaningful than simple social distance measures such as the binary *isFriend*-distance and the shortest-path distance. Due to the computational complexity of computing RWR-distances, we have presented efficient techniques to obtain conservative bounds of RWR-distances which can be used to quickly prune the search space, thus alleviating computational cost significantly as shown by our experimental studies.

As a future step, we want to exploit information given by check-in data provided by users, rather than using their current spatial position only. Such check-in data allows to automatically return for a given a user, who is going to visit a location such as the island of Bali, friends that have recently visited Bali. These friends can be recommended to the user as experts that may help the user to, for example, find good places to visit.

References

1. Amir, A., Efrat, A., Myllymaki, J., Palaniappan, L., Wampler, K.: Buddy tracking - efficient proximity detection among mobile friends. Pervasive and Mobile Computing 3(5), 489–511 (2007)

2. Angles, R., Prat-Pérez, A., Dominguez-Sal, D., Larriba-Pey, J.-L.: Benchmarking database systems for social network applications. In: Proc. GRADES. ACM (2013)
3. Armenatzoglou, N., Papadopoulos, S., Papadias, D.: A general framework for geo-social query processing. In: Proc. VLDB, pp. 913–924. VLDB Endowment (2013)
4. Berkhin, P.: Bookmark-coloring algorithm for personalized pagerank computing. Internet Mathematics 3(1), 41–62 (2006)
5. Borzsony, S., Kossmann, D., Stocker, K.: The skyline operator. In: Proc. ICDE, pp. 421–430. IEEE (2001)
6. Brinkhoff, T.: A framework for generating network-based moving objects. GeoInformatica 6(2), 153–180 (2002)
7. Cho, E., Myers, S.A., Leskovec, J.: Friendship and mobility: user movement in location-based social networks. In: Proc. KDD, pp. 1082–1090. ACM (2011)
8. Deng, K., Zhou, Y., Shen, H.T.: Multi-source skyline query processing in road networks. In: Proc. ICDE, pp. 796–805. IEEE (2007)
9. Fujiwara, Y., Nakatsuji, M., Onizuka, M., Kitsuregawa, M.: Fast and exact top-k search for random walk with restart. In: Proc. VLDB, pp. 442–453. VLDB Endowment (2012)
10. Graf, F., Kriegel, H.-P., Renz, M., Schubert, M.: Memory-efficient A*-search using sparse embeddings. In: Proc. ACM GIS, pp. 462–465. ACM (2010)
11. Guttman, A.: R-Trees: A dynamic index structure for spatial searching. In: Proc. SIGMOD, pp. 47–57. ACM (1984)
12. Hjaltason, G.R., Samet, H.: Distance browsing in spatial databases. ACM Transactions on Database Systems 24(2), 265–318 (1999)
13. Huang, X., Jensen, C.S.: In-route skyline querying for location-based services. In: Kwon, Y.-J., Bouju, A., Claramunt, C. (eds.) W2GIS 2004. LNCS, vol. 3428, pp. 120–135. Springer, Heidelberg (2005)
14. Jang, S.M., Yoo, J.S.: Processing continuous skyline queries in road networks. In: Proc. CSA, pp. 353–356. IEEE (2008)
15. Konstas, I., Stathopoulos, V., Jose, J.M.: On social networks and collaborative recommendation. In: Proc. SIGIR, pp. 195–202. ACM (2009)
16. Kossmann, D., Ramsak, F., Rost, S.: Shooting stars in the sky: an online algorithm for skyline queries. In: Proc. VLDB, pp. 275–286. VLDB Endowment (2002)
17. Lin, X., Yuan, Y., Zhang, Q., Zhang, Y.: Selecting stars: The k most representitive skyline operator. In: Proc. ICDE, pp. 86–95. IEEE (2007)
18. Liu, W., Sun, W., Chen, C., Huang, Y., Jing, Y., Chen, K.: Circle of friend query in geo-social networks. In: Lee, S.-g., Peng, Z., Zhou, X., Moon, Y.-S., Unland, R., Yoo, J. (eds.) DASFAA 2012, Part II. LNCS, vol. 7239, pp. 126–137. Springer, Heidelberg (2012)
19. Morse, M., Patel, J.M., Jagadish, H.: Efficient skyline computation over low-cardinality domains. In: Proc. VLDB, pp. 267–278. VLDB Endowment (2007)
20. Papadias, D., Tao, Y., Fu, G., Seeger, B.: An optimal and progressive algorithm for skyline queries. In: Proc. SIGMOD, pp. 467–478. ACM (2003)
21. Pei, J., Jin, W., Ester, M., Tao, Y.: Catching the best views of skyline: A semantic approach based on decisive subspaces. In: Proc. VLDB, pp. 253–264. VLDB Endowment (2005)
22. Scellato, S., Mascolo, C., Musolesi, M., Latora, V.: Distance matters: geo-social metrics for online social networks. In: Proc. WOSN. USENIX (2010)
23. Tan, K.-L., Eng, P.-K., Ooi, B.C.: Efficient progressive skyline computation. In: Proc. VLDB, pp. 301–310. VLDB Endowment (2001)
24. Tong, H., Faloutsos, C., Pan, J.Y.: Fast random walk with restart and its applications. In: Proc. ICDM, pp. 613–622. IEEE (2006)
25. Zhang, C., Shou, L., Chen, K., Chen, G., Bei, Y.: Evaluating geo-social influence in location-based social networks. In: Proc. CIKM, pp. 1442–1451. ACM (2012)

Reverse-Nearest Neighbor Queries on Uncertain Moving Object Trajectories

Tobias Emrich[1], Hans-Peter Kriegel[1], Nikos Mamoulis[2], Johannes Niedermayer[1],
Matthias Renz[1], and Andreas Züfle[1]

[1] Institute for Informatics, Ludwig-Maximilians-Universität München, Germany
{emrich,kriegel,niedermayer,renz,zuefle}@dbs.ifi.lmu.de
[2] Department of Computer Science, University of Hong Kong
nikos@cs.hku.hk

Abstract. Reverse nearest neighbor (RNN) queries in spatial and spatio-temporal databases have received significant attention in the database research community over the last decade. A reverse nearest neighbor (RNN) query finds the objects having a given query object as its nearest neighbor. RNN queries find applications in data mining, marketing analysis, and decision making. Most previous research on RNN queries over trajectory databases assume that the data are certain. In realistic scenarios, however, trajectories are inherently uncertain due to measurement errors or time-discretized sampling. In this paper, we study RNN queries in databases of uncertain trajectories. We propose two types of RNN queries based on a well established model for uncertain spatial temporal data based on stochastic processes, namely the Markov model. To the best of our knowledge our work is the first to consider RNN queries on uncertain trajectory databases in accordance with the possible worlds semantics. We include an extensive experimental evaluation on both real and synthetic data sets to verify our theoretical results.

1 Introduction

The widespread use of smartphones and other mobile or stationary devices equipped with RFID, GPS and related sensing capabilities made the collection and analysis of spatio-temporal data at a very large scale possible. A wide range of applications benefit from analyzing such data, such as environmental monitoring, weather forecasting, rescue management, Geographic Information Systems, and traffic monitoring. In the past, however, research focused mostly on *certain* trajectory data, assuming that the position of a moving object is known precisely at each point in time without any uncertainty. In reality, though, due to physical limitations of sensing devices, discretization errors, and missing measurements, trajectory data have different degrees of *uncertainty*: GPS and RFID measurements introduce uncertainty in the position of an object. Furthermore, as RFID sensors are usually set up at a certain position, an RFID-based location tracker will only be activated if an object passes near its sensor. Between two consecutive sensor measurements the position of the object remains unknown. This problem of incomplete observations in time is a general problem of trajectory databases, and does not only appear in RFID applications, but also in well-known GPS datasets published for research purposes such as T-Drive [28] and GeoLife [29]. As a consequence,

S.S. Bhowmick et al. (Eds.): DASFAA 2014, Part II, LNCS 8422, pp. 92–107, 2014.

it is important to find solutions for deducting the unknown and therefore uncertain positions of objects in-between discrete (certain) observations. The most straightforward solution for deducting a position between consecutive measurements would be linear interpolation. However, linear interpolation can cause impossible trajectories, such as a bike driving through a lake. Other solutions such as computing the shortest path between consecutive locations produce valid results, but do not provide probabilities for quantifying the quality of the result.

In this paper, we consider a historical database $\mathcal{D}$ of uncertain moving object trajectories. Each of the stored uncertain trajectories consists of a set of observations given at a some (but not all) timesteps in the past. An intuitive way to model such data is by describing it as a time-dependent random variable, i.e., a stochastic process. In this research, we model uncertain objects by a first-order Markov chain. It has been shown recently that even a first-order Markov chain, if augmented with additional observations, can lead to quite accurate results [12]. We address the problem of performing Reverse Nearest Neighbor (RNN) queries on such data. Given a query q, a reverse nearest neighbor query returns the objects in the database having q as one of its nearest neighbors. This query has been extensively studied on certain data [9,15,17]. Recent research has focused on RNN queries in uncertain spatial [3,10] data. Xu et al were the first to address RNN queries on uncertain spatio-temporal data under the Markov model and showed how to answer an "interval reverse nearest neighbor query" [24]. This kind of query has many applications, for example in collaboration recommendation applications. However, as we will see later, the solution presented in [24] does not consider possible worlds semantics (PWS). In this work we fill this research gap by proposing algorithms to answer reverse nearest neighbour queries according to PWS.

The contributions of this work can be summarized as follows:

- We introduce two query definitions for the reverse nearest neighbor problem on uncertain trajectory data, the $P\exists RNNQ(q, \mathcal{D}, T, \tau)$ and $P\forall RNNQ(q, \mathcal{D}, T, \tau)$ query. The queries are consistent with existing definitions of nearest neighbor and window queries on this data.
- We demonstrate solutions to answer the queries we defined efficiently and, most importantly, according to possible worlds semantics.
- We provide an extensive experimental evaluation of the proposed methods both on synthetic and real world datasets.

This paper is organized as follows: In Section 2 we review related work on RNN queries and uncertain spatio-temporal data modeling. Section 3 provides a formal problem definition. Section 4 introduces algorithms for the queries proposed in Section 3. An extensive experimental evaluation follows in Section 5. Section 6 concludes this paper.

2   Related Work

Probabilistic Reverse-Nearest Neighbour Queries. Reverse (k)-Nearest Neighbor queries, initially proposed by Korn et al. [9] on certain data have been studied extensively in the past [15,25,17,1,19]. Many of the early solutions for RkNN queries rely on costly precomputations [9,25,1] and augment index structures such as R-trees or M-trees by additional information in order to speed up query evaluation. Follow-up techniques, such as TPL [17], aim at avoiding costly preprocessing at the cost of a more expensive query evaluation stage; moreover, they do not depend on specialized index structures. Recently, probabilistic reverse nearest neighbor queries have gained significant attention [10,3,2]. The solution proposed by Chen et al. [10] aims at processing PRNN queries on uncertain objects represented by continuous probability density functions (PDFs). In contrast, Cheema et al. [3] provided solutions for the discrete case. In the context of probabilistic reverse nearest neighbor queries, two challenges have to be addressed in order to speed up query evaluation. On the one hand, the I/O-cost has to be minimized; on the other hand, the solution has to be computationally efficient.

Uncertain Spatio-temporal Data. Query processing in trajectory databases has received significant interest over the last ten years. (see for example [18,16,26,23,8]). Initially, trajectories have been assumed to be certain, by employing linear [18] or more complex [16] types of interpolation to handle missing measurements. Later, a variety of uncertainty models and query evaluation techniques has been developed for moving object trajectories (e.g.[11,22,21,7]).

A possible way to approach uncertain data is by providing conservative bounds for the positions of uncertain objects. These conservative bounds (such as cylinders [22,21] or beads [20]) approximate trajectories and can answer queries such as "give me all objects that *could have* (or *definitely have*) the query as a nearest neighbor". However they cannot provide probabilities conforming to possible worlds semantics. For a detailed analysis of this shortcoming, see [7].

Another class of algorithms employ independent probability density functions (pdf) at each point of time. This way of modeling the uncertain positions of an object [4,21,11], can produce wrong results if a query considers more than a single point in time as shown in [7,12], as temporal dependencies between consecutive object positions in time are ignored. A solution to this problem is modeling uncertain trajectories by stochastic processes.

In [13,7,14,24], trajectories are modeled by Markov chains. Although the Markov chain model is still a *model* and can therefore only provide an approximate view of the world, it allows to answer queries according to possible worlds semantics, significantly increasing the quality of results. Recently, [12] addressed the problem of nearest neighbor queries based on the Markov model. Our work builds upon the results from this paper.

Regarding reverse nearest-neighbor processing using the Markov model, to the best of our knowledge, there exists only one work so far which addresses interval reverse nearest neighbor queries [24]. The approach basically computes for each point of time in the query interval separately the probability for each object $o \in \mathcal{D}$ to be the RNN to the query object. Then for each object o, the number of times where o has the highest probability to be RNN is counted. The object with the highest count is returned. Upon investigation, this approach has certain drawbacks. First, the proposed algorithm is not in accordance with possible worlds semantics, since successive points of time are considered independently (a discussion on the outcome of this treatment can be found in Section 3.2). Second, the paper does not show how to incorporate additional observations (besides the first appearance of an object). In this paper, we aim filling this research gap and solving the two said issues by proposing algorithms following possible world semantics that allow incorporating observations.

3   Problem Definition

In this paper, following [7,12], we define a spatio-temporal database $\mathcal{D}$ as a database storing triples $(o_i, time, location)$, with o_i being a unique object identifier, $time \in \mathcal{T}$ a point in time and $location \in \mathcal{S}$ a position in space. Each of these triples describes a certain observation of object o_i at a given $time$ at a given location $location$, e.g. a GPS measurement; the location of an object between two consecutive observations is unknown. Based on this definition an object o_i can be seen as a function $o_i(t) : \mathcal{T} \to \mathcal{S}$ that maps each point in time to a location in space; this function is called $trajectory$.

Following the related literature we assume a discrete time domain $\mathcal{T} = \{0, \dots, n\}$. Furthermore, we assume a discrete state space of possible locations (*states*): $\mathcal{S} = \{s_1, \dots, s_{|\mathcal{S}|}\} \subset \mathbb{R}^d$. Both of these assumptions are necessary to model uncertain trajectories by Markov chains. An object stored in the database can only be located in one of these states at each point in time.

3.1   Uncertain Trajectory Model

The uncertain trajectory model used in this paper has been recently investigated (e.g., by [7,12]) in the context of window queries and nearest neighbor queries. In the following, we recap this model. Let $\mathcal{D}$ be a database containing the trajectories of $|\mathcal{D}|$ uncertain moving objects $\{o_1, \dots, o_{|\mathcal{D}|}\}$. An object $o \in \mathcal{D}$ is represented by a set of observations $\Theta^o = \{\langle t_1^o, \theta_1^o \rangle, \langle t_2^o, \theta_2^o \rangle, \dots, \langle t_{|\Theta^o|}^o, \theta_{|\Theta^o|}^o \rangle\}$ with $t_i^o \in \mathcal{T}$ being the timestamp and $\theta_i^o \in \mathcal{S}$ the state (i.e. location) of observation Θ_i^o. Let $t_1^o < t_2^o < \dots < t_{|\Theta^o|}^o$. This model assumes observations to be certain, however between two certain observations the location of an object is unknown and therefore uncertain. To model this uncertainty we can interpret the uncertain object o as a stochastic process [7]. With this interpretation, the location of an uncertain object o at time t becomes a realization of the random variable $o(t)$. Considering a time interval $[t_s, t_e]$, results in a sequence of uncertain locations of an object, i.e. a stochastic process. With this definition we can compute the probability of a given trajectory.

In this paper, following [7,6,24,12], we investigate query evaluation on a first-order Markov model. The advantage of the first-order Markov model is its simplicity. By employing a Markov model, the position $o(t+1)$ of object o at time $t+1$ only depends on the location of o at time t, i.e. $o(t)$. Therefore, transitions between consecutive points in time can be easily realized by matrix multiplication. However note that by modeling uncertain objects by a Markov chain, the motion of these objects basically degenerates to a random walk, clearly not a realistic motion pattern of objects in real life. The motion of cars for example is better described by shortest paths than by a random walk. Fortunately, as showed in [12], by incorporating a second source of information into the model, namely observations of an object, the Markov chain can be used to accurately describe the uncertainty area of an uncertain object.

Now, let the state space of the Markov chain be given as the spatial domain $\mathcal{S}$, i.e. points in Euclidean space. The *transition probability* $M_{ij}^o(t) := P(o(t+1) = s_j | o(t) = s_i)$ denotes the probability of object o moving from state s_i to s_j at time t. These transition probabilities can be stored in a matrix $M^o(t)$, i.e. the *transition matrix* of o at time t. The transition matrix of an object might change with time, and different objects might have different transition matrices. The first property is useful to model varying motion patterns of moving objects at different times of a day, a month or a year: birds move to the south in autumn and to the north during springtime. Each of these patterns could be described by a different transition matrix. The second property is useful to model different classes of objects such as busses and taxis.

Let $s^o(t) = (s_1, \ldots, s_{|\mathcal{S}|})^T$ be the probability distribution vector of object o at time t, with $s_i^o(t) = P(o(t) = s_i)$. An entry $s_i^o(t)$ of the vector describes the probability of o entering s_i at time t. The state vector $s^o(t+1)$ can be computed from $s^o(t)$ as follows: $s^o(t+1) = M^o(t)^T \cdot s^o(t)$ Note that simple matrix multiplications can only be employed in the absence of observations. In the presence of observations, transition matrices must be adapted, see [12]. Finally note that we assume different objects to be *mutually independent*.

Due to the generality of the Markov model, we can model both continuous space and street networks with this technique. In a continuous space we could sample the set of discrete states randomly. For street networks the states would represent the street crossings (see our experiments, Section 5). Transition probabilities can then be learned from known trajectories. These transition probabilities model the motion patterns of objects. They basically say "given the object was at crossing a at time t, it will move to crossing b at time $t+1$ with probability p".

3.2   Probabilistic Reverse Nearest Neighbor Queries

In the following we define two types of probabilistic time-parameterized RNN queries. The queries conceptually follow the definitions of time-parameterized nearest neighbor and window queries in [12,7]. We assume that the RNN query takes as input a set of timestamps T and either a single state or a (certain) query trajectory q. Still, our definitions and solutions can be trivially extended to consider RNN queries where the input states are uncertain.

Definition 1 (P∃RNN Query). *A probabilistic* ∃ *reverse nearest neighbor query retrieves all objects* $o \in \mathcal{D}$ *having a sufficiently high probability to be the reverse nearest neighbor of* q *for at least one point of time* $t \in T$, *formally:*

$$P\exists RNNQ(q, \mathcal{D}, T, \tau) = \{o \in \mathcal{D} : P\exists RNN(o, q, \mathcal{D}, T) \geq \tau\}$$

where $P\exists RNN(o, q, \mathcal{D}, T) = P(\exists t \in T : \forall o' \in \mathcal{D} \setminus o : d(o(t), q(t)) \leq d(o(t), o'(t)))$

and $d(x, y)$ *is a distance function defined on spatial points, typically the Euclidean distance.*

This query returns all objects from the database having a probability greater τ to have q as their probabilistic ∃ nearest neighbor [12]. In addition to this ∃ query, we consider RNN queries with the ∀ quantifier:

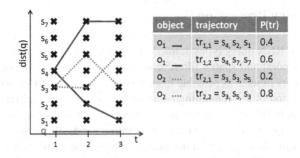

object	trajectory	P(tr)
o_1 —	$tr_{1,1} = s_4, s_2, s_1$	0.4
o_1 —	$tr_{1,2} = s_4, s_7, s_7$	0.6
o_2	$tr_{2,1} = s_3, s_3, s_5$	0.2
o_2	$tr_{2,2} = s_3, s_5, s_3$	0.8

Fig. 1. Example database of uncertatin trajectories

Definition 2 (P∀RNN Query). *A probabilistic* $\forall$ *reverse nearest neighbor query retrieves all objects* $o \in \mathcal{D}$ *having a sufficiently high probability (P∀RNN) to be the reverse nearest neighbor of q for the entire set of timestamps T, formally:*

$$P\forall RNNQ(q, \mathcal{D}, T, \tau) = \{o \in \mathcal{D} : P\forall RNN(o, q, \mathcal{D}, T) \geq \tau\}$$

where $P\forall RNN(o, q, \mathcal{D}, T) = P(\forall t \in T : \forall o' \in \mathcal{D} \setminus o : d(o(t), q(t)) \leq d(o(t), o'(t)))$

The above definition returns all objects from the database which have a probability greater τ to have q as their probabilistic $\forall$ nearest neighbor [12].

Example 1. To illustrate the differences between the proposed queries consider the example in Figure 1. Here for simplicity the query is not moving at all over time and the two objects o_1 and o_2 each have 2 possible trajectories. o_1 follows the lower trajectory $(tr_{1,1})$ with a probability of 0.4 and the upper trajectory $(tr_{1,2})$ with a probability of 0.6. o_2 follows trajectory $tr_{2,1}$ with a probability of 0.2 and trajectory $tr_{2,2}$ with a probability of 0.8. For query object q and the query interval $T = [2, 3]$, we can compute the probability for each object to be probabilistic reverse nearest neighbor of q. Specifically for o_1 the probability $P\exists RNN(o_1, q, \mathcal{D}, T) = 0.4$ since whenever o_1 follows $tr_{1,1}$ then at least at $t = 3$, o_1 is RNN of q. The probability for $P\forall RNN(o_1, q, \mathcal{D}, T)$ in contrast is 0.32 since it has to hold that o_1 follows $tr_{1,1}$ (this event has a probability of 0.4) and o_2 has to follow $tr_{2,2}$ (this event has a probability of 0.8). Since both events are mutually independent we can just multiply the probabilities to obtain the final result probability. Regarding object o_2 we can find no possible world (combination of possible trajectories of the two objects) where o_2 is always ($T = [2, 3]$) RNN, thus $P\forall RNN(o_2, q, \mathcal{D}, T) = 0$. However $P\exists RNN(o_2, q, \mathcal{D}, T) = 0.6$ since whenever o_1 follows $tr_{1,2}$ then o_2 is RNN either at $t = 2$ or at $t = 3$.

An important observation is that it is not possible to compute these probabilities by just considering the snapshot RNN probabilities for each query time stamp individually. For example the probability for o_2 to be RNN at time $t = 2$ is 0.12 (the possible world where objects follow $tr_{1,2}$ and $tr_{2,1}$) and the probability at $t = 3$ is 0.48 (the possible world where objects follow $tr_{1,2}$ and $tr_{2,2}$). However these events are not mutually independent and thus we cannot just multiply the probabilities to obtain the probability that object o_2 is RNN at both points of time (note that the true probability of this event is $P\forall RNN(o_2, q, \mathcal{D}, T) = 0$).

4   PRNN Query Processing

To process the two RNN query types defined in Section 3, we proceed as follows. First, we perform a temporal and spatial filtering to quickly find candidates in the database and exclude as many objects as possible from further processing. In the second step we perform a verification of the remaining candidates to obtain the final result. Although different solutions to this problem are possible, we decided to describe an algorithm that splits the query involving several timesteps into a series of queries involving only a single point in time during the pruning phase. The interesting point in this algorithm is that it shows that spatial pruning *does not introduce errors when disregarding temporal correlations*. However, disregarding temporal correlation during the probability computation phase *does introduce errors*.

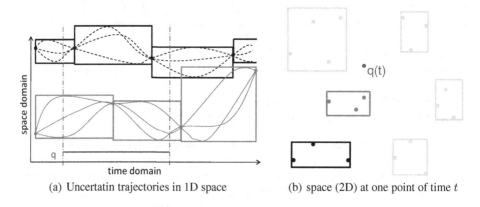

(a) Uncertatin trajectories in 1D space   (b) space (2D) at one point of time t

Fig. 2. Spatio-temporal filtering (only leaf nodes are shown)

4.1   Temporal and Spatial Filtering

In the following we assume that the uncertain trajectory database $\mathcal{D}$ is indexed by an appropriate data structure, like the UST tree [6].[1] For the UST tree, the set of possible (location,time)-tuples between two observations of the same object is conservatively approximated by a minimum bounding rectangle (MBR) (cf Figure 2(a)). All these rectangles are then used as an input for an R*-Tree. For simplicity we only rely on these MBRs and do not consider the more complicated probabilistic approximations of each object.

The pseudo code of the spatio-temporal filter is illustrated in Algorithm 1. The main idea is to (i) perform *candidates search* for each timestamp t in the query interval T separately (cf. Figure 2(a)) and (ii) for each candidate find the set of objects which are needed for the verification step (influence objects S_{ifl}^{cnd}). For each t we consider the R-Tree I_{DB}^{t} which results by intersecting the time-slice t with the R-Tree I_{DB} (cf Figure 2(b)). This can be done efficiently during query processing, by just ignoring pages of

---

[1] Note that the techniques for pruning objects do not rely on this index and thus can also be applied in scenarios where there is no index present.

Algorithm 1. Spatio-Temporal Filter for the P∀RNN query

Require: q, T, I_{DB}

1. $\forall t \in T : S_{cnd}^t = \emptyset$
2. $\forall t \in T : S_{ifl}^{cnd,t} = \emptyset$
3. **for** each $t \in T$ **do**
4.   init min-heap H ordered by minimum distance to q
5.   insert root entry of I_{DB}^t into H
6. $S_{prn} = \emptyset$
7.   **while** H is not empty **do**
8.   de-heap an entry e from H
9.   **if** $\exists e_2 \in H \cup S_{prn} \cup S_{cnd}^t : Dom(e_2, q, e)$ **then**
10. $S_{prn} = S_{prn} \cup \{e\}$
11.   **else if** e is directory entry **then**
12.   **for** each child ch in e **do**
13.   insert ch in H
14.   **end for**
15.   **else if** e is leaf entry **then**
16. $S_{cnd}^t = S_{cnd}^t \cup \{e\}$
17.   **end if**
18.   **end while**
19.   **for** each $cnd \in S_{cnd}^t$ **do**
20.   **if** $\exists le : Dom(le, q, e)$ **then**
21.   continue;
22.   **end if**
23. $S_{ifl}^{cnd,t} = \{le : \neg Dom(le, q, e)\} \wedge \neg Dom(q, le, e)\}$
24.   **end for**
25. **end for**
26. $S_{ref} = \bigcap_{t \in T} S_{cnd}^t$
27. **for** each $cnd \in S_{ref}$ **do**
28. $S_{ifl}^{cnd} = \bigcup_{t \in T} S_{ifl}^{cnd,t}$
29. **end for**
30. **return** $\forall cnd \in S_{ref} : (cnd, S_{ifl}^{cnd})$

---

the index, that do not intersect with the value of t in the temporal domain. For each time t a reverse nearest neighbor candidate search [2] is performed.

Therefore, a heap H is initialized, which organizes its entries by their minimum distance to the query object q. H initially only contains the root node of I_{DB}^t. Additionally we initialize two empty sets. S_{cnd} contains all RNN candidates which are found during query processing and S_{prn} contains objects (leaf entries) or entries which have been verified not to contain candidates. Then, as long as there are entries in H, a best-first traversal of I_{DB}^t is performed. For each entry e, which is de-heaped from H, the algorithm checks whether e can be pruned (i.e., it cannot contain potential candidates) by another object or entry e_2 which has already been seen during processing. This is the case if e_2 dominates q w.r.t. e, i.e. e_2 is definitely closer to e than q which implies that e cannot be RNN of q. To verify spatial domination we adapt the technique proposed in [5], as shown in the following lemma.

Lemma 1 (Spatial Domination [5]). *Let A, B, R be rectangular approximations then the relation*

$$Dom(A, B, R) = \forall a \in A, b \in B, r \in R : dist(a, r) < dist(b, r)$$

can efficiently be checked by the following term

$$Dom(A, B, R) = \sum_{i=1}^{d} \max_{b_i \in \{B_i^{min}, B_i^{max}\}} (MaxDist(A_i, b_i)^2 - MinDist(Q_i, b_i)^2) < 0$$

where X_i ($X \in \{A, B, Q\}$) denotes the projection interval of the rectangular region of X on the ith dimension, X_i^{min} (X_i^{max}) denotes the lower (upper) bound of the interval X_i, and MaxDist(I, p) (MinDist(I, p)) denotes the maximal (minimal) distance between a one-dimensional interval I and a one-dimensional point p.

An entry which is pruned by this technique is moved to the S_{prn} set. If an entry cannot be pruned it is either moved to the candidate set if it is a leaf entry or put into the heap H for further processing.

After the index traversal, for each candidate it can be checked if the candidate is pruned by another object. If the other object it is definitely closer to the candidate than the query, the candidate object can be discarded (see line 21). To find the set of objects that could possibly prune a candidate, we can again use the domination relation (see line 23). An object (leaf entry le) is necessary for the verification step if it might be closer to the candidate than the query, which is reflected by the statement in this line. A more detailed description of this step can be found in [2].

After performing this process for each timestamp t we have to merge the results for each point of time to obtain the final result. In the case of a P∀RNN we intersect the *candidate* sets for each point in time. The only difference for the P∃RNN query is that we have to unify the results in this step.

These *influencing* objects of each candidate have to be unified for each time $t \in T$ to obtain the final set of influencing objects. The algorithm ultimately returns a list of candidate objects together with their sets of influencing objects.

4.2 Verification

The objective of the verification step is to compute, for each candidate c, the probability $P∃RNN(c, q, \mathcal{D}, T)$ ($P∀RNN(c, q, \mathcal{D}, T)$) and compare this probability with the probability threshold τ. An interesting observation is, that we were able to prune objects based on the consideration of each point $t \in T$ separately, however as shown in Section 3.2 it is not possible to obtain the final probability value by just considering the single time probabilities. Thus our approach relies on sampling of possible trajectories for each candidate and the corresponding influence objects, as computing exact result probabilities has a high complexity [12]. For this step, we can utilize the techniques from [12]. On each sample (which then consists of certain trajectories) we then are able to efficiently evaluate the query predicate. Repeating this step often enough we are able to approximate the true probability of $P\exists RNN(c,q,\mathcal{D},T)$ $(P\forall RNN(c,q,\mathcal{D},T))$ by the percentage of samples where the query predicate was satisfied; the reasoning behind this can be found in [12]. The algorithm for the verification of the P∀RNN query is given in Algorithm 2. Note, that it is possible to early terminate sampling of influence objects, when we find a time t where the candidate is not closer to q than to the object trajectory just sampled. For the P∃RNN query we can also implement this early termination whenever we can verify the above for each point of time.

Algorithm 2. Verification for the P∀RNN query

Require: q, T, cnd, S_{ifl}^{cnd}, num_samples
1. num_satisfied = 0
2. **for** $0 \leq i \leq num\_samples$ **do**
3.   num_satisfied = num_satisfied + 1
4. $cnds = sampleTrajectory(cnd)$
5.   **for all** $o \in S_{ifl}^{cnd}$ **do**
6. $os = sampleTrajectory(o)$
7.   **if** $\exists t \in T : dist(cnds(t), os(t)) < dist(cnds(t), q(t))$ **then**
8.   num_satisfied = num_satisfied - 1
9.   break
10.   **end if**
11.   **end for**
12. **end for**
13. **return** num_satisfied/num_samples

---

5   Experiments

In our experimental evaluation, we focus on testing the efficiency of our algorithm for P∀RNNQ and P∃RNNQ, by measuring (i) the number of candidates and influence objects remaining after pruning irrelevant objects based on their spatio-temporal MBRs, and (ii) the runtime of the refinement procedure, i.e. sampling. Similar to [12], we split the sampling procedure into adapting the transition matrices (building the trajecory sampler) of the objects and the actual sampling process, as the adaption of transition matrices can be done as a preprocessing step. All experiments have been conducted in the UST framework that has also been used in [12]. The experimental evaluation has been conducted on a desktop computer with Intel i7-870 CPU at 2.93 GHz and 8GB of RAM.

5.1   Setup

Our experiments are based on both artificial and real data sets. In the artificial dataset states are drawn from a uniform distribution to model a (discretized) *continuous* state space. The real data set uses an OpenStreetMap graph of the city of Beijing to derive the underlying state space, modeling a *street network*. Both of the datasets have also been used in [12], however here we recap their generation for the sake of completeness.

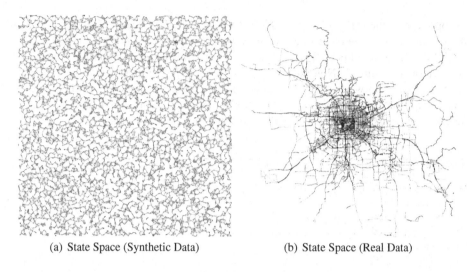

(a) State Space (Synthetic Data)   (b) State Space (Real Data)

Fig. 3. Examples of the models used for synthetic and real data. Black lines denote transition probabilities. Thicker lines denote higher probabilities, thinner lines lower probabilities. The synthetic model consists of 10k states.

Artificial Data. Artificial data was generated in four steps. *First* a state space was generated by randomly drawing N points from the $[0, 1]^2$ space. These points represent the states of the underlying Markov chain. *Second* we built a graph from these neighboring states by connecting neighboring points. Given a reference point p, we selected all points within in a radius of $r = \sqrt{\frac{b}{N*\pi}}$ from p, connecting each of the resulting points with p. Here b denotes the average branching factor of the resulting graph structure. *Third* we weighted each of the graph's edges based on the distance between the corresponding states: The transition probability between two states was set indirectly proportional to the states' distance relative to all outgoing distances to other states. The resulting transition matrix is independent of a specific object and therefore provides a general model for all objects in the database. An example of such a model can be found in 3(a). Finally and *fourth* we had to generate uncertain trajectories from this data for each object o in the database. Each of these uncertain trajectories has a length of 100 timesteps. To generate the uncertain object, we first sampled a random sequence of states from the state space and connected two consecutive states by their shortest path. The resulting paths contain straight segments (shortest paths), modelling the directed motion of objects, and random changes in direction that can appear e.g. when objects move from one destination (e.g. home, work, or disco for humans) to another destination. The resulting trajectories serve as a certain baseline and are made uncertain by considering only a subset of the trajectory points: every l^{th} node of the trajectory, $l = i * v, v \in [0, 1]$ is used as an observation of the corresponding uncertain object. Here, i models the time difference between observations and v is a lag factor, making the object slower. With, for example, $v = 0.8$, the object moves only with 80% of its maximum speed. We randomly distributed the resulting uncertain objects over the time horizon of the database (defaulting 1000 timesteps). The objects can then be indexed by a UST tree [6]. For our experiments we determined candidates based on MBR filtering, i.e. the MBR over all states that can be reached by an object between two consecutive observations. For our experiments we concentrate on the case of the query q beeing given as a query state.

Real Data. As a real dataset we used taxi trajectories in the city of Beijing [27], equivivalent to [12]. After an initial cleaning phase, the taxi trajectories were map-matched to a street graph taken from OpenStreetMap. Then, from the resulting map-matched trajectories, a general model, i.e. the Markov transition matrix and the underlying state space was generated with a time interval of 10s between two consecutive tics. The resulting model is visualized in Figure 3(b). In the model, transition probabilities denote turning probabilities of taxis at street crossings and states model the street crossings. Due to the sparsity of taxi trajectories only a subset of the real street crossings was hit by a taxi. Therefore the resulting network consisting of only 68902 states is smaller than the actual street network. The uncertain object trajectories were taken directly from the trajectory data. Again, uncertain objects have a lifetime of 100 tics, and uncertain trajectories were generated by taking each 8-th measurement as an observation. For a more detailed description of the dataset we refer to [27], for a detailed description of the model generation from the real dataset we refer to [12].

5.2   Evaluation: P∀RNNQ and P∃RNNQ

The default setting for our performance analysis is as follows: We set the number of states to $N = |\mathcal{S}| = 100k$, the database size (number of objects) to $|\mathcal{D}| = 10k$, the average branching factor (synthetic data) to $b = 8$, probability threshold $\tau = 0$. The length of the query interval was set to $|T| = 10$. To compute a result probability, 10k possible worlds where sampled from the candidate objects.

In each experiment, the left plot shows a stacked histogram, visualizing the cost for building the trajectory sampler (TS) and the actual cost for sampling (EX for the $P\exists RNN$ query, FA for the $P\forall RNN$ query). The right plot first visualizes the number of candidates $(C_x(q))$ for the $P\exists RNN$ ($x = E$) and $P\forall RNN$ ($x = A$) query, i.e. the number of objects for which the nearest neighbor has to be computed. Second it visualizes the number of pruners or influence objects $(I_x(q))$, i.e. the number of objects that can prune candidates. Clearly the number of candidates and pruners is different for $P\forall RNN$ and $P\exists RNN$ queries: for the $P\exists RNN$ query objects not totally overlapping the query interval can be candidates, increasing the number of candidates. For the $P\forall RNN$ query, candidates can be definitely pruned if at least at one point of time another object prunes the candidate object.

Varying $|\mathcal{D}|$. Let us first analyze the impact of the database size (see Figure 4), i.e. the number of uncertain objects on the runtime of a probabilistic reverse nearest neighbor query. First of all note that the number of candidates and influence objects increases if the database gets large. This is the case because with more objects, the density of objects increases and therefore more objects become possible results of the RNN query. Also, clearly, the number of influence objects increases due to the higher degree of intersection of MBRs. Second the the probability computation of candidate objects during

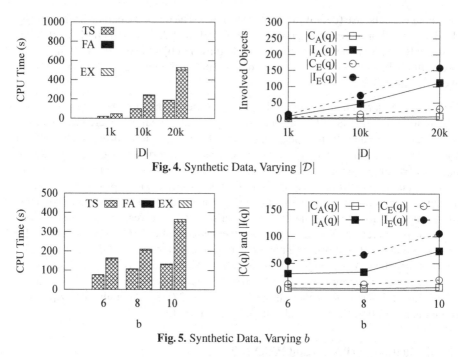

Fig. 4. Synthetic Data, Varying $|\mathcal{D}|$

Fig. 5. Synthetic Data, Varying b

refinement is mostly determined by the computation of the adapted transition matrices, i.e. building the trajectory sampler. This is actually good news, as this step can also be performed offline and the resulting transition matrices can be stored on disk. Last note that evaluating the $P\exists RNN$ query during the actual sampling process (EX and FA for $P\exists RNN$ and $P\forall RNN$ respectively) is also more expensive than the $P\forall RNN$ query, as more samples have to be drawn for each candidate object: possible worlds of $P\forall RNN$-candidates can be pruned if a single object at a single point in time prunes the candidate. For the $P\exists RNN$ query, all points in time have to be pruned which is much less probable. Additionally, more candidates than for the $P\forall RNN$ have to be evaluated, also increasing the complexity of the $P\exists RNN$ query.

Varying b. The effect of varying the branching factor b (see Figure 5) is similar but not as severe as varying the number of objects. Increasing the branching factor increases the number of states that can be reached during a single transition. In our setting, this also increases the uncertainty area of an uncertain object, making pruning less effective, and therefore increasing the number of objects that have to be considered during refinement. As a result, the computational complexity of the refinement phase increases with increasing branching factor.

Varying $|\mathcal{S}|$. Increasing the number of states in the network (see Figure 6) while keeping the branching factor b constant shows an opposite effect on the number of candidates: as the number of states increases, objects become more sparsely distributed such that pruning becomes more effective. Note that for small networks especially the sampling process of the $P\exists RNN$ query becomes very expensive. This effect diminishes with increasing size of the network. Although less objects are involved, building the

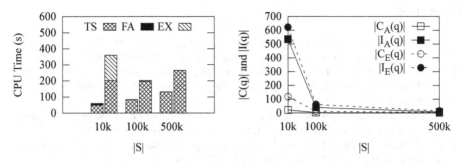

Fig. 6. Synthetic Data, Varying $|\mathcal{S}|$

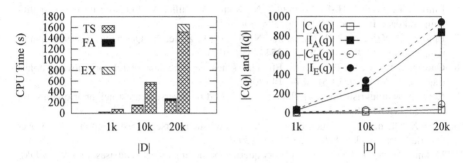

Fig. 7. Real Data, Varying $|\mathcal{D}|$

adapted transition matrices becomes more expensive, as for example matrix operations become more expensive with larger state spaces.

Real Dataset. Last but not least we show results of the $P\forall RNN$ and $P\exists RNN$ queries on the real dataset (see Figure 7). We decided to vary the number of objects as the number of states and the branching factor is inherently given by the underlying model. The results are similar to the synthetic data, however more than twice as many objects have to be considered during refinement. We explain this exemplarily by the non-uniform distribution of taxis in the network. A part of this performance difference can also be explained by the slightly smaller network consisting of about 70k states instead of 100k states in the default setting.

6   Conclusion

In this paper we addressed the problem of probabilistic RNN queries on uncertain spatio-temporal data following the possible worlds semantics. We defined two queries, the $P\exists RNNQ(q,\mathcal{D},T,\tau)$ and the $P\forall RNNQ(q,\mathcal{D},T,\tau)$ query and proposed pruning techniques to exclude irrelevant objects from costly propability computations. We then used sampling to compute the actual $P\exists RNNQ(q,\mathcal{D},T,\tau)$ and $P\forall RNNQ(q,\mathcal{D},T,\tau)$ probabilities for the remaining candidate objects. During an extensive performance analysis on both synthetic and real data we empirically evaluated our theoretic results.

References

1. Achtert, E., Böhm, C., Kröger, P., Kunath, P., Pryakhin, A., Renz, M.: Efficient reverse k-nearest neighbor search in arbitrary metric spaces. In: Proc. SIGMOD (2006)
2. Bernecker, T., Emrich, T., Kriegel, H.-P., Renz, M., Zankl, S., Züfle, A.: Efficient probabilistic reverse nearest neighbor query processing on uncertain data. In: Proc. VLDB, pp. 669–680 (2011)
3. Cheema, M.A., Lin, X., Wang, W., Zhang, W., Pei, J.: Probabilistic reverse nearest neighbor queries on uncertain data. IEEE Trans. Knowl. Data Eng. 22(4), 550–564 (2010)
4. Cheng, R., Kalashnikov, D., Prabhakar, S.: Querying imprecise data in moving object environments. In: IEEE TKDE, vol. 16, pp. 1112–1127 (2004)
5. Emrich, T., Kriegel, H.-P., Kröger, P., Renz, M., Züfle, A.: Boosting spatial pruning: On optimal pruning of mbrs. In: Proc. SIGMOD (2010)
6. Emrich, T., Kriegel, H.-P., Mamoulis, N., Renz, M., Züfle, A.: Indexing uncertain spatio-temporal data. In: Proc. CIKM, pp. 395–404 (2012)
7. Emrich, T., Kriegel, H.-P., Mamoulis, N., Renz, M., Züfle, A.: Querying uncertain spatio-temporal data. In: Proc. ICDE, pp. 354–365 (2012)
8. Güting, R.H., Behr, T., Xu, J.: Efficient k-nearest neighbor search on moving object trajectories. VLDB J. 19(5), 687–714 (2010)
9. Korn, F., Muthukrishnan, S.: Influenced sets based on reverse nearest neighbor queries. In: Proc. SIGMOD (2000)
10. Lian, X., Chen, L.: Efficient processing of probabilistic reverse nearest neighbor queries over uncertain data. VLDB J. 18(3), 787–808 (2009)
11. Mokhtar, H., Su, J.: Universal trajectory queries for moving object databases. In: Proc. MDM, pp. 133–144 (2004)
12. Niedermayer, J., Züfle, A., Emrich, T., Renz, M., Mamoulis, N., Chen, L., Kriegel, H.-P.: Probabilistic nearest neighbor queries on uncertain moving object trajectories. In: Proc. VLDB (to appear, 2014)
13. Qiao, S., Tang, C., Jin, H., Long, T., Dai, S., Ku, Y., Chau, M.: Putmode: Prediction of uncertain trajectories in moving objects databases. Appl. Intell. 33(3), 370–386 (2010)
14. Ré, C., Letchner, J., Balazinksa, M., Suciu, D.: Event queries on correlated probabilistic streams. In: Proc. SIGMOD, pp. 715–728 (2008)
15. Stanoi, I., Agrawal, D., Abbadi, A.E.: Reverse nearest neighbor queries for dynamic databases. In: Proc. DMKD (2000)
16. Tao, Y., Faloutsos, C., Papadias, D., Liu, B.: Prediction and indexing of moving objects with unknown motion patterns. In: Proc. SIGMOD, pp. 611–622 (2004)
17. Tao, Y., Papadias, D., Lian, X.: Reverse kNN search in arbitrary dimensionality. In: Proc. VLDB (2004)
18. Tao, Y., Papadias, D., Shen, Q.: Continuous nearest neighbor search. In: Proc. VLDB, pp. 287–298 (2002)
19. Tao, Y., Yiu, M.L., Mamoulis, N.: Reverse nearest neighbor search in metric spaces. IEEE TKDE 18(9), 1239–1252 (2006)
20. Trajcevski, G., Choudhary, A.N., Wolfson, O., Ye, L., Li, G.: Uncertain range queries for necklaces. In: Proc. MDM, pp. 199–208 (2010)
21. Trajcevski, G., Tamassia, R., Ding, H., Scheuermann, P., Cruz, I.F.: Continuous probabilistic nearest-neighbor queries for uncertain trajectories. In: Proc. EDBT, pp. 874–885 (2009)
22. Trajcevski, G., Wolfson, O., Hinrichs, K., Chamberlain, S.: Managing uncertainty in moving objects databases. ACM Trans. Database Syst. 29(3), 463–507 (2004)
23. Xiong, X., Mokbel, M.F., Aref, W.G.: Sea-cnn: Scalable processing of continuous k-nearest neighbor queries in spatio-temporal databases. In: Proc. ICDE, pp. 643–654 (2005)

24. Xu, C., Gu, Y., Chen, L., Qiao, J., Yu, G.: Interval reverse nearest neighbor queries on uncertain data with markov correlations. In: Proc. ICDE (2013)
25. Yang, C., Lin, K.-I.: An index structure for efficient reverse nearest neighbor queries. In: Proc. ICDE (2001)
26. Yu, X., Pu, K.Q., Koudas, N.: Monitoring k-nearest neighbor queries over moving objects. In: Proc. ICDE, pp. 631–642 (2005)
27. Yuan, J., Zheng, Y., Xie, X., Sun, G.: Driving with knowledge from the physical world. In: Proc. KDD, pp. 316–324 (2011)
28. Yuan, J., Zheng, Y., Zhang, C., Xie, W., Xie, X., Huang, Y.: T-drive: Driving directions based on taxi trajectories. In: Proc. ACM GIS (2010)
29. Zheng, Y., Li, Q., Chen, Y., Xie, X., Ma, W.: Understanding mobility based on gps data. In: Proc. Ubicomp, pp. 312–312 (2008)

Selectivity Estimation of Reverse k-Nearest Neighbor Queries

Michael Steinke, Johannes Niedermayer, and Peer Kröger

Institute for Computer Science
Ludwig-Maximilians-Universität München, Germany
{steinke,niedermayer,kroeger}@dbs.ifi.lmu.de

Abstract. This paper explores different heuristics to estimate the selectivity of a reverse k-nearest neighbor query. The proposed methods approximate the number of results for a given RkNN query, providing a key ingredient of common (relational) query optimizers. A range of experiments evaluate the quality of these estimates compared to the true number of results on both real and synthetic data, analyzing the potentials of the proposed approximations to make accurate predictions that can be used to generate efficient query execution plans.

1 Introduction

Storing and managing multidimensional feature vectors are basic functionalities of a database system in applications such as multimedia, CAD, molecular biology, etc. In such applications, efficient and effective methods for content based similarity search and data mining are required that are typically based on multidimensional feature vectors and on query predicates consisting of metric distance functions such as any L_p-norm. In order to give full support to applications dealing with multidimensional feature vectors, the database system needs efficient and effective techniques for query optimization. Often, a key step during query optimization is estimating the selectivity of queries before actually executing them. Selectivity estimation aims at predicting the number of objects that do meet the query predicate. Knowing about the selectivity of a given query would help a physical query optimizer to decide which particular algorithm to use. Also, logical query optimization uses such estimates to minimize intermediate results in order to find optimal query plans.

While there have been many different proposals how to accurately guess the selectivity of "classical" similarity queries on multidimensional vector data like range queries and window queries, to the best of our knowledge, there is no systematic approach for the concept of RkNN queries. RkNN queries retrieve for a given query object o all database objects that have o amongst their k-nearest neighbors (kNN). It is worth noting that RkNN queries and kNN queries are not symmetric and the number of objects that qualify for a given query are not known beforehand. Due to the importance of RkNN queries in numerous applications, a lot of work has been published for supporting this query predicate following different algorithmic paradigms. This variety of approaches have very different properties in different use cases [1], offering a huge potential for case-based ad-hoc physical query optimization. Let RkNN(q) denote the

S.S. Bhowmick et al. (Eds.): DASFAA 2014, Part II, LNCS 8422, pp. 108–123, 2014.

set of RkNNs of object q. Then the aim of this paper is to compute $RkNN_{est}(q_i)$, i.e. estimate $|RkNN(q)|$, the number of RkNN of the query point q.

Many methods for selectivity estimation in the context of range queries and nearest neighbor queries are based on sampling or histogram-based techniques. In this paper, we apply these two techniques for estimating the selectivity of RkNN queries. The first approach and at the same time a baseline for other solutions is simply taking k as a selectivity estimate. Second we apply sampling to RkNN selectivity estimation. Third we evaluate the applicability of histogram-based selectivity estimation in the context of the RkNN query. For this technique, we employ common RkNN query techniques on rough approximations of the data space. As an additional technique for selectivity estimation, we consider a relationship between the RkNN selectivity and an object's kNN sphere.

The reminder is organized as follows: In Section 2 we review related work on both RkNN queries and selectivity estimation. Then, in Section 3 we propose our selectivity estimation techniques. Finally, in Section 4 we extensively evaluate the proposed approaches. Section 5 concludes this work.

2   Related Work

In this section we review related work on RkNN queries and selectivity estimation.

RkNN Query Processing. RkNN query algorithms can be roughly cathegorized into *self-pruning* and *mutual-pruning* approaches. In self-pruning techniques [2,3,4,5], objects, i.e. data objects or entries of an index structure such as an R-tree, prune parts of the data space based on information stored in the object itself. Therefore, these techniques often involve costly precomputations. The precomputed information is then stored in specialized index structures and accessed during query time. Because of these precomputations, the computational complexity of such solutions is moved to a preprocessing stage such that the actual query can be evaluated with high speed. However, this preprocessing makes updates of the data base costly, because preprocessed information has to be updated as well if items are inserted or deleted.

Mutual-pruning approaches on the other hand do not rely on precomputations and therefore can be used with traditional index strucutures such as the R*-tree or M-tree. However by avoiding precomputations, query evaluation can be slower than with self-pruning approaches. A famous example of such mutual-pruning techniques is TPL [6], but there exist also other approaches such as [7,8,9].

There exist also approximate solutions aiming at reducing the time of query evaluation for the cost of accuracy [8,10], and solutions for uncertain data [11,12].

There exists also work on reverse nearest neighbor aggregates on data streams where aggregates over the bichromatic reverse nearest neighbors, such as the count and the maximum distance are computed in a stream environment [13]; we however address the monochromatic case. The number of RkNNs has also been used to identify boundary points in [14].

Selectivity Estimation Approaches. For range and window queries, three different paradigms of data modelling for selectivity estimation in general can be distinguished: Histograms, sampling, and parametric techniques.

Many different sampling methods have been proposed. They share the common idea to evaluate the predicate on top of a small subset of the actual database objects and to extrapolate the observed selectivity. The well-known techniques differ in the way how the sample is drawn as well as in the determination of the suitable size of the sample. The general drawback of sampling techniques is that the accuracy of the result is strictly limited by the sample rate. To get an accurate estimation of the selectivity, often a large sample of the database is required. To evaluate the query on top of the large sample is not much cheaper than to evaluate it on the original data set which limits its usefulness for query optimization. Examples of sampling based methods for selectivity estimations are [15,16].

Histogram techniques, the most prevalent paradigm to model the data distribution in the one-dimensional case, have a different problem. This concept is very difficult to be carried over to the multidimensional case, even for low or moderate dimensional data. One way to adapt one-dimensional histograms to multidimensional data is to describe the distribution of the individual attributes of the vectors independently by usual histograms. These histograms are sometimes called marginal distributions. In this case, the selectivity of multidimensional queries can be determined easily provided that the attributes are statistically independent, i.e. neither correlated nor clustered. Real-world data sets, however, rarely fulfill this condition. Another approach is to partition the data space by a multidimensional grid and to assign a histogram bin to each grid cell. This approach may be possible for two- and three-dimensional spaces. However, for higher dimensional data this method becomes inefficient and ineffective since the number of grid cells is exponential in the dimensionality. Techniques of dimensionality reduction such as Fourier transformation, wavelets, principal component analysis or space-filling curves (Z-ordering, Hilbert) may reduce this problem to some extent. The possible problem reduction, however, is limited by the intrinsic dimensionality of the data set. Representative methods include e.g. MHIST [17] and STHoles [18].

The idea of parametric techniques is to describe the data distribution by curves (functions) which have been fitted into the data set. In most cases Gaussian functions (normal distributions) are used. Instead of using one single Gaussian, a set of multivariate Gaussians can be fitted into the data set which makes the technique more accurate. Each Gaussian is then described by three parameters (mean, variance and the relative weight of the Gaussian in the ensemble). This approach can exemplarily be transferred into the multidimensional case as follows. Like described above for histograms, the marginal distribution of each attribute can be modelled independently by a set of Gaussians. The multidimensional query selectivity can be estimated by combining the marginal distributions. This approach leads to similar problems like marginal histograms. A popular example for this class of techniques is ASE [19].

3   Selectivity Estimation for RkNN Queries

In this section, we introduce four different approaches for estimating the selectivity of RkNN queries. The first approach simply considers k as an estimate of selectivity. Next we introduce sampling and histograms, and finally we aim at exploiting the correlation between kNN distances and the number of RkNN of a query point for selectivity estimation.

3.1 Derivation from k

A simple approach – our baseline – of estimating the RkNN-selectivity of a query q is based directly on the parameter k, exploiting the property that on average the selectivity of q, $RkNN_{est}(q)$ equals k.

For basic kNN queries, the selectivity can be easily estimated, as a kNN query, based on its definition, returns *exactly* k or *at least* k query results, depending on the handling of ties during query evaluation. For an RkNN query, we can exploit this property to deduce that the expected number of RkNN of a query q is k: Let a dataset $\mathbb{M}$ consisting of $m = |\mathbb{M}|$ points be given. Let us assume that ties of a kNN query have a low probabiltiy given the underlying data and thus can be neglected. In general this is the case, as most distance functions have a real-valued range, and therefore the probability of two points having the same distance to a query point is close to zero, as long as the database does not contain duplicates. In such a setting, posing a kNN query of a point $p \in \mathbb{M}$ onto the dataset $\mathbb{M}$ returns by definition exactly k kNNs. As a result, the resulting k points from $\mathbb{M}$ have p as one of their RkNN. Now, if we pose a kNN query for all points $p \in \mathbb{M}$, each of these points has k nearest neighbors, resulting in $m * k$

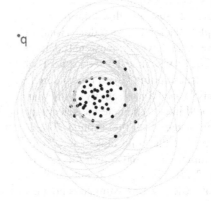

Fig. 1. Selectivity Estimation with k ($k = 50$)

RkNN for the dataset $\mathbb{M}$. The average number of RkNN over all points in $\mathbb{M}$ therefore is $\frac{m*k}{m} = k$ RkNN, concluding the assumption.

Clearly, the advantage of this approach is its constant runtime complexity. However, only considering the expected number of RkNN does not necessarily provide good estimates as the actual distribution of the underlying dataset is not considered.

Figure 1 visualizes this problem. It shows a clustered dataset consisting of 55 points (black) in two-dimensional Euclidean space and a query point q (red). Now let us assume that k is set to 50. The kNN spheres for $k = 50$ are also drawn in the figure. The query q which lies outside the cluster is not contained in any kNN spheres of the data points and therefore does not have any RkNN. However, as the selectivity estimate is simply k, the result of the selectivity estimation is 50, resulting in a very high error.

3.2 Sampling

As we have seen, while simply taking k as a selectivity estimate provides good results on average, the absolute error of these selectivity estimates can be high. Another choice to solve the problem is based on sampling. The idea of this approach is to precompute the number of RkNN results for a set of sample points from the dataset $\mathbb{M}$. During query evaluation, the selectivity of a query point is then estimated based on the sample point closest to the query.

During the preprocessing stage, a sample $\mathbb{S}$ of size $n, 0 < n \leq |\mathbb{M}|$ is drawn from $\mathbb{M}$. While in general the accuracy of selectivity estimation increases with increasing sample size, the sample size still has to be kept low in order to minimize the runtime complexity of this approach, so there is clearly a trade-off. For each $p \in \mathbb{S}$, an RkNN query is posed on the dataset $\mathbb{M}$, and the number of RkNN results is stored as tuples $(p, |RkNN(p)|)$. To increase the performance at runtime, these result tuples can be stored in an index structure that provides fast nearest neighbor queries, such as an R-tree.

During query evaluation, given a query point q and the sample set $\mathbb{S}$, the selectivity $|RkNN(q)|$ has to be estimated. To achieve this, the nearest neighbor of q in $\mathbb{S}$, i.e. $NN(q, S)$ is determined, and the selectivity of of the result point is used as an estimate of the selectivity of q.

Figure 2 illustrates this solution by an example: Black and green dots visualize an exemplary two-dimensional dataset $\mathbb{M}_B$. Circles visualize the 1NN spheres of each database point. Green points denote the sample used for selectivity estimation. We would like to estimate the selectivity of point q based on this sample. Therefore we find the point closest to q from the sample, i.e. p and return the size of its $|RkNN(p, \mathbb{M}_B)| = 1$ result set as an estimate. Clearly, this estimate is not perfect, as the actual result for q is 2.

Furthermore note that the accuracy of an estimate depends significantly on $dist(q, NN(q, \mathbb{S}))$ - i.e. the distance between q and its nearest neighbor from the sample set. If the actual dataset is taken as a basis of the sample set $\mathbb{S}$, we face the

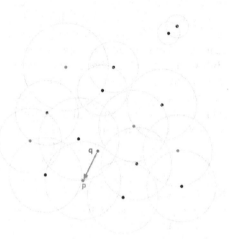

Fig. 2. Selectivity Estimation via Sampling

problem that query points taken from a different distribution will often have a high distance to their closest sample point such that the estimate for these points is often insufficient. However, if the query point q lies in a dense area of the database, the selectivity estimate of q is usually good. On the other hand the sample could be drawn from a different distribution, such as a regular grid, or a uniform distribution. The problem with this approach is that it might draw samples from volumes in the dataset that are irrelevant. In conclusion, the choice of the underlying sampling strategy depends significantly on the user's needs, and on the properties of both the database and the expected queries. Another solution to solve the issue of sample selection would be learning from previous queries. However, we do not focus on this aspect here, but follow the approach of drawing samples directly from the dataset in our experiments.

3.3 Selectivity Estimation Based on kNN-RkNN Correlations

The next method considers an idea similar to [20]. The authors of [20] exploit a (low) correlation between the sets $NN(q)$ and $RNN(q)$ to answer reverse nearest neighbor queries. Based on this idea, but slightly different, we could build a relationship between the diameter of a kNN sphere of a point q and the cardinality of its RkNN set. The intention of this approach is that the computational complexity of kNN queries is lower than the complexity of RkNN queries: Without an index structure, the complexity of a kNN query is in $O(|\mathbb{M}|)$, while for the RkNN query it is $O(|\mathbb{M}|^2)$

To exploit this property, during a preprocessing stage we could compute the kNN-diameters and |RkNN| cardinalities of each point (or a sample set) in the dataset. During query evaluation, given a query point q we could then compute the diameter of its kNN sphere, and estimate its RkNN selectivity by employing the kNN spheres of other points in the database. To estimate the selectivity, we could use one of three approaches:

1. Take $|RkNN(p, \mathbb{M})|$ from the point $p \in \mathbb{M}$ with the kNN diameter closest to the kNN diameter of q by running a NN query on the one-dimensional feature space of kNN distances.
2. Do not only consider the most similar point concerning the kNN diameter, but instead the x most similar points by performing an xNN query. To estimate the selectivity, use the median or average selectivity of the resulting objects.
3. Select the closest objects based on a range query instead of a nearest neighbor query, and compute the result as the median or the average selectivity of the result set. As with range queries in general, the problem of this solution is that the result set might either be empty or very large.

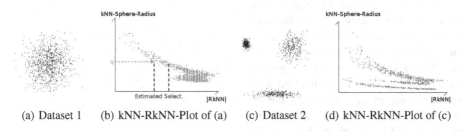

(a) Dataset 1   (b) kNN-RkNN-Plot of (a)   (c) Dataset 2   (d) kNN-RkNN-Plot of (c)

Fig. 3. RkNN-Number-/kNN-Sphere-Relationship ($k = 50$)

A visualization of this approach based on the dataset from Figure 3 (a), showing several problems of this idea, can be found in Figure 3 (b) (k=50). First of all note that the shown diagrams contain a long very flat tail (see Figure 3 (b),(d)). Therefore a high number of points with a similar kNN diameter shows a very large variation in its selectivity. As a result, the selectivity estimate of a query point q can be very imprecise. The optimal case for our goal of selectivity estimation would be that a kNN diameter precisely determines $RkNN(q, \mathbb{M})$. Second note that a dataset often does not consist of a simple linearly decreasing curve, but rather of a set of curves with different properties. The clustered dataset from Figure 3 (c) returns for $k = 50$ the RkNN/kNN relationship shown in Figure 3 (d). Each of the three curves represents a single cluster. Especially in the lower area, these curves overlap. As the RkNN/kNN relationship varies with the cluster-membership, this makes good selectivity estimates difficult to achieve. A possible solution to solve this problem would be to run a clustering algorithm, dividing the dataset into several clusters with similar properties. Then, given a query point q, we could first determine the cluster q falls into, and then compute the selectivity of q only based on the points contained in the same cluster. However, this solution would not solve the first problem of the long tail. As this solution is very data-specific, we will not consider it during our experimental evaluation.

3.4   Selectivity Based on Histograms

In this section we want to investigate the applicability of histograms for RkNN selectivity estimation. Based on the RkNN algorithm from [21], the minimum distance (*MinDist*) and maximum distance (*MaxDist*) of objects are used to narrow down the selectivity of an RkNN query, resulting in a *lower Bound* and an *upper Bound* of the real selectivity; these bounds are conservative, i.e. the actual selectivity is somewhere between the lower and the upper bound, but never lies outside the corresponding interval.

To reduce the complexity of this algorithm, instead of computing the exact RkNN result set we split the data space into a equally-sized regions in each dimension, for each resulting bucket (or bin) storing the number of points falling into them, similar to a histogram. An example for a two-dimensional Euclidean space can be found in Figure 4; each dimension is split into $a = 4$ equally sized regions. Note that with increasing dimensionality d, the number of buckets increases exponentially (a^d), a disadvantage of this solution.

The resulting bins of a histogram can be easily linearized since the coordinates of each corner can be reconstructed based on a bins position in the linearized array. As a result, the storage complexiy for each bin reduces to a single integral value, i.e. the number of points falling into the bin. The position of a bucket in the linearized list of all buckets can be computed as:

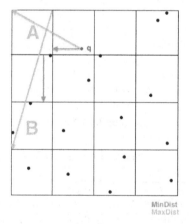

$$binPos_p = \sum_{i=0}^{d-1} a^i * \left\lfloor \frac{(p.x_i - x_i^{min})}{len/a} \right\rfloor \tag{1}$$

Here, $p.x_i$ stands for the coordinate value x_i of point p, x_i^{min} describes the smallest coordinate value in the data set and $len = x_i^{max} - x_i^{min}$ (with x_i^{max} the largest value in the data set). The runtime

Fig. 4. Min/Max-Approach

of inserting a point into the list is therefore linear in the number of dimensions of a data point. Now let $upperBound(q)$ and $lowerBound(q)$ be methods for computing the upper and lower bound of the selectivity of a query point q. Let $minDist(bin_1, bin_2)$ denote the minimum distance, $maxDist(bin_1, bin_2)$ the maximum distance between two bins bin_1 and bin_2.

Upper Bound. Estimating the selectivity upper bound is based on the exclusion of bins from the result. A bin is excluded if not a single point in the bin, independent of its position, could have q as one of its nearest neighbors. To find out whether a point in bin $B_i \in \mathbb{L}_{bin}$ could possibly have q as one of its nearest neighbors we first compute $minDist(B_i, q)$. Then, the maximum distance $maxDist(B_i, B_j)$ of B_i to every bin $B_j \in \mathbb{L}_{bin}$, including B_i, is computed. If $maxDist(B_i, B_j) < minDist(B_i, q)$ holds, every point in B_j is closer to every point in B_i than q; therefore the prune count $prune\_cnt_{B_i}$ can be increased by the number of values in B_j, i.e. $data\_cnt_{B_j}$.

If $prune\_cnt_{B_i} \geq k$, the bucket can not possibly contain a RkNN of q as at least k points are closer to B_i than to q. Therefore the bucket can be discarded. Otherwise, if each bucket B_j has been tested against B_i and $prune\_cnt_{B_i} < k$, all points in B_i are potential RkNN of q. The upper bound then results from the following formula:

$$\sum_{\{B_i : prune\_cnt(B_i) < k\}} data\_cnt(B_i)$$

Lower Bound. The lower bound $B_i \in \mathbb{L}_{bin}$ returns the number of points in the data set that are definitely RkNN of a given query point q. To compute this lower bound, first we compute the maximum distance between q and B_i, $maxDist(B_i, q)$. Then for each other bin $B_j \in \mathbb{L}_{bin}$, $minDist(B_i, B_j)$ is computed. If $maxDist(B_i, q) \geq minDist(B_i, B_j)$ points in B_j might be closer to points in B_i than q; hence $prune\_cnt_{B_i}$ is increased by $data\_cnt_{B_j}$. On the other hand, if $maxDist(B_i, q) < minDist(B_i, B_j)$, $prune\_cnt_{B_i}$ is not increased as q is closer to each point from B_i than any point from B_j. Grid cell B_i can be pruned as soon as its prune count $prune\_cnt(B_i) \geq k$. The resulting selectivity estimate follows again as

$$\sum_{\{B_i : prune\_cnt(B_i) < k\}} data\_cnt(B_i)$$

The accuracy of this approach depends directly on the number of bins used for constructing the histograms, and an exact selectivity computation becomes possible as soon as the number of buckets becomes large enough. However increasing a increases the number of buckets exponentially in a^d. Therfore this solution performs significantly worse for high-dimensional data.

To reduce errors and tighten the selectivity bounds, we could not only increase the grid resolution (which increases the computational complexity exponentially), but also replace the $MinDist/MaxDist$ approach by the decision criterion from [22] that is optimal in the sense of pruning irrelevant grid cells based on their bounding box. Another approach to increase the tightness of the bounds would be to use minimum bounding rectangles for each of the cells, however this would increase the storage complexity of the algorithm, as we would have to store the MBR coordinates for each grid cell. In our solution, the bounds of a grid cell can be derived from its position in the underlying list, decreasing the memory complexity of the problem.

3.5   Extension to Metric Space

A metric space is defined as a pair $(\mathbb{X}, d)$, with $\mathbb{X}$ a domain and $d : \mathbb{X} \times \mathbb{X} \to \mathbb{R}$ a metric defined on $\mathbb{X}$. By definition, the metric d must be *positive definite*, *symmetric*, and it must fulfill the *triangle inequality* [23]. Data types involving metric spaces exemplarily include graphs and genetic data. The approaches from Section 3.1 and Section 3.2 can be trivially extended to metric spaces. More difficult is the MIN/MAX approach from Section 3.4 that can also be used in metric spaces with some modifications: the division of the data space into bins must be modified. Instead of laying a grid over the data space, a set of anchor points $\mathbb{X}$ can be sampled from the data itself. The remaining points are then assigned to these anchor points based on heuristics such as the distance to an anchor point and the number of points already assigned to this point. This assignment is similar to the M-tree [24] and has been applied to RkNN query processing, recently [21]. The remaining algorithm is equivalent to vector spaces, however the $minDist-$ und $maxDist-$ functions have to be defined according to the underlying data.

4   Evaluation

In our experimental evaluation, we compare the sampling based selectivity estimation approach with the MIN/MAX estimation approach and the simple solution of taking k as an estimate of selectivity.

4.1   Experimental Setup

The approaches have been evaluated on three synthetic datasets: an equi-distributed one, a clustered one and a normally distributed dataset. The clustered dataset consists of four normally distributed clusters in a square with an additional cluster in the center of the outer clusters. A visualization of this dataset can be found in Figure 5. By default, experiments are conducted on the normally distributed dataset. Furthermore we studied the three approaches on a real dataset.

During our experiments, we varied the parameter k and the size of the data set n, the dimensionality d of the underlying vector space, the distribution of the underlying data, the sample size s for the sampling approach, and the number of buckets b per dimension for the MIN/MAX-based approach. The overall number of buckets is therefore b^d and hence grows exponentially with the dimensionality of the data space. We followed the approach of keeping all but one variable fixed during an experiment to be able to give insights into the behaviour of our algorithms under a specific change. An overview over the variable values used during our experiments can be found in Table 1. Default values are denoted in bold font. Note that the number of bins for the MIN/MAX approach is relatively low. We chose this setup because this algorithm generally has to approximate a real RkNN query, which is expensive. Therefore, by using many buckets, the approach would degenerate to an exact RkNN query which has to be avoided. For the sampling-based approach we chose the sample size to be 10% of the dataset size by default. The samples were drawn directly from the database, i.e. follow the distribution of the data.

For the query points we decided to use the same distribution as the underlying data mixed with 20% uniform noise by default. We chose this approach because it is possible that queries are posed to areas in the database where the actual density of database values is low. This is especially of interest for our selectivity estimation, since sparse areas in the database affect the performance of the sampling based approach.

Table 1. Experimental Setup: Parameter Values

Parameter	Values
k	1, 10, **50**, 100
d	**2**, 4, 6, 8
n	1.000, **10.000**, 25.000, 70.000, 100.000
Distribution	clustered, **normal**, uniform
s	1%, 5%, **10%**, 50%
b	**8**, 10, 20, 50, 100, 500

As a result of the test runs we computed the average error over the 1000 RkNN estimates for the sampling based and k-based approaches, i.e.

$$\sum_{i<1000} \frac{||RkNN(q_i)| - RkNN_{est}(q_i)|}{1000}$$

For the MIN/MAX approach we did not compute the average error but instead plotted the average over the lower and upper bound of the estimate in order to visualize the large gap between these two bounds.

4.2 RkNN Distribution

Before stepping to the actual evaluation, let us first take a look at the actual RkNN distribution in the evaluated datasets. Figure 5 visualizes the distribution of the number of RkNNs with varying k (50,300,1000)in form of a heat-map. For each point in the database, an RkNN query was performed and the number of results was counted. Blue values visualize the lowest number of RkNN, and red values visualize the highest number of RkNN in the database. Remaining values are interpolated between blue and red. The visualization shows that especially in the clustered dataset the number of RkNN diminishes at the cluster boundaries (this has also been realized by [14]), indicating that the selectivity estimation using k only can not model the dataset sufficiently; this estimate assumes the same number of RkNN for every point in the dataset, clearly a major simplification.

4.3 Evaluation Results

Varying k. In the following we will visualize the errors of the sampling and k estimate in a single plot, and the MIN/MAX estimate separately, because results of the MIN/MAX approach are significantly larger and would therefore obscure the details of the other approaches if plotted in the same figure.

With varying k, the estimates based on k and sampling behave similar, compare Figure 6(left). With small k ($k = 2$ and $k = 10$), the errors of these approaches are nearly equivalent and relatively low. However with larger valuse of k, i.e. $k = 50$ and $k = 100$, the error of both approaches increases. Note that the sampling approach shows a lower increase of the error with increasing k. The observation that errors based on the value of k increase faster can be explained by the fact that with increasing k the number

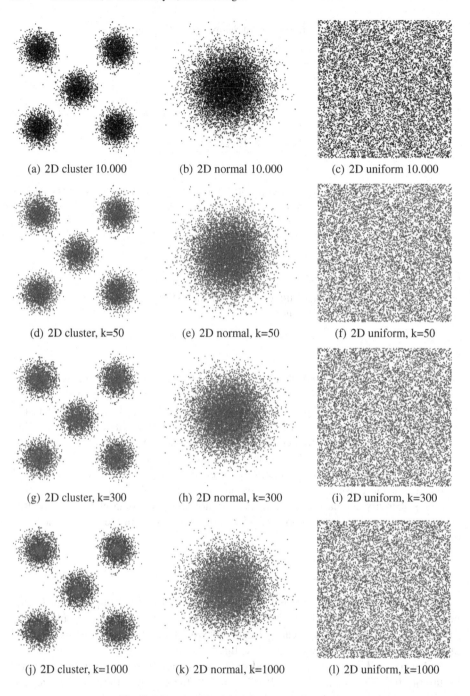

(a) 2D cluster 10.000   (b) 2D normal 10.000   (c) 2D uniform 10.000

(d) 2D cluster, k=50   (e) 2D normal, k=50   (f) 2D uniform, k=50

(g) 2D cluster, k=300   (h) 2D normal, k=300   (i) 2D uniform, k=300

(j) 2D cluster, k=1000   (k) 2D normal, k=1000   (l) 2D uniform, k=1000

Fig. 5. Data visualization (best viewed in color)

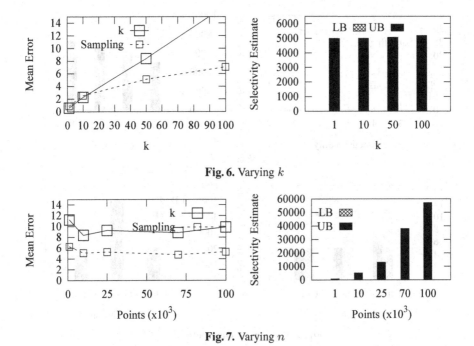

Fig. 6. Varying k

Fig. 7. Varying n

of RkNN can vary more than with lower k. Exemplarily in 2D space, the kissing number problem reduces the number of R1NN for all $p \in \mathbb{M}$ to $0 \leq RNN(p) \leq 6$ [25]. With higher k, this value becomes significantly higher, such that the estimate based on k becomes worse. In the sampling-based approach the aberrance of using k is reduced by incorporating actual information retrieved from the data set.

Figure 6(right) shows the results of the MIN/MAX estimate with varying k. The upper bound of the selectivity estimated with this aproach is around 6500 while the lower bound is always 0 in the default setting. Clearly, these results are inappropriate for estimating the selectivity of RkNN queries due to their high error. Note that the estimate of this approach also depends on k as the other solutions, however this effect is hidden by the generally high error. With a higher number of bins used to discretize the data space, the estimates of the MIN/MAX approach could be increased, however at the cost of computational complexity. We will address this in one of the following experiments.

Varying n. When varying the number of points in the database (Figure 7(left)), the estimate based on k stays worse than the estimate based on sampling. The MIN/MAX approach (Figure 7(right)) depends significantly on the number of data points in the database, because with a higher number of points in the database the number of values aggregated in a single bucket increases as well. As a result, less values can be pruned, and therefore the upper bound grows with increasing n.

Varying d. Figure 8 shows the impact of changing the dimensionality of the underlying data on the performance of the proposed solutions. With increasing dimensionality

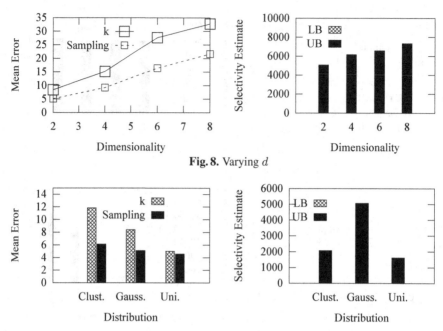

Fig. 8. Varying d

Fig. 9. Varying the Distribution

the error for both the estimate based on k (Figure 8(left)) and the error based on sampling increases. The increasing imprecision with higher dimensionality could be, as described in [26], related to the *Curse of Dimensionality*. As a result, to keep the accuracy constant in high-dimensional spaces, the sample size would have to be increased significantly. Note that a similar behaviour is visible for the MIN/MAX approach as well (Figure 8(right)). With increasing dimensionality the upper bound of this approach becomes significantly larger.

Varying the Data Distribution. Let us now analyze the impact of the data distribution on the accuracy of the selectivity estimates, see Figure 9. The estimates based on k and sampling yield the best results if data is normally or equi-distributed. On clustered data the precision of these solutions is lower. This behaviour can be explained by the fact that with equi-distributed data, sparse areas as the ones between different clusters in the clustered dataset do not apear. However in these sparse areas, the estimates of both the sampling-based approach as well as the solution based on k produce inaccurate estimates. Taking k as an estimate would often introduce an error of about k, because the actual number of RkNN in empty areas is zero. The same holds for the sampling-based approach. For this approach we have drawn samples from the database. As a result there exist barely samples in sparse areas. Therefore the closest sample found for a given query is often contained in the closest cluster, and therfore does not correctly estimate the number of RkNN of a given query point.

The upper-bound of the MIN/MAX approach (Figure 9(right)) however works best for the clustered data and the equi-distribution, indicating that for these datasets more values can be pruned than for the normally distributed data.

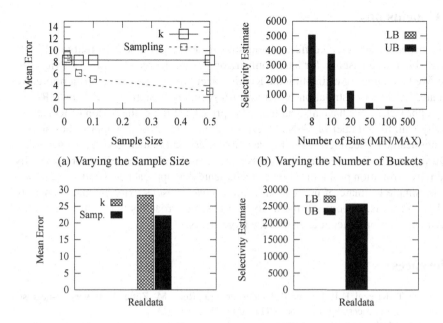

Fig. 10. Real Dataset

Varying s. As shown in Figure 10(a) the sample size has a high impact on the precision on the sampling-based selectivity estimation. The larger the size of a sample, the lower the error. Note that with a high enough number of samples the sampling-based approach becomes better than the approach based on k.

Varying b. Similar to the sampling-based approach, the precision of the MIN/MAX estimate increases with increasing number of buckets, see Figure 10(b). However, unfortunately the runtime of this approach increases significantly with the number of buckets, therefore only a very rough disretization of the data space is feasible in reality. If we would compute the selectivity estimate with a high number of buckets, we could directly run an RkNN query, and therefore there would be no utility in estimating selectivity. Note that in general the number of bins needed to discretize the data space grows exponentially to the number of splits in one dimension, such that for high-dimensional spaces the number of bins needed to model the data space is even higher.

Real Dataset. We also ran experiments on a real dataset, containing the geographic positions (coordinates) of 123593 postal addresses in the north eastern united states. As the dataset contains three conurbations (New York, Philadelphia, Boston), it contains three larger clusters and a lot of additional noise. The error with this real dataset is higher than with the synthetic data, however the ordering of the different approaches is similar: the MIN/MAX approach performs worse, k is similar to the sampling approach and the sampling approach shows the lowest error.

5   Conclusions

In this paper we addressed different heuristics for estimating the selectivity of RkNN queries. First, as a baseline for later comparison, we used k as a trivial selectivity estimate. We then aimed at estimating the RkNN selectivity by sampling. Furthermore, we evaluated selectivity estimation based on histograms by adapting well-known RkNN query approaches. We have also addressed approaches linking the notion of RkNN selectivity with the number of kNNs of a query point. The proposed approaches are not only applicable to Euclidean data, but can also be applied to metric spaces in general.

During an extensive performance analysis we compared our solutions for the RkNN selectivity estimation problem. The results indicate that approaches based on sampling and the simple estimate of using k as location-invariant selectivity estimate generally yield good results. In contrast, the histogram-based approach is not applicable for selectivity estimation, since it leads to very large errors.

References

1. Emrich, T., Kriegel, H.P., Kröger, P., Niedermayer, J., Renz, M., Züfle, A.: Reverse-k-nearest-neighbor join processing. In: Proc. SSTD, pp. 277–294 (2013)
2. Korn, F., Muthukrishnan, S.: Influenced sets based on reverse nearest neighbor queries. In: Proc. SIGMOD, pp. 201–212 (2000)
3. Yang, C., Lin, K.I.: An index structure for efficient reverse nearest neighbor queries. In: Proc. ICDE, pp. 485–492 (2001)
4. Achtert, E., Böhm, C., Kröger, P., Kunath, P., Pryakhin, A., Renz, M.: Efficient reverse k-nearest neighbor search in arbitrary metric spaces. In: Proc. SIGMOD, pp. 515–526 (2006)
5. Tao, Y., Yiu, M.L., Mamoulis, N.: Reverse nearest neighbor search in metric spaces. IEEE TKDE 18(9), 1239–1252 (2006)
6. Tao, Y., Papadias, D., Lian, X.: Reverse kNN search in arbitrary dimensionality. In: Proc. VLDB, pp. 744–755 (2004)
7. Stanoi, I., Agrawal, D., Abbadi, A.E.: Reverse nearest neighbor queries for dynamic databases. In: Proc. DMKD, pp. 44–53 (2000)
8. Singh, A., Ferhatosmanoglu, H., Tosun, A.S.: High dimensional reverse nearest neighbor queries. In: Proc. CIKM, pp. 91–98 (2003)
9. Cheema, M.A., Lin, X., Zhang, W., Zhang, Y.: Influence zone: Efficiently processing reverse k nearest neighbors queries. In: Proc. ICDE (2011)
10. Xia, C., Hsu, W., Lee, M.L.: ERkNN: Efficient reverse k-nearest neighbors retrieval with local kNN-distance estimation. In: Proc. CIKM, pp. 533–540 (2005)
11. Cheema, M.A., Lin, X., Wang, W., Zhang, W., Pei, J.: Probabilistic reverse nearest neighbor queries on uncertain data. IEEE TKDE 22(4), 550–564 (2010)
12. Bernecker, T., Emrich, T., Kriegel, H.P., Renz, M., Zankl, S., Züfle, A.: Efficient probabilistic reverse nearest neighbor query processing on uncertain data. In: Proc. VLDB, pp. 669–680 (2011)
13. Korn, F., Muthukrishnan, S., Srivastava, D.: Reverse nearest neighbor aggregates over data streams. In: Proc. VLDB, pp. 814–825 (2002)
14. Xia, C., Hsu, W., Lee, M.L., Ooi, B.C.: Border: Efficient computation of boundary points. IEEE TKDE 18(2), 289–303 (2006)
15. Lipton, R., Naughton, J., Schneider, D.: Practical selectivity estimation through adaptive sampling. In: Proc. SIGMOD (1990)

16. Hou, W.C., Ozsoyoglu, G., Dodgu, E.: Error-constrained count query: Evaluation in relational databases. In: Proc. SIGMOD (1991)
17. Poosala, V., Ioannidis, Y.E.: Selectivity estimation without the attribute value independence assumption. In: Proc. VLDB (1997)
18. Bruno, N., Chaudhuri, S., Gravan, L.: Stholes: A multidimensional workload-aware histogram. In: Proc. SIGMOD (2001)
19. Chen, C.M., Roussopoulos, N.: Adaptive selectivity estimation using query feedback. In: Proc. SIGMOD (1994)
20. Singh, A., Ferhatosmanoglu, H., Tosun, A.S.: High dimensional reverse nearest neighbor queries. In: Proc. CIKM, pp. 91–98 (2003)
21. Achtert, E., Kriegel, H.P., Kröger, P., Renz, M., Züfle, A.: Reverse k-nearest neighbor search in dynamic and general metric databases. In: Proc. EDBT, pp. 886–897 (2009)
22. Emrich, T., Kriegel, H.P., Kröger, P., Renz, M., Züfle, A.: Boosting spatial pruning: On optimal pruning of MBRs. In: Proc. SIGMOD, pp. 39–50 (2010)
23. Chávez, E., Navarro, G., Baeza-Yates, R.A., Marroquín, J.L.: Searching in metric spaces. ACM Comput. Surv. 33(3), 273–321 (2001)
24. Ciaccia, P., Patella, M., Zezula, P.: M-tree: An efficient access method for similarity search in metric spaces. In: Proc. VLDB, pp. 426–435 (1997)
25. Conway, J., Sloane, N., Bannai, E.: Sphere Packings, Lattices and Groups. Grundlehren der mathematischen Wissenschaften. Springer (1999)
26. Houle, M.E., Kriegel, H.-P., Kröger, P., Schubert, E., Zimek, A.: Can shared-neighbor distances defeat the curse of dimensionality? In: Gertz, M., Ludäscher, B. (eds.) SSDBM 2010. LNCS, vol. 6187, pp. 482–500. Springer, Heidelberg (2010)

Efficient Sampling Methods for Shortest Path Query over Uncertain Graphs

Yurong Cheng, Ye Yuan, Guoren Wang, Baiyou Qiao, and Zhiqiong Wang

College of Information Science and Engineering, Northeastern University, China
cyrneu@gmail.com

Abstract. Graph has become a widely used structure to model data. Unfortunately, data are inherently with uncertainty because of the occurrence of noise and incompleteness in data collection. This is why uncertain graphs catch much attention of researchers. However, the uncertain graph models in existing works assume all edges in a graph are independent of each other, which dose not really make sense in real applications. Thus, we propose a new model for uncertain graphs considering the correlation among edges sharing the same vertex. Moreover, in this paper, we mainly solve the shortest path query, which is a funduemental but important query on graphs, using our new model. As the problem of calculating shortest path probability over correlated uncertain graphs is #P-hard, we propose different kinds of sampling methods to efficiently compute an approximate answer. The error is very small in our algorithm, which is proved and further verified in our experiments.

1 Introduction

In recent decades, graph has emerged to be a popular structure to model data in database community. The road network data, social network data, biological network data, RDF data, etc., are well-known graph data. However, these data are usually uncertain because noise and incompleteness inevitably exist when collecting data in real applications. For example, researchers usually use sensors in road networks to test all kinds of data such as vehicle speed. Due to the limitation of hardware, sensors may delay or miss some data messages, which causes some inaccuracy to the collected data. Given another example, in the *Protein-Protein Interaction (PPI)* network, the proteins obtained from experiments may either contain some nonexisting protein interactions, or miss some existing ones. So this is the reason why uncertain graphs catch a lot of researches' attentions in recent years [1][2][15][17].

As a fundamental and important query on graphs, shortest path query has a wide application. Taking an example, when a tourist arrives in a new city, he would like to know how to get to a place of interest from his current location using the least time. Though the shortest path query has been studied for a long time, there are few works that cope with it over uncertain graphs.

The existing ones consider the edges of an uncertain graph are independent of each other [26][30]. Specifically, given an uncertain graph $\mathcal{G}(V, E, W, Pr)$, its vertices are deterministic, and each edge $e \in E$ is labeled by a two-tuple pair (w, pr), where $w \in W$

S.S. Bhowmick et al. (Eds.): DASFAA 2014, Part II, LNCS 8422, pp. 124–140, 2014.

is the weight of the edge and $pr \in Pr$ is the existence probability of the edge. A *possible world graph* g is an instance of $\mathcal{G}$. The probability of a *possible world graph* equals to,

$$Pr(g) = \prod_{e \in E_i} pr_i \prod_{e \notin E_i} (1 - pr_i)$$

The shortest path probability SPr of a path is the sum of the probability of all the *possible world graphs* in which it is shortest between the query points. The follows is an example of calculating the shortest path probability under the independent uncertain graph model.

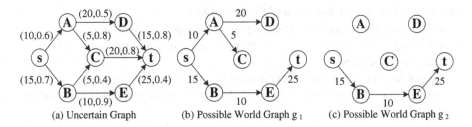

(a) Uncertain Graph   (b) Possible World Graph g_1 (c) Possible World Graph g_2

Fig. 1. Independent Uncertain Graph Model

Example 1. Fig. 1(a) shows an uncertain graph $\mathcal{G}$ under the independent uncertain graph model, and Fig. 1(b)-(c) are two *possible world graphs* of $\mathcal{G}$, g_1 and g_2 respectively. The probability of g_1 equals to

$$Pr(g_1) = 0.6 \times 0.7 \times 0.5 \times 0.8 \times (1-0.8) \times (1-0.4) \times 0.9 \times (1-0.8) \times 0.4 = 0.00145$$

Among all the *possible world graphs*, there are some ones in which the path P_{sBEt} is the shortest path, such as g_1, g_2, etc. Thus, the shortest path probability of P_{sBEt} is $SPr(P_{sBEt}) = Pr(g_1) + Pr(g_2) + \cdots$

However, in real applications, the edges are usually not independent of each other. In other words, there are always some hidden correlationships among the edges sharing the same vertex. For example, assuming Fig. 1(a) represents a road network, if a car is known to travel fluently on one road e_{sB}, it will be more possible to still travel fluently after it goes across an intersection B to another road e_{BE}. Therefore, whether the car travels fluently on edge BE, to some extent, depends on whether it travels fluently on edge sB. However, the independent model cannot present such correlation among edges such as BE and sB. So the independent uncertain graph model is not good enough to describe the real life uncertain graphs.

Is There a More Reasonable Model to Describe the Uncertain Graphs? In this paper, we propose a new uncertain graph model. This new model can show the correlationship among the edges sharing the same vertex, which overcomes the shortcoming of the existing uncertain graph model.

How Can We Get the Query Answers Efficiently and Effectively? It can be seen that calculating the shortest path probability over correlated uncertain graph is a #P-hard problem (in Section 2.2). A naive method is to find out all the *possible world graphs*, compute the shortest path in each *possible world graph*, and sum up the corresponding probabilities of the *possible world graphs* whose shortest path is the same. This method is apparently infeasible since the number of *possible world graphs* is exponential, which is impossible to be enumerated.

Thus, in this paper, we first propose a baseline sampling algorithm based on *Monte Carlo* theory, and then design an improved sampling algorithm based on *Unequal Probability Sampling*. The improved algorithm has a less processing time and can provide a more accurate result.

Summery of Our Contributions and Organization:

- We propose a new Uncertain Graph Model for the shortest path query.
- We propose a more effective and efficient sampling algorithm for shortest path query under our model.
- We conduct a comprehensive experiment and verify that our improved algorithm can answer the shortest path query more efficiently and accurately than the baseline algorithm.

In the rest of our paper, we introduce our new model and define the problem formally in Section 2. In Section 3, we introduce the baseline sampling algorithm. In Section 4, we describe our improved sampling method. In Section 5, we report our experiment results. Finally, we briefly review the related works in Section 6 and make a conclusion in Section 7. Some frequently used symbols in this paper are listed in Table 1.

Table 1. Main Symbols

Symbols	Description	Symbols	Description
DAG	*Directed Acyclic Graph*	$\mathcal{G}$	uncertain graph
g_c	deterministic graph ignoring uncertainty	P	path
pmf	*probability mass function*	SPr	shortest path probability
CPr	conditional probability	*CPT*	Conditional Probability Table
s	the source query vertex	t	the terminal query vertex
e_s	the edge outgoing s	v_z	a vertex whose in-degree is zero
s_z	a source vertex whose in-degree is zero	*MSE*	*Mean Square Error*

2   The New Uncertain Graph Model and Problem Statement

In this paper, we use directed graphs due to the wide applications in real life. For example, in a road network, it is common to see vehicles stuck in one direction, while the opposite direction has a very light traffic. What's more, in other domains, such as semantic Web (RDF), social networks, and bioinformations, graphs are often directed. Specifically, we take *DAG(Directed Acyclic Graph)* typically in our model, since the cyclic graphs can equivalently be changed into *DAG*s by merging the strong connected components into a single vertices [29]. We give the definition of *DAG* as follows.

Definition 1 (Directed Acyclic Graph (DAG)). *A Directed Acyclic Graph is denoted as $DAG(V, E, W)$, where V is the vertex set, E is the edge set, and each edge is given a weight $w \in W$. For any path P from v_1 to v_2 in a DAG, there should be no such paths $P's$ from v_2 to v_1. If there is an edge $e(v_1, v_2) \in E$, it is called **outgoing** v_1 and **injecting** v_2, and v_1 is called a **parent vertex** of v_2. If there are two edges $e_1(v_1, v_2)$ and $e_2(v_2, v_3)$, e_1 is called a **parent edge** of e_2. The total number of edges outgoing a vertex is called the **out-degree** of this vertex, and the total number of edges injecting a vertex is called its **in-degree**. The vertices with zero in-degree are called **root vertices**.*

In this section, we first describe our new uncertain graph model in Section 2.1. Then, in Section 2.2, we formally define our problem.

2.1 Uncertain Graph Model

Definition 2 (Uncertain Graph Model). *An Uncertain Graph is denoted as $\mathcal{G}(g_c, Pr)$, where $g_c(V, E, W)$ is a DAG and Pr is a* probability mass function (pmf) *showing the correlation among edges and their parent edges. Similar to* Bayesian Network, *the pmf is represented by a conditional probability CPr_e that edge e exists (or not) on the condition that its parent edges exist (or not). That is,*

$$CPr_{state(e)} = Pr(state(e)|state(e_1), \ldots, state(e_i), \ldots, state(e_{in})) \qquad (1)$$

where $e_1 \ldots e_{in}$ are the parent edges of e, and $state(e)$ is the function presenting whether e exists. If e exists, $state(e) = 1$. Otherwise, $state(e) = 0$.

A conditional probability table CPT is used on e to list all its $CPrs$.

Example 2 (Uncertain Graph Model). Fig. 2 shows an uncertain graph, and Gi. 2(a) is its g_c. As there are 9 edges in g_c, there should be 9 *CPT*s in total, and Fig. 2(b)-(f) show 5 of them. The edge e_{sA} and e_{sB} are outgoing a root vertex s, and their conditional probabilities are shown in Fig. 2(b) and (c). The state of edge e_{BE} depends on the state of the edge e_{sB}. Similarly, the state of edge e_{Ct} depends on the states of both e_{AC} and e_{BC}. For instance in Fig 2(e), the value 0.2 means the probability that e_{BE} exists on the condition that e_{sS} exists.

This correlated uncertain graph model can be used in road network to show the fluency correlationship among neighbor roads. For example, in Fig. 2(a), road e_{Ct} is more possible to be fluent (i.e., e_{Ct} exists) if both the roads e_{AC} and e_{BC} are fluent (i.e., e_{AC} and e_{BC} both exist). Besides, if this model is used in website society, for example, in Fig. 2(a), if a user visit a website A from its similar website s, he is more probable to continue visit its another similar website C or D than not to visit them. These correlationships can only be presented by our correlated uncertain graph model other than the existing independent one.

Given the above uncertain graph model, how to calculate the probability of a *possible world graph*? This would apply the following definition of *Conditional Independent*.

Definition 3 (Conditional Independent). *X, Y, Z are three sets of random variables. Y is conditionally independent of Z given X (denoted as $Y \perp Z|X$) in distribution Pr if $\forall x \in X, y \in Y$ and $z \in Z$, that is,*

$$Pr(Y = y, Z = z|X = x) = Pr(Y = y|X = x) \times Pr(Z = z|X = x)$$

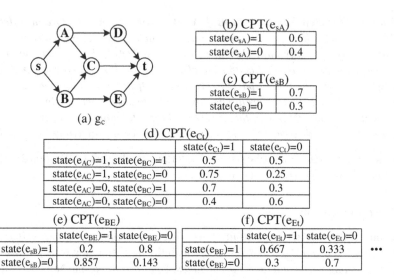

(b) CPT(e_{sA})

state(e_{sA})=1	0.6
state(e_{sA})=0	0.4

(c) CPT(e_{sB})

state(e_{sB})=1	0.7
state(e_{sB})=0	0.3

(a) g_c

(d) CPT(e_{Ct})

	state(e_{Ct})=1	state(e_{Ct})=0
state(e_{AC})=1, state(e_{BC})=1	0.5	0.5
state(e_{AC})=1, state(e_{BC})=0	0.75	0.25
state(e_{AC})=0, state(e_{BC})=1	0.7	0.3
state(e_{AC})=0, state(e_{BC})=0	0.4	0.6

(e) CPT(e_{BE})

	state(e_{BE})=1	state(e_{BE})=0
state(e_{sB})=1	0.2	0.8
state(e_{sB})=0	0.857	0.143

(f) CPT(e_{Et})

	state(e_{Et})=1	state(e_{Et})=0
state(e_{BE})=1	0.667	0.333
state(e_{BE})=0	0.3	0.7

⋯

Fig. 2. Uncertain Graph Model

From Definition 3, the probability of a *possible world graph* g equals to

$$Pr(g) = \prod_{1 \le i \le m} CPr(state(e_i)) \tag{2}$$

where m is the number of edges in a graph.

For example, the probability of g_c shown in Fig. 2(a) equals to

$$Pr(g_1) = CPr(state(e_{sB}) = 1) \times CPr(state(e_{BE}) = 1) \times \cdots$$
$$= Pr(state(e_{sB}) = 1) \times Pr(state(e_{BE}) = 1)|state(e_{sB}) = 1) \times \cdots$$
$$= 0.7 \times 0.2 \times \cdots \tag{3}$$

2.2 Problem Statement

Definition 4 (Shortest Path Probability (SPr)). *For any path P of g_c, the probability of the event that P is the shortest path in the uncertain graph $\mathcal{G}(g_c(V, E, W), Pr)$ is called its Shortest Path Probability, which is calculated as*

$$SPr(P) = \sum_{\substack{P \text{ is the shortest path in } g_i \ (1 \le i \le n)}} Pr(g_i) \tag{4}$$

Definition 5 (Threshold-based Shortest Path Query). *Given a correlated uncertain graph $\mathcal{G}(g_c(V, E, W), Pr)$, two vertices $s, t \in V$, and a probabilistic threshold τ, the threshold-based shortest path query returns a path set $SP = \{P_1, P_2, \ldots, P_n\}$ in which each path P_i has a $SPr(P_i)$ larger than the threshold τ.*

Apparently, if we want to quickly answer the *threshold-based shortest path query*, we have to efficiently calculate the SPr. However, the calculation of SPr under the correlated uncertain graph model is #P-hard. This is because calculating the reachability probability under the independent uncertain graph model is #P-complete [11][23].

When considering the correlation among edges, the calculation of the reachability probability under our model will be harder, whose complexity is at least #P-complete. Apparently, the calculation of the shortest path probability under our model is even harder than calculating the reachability probability. So our problem is at least #P-complete, which is #P-hard.

3 Baseline Sampling Algorithm

As discussed in the last section, calculating SPr is a #P-hard problem, so there is no exact solution in polynomial time unless P=NP. Thus, in the rest of this paper, we use sampling methods to get an approximate result. In this section, we give a baseline sampling method based on *Monte Carlo* theory.

During the sampling process, we sample N *possible world graphs*, $G_1, G_2, \ldots, G_N$, according to the *pmf* of each edge. Then, on each sampled possible world graph, we run a shortest path algorithm over deterministic graphs to find out its shortest path. Finally, we set a flag y_{P_i} for each path, so that

$$y_{P_i} = \begin{cases} 1 \text{ if } P_i \text{ is the shortest path in the sampled } possible\ world\ graph \\ 0 \text{ otherwise} \end{cases}$$

Thus, the estimator $\widehat{SPr_B}$ equals to,

$$SPr_B \approx \widehat{SPr_B} = \frac{\sum_{i=1}^{N} y_{P_i}}{N} \tag{5}$$

For any sampling method, The *Mean Square Error (MSE)* incorporates both the bias and the precision of an estimator into a measure of overall accuracy. It is calculated as,

$$MSE(\widehat{\theta}) = E[(\widehat{\theta} - \theta)^2] = Var(\widehat{\theta}) + Bias(\widehat{\theta}, \theta)$$

The bias of an estimator is given by,

$$Bias(\widehat{\theta}) = E(\widehat{\theta}) - \theta$$

An estimator of θ is unbiased if its bias is 0 for all values of θ, that is, $E(\widehat{\theta}) = \theta$ As the estimator of *Monte Carlo* method is unbiased [6], thus,

$$MSE(\widehat{SPr_B}) = Var(\widehat{SPr_B}) = \frac{1}{N} SPr_B(1 - SPr_B) \approx \frac{1}{N} \widehat{SPr_B}(1 - \widehat{SPr_B}) \ (6)$$

The pseudo code is shown in Algorithm 1.

If the sampling number N is far smaller than the number of *possible world graphs*, the *Baseline Sampling Algorithm* is more efficient than the naive exact method discussed in Section 1. But as a consequence, the calculation result is less accurate. According to our experiment result in Section 5, if we want to get a more accurate result, we need to spend many hours to calculate the result using the *Baseline Sampling* algorithm. Whereas, if we want to quickly access the result in seconds, the error may be as large as more than 20%.

Algorithm 1. Baseline Algorithm

Input: Start Point s, Termination Point t
Output: $\widehat{SPr_B}$, $Var(\widehat{SPr_B})$
1 **for** i *from 1 to N* **do**
2 | *Sample a possible graph*;
3 | $Dijkstra(PossibleGraph\ PG)$;
4 | $C_{P_i}++$;
  | // to count the number of 1 in y_{P_i}
5 return $(\widehat{SPr_B}, Var(\widehat{SPr_B}))$ according to formula (5) and (6);

The total time cost t_{total} of the sampling algorithm equals to $t_{total} = t_{once} \times N$ (t_{once} is the time cost in one sampling). **So, there are two bottlenecks we need to break. The first one is how to get a more accurate result using a smaller sampling size (reduce N). The other one is whether we can reduce the time cost in one sampling (reduce t_{once}).** Taking the above consideration, we propose an improved sampling method based on *Unequal Probability Sampling*.

4 Improved Sampling Algorithm

In this section, we first introduce some knowledge on *Unequal Probability Sampling* in Section 4.1. Then, in Section 4.2, we illustrate our improved sampling algorithm, which is called *DijSampling* algorithm.

4.1 Unequal Probability Sampling

Unequal probability sampling is when some units in the population have probabilities of being selected from others. Suppose a sample of size N is selected randomly from a population S but that on each draw, unit i is sampled according to any probability q_i, where $\sum_{i=1}^{S} q_i = 1$ [14]. Let y_i be the response variable measured on each unit selection, i.e.,

$$y_i = \begin{cases} 1 \text{ if } i \text{ is selected in one sampling} \\ 0 \text{ otherwise} \end{cases}$$

Note that if a unit is selected more than once, it is used as many times as it is selected.

There are two classical unbiased estimators, *Hansen-Hurwitz (H-H)* and *Horvitz-Thompson (H-T)* in unequal probability sampling method. The *H-H* estimator is for random sampling with replacement, while the *H-T* estimator is a more general estimator, which can be used for any probability sampling plan, including both sampling with and without replacement [14]. For *H-H* estimator, an unbiased estimator of the population total $\xi = \sum_{i=1}^{S} y_i$ is given by

$$\hat{\xi}_q = \frac{1}{N} \sum_{i=1}^{N} \frac{y_i}{q_i}$$

The variance of *H-H* estimator is

$$\widehat{Var}(\hat{\xi}_q) = \frac{1}{N(N-1)} \sum_{i=1}^{N} q_i (\frac{y_i}{q_i} - \hat{\xi}_q)^2$$

For *H-T* estimator, the population total $\xi = \sum_{i=1}^{S} y_i$ is estimated by

$$\hat{\xi}_\pi = \frac{1}{d} \sum_{i=1}^{N} \frac{y_i}{\pi_i}$$

where π_i is the probability that the ith unit of the population is included in the sample (*inclusion probability*), and d is the distinct unit number in the sample, which is called *effective sample size*. In addition, the variance of *H-T* estimator is,

$$\widehat{Var}(\hat{\xi}_\pi) = \sum_{i=1}^{d} (\frac{1-\pi_i}{\pi_i^2}) y_i^2 + \sum_{i=1}^{d} \sum_{j=1, j\neq i}^{d} (\frac{\pi_{ij} - \pi_i \pi_j}{\pi_i \pi_j}) \frac{y_i y_j}{\pi_{ij}}$$

where π_{ij} is the *joint inclusion probability* of units i and j (the probability that i and j are sampled at the same time).

As both of the above estimators are unbiased, i.e., $Bias(\widehat{\theta}, \theta) = 0$, the variance can be used to evaluate the accuracy of the estimators.

4.2 DijSampling Algorithm

Reviewing the two bottlenecks in Section 3, the underlying cause of the first one is that the *Baseline Sampling* algorithm only samples one instance of the sample space in one sampling. **We consider a method to sample a lot of *possible world graphs* together in one sampling in order to reduce** N. Then, for the second bottleneck, when sampling once, the *Baseline Sampling* algorithm samples the whole *possible world graph*, and the followed *Dijkstra* algorithm is run on the whole *possible world graph*. Thus, **we should think of a method that only samples a part of a *possible world graph* each time to reduce** t_{once}. These are the main purposes of our *DijSampling* algorithm. We use an example to explain these advantages in detail.

Example 3 (Advantages of Improved Sampling Algorithm). Fig. 3 shows 4 *possible world graphs*. These *possible world graphs* have the same shortest path, P_{sBEt}. (Actually, there are 16 such *possible world graphs*, and we only show 4 of them.) We apply shortest path algorithm to sample edges e_{sB}, e_{BE}, and e_{Et}, and the 16 *possible world graphs* are totally the *possible world graphs* that contain P_{SBEt} as the shortest path. Thus, our *DijSampling* algorithm only needs one sampling and can sample all these 16 *possible world graphs* together. But in *Baseline Sampling* algorithm, it takes 16 times to sample them. In other words, sampling once in *DijSampling* algorithm has the same effect with sampling 16 times in *Baseline Sampling*. If the *Baseline Sampling* needs 1600 times of sampling to get an accurate result, the *DijSampling* algorithm only needs 100 times of sampling. Thus, by together sampling the *possible world graphs* containing the same shortest path, the *DijSampling* algorithm can effectively reduce the sample size N. Moreover, since the 4 edges, e_{AC}, e_{AD}, e_{Ct} and e_{Dt}, are not sampled in one sampling of *DijSampling* algorithm, the *DijSampling* algorithm has a smaller t_{each} than the *Baseline Sampling* algorithm.

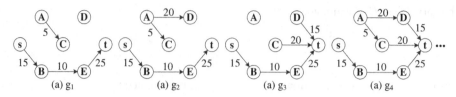

Fig. 3. Possible World Graphs in which P_{sBEt} is the shortest path

As indicated in the above example, the main idea of our *DijSampling* algorithm is to sample the edges at the same time with the process of *Dijkstra* algorithm. The pseudocode is shown in Algorithm 2. The *DijSampling* is proceeded during the process of *Dijkstra* algorithm. When the *Dijkstra* algorithm needs the messages of an edge such as its existence or weight, we sample it according to its *pmf* (Line 3). Note that if this edge is outgoing the source query point s, we need to determine whether its in-degree is zero. If it is so ($s = s_z$), we just sample it according to its *pmf* (Line 5). Otherwise (Line 7), we sample this edge with the probability estimated by the following formula

$$Pr(state(e_s) = 1) = \frac{\sum_{1 \le i \dots \le m_{in}; k=0,1} CPr(state(e_s) = 1 | state(e_i) = k, \cdots)}{\sum_{1 \le i \dots \le m_{in}; k=0,1} CPr(state(e_s) = k | state(e_i) = k, \cdots)} \quad (7)$$

This formula means we sample the edge e_s using the average probability when $state(s) = 1$ in its *CPT* as its existence probability, while using the average probability when $state(s) = 0$ as its nonexistence probability. The reason why we treat the e_s specially will be explained latter in this subsection.

The following is an example illustrating the process of *DijSampling* algorithm.

Algorithm 2. DijSampling Algorithm

Input: Start Point s, Termination Point t, sampling times N
Output: $\widehat{SPr_{H-H}}, Var(\widehat{SPr_{H-H}}), \widehat{SPr_{H-T}}, Var(\widehat{SPr_{H-T}})$

1 **for** i *from* 1 *to* N **do**
2 | running *Dijkstra* **begin**
3 | | **if** Dijkstra *needs the messages of an edge* **then**
4 | | | **if** $s == s_z$ **then**
5 | | | | Sample the edges outgoing s according to their *CPT*s ;
6 | | | **else**
7 | | | | Sample the edges outgoing s according to formula (7) ;
8 | | | Sample the other edges according to their *CPT*s ;
9 | | **else**
10 | | | continue the *Dijkstra* algorithm ;
11 | | return (result shortest path P_i) ;
12 | y_{P_i}++ ;
13 calculate $\widehat{SPr_{H-H}}, Var(\widehat{SPr_{H-H}}), \widehat{SPr_{H-T}}, Var(\widehat{SPr_{H-T}})$;

Example 4. Fig. 4 shows the process of *DijSampling* algorithm. In this figure, we use dashed lines to denote the unsampled edges, solid lines to denote the existing edges after sampling, and no lines (empty) to denote the nonexisting edges after sampling. Specifically, Fig. 4(a) is the g_c of the uncertain graph in Fig. 2 before sampling, and Fig. 4(b)-(d) are the graphs after sampling the designated edges.

Assume the start query point is s and terminal query point is t. We first sample e_{sA} and e_{sB} according to Fig. 2(b) and Fig. 2(c) respectively. For example, we sample the e_{sB} with probability 0.7 existing and 0.3 nonexisting. After that, we assume that e_{sA} does not exists and e_{sB} exists (Fig. 4(b)). Then, we sample edge e_{BE} and e_{BC}. For example, for edge e_{BE}, as we have sampled edge e_{sB} exists, then $state(e_{sB}) = 1$. So we sample e_{BE} with probability 0.2 existing, and probability 0.8 nonexisting (Fig. 4(c)). Repeat the above process, after we sample to Fig. 4(d), we can know P_{sBEt} is the shortest path. Thus, there is no need to sample the other edges (dashed edges in Fig. 4(d)).

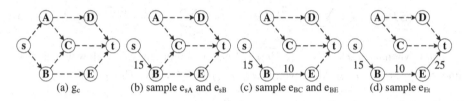

(a) g_c (b) sample e_{sA} and e_{sB} (c) sample e_{BC} and e_{BE} (d) sample e_{Et}

Fig. 4. *DijSampling* Process

In addition, if the query is B to t instead, the first edges should be sampled are e_{BE} and BC. The sampling existence probability of e_{BE} is $(0.2 + 0.857)/(0.2 + 0.857 + 0.8 + 0.143) = 0.5285$, and similarly the nonexistence probability is 0.4715. Then, the remained sampling steps are the same with those mentioned above (from s to t).

After describing the algorithm, let us see how to estimate the $SPrs$. According to Section 4.1, the H-H estimator of SPr can be calculated as

$$SPr_{H-H} \approx \widehat{SPr_{H-H}} = \frac{1}{N} \sum_{i=1}^{N} \frac{Pr_{P_i} y_{P_i}}{q_i} \qquad (8)$$

where N is the sampling times, and Pr_{P_i} is the probability of the sampled *possible world graph* that P_i is the shortest path in it, which can be calculated as

$$Pr_{P_i} = \prod_{1 \leq i \leq m_{samp}} CPr(state(e_i)) \qquad (9)$$

where m_{samp} is the number of sampled edges in one sampling process.

The variance of this estimator equals to

$$Var(\widehat{SPr_{H-H}}) = \frac{1}{N(N-1)} \sum_{i=1}^{N} q_i (\frac{Pr_{P_i}}{q_i} - \widehat{SPr_{H-H}})^2 \qquad (10)$$

When $q_i = Pr_{P_i}$, i.e., we sample each edge according to its *pmf* [11], it has been proved that, the variance in formula (10) is minimum. At this moment, the $\widehat{SPr_{H-H}}$ in formula (8) equals to

$$min(\widehat{SPr_{H-H}}) = \frac{\sum_{i=1}^{N} y_{P_i}}{N} \qquad (11)$$

and its corresponding variance is

$$min(Var(\widehat{SPr_{H-H}})) = \frac{1}{N}\widehat{SPr_{H-H}}(1 - \widehat{SPr_{H-H}}) \tag{12}$$

The H-H estimator and its variance equals to those of the baseline algorithm.

Similarly, we give the formula to compute H-T estimator. Suppose, in the N times sampling, there are v different sampling results. Thus the probability that each one of the v results may be sampled is $\pi_i = 1 - (1 - q_i)^N$. So the estimator $\widehat{SPr_{H-T}}$ is,

$$SPr_{H-T} \approx \widehat{SPr_{H-T}} = \frac{1}{N}\sum_{i=1}^{N}\frac{Pr_{P_r}y_{P_i}}{\pi_i} \tag{13}$$

This estimator is unbiased as well [11][14]. $\pi_{ij} = 1 - (1 - q_i)^n - (1 - q_j)^N - (1 - q_i - q_j)^N$ is the probability that π_i and π_j are in the result set at the same time, and the variance of this estimator is,

$$Var(\widehat{SPr_{H-T}}) = \sum_{i\in v}(\frac{1-\pi_i}{\pi_i})(Pr_{P_i})^2 + \sum_{i,j\in v, i\neq j}(\frac{\pi_{ij}-\pi_i\pi_j}{\pi_i\pi_j})Pr_{P_j}Pr_{P_j} \tag{14}$$

Again, when $q_i = Pr_{P_i}$, $Var(\widehat{SPr_{H-T}})$ gets its minimum value. However, at this time, $\widehat{SPr_{H-T}} \neq \widehat{SPr_B}$. Moreover, $Var(\widehat{SPr_{H-T}}) \leq Var(\widehat{SPr_B})$ [11].

The Special e_s in Line 7 of Algorithm 2. The reason why we treat the e_s specially is that the edges outgoing a vertex always depend its parent edges under the uncertain graph model introduced in Section 2.1. Just like Example 4, we cannot know the sampling probability of e_{BE} if the state of e_{sB} is unknown. Actually, we can just randomly choose a root vertex and sample from it. After it reach the source query point, we continue our algorithm. However, this method is apparently uneconomical since the long journey before the source query point has no contribution to the result set except for a start probability. Thus, we use formula (7) to quickly estimate this probability instead.

Our algorithm can achieve $q_i = Pr_{P_i}$ except for this step in Line 7 of Algorithm 2. In this situation, if we still use the estimators with the minimum variance, they are no longer unbiased. But fortunately, we can prove that this bias is close to zero when the number of sampled edges are large, which is Theorem 1 in follows.

Theorem 1. *The bias of estimators equals nearly to zero when the number of sampled edges are large.*

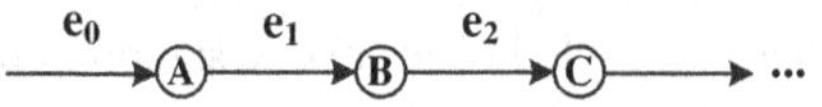

Fig. 5. *Dependency Schematic Diagram for Derivation of Bias*

Proof. We take the situation in which s has only one injection edge as an example to derivate the bias. The derivation when s has more than one injection edges are similar. A schematic diagram is shown in Fig. 5. e_1 is the first edge we want to sample. Let $Pr(e_i|e_j) = Pr(state(e_i) = 1|state_{e_j} = 1)$, $Pr(e_i|\overline{e_j}) = Pr(state(e_i) = $

$1|state_{e_j} = 0)$, $Pr(\overline{e_i}|e_j) = Pr(state(e_i) = 0|state_{e_j} = 1)$, and $Pr(\overline{e_i}|\overline{e_j}) = Pr(state(e_i) = 0|state_{e_j} = 0)$, then the real sampling probability of e_1 equals to

$$Pr(e_1) = Pr(e_1|e_0) \times Pr(e_0) + Pr(e_1|\overline{e_0}) \times Pr(\overline{e_0})$$
$$= Pr(e_1|e_0) \times Pr(e_0) + Pr(e_1|\overline{e_0}) \times (1 - Pr(\overline{e_0})) \qquad (15)$$

But our estimated probability of e_1 equals to,
$$\widehat{Pr(e_1)} = Pr(e_1|e_0) \times \widehat{Pr(e_0)} + Pr(e_1|\overline{e_0}) \times (1 - \widehat{Pr(\overline{e_0})})$$

Thus, the bias after sampling the first edge $Bias^{(1)}$ equals to,
$$Bias^{(1)} = \widehat{Pr(e_1)} - Pr(e_1) = (Pr(e_1|e_0) - Pr(e_1|\overline{e_0})) \times (Pr(e_0) - \widehat{Pr(e_0)})$$

Similarly, we can get the bias after sampling the second edge,
$$Bias^{(2)} = \widehat{Pr(e_2)} - Pr(e_2)$$

$$= (Pr(e_2|e_1) - Pr(e_2|\overline{e_1})) \times (Pr(e_1|e_0) - Pr(e_1|\overline{e_0})) \times (Pr(e_0) - \widehat{Pr(e_0)})$$

Recursively, we can get the bias after sampling m_{sample} edges,
$$Bias^{(m_{sample})} = \widehat{Pr(e_{m_{sample}})} - Pr(e_{m_{sample}})$$

$$= \prod_{i=1}^{m_{sample}} (Pr(e_i|e_j)^{(i)} - Pr(e_i|\overline{e_j})^{(i)}) \times (Pr(e_0) - \widehat{Pr(e_0)})$$

where $Pr(e_i|e_j)^{(i)}$ and $Pr(e_i|\overline{e_j})^{(i)}$ are conditional probabilities of ith sampled edge. Apparently, $Bias^{(m_{sample})} \to 0$ when m_{sample} is large enough. $\qquad \square$

5 Performance Evaluations

In this section, we will report and analyze our experiment results. All the experiments are proceeded on a PC with CPU Inter(R) Core(TM)i7-2600, frequency 3.40GHz, memory 8.00GB, hard disk 500GB. The Operation System is Microsoft Windows 7 Enterprise Edition. The development software is Microsoft Visual Studio 2010, using language C++ and its standard template library (STL).

The Dataset. The real data used in this paper can be classified into two categories. One is sparse dataset such as Road Network data[1], and the other one is Social Network data[2]. The numbers of vertices and edges are listed in Table 2.

In real datasets, the edges are certain, so we need to change the certain graphs into uncertain graphs. In Road Network Datasets, we use the method which is introduced in [10]. Use Normal Distribution $N(\mu, \sigma)$ to generate the weights on each edge. Here, μ is the edge weight in original datasets, and σ is the variance of the generated weights. σ is different according to different edges, which is normally distributed as $N(\mu_\sigma, \sigma_\sigma)$. Here, $\mu_\sigma = xR$, and the value range of x is $[1\%, 5\%]$. In default condition, $\mu_\sigma = 1\%R$, and R is the value range of all weights in original datasets. In the same way, we generate the weights and probabilities on edges in Social Network datasets.

For all the datasets, we choose 100 pairs of vertices as starting points and termination points. After the 100 tests, we calculate the average time cost, memory cost, *Mean Square Error (MSE)* and relative error as the experiment results. We set the threshold defaulted to be 0.5 and compare the proformance of baseline sampling algorithm (denoted as *BS*) and our improved *DijSampling* algorithm (denoted as *DS*).

---

[1] http://www.cs.utah.edu/~lifeifei/SpatialDataset.htm
[2] http://snap.stanford.edu/data

Table 2. Real Dataset Parameters

Name of Dataset	Vertex Number	Edge Number
OLdenburg (OL) Road Network	6,105	7,035
San Francisco Road Network (SF)	174,956	223,001
wiki-Vote	7,115	103,689

Running Time. Fig. 6 shows the running time vs sample size for the two sampling algorithms on different real datasets. From the results, we can observe that with the increase of sample size, the time cost for the two algorithms all increases. The time cost of *BS* is always largest than that of *DS*. In addition, we observe that the larger the graph is, the more time will be cost.

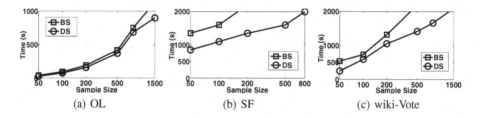

(a) OL   (b) SF   (c) wiki-Vote

Fig. 6. Running time vs Sample Size

The results above are reasonable. The running time depends on the number of sampled edges. The more edges sampled, the more time an algorithm will cost. As the *BS* Algorithm samples more edges, the whole possible graph, its running time is longer.

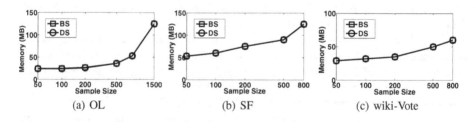

(a) OL   (b) SF   (c) wiki-Vote

Fig. 7. Memory Cost vs Sample Size

Memory Cost. Fig. 7 shows the memory cost vs sample size on different datasets. It can be seen that with the increase of sample times, the memory cost of two algorithms increases. The memory cost of the two algorithms is almost the same.

The phenomena above is reasonable that as sample size increases, the program needs to explore more space to find the shortest path and calculate estimator with corresponding variance. Thus, the memory cost increases with the increase of sample times. Moreover, both the algorithms need to save the structures of graph data and some queues for

Dijkstra algorithm, which are the same. The only difference between the two methods are the flags showing whether each edge is sampled, which takes little memory cost. Thus, the memory cost of the two algorithms is nearly the same.

Accuracy. We test *Mean Square Error (MSE)* and relative error of the estimators to show the accuracy of different algorithms. As the bias caused by formula (7) is always 0 in the experiment result, we do not show it in our result figures. Since the variance of *BS* estimator is the same as the *H-H* estimator of *DS*, we show them using the same line in result figure. The method of calculating *relative error* is the same as that in [11].

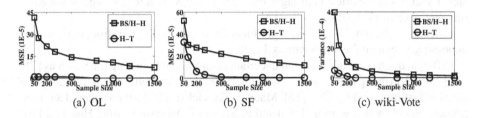

(a) OL   (b) SF   (c) wiki-Vote

Fig. 8. Mean Square Error vs Sample Size

From Fig. 8, it can be seen that no matter how the size of datasets change, the variance of estimators decreases as sample size increases. Moreover, the variance of $\widehat{SPr}_{H-T}$ always keeps smaller than that of $\widehat{SPr}_{H-H}$. This result means the $\widehat{SPr}_{H-T}$ estimator is better than $\widehat{SPr}_{H-T}$, which verifies the discussion in Section 4.1.

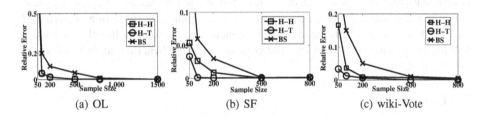

(a) OL   (b) SF   (c) wiki-Vote

Fig. 9. Relative Error vs Sample Size

From Fig. 9, the relative error of *BS* is always the largest, and *H-T* of *DS* is always the smallest. This means the stability and accuracy of $\widehat{SPr}_{H-T}$ is strongest. This result verifies our analysis in Section 4.2. That is, we need fewer sampling times by applying *DijSampling* Algorithm to get the same accuracy as baseline algorithm.

As we cannot get the distribution of unequal probability sampling, the error cannot be bounded. However, from Figure 9, in the 4 datasets used in our experiments, the error of both $\widehat{SPr}_{H-T}$ and $\widehat{SPr}_{H-H}$ is very low. Moreover, with the increase of sample times, the error fluctuates very gently. Thus, the estimators can approximate the exact answer well.

6   Related Works

The most popular shortest path algorithms over deterministic graphs are *Dijkstra* [3] and A^* [7]. These algorithms has a large time complexity. Thus, there are large number of works focusing on building indexes to speed up the algorithms. To accelerate the exact shortest path query, Cohen et al. [5] proposed a 2-hop labeling method, and F.Wei [24] proposed a tree-width decomposition index. Moreover, authors in [9] proposed a landmark encoding method to provide an approximate answer for shortest path query. In addition, there are also some shortest path algorithms over deterministic graphs designed typically for road networks such as [20][19][18][13], and [25] is a good summary. In recent years, some shortest path algorithms are also designed to for large graph environment, such as [12][4][8].

Essentially, the models of these algorithms above are all different from ours, so it is impossible to extend these algorithms directly.

Different queries are addressed in uncertain environment for a long time such as [21][22][28][27]. Among them, shortest path query over uncertain graph is important. This query is first proposed by Loui [16]. Many works such as [26][30] considered an independent model, which is argued in detail in Section 1. Moreover, Ming Hua et.al [10] built a simple correlated model. It modeled simple correlation in uncertain graphs between each two edges sharing the same vertex. But when there are more than two edges sharing the same vertex, it would be confusing.

In addition, ruoming Jin et.al [11] applied *Unequal Probability Sampling* method, which solved the uncertain reachability problem efficiently and effectively. However, their algorithms cannot be applied into our problem directly for three reasons. First, their model was independent uncertain graph model, and their algorithms cannot handle the correlated uncertain graph model. Secondly, their query was reachability, which is different from our shortest path query. As discussed in Section 2.2, our problem is harder than theirs. Thirdly, they applied their sampling algorithm in a *divided and conquer* framework. If this framework is extended to our problem directly, we cannot find out the shortest path unless the last edge is sampled. Then, their algorithms would lose the advantages.

7   Conclusion

In this paper, we first propose a new uncertain graph model, which considers the hidden correlation among edges sharing the same vertex. As calculating the *shortest path probability* is a #P-hard problem, we use sampling methods to approximately compute it. We propose a baseline algorithm and an improved algorithm. Our improved algorithm is more efficient than the baseline algorithm with more accurate answers when sampling the same times. Moreover, we preform comprehensive experiments to verify the efficiency and accuracy of our algorithms.

Acknowledgments. Yurong Cheng, Ye Yuan and Guoren Wang were supported by the NSFC (Grant No.61025007, 61328202, 61202087, and 61100024), National Basic Research Program of China (973, Grant No.2011CB302200-G), National High Technology Research and Development 863 Program of China (Grant No.2012AA011004), and the Fundamental Research Funds for the Central Universities (Grant No. N110404011). Baiyou qiao was supported by the NSFC (Grant No.61073063).

References

1. Adar, E., Ré, C.: Managing uncertainty in social networks. IEEE Data Eng. Bull. 30(2), 15–22 (2007)
2. Asthana, S., King, O.D., Gibbons, F.D., Roth, F.P.: Predicting protein complex membership using probabilistic network reliability. Genome Research 14(6), 1170–1175 (2004)
3. Bast, H., Funke, S., Matijevic, D.: Transitultrafast shortest-path queries with linear-time pre-processing. In: 9th DIMACS Implementation Challenge [1] (2006)
4. Cheng, J., Ke, Y., Chu, S., Cheng, C.: Efficient processing of distance queries in large graphs: A vertex cover approach. In: SIGMOD, pp. 457–468. ACM (2012)
5. Cohen, E., Halperin, E., Kaplan, H., Zwick, U.: Reachability and distance queries via 2-hop labels. SIAM Journal on Comp 32(5), 1338–1355 (2003)
6. Fishman, G.S.: A monte carlo sampling plan based on product form estimation. In: Proceedings of the 23rd Conference on Winter Simulation, pp. 1012–1017. IEEE Computer Society (1991)
7. Fu, L., Sun, D., Rilett, L.R.: Heuristic shortest path algorithms for transportation applications: State of the art. Computers & Operations Research 33(11), 3324–3343 (2006)
8. Gao, J., Jin, R., Zhou, J., Yu, J.X., Jiang, X., Wang, T.: Relational approach for shortest path discovery over large graphs. PVLDB 5(4), 358–369 (2011)
9. Gubichev, A., Bedathur, S., Seufert, S., Weikum, G.: Fast and accurate estimation of shortest paths in large graphs. In: CIKM, pp. 499–508. ACM (2010)
10. Hua, M., Pei, J.: Probabilistic path queries in road networks: Traffic uncertainty aware path selection. In: EDBT, pp. 347–358. ACM (2010)
11. Jin, R., Liu, L., Ding, B., Wang, H.: Distance-constraint reachability computation in uncertain graphs. PVLDB 4(9), 551–562 (2011)
12. Jin, R., Ruan, N., Xiang, Y., Lee, V.: A highway-centric labeling approach for answering distance queries on large sparse graphs. In: SIGMOD, pp. 445–456. ACM (2012)
13. Jing, N., Huang, Y.W., Rundensteiner, E.A.: Hierarchical encoded path views for path query processing: An optimal model and its performance evaluation. TKDE 10(3), 409–432 (1998)
14. Thompson, S.K.: Sampling the Third Edition. Wiley Series In Probability And Statistics. Wiley (2012)
15. Lian, X., Chen, L.: Efficient query answering in probabilistic rdf graphs. In: SIGMOD, pp. 157–168. ACM (2011)
16. Loui, R.P.: Optimal paths in graphs with stochastic or multidimensional weights. CACM 26(9), 670–676 (1983)
17. Nierman, A., Jagadish, H.: Protdb: Probabilistic data in xml. In: Proceedings of the 28th International Conference on Very Large Data Bases, pp. 646–657. VLDB Endowment (2002)
18. Rice, M., Tsotras, V.J.: Graph indexing of road networks for shortest path queries with label restrictions. PVLDB 4(2), 69–80 (2010)
19. Samet, H., Sankaranarayanan, J., Alborzi, H.: Scalable network distance browsing in spatial databases. In: SIGMOD, pp. 43–54. ACM (2008)
20. Sankaranarayanan, J., Samet, H., Alborzi, H.: Path oracles for spatial networks. PVLDB 2(1), 1210–1221 (2009)
21. Tong, Y., Chen, L., Cheng, Y., Yu, P.S.: Mining frequent itemsets over uncertain databases. PVLDB 5(11), 1650–1661 (2012)

22. Tong, Y., Chen, L., Ding, B.: Discovering threshold-based frequent closed itemsets over probabilistic data. In: ICDE, pp. 270–281. IEEE (2012)
23. Valiant, L.G.: The complexity of enumeration and reliability problems. SIAM Journal on Comp. 8(3), 410–421 (1979)
24. Wei, F.: Tedi: Efficient shortest path query answering on graphs. In: Proceedings of SIGMOD, pp. 99–110. ACM (2010)
25. Wu, L., Xiao, X., Deng, D., Cong, G., Zhu, A.D., Zhou, S.: Shortest path and distance queries on road networks: an experimental evaluation. PVLDB 5(5), 406–417 (2012)
26. Yuan, Y., Chen, L., Wang, G.: Efficiently answering probability threshold-based shortest path queries over uncertain graphs. In: Kitagawa, H., Ishikawa, Y., Li, Q., Watanabe, C. (eds.) DASFAA 2010. LNCS, vol. 5981, pp. 155–170. Springer, Heidelberg (2010)
27. Yuan, Y., Wang, G., Chen, L., Wang, H.: Efficient keyword search on uncertain graph data. IEEE Transactions on Knowledge and Data Engineering 25(12), 2767–2779 (2013)
28. Yuan, Y., Wang, G., Wang, H., Chen, L.: Efficient subgraph search over large uncertain graphs. In: International Conference on Very Large Data Bases (2011)
29. Zhang, Z., Yu, J.X., Qin, L., Chang, L., Lin, X.: I/o efficient: Computing sccs in massive graphs. In: Proceedings of the 2013 International Conference on Management of Data, pp. 181–192. ACM (2013)
30. Zou, L., Peng, P., Zhao, D.: Top-k possible shortest path query over a large uncertain graph. In: Bouguettaya, A., Hauswirth, M., Liu, L. (eds.) WISE 2011. LNCS, vol. 6997, pp. 72–86. Springer, Heidelberg (2011)

Exploiting Transitive Similarity and Temporal Dynamics for Similarity Search in Heterogeneous Information Networks

Jiazhen He[1,2], James Bailey[1,2], and Rui Zhang[1]

[1] Department of Computing and Information Systems, The University of Melbourne
[2] Victoria Research Laboratory, National ICT Australia

Abstract. Heterogeneous information networks have attracted much attention in recent years and a key challenge is to compute the similarity between two objects. In this paper, we study the problem of similarity search in heterogeneous information networks, and extend the meta path-based similarity measure *PathSim* by incorporating richer information, such as transitive similarity and temporal dynamics. Experiments on a large DBLP network show that our improved similarity measure is more effective at identifying similar authors in terms of their future collaborations.

Keywords: similarity search, heterogeneous network, meta path.

1 Introduction

Heterogeneous information networks are ubiquitous in many real-world applications, such as bibliographic networks and healthcare networks. Different from homogeneous information networks (which only consider one type of object and link), heterogeneous information networks involve multiple types of objects and links. For example, heterogeneous bibliographic networks contain authors as well as other types of objects, such as papers, venues, and terms. In addition, heterogeneous information networks contain rich semantic information. For example, two objects can be connected through different links with different semantic meanings (i.e. two authors can be connected by co-authoring a paper or publishing different papers on a same venue). Such networks can more accurately model complex network data.

Heterogeneous information networks have been studied in many data mining tasks [6,16,15]. In this paper, we focus on the problem of similarity search in these networks. Similarity search aims to discover the most relevant objects with respect to a given query object. In heterogeneous information networks where multiple types of objects are available, we focus on identifying similar objects of the same type considering rich semantic information. For example, in a heterogeneous bibliographic network, given a query author, we can discover similar authors based on the diversified semantic meanings, such as co-author relationships and venues of publication.

S.S. Bhowmick et al. (Eds.): DASFAA 2014, Part II, LNCS 8422, pp. 141–155, 2014.
© Springer International Publishing Switzerland 2014

Intuitively, two objects are similar if there many paths between them. A major challenge for similarity search in heterogeneous information networks is how to exploit the diversified semantic meanings under different paths. Existing similarity measures for *homogeneous information networks* cannot effectively capture such meanings since they treat all the paths between two objects equally without distinguishing the different semantic meanings. Some existing studies have recognised this problem and tackled similarity search in *heterogeneous information networks* based on the concept of meta paths[19,14]. A *meta path* is a sequence of links between object types, which can capture a particular semantic meaning between its starting type and ending type. The meta path-based similarity measures treat the concrete paths following a given meta path equally. However, the impacts of the paths connected through different objects can vary. The challenge is how to model such impacts. In addition, heterogeneous information networks evolve over time, and contain rich temporal information. For example, the link between two objects is generally formed with a timestamp. The challenge is how to exploit this temporal information for similarity search.

In this paper, we extend the meta path-based similarity measure $PathSim$[14] by incorporating transitive similarity and temporal information. A meta path can be concatenated by multiple short meta paths. Given a meta path, we first decompose it into multiple short meta paths with the start type and end type of the same type. For example, meta path "*author-paper-author-paper-author*" ($APAPA$) describing two authors share same co-authors can be decomposed into two meta paths APA and APA. Then we add weights to the paths following a short meta path, according to the similarity between the two end objects of the short meta path, which is called transitive similarity. The transitive similarity between two objects can be obtained based on the different meta paths between them with different semantic meanings. The higher the transitive similarity between two objects, the more important the paths between them. For example, suppose two end authors x and y of $APAPA$ are connected through two common co-authors z_1 and z_2, if z_1 is more similar to x and y compared with z_2, the paths between x and z_1, and the ones between y and z_1 should be more important.

In addition, the paths between two objects are generally associated with temporal information, i.e., the building time. Intuitively, the recent paths should be more important than old ones. The paths are generally built as a result of an event. For example, the path "$Tom - P_1 - SIGKDD$" with building time 2012 following the meta path "*author-paper-venuer*" is built due to the event that Tom published paper P_1 in $SIGKDD$ in 2012. To differentiate the importance of different paths, we first decompose a meta path into multiple short meta paths with the maximum length that an event can affect, for example, meta path "*author-paper-venue-paper-author*" can be decomposed into "*author-paper-venuer*" and "*venue-paper-author*". Then we add weights to the paths following the short meta paths according to their building time.

On the other hand, evaluating a new similarity measure is difficult, since it is difficult to obtain ground truth. We approach this challenge by assuming that similar objects will exhibit their similarity by their future behaviour. For example, in the Flickr image network, similar images are more likely to share the same tags or be in the same categories in the future. In bibliographic networks, similar authors are more likely to have collaborations in the future. Under this assumption, we can obtain a ground truth to evaluate our extended similarity measure and compare it against existing methods.

The contributions of this paper are summarized as follows:

- We develop a new method that incorporates transitive similarity to capture the impacts of different paths between two objects given a meta path.
- We incorporate temporal information for similarity search in heterogeneous information networks, by assigning different weights for the paths with different building time.
- Experiments on DBLP network data demonstrate the effectiveness of our proposed methods.

The rest of the paper is organized as follows. Section 2 presents related work, then preliminary concepts and a problem definition are given in Section 3. Section 4 introudces our proposed methods, and Section 5 presents the experimental results. Finally, Section 6 concludes the paper.

2   Related Work

The key basis for similarity search is a similarity measure, which measures the similarity between two objects. Similarity measures for traditional data types have been widely studied, for example the Jaccard coefficient and cosine similarity. For graph data, a number of studies utilize link information to measure the similarity between two objects. Early similarity measures include co-citation[11] and co-coupling[7], which were developed for scientific papers. Other similarity measures based on random walks have also been developed, such as SimRank[4] and Personalized PageRank [5]. SimRank measures the similarity between two objects recursively, by averaging the similarity of their neighbours. Personalized PageRank measures the similarity between two objects by the probability of a random walk with restart starting from source object to target object.

The similarity measures defined in homogeneous networks ignore the different types of semantic information that is available under different paths in heterogeneous networks. There are several works on similarity search in heterogeneous information networks. In [14], a meta path framework was proposed for heterogeneous information networks, where a meta path corresponds to a sequence of links between the objects. Based on the framework, a similarity measure called *PathSim* was proposed, which aims to find similar objects with the same type. In [19], the similarity query ambiguity problem was studied, arising from the diversified semantic meanings in heterogeneous information networks. For a query object, users can provide example similar objects for the query as guidance for choosing related objects. Recently, relevance search in heterogeneous networks was studied in [10]. A relevance measure called *HeteSim*, was proposed to measure the relatedness of the objects in heterogeneous networks, either of the same or different type. Overall, these works are based on the meta path framework and can capture semantic information under a meta path. However, they do not differentiate the impacts of concrete paths given a meta path, which can affect the similarity between two objects.

Another line of work related to our problem is link prediction, as the similarity between two objects can be used to predict the existence of a link between them (i.e., friendships and co-authorship). In addition, since we evaluate the similarity measures considering the future behaviour between two similar objects, and such behaviour can be that a link will be formed between them in the future, our problem is similar to link prediction. However, we focus on developing similarity measures and the future information is only used for evaluation, while link prediction aims at developing methods to predict the existence of a link between two objects. The methods for link prediction can be directly using similarity measures[8] or more sophisticated such as using supervised learning[2].

There are several works on link prediction in heterogeneous information networks[12,18,13]. The most related work to our problem is co-author relationship prediction in heterogeneous networks. Sun et al.[12], considering heterogeneous meta path-based features, used a logistic regression-based co-author relationship prediction model, to predict future co-author relationships. Our similarity measure can actually serve as a heterogeneous feature for their link prediction model.

3   Preliminaries and Problem Statement

In this section, we briefly introduce concepts related to heterogeneous information networks and define the problem.

A **Heterogeneous information network** is defined as a graph $G = (V, E, \mathcal{T}, \mathcal{R})$ where V is a set of objects, E is a set of links, $\mathcal{T}$ is a set of object types and $\mathcal{R}$ is a set of link types between object types. Since a heterogeneous information network contains multiple types of objects and links, $|\mathcal{T}| > 1$ and $|\mathcal{R}| > 1$. Each object $v \in V$ is associated with a particular type $T_i \in \mathcal{T}$, and each link $e \in E$ is associated with a particular type $R_j \in \mathcal{R}$.

The concept of **network schema**[14] has been proposed to describe the meta structure of a heterogeneous network for better understanding. It is a graph defined as $S_G = (\mathcal{T}, \mathcal{R})$ where each object is an object type and each link is a link type between object types.

For example, Fig. 1(a) shows the network schema for a bibliographic information network. There are four types of objects: papers (P), venues(conferences/journals) (C), authors (A) and terms (T) which are the words appearing in the paper title. Also there are different links between the objects. For example, the links between authors and papers denote the writing or written-by relations.

A **meta path** $\mathcal{P}$ is a path defined over network schema, and is formalized as $T_1 \xrightarrow{R_1} T_2 \xrightarrow{R_2} \cdots \xrightarrow{R_l} T_{l+1}$, which defines a composite relation between type T_1 and T_{l+1}. The length of $\mathcal{P}$ is the number of relations in it. The objects can be connected through different meta paths. Two examples of meta path are shown

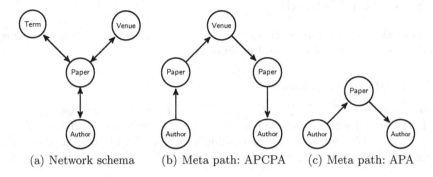

(a) Network schema   (b) Meta path: APCPA   (c) Meta path: APA

Fig. 1. (a) A bibliographic network schema; (b) meta path "author-paper-venue-paper-author" (APCPA) describing authors publish papers in the same conferences; (c) meta path "author-paper-author" (APA) describing co-author relationship

in Fig. 1(b) and Fig. 1(c). For simplicity, the meta path is denoted by the names of object types.

PathSim[14] is a meta path-based similarity measure, which aims at finding similar peer objects for a query object, such as finding similar authors in terms of research area and reputation. Given a symmetric meta path $\mathcal{P}$, *PathSim* computes the similarity between two objects x and y according to

$$s(x, y) = \frac{2 \times |\mathcal{P}_{x \rightsquigarrow y}|}{|\mathcal{P}_{x \rightsquigarrow x}| + |\mathcal{P}_{y \rightsquigarrow y}|} \tag{1}$$

where $\mathcal{P}_{x \rightsquigarrow y}$ is the set of paths between x and y following $\mathcal{P}$, $\mathcal{P}_{x \rightsquigarrow x}$ is that between x and x, and $\mathcal{P}_{y \rightsquigarrow y}$ is that between y and y. The intuition behind *PathSim* is that two similar peer objects should not only be strongly connected, but also share comparable visibility. Their connectivity is defined as the number of paths between them following $\mathcal{P}$, and the visibility is defined as the number of paths between themselves[14].

Given a symmetric meta path $\mathcal{P} = T_1 T_2 \cdots T_l$, *PathSim* similarity between two objects $x_i \in T_1$ and $x_j \in T_l$ with the same type $s(x_i, x_j)$, can be computed through the **commuting matrix** M, which is defined as $M = W_{T_1 T_2} W_{T_2 T_3} \cdots W_{T_{l-1} T_l}$, where $W_{T_i T_j}$ is the adjacency matrix between type T_i and type T_j. M_{ij} denotes the number of paths between object $x_i \in T_1$ and objects $y_j \in T_l$ following meta path $\mathcal{P}$, and $M_{ij} = |\mathcal{P}_{x_i \rightsquigarrow x_j}|$. Similarly, $M_{ii} = |\mathcal{P}_{x_i \rightsquigarrow x_i}|$ and $M_{jj} = |\mathcal{P}_{x_j \rightsquigarrow x_j}|$.

Problem Statement: The problem studied in this paper is as follows. Given a heterogeneous information network and a query object, the goal is to find the top-k objects with the same type and the highest similarity with respect to the query object.

4   Proposed Methods

In this section, we introduce our methods to extend *PathSim* by incorporating transitive similarity and temporal information.

4.1   Transitive Similarity

Given a meta path $\mathcal{P} = T_1 T_2 \cdots T_l$, where T_1 and T_l are the same type ($T_1 = T_l$), $\mathcal{T}_m$ is the set of intermediate types which are the same as T_1 and T_l, $\mathcal{T}_m = (T_{m1}, T_{m2}, \cdots, T_{md})$ where d is the cardinality of $\mathcal{T}_m$. Therefore, $\mathcal{P}$ can be concatenated by multiple meta paths $\mathcal{P}_i (i = 1, \cdots, d+1)$, which is shown in Eq.(2).

$$\mathcal{P} = \underbrace{T_1 \cdots T_{m1}}_{\mathcal{P}_1} \underbrace{\cdots T_{m2}}_{\mathcal{P}_2} \cdots \underbrace{T_{md} \cdots T_l}_{\mathcal{P}_{d+1}} \tag{2}$$

PathSim [14] treats all the paths between object $x \in T_1$ and $y \in T_l$ connected through different transitive objects $z \in T_{mh}$ equally. However, intuitively, we are more likely to trust the paths betweens the objects which are more similar to each other. We can put different weights on the paths following $\mathcal{P}_i$ considering the transitive similarity between the start type and the end type of $\mathcal{P}_i$. A simple way of obtaining the transitive similarity is to utilize *PathSim* over different meta paths with different semantic meanings. Therefore, for meta path $\mathcal{P}$, its commuting matrix can be computed as

$$M_{\mathcal{P}} = M_{\mathcal{P}_1}^s M_{\mathcal{P}_2}^s \cdots M_{\mathcal{P}_{d+1}}^s \tag{3}$$

where $M_{\mathcal{P}_i}^s$ is the commuting matrix for meta path $\mathcal{P}_i$ with transitive similarity incorporated, and can be computed as

$$M_{\mathcal{P}_i}^s = M_{\mathcal{P}_i} \cdot S_{\mathcal{P}'} \tag{4}$$

where $M_{\mathcal{P}_i}$ denotes the commuting matrix of $\mathcal{P}_i$, with each element representing the number of paths between object $x \in T_s(\mathcal{P}_i)$ and object $y \in T_e(\mathcal{P}_i)$, where $T_s(\mathcal{P}_i)$ and $T_e(\mathcal{P}_i)$ represents the start type and the end type of $\mathcal{P}_i$ respectively. $S_{\mathcal{P}'}$ denotes a transitive similarity matrix computed on meta path $\mathcal{P}'$. $\mathcal{P}'$ can be different meta paths such that $T_s(\mathcal{P}') = T_e(\mathcal{P}') = T_s(\mathcal{P}) = T_e(\mathcal{P})$. $S_{\mathcal{P}'}$ allows us to incorporate different meta paths with different semantic meanings.

To better illustrate our method, we give an example in bibliographic networks. Fig. 2 shows the paths between *Rao Kotagiri*(*Rao*) and *Jian Pei*(*Jian*) following meta path *APAPA*, and the one between *Rao* and *Kim Marriott* (*Kim*) according to DBLP between 1990 and 2007. *Rao* and *Jian* (*Kim*) are not co-authors between 1990 and 2007. But they are connected through their common co-authors. Suppose *Rao* is the query author, the *PathSim* similarity between *Rao* and *Jian* according to Eq.(1) is,

$$\begin{aligned} s(Rao, Jian) &= \frac{2 \times |APAPA_{Rao \rightsquigarrow Jian}|}{|APAPA_{Rao \rightsquigarrow Rao}| + |APAPA_{Jian \rightsquigarrow Jian}|} \\ &= \frac{2 \times (9 \times 2 + 1 \times 2 + 18 \times 11)}{21280 + 15333} = 0.0119 \end{aligned}$$

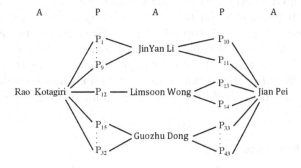

(a) The paths between *Rao Kotagiri* and *Jian Pei* following *APAPA*

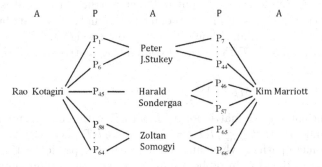

(b) The paths between *Rao Kotagiri* and *Kim Marriott* following *APAPA*

Fig. 2. Example of paths following *APAPA* with *Rao Kotagiri* as the query author and two candidate authors

where the process of computation of $|APAPA_{Rao \rightsquigarrow Rao}| = 21280$ is not shown due to the space limitation, and the same for *Jian* (15333). Similarly, $s(Rao, Kim) = 0.0134$. However, according to our improved similarity measure,

$$s'(Rao, Jian) = \frac{2 \times \sum_{c \in Co}(|APA_{Rao \rightsquigarrow c}| \times S_{Rao,c} + |APA_{c \rightsquigarrow Jian}| \times S_{c,Jian})}{19357.04 + 12594.43}$$

$$= \frac{2 \times 3.59}{19357.04 + 12594.43} = 2.25E - 04$$

where c denotes a common co-author of *Rao* and *Jian*, $Co = \{JinYan\ Li,\ Limsoon\ Wong, Guozhu\ Dong\}$ denotes the set of common co-authors of *Rao* and $Jian$, $S_{Rao,c}$ denotes the transitive similarity between Rao and c (in this example, S is computed based on APA), and similarly for $S_{c,Jian}$. The number of paths (weighted) between Rao and Rao (19357.04) is given directly due to the space limitation, and the same for $Jian$ (12594.43). Similarly, $s'(Rao, Kim) = 1.43E - 04$. We assume that more similar authors are more likely to collaborate with the query author in future. In this example, based on the DBLP data between 2008 and 2013, $Jian$ has collaboration with Rao, while Kim does not. We can see that our improved similarity measure can rank $Jian$ higher compared with Kim.

4.2   Temporal Dynamics

Heterogenous information networks evolve over time, and also the similarity between two objects can change over time. We are more interested in finding similar objects now or even in the future. Intuitively, two objects are more similar if there are more recent connections between them. Instead of treating the paths given a single snapshot equally, we differentiate the impacts of paths formed at different timestamps. A simple way is to put different weights on the paths formed in different timestamps. Essentially, the older paths make less contribution to similarity than recent ones, and should be given lower weights.

Given a meta path $\mathcal{P} = T_1 T_2 \cdots T_l$, its commuting matrix can be computed as

$$M_{\mathcal{P}} = M_{\mathcal{P}_1}^t M_{\mathcal{P}_2}^t \cdots M_{\mathcal{P}_g}^t \tag{5}$$

where $M_{\mathcal{P}_i}^t$ is the commuting matrix for meta path $\mathcal{P}_i$ with temporal information incorporated, and such that $\sum_{i=1}^{g} l(P_i) = l(\mathcal{P})$, where $l(\mathcal{P}_i)$ is the length of meta path $\mathcal{P}_i$. $\mathcal{P}_i$ is a meta path on which an event happens in a particular timestamp. For example, it can be APC in bibliographic networks which represents author publish paper in conference in a particular year. $M_{\mathcal{P}_i}^t$ can be computed as

$$M_{\mathcal{P}_i}^t = M_{\mathcal{P}_i} \cdot Y_{\mathcal{P}_i} \tag{6}$$

where $Y_{\mathcal{P}_i}$ is the temporal matrix on $\mathcal{P}_i$, with each element represents the weight of the path between object $x \in T_s(\mathcal{P}_i)$ and object $y \in T_e(\mathcal{P}_i)$. The weight can be assigned according to the timestamp of the path formed. Here, we define a function $f(t)$ of timestamp t to decide the weights,

$$f(t) = \alpha^{(t_1-t)}(t_0 \le t \le t_1) \tag{7}$$

where t_0 and t_1 represent the start time and end time of the data used for computing similarities. $\alpha(0 < \alpha < 1)$ can be varied. The path formed most recently in t_1 has the largest weight 1. The smaller α is, the more rapidly the weight of the less recent path drops. Different $f(t)$ can be defined. In this paper, we focus on the importance of incorporating temporal information instead of studying the impacts of different $f(t)$.

Based on the above proposed methods, we can improve $PathSim$ by incorporating transitive similarity and/or temporal dynamic, and find the top-k similar objects for a give query object based on our improved similarity measure.

5   Experiments

In this section, we compare the effectiveness of our improved similarity measure using the *PathSim* measure as a baseline.

5.1   Evaluation Measure

Assessing similarity is challenging since it is difficult to obtain ground truth providing a quantitative measure for the similarity between two objects. Most existing methods to evaluate the performance of similarity measures rely on user studies or on an reliable external measure of similarity. The study in [14] used case studies and manually labeled the results for a handful of queries, evaluating using domain knowledge based on these queries. In this paper, since we assume that similar objects will show similar behaviour in some way in the future, we can obtain ground truth to evaluate the similarity measure and provide a comprehensive experimental assessment using thousands of test queries.

We use NDCG (Discounted Normalised Cumulative Gain), a widely used measure in information retrieval [1][3], to evaluate the ranking performance. It rewards relevant objects in the top ranked results more heavily than those ranked lower. In particular, we use NDCG@n, which computes NDCG over the top n ranked objects, and which can be computed as

$$NDCG@n = \frac{DCG@n}{IDCG@n}$$
$$DCG@n = rel(1) + \sum_{i=2}^{n} \frac{rel(x_i)}{log_2(i)} \tag{8}$$

where $IDCG@n$ denotes the Ideal DCG for a perfect ranking and $rel(x_i)$ denotes the relevance score for an object x_i at position i.

5.2   Experiment Setup

The DBLP dataset downloaded on 25th April 2013 is used in our experiments. The network schema of DBLP network is same as Fig. 1(a). The data from 1990 to 2007 (denoted as $T_{1990-2007}$) is used to compute similarity, while the data from 2008 to 2013(denoted as $T_{2008-2013}$) is used for evaluation. The number of authors, papers, conferences (including journals) and terms (after removing stopwords in paper titles) between 1990 and 2007 are shown in Table. 1.

Table 1. DBLP data between 1990 and 2007

Data	Author	Paper	Conference	Term
1990-2007	698,507	1,114,726	4,949	139,613

We focus on computing the similarity between two authors given a meta path between them. In particular, we use meta path $APAPA$ which implies two authors share the same co-authors. Given a query author q, the top n similar authors are returned with similarity computed based on the data in $T_{1990-2007}$. We assume that similar authors will exhibit their similarity by their future behaviour. For meta path $APAPA$, two similar authors might collaborate in the future ($T_{2008-2013}$). To easily capture such behaviour for evaluation, we only return the top n similar authors who have not collaborated with the query author in $T_{1990-2007}$. To evaluate the ranking performance, we need the relevance score $rel(x_i)$ for each returned similar author w.r.t. q. According to the number of co-authored publications between x_i and q in $T_{2008-2013}$, $rel(x_i)$ can be set as

$$rel(x_i) = \begin{cases} 0 & \text{if } N(q, x_i)=0 \\ \varphi\left(N(q, x_i)\right) & \text{if } N(q, x_i) \neq 0 \end{cases} \quad (9)$$

where $N(q, x_i)$ denotes the number of papers that q and x_i publish together in $T_{2008-2013}$. We use $\mathcal{C}$ to denote the set of all the candidate authors. The candidate authors are ranked in ascending order according to $N(q, x)(x \in \mathcal{C})$, and each candidate is assigned a ranking value according to its ranking position. For those who have same value of $N(q, x)$, the same ranking value will be assigned. $\varphi(\cdot)$ is a mapping function from $N(q, x_i)$ to the ranking value for x_i.

The query authors can be chosen from the set of authors who exist in $T_{1990-2007}$, and have new collaborations with authors exist in $T_{1990-2007}$ in future time interval $T_{2008-2013}$. We randomly select 3000 authors as query authors, and compute the averaged results over the 3000 authors. We compare our improved similarity measure with $PathSim$ using paired t-test with $p = 0.05$. This process is repeated 10 times, and the results reported in this paper are the averaged results over 10 runs. In addition, we show the effectiveness of our similarity measure on two sets of query authors, highly productive authors with more than 15 publications in $T_{1990-2007}$ (denoted as HP), and less productive authors with between 5 and 15 publications in $T_{1990-2007}$ (denoted as LP).

5.3   Experimental Results

Transitive Similarity Incorporated. In this group of experiments, we incorporate different kinds of transitive similarity into meta path $APAPA$. We compare our methods with the baseline method, $PathSim$ applied on $APAPA$. The results are shown in Fig.3, where $(APA)^2$ represents the baseline method, and $(APA)^2 - S_{APA}$, $(APA)^2 - S_{APCPA}$ and $(APA)^2 - S_{APTPA}$ represents our methods on $APAPA$ with incorporated transitive similarity based on APA, $APCPA$ and $APTPA$ respectively. All the results have statistical significance with p-value<0.05.

It can be seen from Fig.3 that after incorporating different similarity information, the performances of our methods are improved over all the varying n on both HP and LP queries. Basically, the similarity incorporated based on APA gives better performance compared with $APCPA$ and $APTPA$. In addition, the

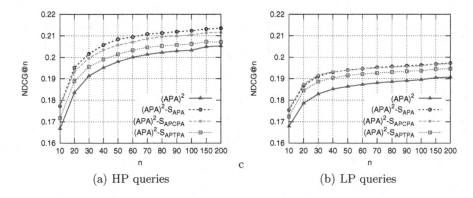

Fig. 3. NDCG@n of $(APA)^2$ denoting the baseline method $(PathSim)$ on $APAPA$ and our methods $(APA)^2 - S_{APA}$, $(APA)^2 - S_{APCPA}$ and $(APA)^2 - S_{APTPA}$ denoting $APAPA$ with incorporated transitive similarity based on APA, $APCPA$ and $APTPA$ respectively, for (a)HP queries and (b)LP queries

performances of all the similarity measures in terms of NDCG@N are low. The main reason is that ranking is generally difficult, especially in the case of similar authors in terms of future collaborators, and only using the raw similarity produced by the similarity measures. Actually, two authors can collaborate due to many external factors that cannot be captured using the similarity measures in this paper. Another reason is that for each run, among the 3000 queries, there are a number of queries with 0 for NDCG@n , which degrade the average results. Such queries do not have future collaborations with their 2-hop authors.

In addition, the overall performance of both the baseline method and our methods on LP queries is worse than that on HP queries. The reason is that for each run, among the 3000 queries, only about 1500 queries have new collaborations with their 2-hop authors for LP queries, while about 2200 for HP queries. Meanwhile, it indicates that HP authors are more likely to collaborate with their 2-hop authors compared with LP authors.

Since the absolute improvements can be misleading, we mainly report the relative improvements of NDCG@n (which is also used in studies in information retrieval[9,17]) in the following experiments. The relative improvements of our methods over $PathSim$ on meta path $APAPA$ are given in Fig. 4. We can see that the relative improvements of our method with transitive similarity S_{APA} and S_{APCPA}, are more than 4% and 3% respectively over all the values of varying n on HP queries. Furthermore, the relative improvements for S_{APTPA} on HP queries is less than that on LP queries. The reason might be that HP authors are generally active in diverse research topics, which yields diverse terms.

Temporal Information Incorporated. In this group of experiments, we show the effectiveness of incorporating temporal information. We incorporate

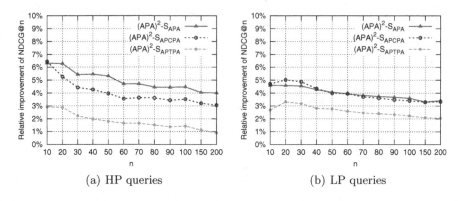

(a) HP queries   (b) LP queries

Fig. 4. Relative improvements of our methods $(APA)^2 - S_{APA}$, $(APA)^2 - S_{APCPA}$ and $(APA)^2 - S_{APTPA}$ over $PathSim$ on $APAPA$

temporal information into meta path $APAPA$, and use Eq.(7) to decide the weights of the paths following APA. Here, $t_0 = 1990$, $t_1 = 2007$.

First we study the impact of parameter α. Fig.5 shows the relative improvements of our method $(APA)^2\_T_\alpha$ with varying α over $PathSim$ on $APAPA$, where $(APA)^2\_T_\alpha$ denotes incorporating the temporal information (with varying α) into $APAPA$. It can be seen that when $\alpha = 0.8$, our method can yield good performance on both HP and LP queries. In addition, the relative improvements on HP queries are much higher than LP queries. The reason might be that the links associated with LP authors are relatively sparse, and are formed in a relatively short time interval, which do not contain much diversified temporal information to be exploited.

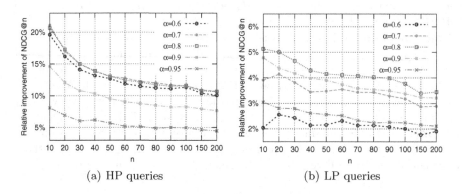

(a) HP queries   (b) LP queries

Fig. 5. Relative improvements of our method $(APA)^2\_T_\alpha$ denoting the temporal information (with varying α) incorporated to $APAPA$ over $PathSim$ on $APAPA$

Furthermore, we compare the relative improvements over $PathSim$ when incorporating temporal information and/or transitive similarity into $APAPA$. Fig. 6 shows the results when incorporating only transitive similarity ($(APA)^2\_S_{APA}$), only temporal information ($(APA)^2\_T_{0.8}$), and both of them ($APAPA\_T_{0.8} - S_{APA}\_T_{0.8}$) to $APAPA$.

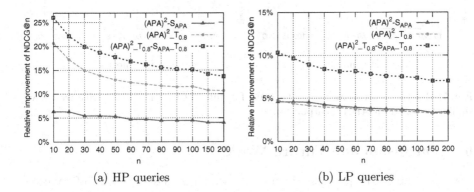

(a) HP queries   (b) LP queries

Fig. 6. Relative improvements of our method $(APA)^2\_S_{APA}$, $(APA)^2\_T_{0.8}$ and $APAPA\_T_{0.8} - S_{APA}\_T_{0.8}$ over $PathSim$ on HP queries and LP queries

It can be seen that there is little difference for the relative improvements of incorporating transitive similarity on HP queries and LP queries. But incorporating temporal information makes huge differences, and basically it works better for HP queries. In addition, the more information incorporated, the higher the performance is, which can be seen from Fig.6 that, $APAPA\_T_{0.8} - S_{APA}\_T_{0.8}$ achieves the best performance with relative improvements more than 15% on HP queries and more than 7% on LP queries.

Impacts on Different Length of Meta Path. In this group of experiments, we check the impacts of transitive similarity on different length of meta path. Fig. 7 shows the relative improvements of incorporating transitive similarity (based on APA) into different length of meta path APA over $PathSim$ applied on corresponding length of meta path APA, where $(APA)^4 - S_{APA}$ represents the relative improvements of incorporating transitive similarity (based on APA) into $(APA)^4$ over $PathSim$ on $(APA)^4$, and similarly for $(APA)^3 - S_{APA}$ and $(APA)^2 - S_{APA}$.

It can be seen that the relative improvement on longer paths is much higher than shorter paths. This is because $PathSim$ does not distinguish the importance of different paths given a meta path. When increasing the length of a meta path, $PathSim$ will treat more remote (and possibly irrelevant) neighbours as similar, whilst our methods which take into account transitive similarity can alleviate this effect.

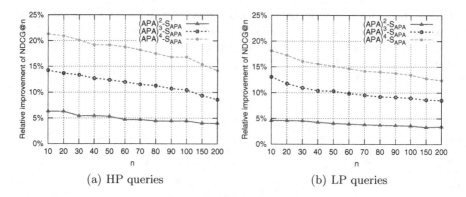

(a) HP queries   (b) LP queries

Fig. 7. Relative improvement on NDCG@n for different length of APA with transitive similarity based on APA incorporated

6   Conclusion and Future Work

We have studied the problem of similarity search in heterogeneous information networks and we have proposed an improved meta path-based similarity measure which incorporates transitive similarity and temporal information. Experimental results show that our improved similarity measures outperforms the baseline existing method. We also found that using temporal information can provide greater gains on highly productive authors than less productive authors. Furthermore, using transitive similarity and temporal information simultaneously can produce the best performance. In future, we plan to consider in more detail other types of objects and networks.

References

1. Agichtein, E., Brill, E., Dumais, S.: Improving web search ranking by incorporating user behavior information. In: Proceedings of the 29th Annual International ACM SIGIR Conference on Research and Development in Information Retrieval, pp. 19–26. ACM (2006)
2. Al Hasan, M., Chaoji, V., Salem, S., Zaki, M.: Link prediction using supervised learning. In: Proceedings of the Sixth SIAM Data Mining Workshop on Link Analysis, Counter-Terrorism and Security (2006)
3. Balasubramanian, N., Kumaran, G., Carvalho, V.R.: Predicting query performance on the web. In: Proceedings of the 33rd International ACM SIGIR Conference on Research and Development in Information Retrieval, pp. 785–786. ACM (2010)
4. Jeh, G., Widom, J.: Simrank: A measure of structural-context similarity. In: Proceedings of the Eighth ACM SIGKDD International Conference on Knowledge Discovery and Data Mining, KDD 2002, pp. 538–543. ACM, Edmonton (2002)
5. Jeh, G., Widom, J.: Scaling personalized web search. In: Proceedings of the 12th International Conference on World Wide Web, WWW 2003. ACM, Budapest (2003)
6. Ji, M., Han, J., Danilevsky, M.: Ranking-based classification of heterogeneous information networks. In: Proceedings of the 17th ACM SIGKDD International Conference on Knowledge Discovery and Data Mining, pp. 1298–1306. ACM (2011)

7. Kessler, M.M.: Bibliographic coupling between scientific papers. American Documentation 14(1), 10–25 (1963)
8. Liben-Nowell, D., Kleinberg, J.: The link prediction problem for social networks. In: Proceedings of the Twelfth International Conference on Information and Knowledge Management, CIKM 2003, pp. 556–559. ACM, New Orleans (2003)
9. Qin, T., Zhang, X.D., Wang, D.S., Liu, T.Y., Lai, W., Li, H.: Ranking with multiple hyperplanes. In: Proceedings of the 30th Annual International ACM SIGIR Conference on Research and Development in Information Retrieval, pp. 279–286. ACM (2007)
10. Shi, C., Kong, X., Yu, P.S., Xie, S., Wu, B.: Relevance search in heterogeneous networks. In: Proceedings of the 15th International Conference on Extending Database Technology, EDBT 2012, pp. 180–191. ACM, Berlin (2012)
11. Small, H.: Co-citation in the scientific literature: A new measure of the relationship between two documents. Journal of the American Society for information Science 24(4), 265–269 (1973)
12. Sun, Y., Barber, R., Gupta, M., Aggarwal, C.C., Han, J.: Co-author relationship prediction in heterogeneous bibliographic networks. In: Proceedings of the 2011 International Conference on Advances in Social Networks Analysis and Mining, ASONAM 2011, IEEE Computer Society, Washington, DC (2011)
13. Sun, Y., Han, J., Aggarwal, C.C., Chawla, N.V.: When will it happen?: relationship prediction in heterogeneous information networks. In: Proceedings of the Fifth ACM International Conference on Web Search and Data Mining, WSDM 2012, pp. 663–672. ACM, Seattle (2012)
14. Sun, Y., Han, J., Yan, X., Yu, P.S., Wu, T.: Pathsim: Meta path-based top-k similarity search in heterogeneous information networks. Proceedings of the VLDB Endowment 4(11) (2011)
15. Sun, Y., Tang, J., Han, J., Gupta, M., Zhao, B.: Community evolution detection in dynamic heterogeneous information networks. In: Proceedings of the Eighth Workshop on Mining and Learning with Graphs, pp. 137–146. ACM (2010)
16. Sun, Y., Yu, Y., Han, J.: Ranking-based clustering of heterogeneous information networks with star network schema. In: Proceedings of the 15th ACM SIGKDD International Conference on Knowledge Discovery and Data Mining, pp. 797–806. ACM (2009)
17. Yeh, J.Y., Lin, J.Y., Ke, H.R., Yang, W.P.: Learning to rank for information retrieval using genetic programming. In: Proceedings of SIGIR 2007 Workshop on Learning to Rank for Information Retrieval, LR4IR 2007 (2007)
18. Yu, X., Gu, Q., Zhou, M., Han, J.: Citation prediction in heterogeneous bibliographic networks. In: Proceedings of the Twelfth SIAM International Conference on Data Mining, Anaheim, California, USA, pp. 1119–1130 (2012)
19. Yu, X., Sun, Y., Norick, B., Mao, T., Han, J.: User guided entity similarity search using meta-path selection in heterogeneous information networks. In: Proceedings of the 21st ACM International Conference on Information and Knowledge Management, CIKM 2012, pp. 2025–2029. ACM, Maui (2012)

Top-k Similarity Matching in Large Graphs
with Attributes

Xiaofeng Ding[1,2], Jianhong Jia[1], Jiuyong Li[2], Jixue Liu[2], and Hai Jin[1]

[1] SCTS & CGCL, School of Computer Science and Technology
Huazhong University of Science and Technology, Wuhan, China
{xfding,jhjia,hjin}@hust.edu.cn
[2] School of Information Technology & Mathematical Sciences
University of South Australia, Adelaide
{xiaofeng.ding,jiuyong.li,jixue.liu}@unisa.edu.au

Abstract. Graphs have been widely used in social networks to find interesting relationships between individuals. To mine the wealthy information in an attributed graph, effective and efficient graph matching methods are critical. However, due to the noisy and the incomplete nature of real graph data, approximate graph matching is essential. On the other hand, most users are only interested in the top-k similar matching, which proposed the problem of top-k similarity search in large attributed graphs. In this paper, we propose a novel technique to find top-k similar subgraphs. To prune unpromising data nodes effectively, our indexing structure is established based on the nodes degrees and their neighborhood connections. Then, a novel method combining graph structure and node attributes is used to calculate the similarity of matchings to find the top-k results. We integrate the adapted TA into the procedure to further enhance the similar graph search. Extensive experiments are performed on a social graph to evaluate the effectiveness and efficiency of our methods.

1 Introduction

With the fast development of online social networks, an increasing number of people are getting involved to communicate with friends and colleagues, which leads to more and more relationships being recorded in various networks, such as Email networks like *Gmail*, online social networks like *Facebook*, and role play game networks like *WOW*, etc. Graph has been widely used to model the structure of online social networks, where nodes standing for individuals and edges represent the relationships among individuals, and attributes of individuals such as address, occupation, etc are also attached to nodes in graph. Therefore, graphs abstracted from online social networks fascinate many researchers from different fields such as computer science, sociology and psychology. Generally speaking, there exists a common problem for all researchers is to find a substructure in a large graph, where nodes with attributes are connected with others. Nevertheless, due to the noisy or imprecise information in a query, approximate graph matching is of paramount important rather than exact matching. Furthermore, top-k answers are interested by users in most situations. Therefore, it is a crucial problem to find top-k similarity matching of a query in a large graph with attributes.

S.S. Bhowmick et al. (Eds.): DASFAA 2014, Part II, LNCS 8422, pp. 156–170, 2014.

To calculate the similarity between two graphs, many measurements have been proposed, and most of them are based on the mismatch of nodes or edges. [25] proposed the maximum common subgraph between a graph and its matching which is used in [17], and it is further enhanced to connected substructure in [24] [18]. Graph edit distance, the number of edit operations making a graph reach another graph, is used in [21] [20] [19]. Random walk is also employed as the similarity measurement in [15] [14]. All of the above works not only proposed the reasonable measurements, but also discussed the approximate matching in different situations. Most existing works, G-Ray [14], TALE [23], TraM [15], TreeSpan [26] and the star representations of a graph [21], build an index structure to accelerate the approximate matching in labeled graphs. In all works, however, a node in the query graph either match or mismatch a node in the data graph, is boolean.

In real social graph, a query node may similar to a data node based on a given measurement, which is neither match nor mismatch with each other. As an example, one researcher attempts to find a substructure consisting of two linked individuals, $(12, F)$ and $(13, M)$, in the graph G in Fig. 1. Clearly, there does not exist an exact matching for that query in G. However, it is very reasonable to return an approximate matching, $\{(17, F), (15, M)\}$. Therefore, it is much more interesting and practical to find the top-k similarity matching in a large attributed graphs for a query graph.

It is challenging to find top-k similarity matching since find one matching (subgraph isomorphism) of a query graph is well known as NP-Complete [12], which can be treated as a special case of approximate matching. A straightforward method, which is called $Bruteforce$, is to search all matches for the query in a data graph, calculate the similarity for each pair and find the top-k similarity matching. However, the naive method is too costly and inefficient. To overcome this drawback, we employ the index structure, NH-index, which was used in TALE [23] to prune unpromising data nodes for each query node and the theory of TA [11] to find the top-k similarity matching in a large attributed graph.

As far as we are concerned, this is the first paper that focuses on the top-k similarity matching in a large graph with attributes. The main contributions of this paper can be summarized as follows:

– The top-k similarity matching problem which combines structural information and attribute information in large attributed graph is firstly motivated and formalized.
– We employ the index structure, NH-index, to prune unpromising nodes of data graph according to the query node degree and its neighborhood connections. Meanwhile, we propose the search algorithm adapted from TA [11] to dramatically reduce search space to improve efficiency.
– Comprehensive experiments are performed on a large graph with attributes, the DBLP dataset. The results verified the effectiveness and efficiency of our index structure and the search algorithm.

The rest of this paper is organized as follows. In Section 2, we introduce the concepts about approximate matching in a large attributed graph. Section 3 presents the index structure, NH-index, which is followed by the search algorithm adapted from TA in Section 4. Experimental results are provided in Section 5 to reveal the performance of our techniques. We present related work in Section 6. Finally, we conclude the paper in section 7.

2 Problem Statement

Based on the definition of social networks [1], we define a social graph as follows,

Definition 1. *Let $A_1, ..., A_I$ is a collection of attributes. A **social graph** is defined as $G = (V, E, R)$, where $V = \{v_1, ..., v_N\}$ is the set of individuals, $E \subseteq \binom{V}{2}$ is the edge set of G describing the relationships among individuals in V, and $R = \{R_1, ..., R_N\}$, where $R_n \in A_1 \times ... \times A_I$, $1 \leq n \leq N$, and A_i, $1 \leq i \leq I$, which denotes the value of i-th attribute, are the description of individuals in V.*

Here, $R_n = \{val(A_1)_n, ..., val(A_I)_n\}$, where $val(A_i)_n$ represents the value of the i-th attribute of R_n, which is *null* or an element in A_i. We define $R_n \subseteq R_m$ if and only if $val(A_i)_n = val(A_i)_m$ or $val(A_i)_n = null$. We say $R_n = R_m$ if $val(A_i)_n = val(A_i)_m$. Hereafter, we denote a social graph as an undirected node-attributed connected graph, a graph for short. Besides the attributes, we assign an unique ID to each node to distinguish them.

Definition 2. *Given two graphs $g_1 = (V_1, E_1, R_1)$ and $g_2 = (V_2, E_2, R_2)$, a **subgraph mapping** f from g_1 to g_2 is an injective function $f : V_1 \rightarrow V_2$ such that:*

- *For any $u \in V_1$, there is a $v = f(u)$, such that $v \in V_2$;*
- *For any $(u, u') \in E_1$, $(f(u), f(u')) \in E_2$.*

If there is a subgraph mapping from g_1 to g_2, we denote $g_1 \subseteq g_2$. Note that the attributes of vertices are not taken into consideration during the subgraph mapping calculation.

Given a data graph G and a query graph q, there is a subgraph mapping f from the node set of q to the node set of G. $f(q) = (V_{f(q)}, E_{f(q)})$ is called a matching of q in G, where $V_{f(q)} = \{f(u)|u \in V_q\}$ and $E_{f(q)} = \{(f(u), f(v)) \in E_G|u \in V_q \cap v \in V_q\}$. The matching is an induced subgraph consisting of the vertices mapped from q. Based on the graph edit distance in [21] and the subgraph mapping in Definition 2, two basic concepts about the matching are presented as follows. In this paper, an edit operation is an insertion or deletion of an edge not including others such as node or edge relabelling. Based on the description about edit operations, the structure similarity is defined.

Definition 3. *The **structure similarity** between q and G under a subgraph mapping f is defined as as*

$$SIM_{str}(q, G, f) = \frac{1}{|E(f(q)) - E(q)| + |E(q) - E(f(q))| + 1} \tag{1}$$

Here, $f(q)$, an induced subgraph of G, is a matching of q in G under mapping f. The structure similarity between a query graph and its matching in the data graph is defined based on the number of missing edges. We add 1 to the number of missing edges to avoid $|E(f(q)) - E(q)| + |E(q) - E(f(q))| = 0$.

Definition 4. *The **attribute similarity** between q and G under a subgraph mapping f is calculated by*

$$SIM_{attr}(q, G, f) = \frac{1}{\sum_{u \in V(q)} dist(attr(u), attr(f(u))) + 1} \qquad (2)$$

Here, $attr(u)$ represents the attributes of vertex u and $dist(attr(u), attr(f(u)))$ is the attribute distance between u and $f(u)$. We convert the attributes of a node into a vector, then the distance can be calculated by vector distances such as Euclidean distance or Cosine distance. Similar to Eqn. 1, we also add 1 to the attribute distance to avoid $\sum_{u \in V(q)} dist(attr(u), attr(f(u))) = 0$. In this paper, the Euclidean distance is employed.

Based on the definitions of structure similarity and attribute similarity, we define the matching similarity between a data graph and a query graph as below.

Definition 5. *A **matching similarity** between q and G under a subgraph mapping f is defined by*

$$SIM(q, G, f) = \beta \times SIM_{attr}(q, G, f) + (1 - \beta) \times SIM_{str}(q, G, f) \qquad (3)$$

Here, β is a parameter to adjust the weight between structure similarity and attribute similarity.

Problem Definition. Given a data graph G, a query graph q, a structure similarity threshold θ, the problem is to identify k subgraphs of G, whose matching similarity is the maximum among all possible matchings of q in G. In other words, we find the top-k approximate matching measured by the matching similarity within a structure similarity threshold. Here, k is a parameter assigned by users.

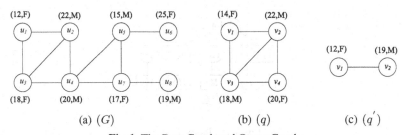

(a) (G) (b) (q) (c) (q')

Fig. 1. The Data Graph and Query Graphs

Example 1. Given a data graph G and a query graph q, as shown in Fig. 1, u_i is the ID of data node and v_i is the ID of query node. The attributes are also attached to the nodes in all graphs.

Based on the above definitions, we can find serval mappings from the query to the data graph, such as $f_0 = \{v_1 \to u_1, v_2 \to u_2, v_3 \to u_3, v_4 \to u_4\}$, $f_1 = \{v_1 \to u_5, v_2 \to u_2, v_3 \to u_7, v_4 \to u_4\}$, $f_2 = \{v_1 \to u_5, v_2 \to u_6, v_3 \to u_7, v_4 \to u_8\}$, and $f_3 = \{v_1 \to u_6, v_2 \to u_5, v_3 \to u_8, v_4 \to u_7\}$. Note that, different mappings may generate the same induced subgraph as matchings such as f_2 and f_3, both of them will be returned as matchings because of different mappings.

It is obvious that the structure similarity between q and G under f_0 is 1. The structure similarity under f_1 is 0.25 since one edge (u_4, u_5) in G and two edges (v_1, v_2), (v_2, v_3) in q are missed. Similarly, we get the structure similarity 0.33 for f_2 and 0.33 for f_3. On the other hand, the attribute similarity under f can be calculated as $SIM_{attr}(q, G, f_0) = 0.2$ for mapping f_0, $SIM_{attr}(q, G, f_1) = 0.207$ for mapping f_1, $SIM_{attr}(q, G, f_2) = 0.065$ for f_2 and $SIM_{attr}(q, G, f_3) = 0.071$ for f_3. Given the parameter $\beta = 0.5$, we calculate the subgraph similarity of the four given mappings by Eqn 3, $SIM(q, G, f_0) = 0.6$, $SIM(q, G, f_1) = 0.229$, $SIM(q, G, f_2) = 0.198$ and $SIM(q, G, f_3) = 0.2$. If we set the parameter, $k = 1$, the subgraph of G, consisting of $\{u_1, u_2, u_3, u_4\}$, and the mapping f_0 will be returned among the four mappings.

3 Indexing Structure

In this section, we introduce the index structure to prune unpromising data nodes for approximate matching in a large data graph. We construct the index based on NH-index which was proposed in [23].

In particular, the node degree is used to denote the number of neighbors in a graph, and the neighbor connection of a node is used to denote the counts of edges existing among the neighbors of the node. Both degree and neighbor connection of nodes capture the feature of a graph structure in some extents. Similar to B$^+$-tree, the NH-index was build on degree and neighbor connection to prune these unpromising nodes, which are impossible to be the matching nodes. In our index, each key is a pair, $<$ $degree, neighborconnection >$, and each key in the leaf node points to a list consisting of nodes whose degree and neighbor connection equal to the key.

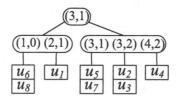

Fig. 2. An Index Tree

The index tree of data graph G in Fig. 1 is shown in Fig. 2. As an example, the graph node, u_2, has 3 neighbors, $\{u_1, u_3, u_4\}$, and there exists 2 edges among its neighbors, $\{(u_1, u_3), (u_3, u_4)\}$. Consequently, u_2 is pointed by the key $(3, 2)$. Then, the key is used to establish the index tree.

We define the partial order for the set of key pairs. Given two key pairs, P_1 and P_2, we say $P_1 \leq P_2$ if and only if

- $P_1.degree \leq P_2.degree$;
- $P_1.degree = P_2.degree$ and $P_1.connection \leq P_2.connection$;

Based on the partial order on the set of key pairs, we establish a NH-index for a data graph. Firstly, we count the degree and neighbor connection for each data node. Nodes with same degree and same neighbor connection are gathered together and pointed by the same pair, $< degree, neighbor connection >$. Based on the set of key pairs, we build the NH-index by hierarchically clustering [13]. Other operations, such as insertion, deletion, update, are the same as the operations on B^+-tree.

4 Search Procedure

In this section, we describe how to pruning unpromising data nodes using NH-index and how TA strategy is applied to find the top-k similarity matching.

4.1 Algorithm Framework

It is too expensive to calculate the candidate list for every query node. Firstly, we introduce some observations as [23] describes:

- **Observation 1**. Some nodes play more important roles in the graph structure than others. In other words, these nodes capture the mainly feature of the graph structure.
- **Observation 2**. A good approximate matching should be more tolerant to missing unimportant nodes in the query than missing important nodes. In other words, we should match the important nodes and satisfy their constraints firstly. Then, unimportant nodes are taken into consideration.

Generally speaking, in social graphs, users submitting queries are more concerned about node attributes than graph structure in most cases. In other words, users pay more attentions to how individuals with certain attributes connect with each other, rather than what the attributes are attached to individuals with certain connections. It can be treated as **Observation 3**.

Based on the former two observations, we can reduce the constraints of a query to shrink the search space. The third observation leads us to divide the combination of attribute similarity and structural similarity into two steps: Firstly, we find the most top-k similar matching under the condition that the structure similarity is no more than the threshold. Then, the k results are returned which are ordered by the combination of structure similarity and attribute similarity.

Given a query, we retrieve candidate data nodes for each important query node using the NH-index. The candidate nodes are listed by descending order according to attribute similarity. Then, the principle of TA [11] is employed to find the top-k similarity matching for the query.

4.2 Step 1: Structural Pruning

To reduce the search space, we prune unpromising data nodes for each important query node using NH-index.

As in [23], we introduce a parameter, q_{imp}, the fraction of the important nodes in a query. Given q_{imp}, we select the important nodes according to a certain measurement. In this paper, degree is used as the measurement.

After getting the set of important nodes, we compute candidate nodes for each important node based on its degree and neighborhood connection using NH-index.

Given a query node, the number of missing neighbors is denoted by Deg_{miss}. Then, the maximum neighbor connection missing can be calculated by $NC_{miss} = Deg_{miss} \times (Deg_{miss} - 1)/2 + (Deg(n_q) - Deg_{miss}) \times Deg_{miss}$, where $Deg(n_q)$ is the degree of a query node n_q. The maximum neighbor connection missing happens when Deg_{miss} neighbors connect with each other and Deg_{miss} neighbors connect with all the remaining $Deg(n_q) - Deg_{miss}$ neighbors. Then, we can prune some data nodes according to Deg_{miss} and NC_{miss}. Based on the relationship between the number of missing neighbors and their connection, we can search the data nodes for each query node on the index tree according to the structure similarity threshold.

After getting the candidate nodes for a query node, we sort these nodes based on the attribute similarity by descending order. Then, TA strategy is applied for retrieving the top-k similarity matching.

4.3   Step 2: TA Strategy

In this section, we present how to find the top-k similarity matching using the principle of TA strategy on the candidate node list.

Based on the structure of data graph and query graph, we join the candidate nodes from different candidate lists retrieved by different important query nodes. According to *observation 3*, we select the most similar data node for each query node in parallel, and extend the mapping until obtaining a matching for the query or proving this mapping fails. Repeat this procedure until the search halts.

Which data node is selected for a query node is crucial to get maximum attribute similarity during our extension. We select a data node which is the most similar to the unmatched query node. If the data node has been matched with another query node, we select the better one from these two pairs. The extension was stopped as TA strategy, i.e. when the top-k similarity matching have been found with the attribute similarity above the attribute similarity summation in the current row.

Given a data graph G and a query graph q in Fig. 1, we select the fraction, q_{imp}, as 0.5. The set of important query nodes can be calculated based on the node degree, $\{v_2, v_3\}$. We use NH-index to prune unpromising data nodes for each important query node based on the structure of query graph, the degree of query node and its neighborhood connection, and the given structure similarity threshold, 1 edge missing. Then, two data nodes, u_6 and u_8, are pruned for both important quer nodes. We sort the candidate nodes based on the attribute similarity by descending order. A matrix can be calculated in Fig. 3, where each entry is the attribute similarity between the query node and the data node. TA strategy is applied to calculated the top-k similarity matching based on the matrix.

To find the top-k matching, we extend the partial match beginning from the first row in parallel. Firstly, we select candidate nodes based on the attribute similarity with the constraint of structure similarity threshold for every other query nodes

V_2	V_3
u_2 (1)	u_3(0.5)
u_4(0.33)	u_7 (0.41)
u_3 (0.2)	u_4 (0.33)
u_7(0.16)	u_5(0.25)
u_5(0.13)	u_2 (0.2)
u_1 (0.09)	u_1 (0.14)

Fig. 3. TA Matrix

according to the matrix. After computing the TA matrix for G and q in Fig. 1, we extend the partial match (v_2, u_2) and (v_3, u_3), then we get a complete mapping $f = \{(v_1, u_1), (v_2, u_2), (v_3, u_3), (v_4, u_4)\}$, since both of them calculate the same mapping f. After getting the matching for a given pair, we adjust the top-k heap. After all pairs in a row were computed, we stop search if k matchings have been found, and the minimum attribute similarity in the top-k heap is above the attribute similarity summation in the current row.

The procedure of TA strategy to find the top-k similarity matching for a query G_q in a data graph G_d is presented in Algorithm 1.

Algorithm 1. extendMatch(G_q, G_d)

Require: A query graph G_q, A data graph G_d
Ensure: TK is a heap containing the top-k similarity matching.
1. calculate TA matrix; //pruning unpromising data nodes for important nodes using NH-index and ordering candidate nodes based on the attribute similarity;
2. M_c is a temporary variable containing matched pairs;
3. **for** each row of TA matrix **do**
4.   **for** each entry (N_q, N_d) in a row **do**
5.   put (N_q, N_d) into M_c;
6.   extendMatch(G_q, G_d, M_c);
7.   heapAdjust(M_c); //adjusting the top-k heap
8.   clear M_c;
9.   calculate τ which is the summation of attribute similarity in the current row;
10.   **if** getHeapSize()$\geq k$ and getMinSim()$> \tau$ **then**
11.   break;
12. **return** TK;

In Algorithm 1, $extendMatch(G_q, G_d, M_c)$ was called to extend similar matching based on the current mapping in M_c which is implemented in Algorithm 2. $heapAdjsut(M_c)$ was employed to adjust the heap to calculate the top-k similarity matching and $getHeapSize()$ returns the number of elements in the heap TK. $getMinSim()$ returns the minimum attribute similarity in the heap.

Algorithm 2. extendMatch(G_q, G_d, M_c)

Require: A query graph G_q, A data graph G_d, M_c contains the match pair on which the extension is based

Ensure: M_c presents a similar matching for important query nodes

1. put the given match (N_q, N_d) into a priority queue Q ordered by attribute similarity;
2. **while** Q is not empty **do**
3.   pop up the head match (N_q, N_d) from Q;
4.   **if** M_c' satisfies the threshold //M_c' is calculated by trying to push (N_q, N_d) into M_c **then**
5.   push (N_q, N_d) into M_c;
6.   extendNodesBFS($G_q, G_d, N_q, N_d, M_c, Q$); //extending the matching based on the current matching.
7. **return** M_c;

Algorithm 2 describes the extension match based on given matches contained in M_c. Akin to breath first search to a graph, a queue is employed to extend the current matching. In each loop, a best match is selected from candidate match queue. Then, we will continue to extend the last match from the nearby nodes of N_q and N_d. This algorithm is similar to the extension algorithm in [23]. We present the $extendNodesBFS()$ in Algorithm 3.

Algorithm 3. extendNodesBFS($G_q, G_d, N_q, N_d, M_c, Q$)

Require: A query graph G_q, A data graph G_d, (N_q, N_d) present the basic match, M_c contains the current matches, Q is the queue of candidate matches

Ensure: M_c presents a similar matching for important query nodes

1. calculate $NB1_q$, the immediate neighbors of N_q not in M_c;
2. calculate $NB2_q$, the neighbors two hops away from N_q not in M_c;
3. calculate $NB1_d$, the immediate neighbors of N_d not in M_c or Q;
4. calculate $NB2_d$, the neighbors two away from N_d not in M_c or Q;
5. MatchNodes($G_q, G_d, NB1_q, NB1_d, M_c, Q$);
6. MatchNodes($G_q, G_d, NB1_q, NB2_d, M_c, Q$);
7. MatchNodes($G_q, G_d, NB2_q, NB1_d, M_c, Q$);

Both Algorithm 3 and Algorithm 4 are the same with extension algorithms in [23].

In Algorithm 3, we first calculate the immediate and two-hop-away neighbors of a given query node and its corresponding data node. Based on the neighborhoods, we compute the candidate mappings. It is reasonable that two-hop neighborhoods are taken into consideration. As an example, for the query graph q' in Fig. 1, $\{u_1, u_8\}$ is the best matching according to the attribute similarity under the mapping, $\{(v_1, u_1), (v_2, u_8)\}$, with the constraint of 1 edge missing. However, u_1 is too far away from u_8 in the data graph G. $\{u_1, u_4\}$ under the mapping $\{(v_1, u_1), (v_2, u_4)\}$ and $\{u_1, u_3\}$ under the mapping $\{(v_1, u_1), (v_2, u_3)\}$ are returned as the results of top-2 similarity matching.

Algorithm 4 is employed to calculate candidate mappings based on the neighbors computed in 3. For each candidate query node N_q, find the best match data node N_d. If N_d has not been matched, add mapping (N_q, N_d) to the candidate queue. Otherwise,

---

Algorithm 4. extendNodesBFS($G_q, G_d, S_q, S_d, M_c, Q$)

Require: A query graph G_q, A data graph G_d, S_q is the set of query nodes, S_d is the set of data nodes, M_c contains the current matches, Q is the queue of candidate matches
Ensure: M_c presents a similar matching for important query nodes
1. **for** each node N_q in S_q **do**
2.   calculate N_d, the best mapping of N_q in S_d;
3.   **if** N_q has not been matched in Q **then**
4.   put (N_q, N_d) into Q;
5.   remove N_d from S_d;
6.   **else**
7.   **if** (N_q, N_d) is a better match except for $M_c[0]$ **then**
8.   remove the existing match of N_q from Q;
9.   put (N_q, N_d) into Q;
10.   remove N_d from S_d;

---

if there is another pair (N_q, N_d') for N_q and (N_q, N_d) is a better match than (N_q, N_d'), we replace (N_q, N_d') by (N_q, N_d) except that (N_q, N_d') is in $M_c[0]$ which is the initial pair. Repeat this procedure until each query node in S_q have a match data node.

5 Experimental Evaluation

In this section, we show the experimental results to reveal the effectiveness and the efficiency of our method on a social graph. Our technique is implemented by C++ and run on an Intel Celeron 2.4GHz machine with 4GB RAM.

Data Graph. DBLP collaboration graph is used in our experiments. It consists of 356K distinct authors and 1.9M co-author edges among the authors, can be downloaded from the internet[1]. We abstract a subgraph making up to 100K vertices from DBLP graph.

Attribute Data. Besides the graph structure, we also need an attribute database. We choose adult database published in internet[2], consisting of 49K items, as the attribute database to describe the vertices. We extract the descriptive data from the adult dataset and make it consist of 5 attributes, including age, salary, education level, gender and work hour. For each vertex in the data graph, we assign an item from adult database to a vertex randomly. For each remaining vertex which has not been assigned the attributes, it was allocated with attributes adding a random number to the attribute selected from the dataset. Up to now, a social graph with attribute is obtained.

Query Graphs. We select a random node from the real dataset as a seed. Breadth first search is employed to obtain query graphs according to the size and the number of vertices. We range the size from 5 to 15 with an interval of 1. For each query size, we generate a set of 50 queries. During the generation of queries, we add a random integer to the attributes of data nodes to get the attributes of queries. At the same time, one edge is deleted or added randomly. The mainly used parameters were in Table 1.

---

[1] http://www.informatik.uni-trier.de/~ley/db/
[2] http://archive.ics.uci.edu/ml/datasets.html

Table 1. The Description of Parameters

Parameter	Description
k	the number of returned results
TH	the threshold of missing edges
β	the weight of structure similarity
q_{imp}	the fraction of important nodes in a query
$|V|$	the number of nodes in a graph
$|q|$	the number of nodes in a query

5.1   Effectiveness

In our experiments, we set the parameter $\beta = 0.5$, to make the weights of structure similarity and attribute similarity equal. To demonstrate the effectiveness of our methods, experiments are randomly conducted on a $|V| = 10K$ graph with constraints of 2 missing edges. We set $q_{imp} = 50\%$. A sample query and its top-2 results were shown in Fig. 4, where attributes are attached to nodes and important nodes are in black. Since our random modification, no exact matchings happen in the data graph. Two approximate matchings are retrieved based on our measurements.

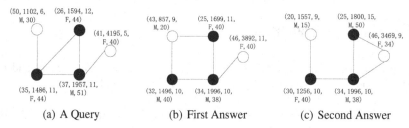

(a) A Query   (b) First Answer   (c) Second Answer

Fig. 4. A Sample Query and Top-2 Answers

5.2   Efficiency

We use various subsets of the whole DBLP database to test the scalability of our methods. In the following experiments, the parameters are set at $\beta = 0.5$, $|V| = 10K$, $k = 2, TH = 2$ and $q_{imp} = 50\%$ by default. For each subgraph, we return top-2 results according to the set of query graphs. We run the experiments serval times and report average time for each type of queries.

Fig. 5 shows the response time increases with the increasing number of query size (the number of nodes), which ranges from 10 to 20 with an interval of 2. For a larger query size, it takes more time since more constraints contained in the query need to be examined. Note that, $NH - TA$ in Fig. 5 describes the performance of our method using NH-index to prune unpromising nodes and TA-strategy to halt the search procedure, BF represents the $Bruteforce$ method.

Fig. 6(a) describes the time to build NH-index for different graph size. It takes more time for a larger data graph since it clusters more pairs, more degree and neighborhood connections, into the NH-index. In Fig. 6(b), response time increases along with the

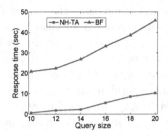

Fig. 5. Response Time vs. Query Size

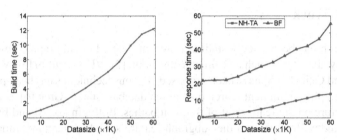

(a) Index Building Time vs. Datasize   (b) Response Time vs. Datasize

Fig. 6. Index Build Time and Response Time

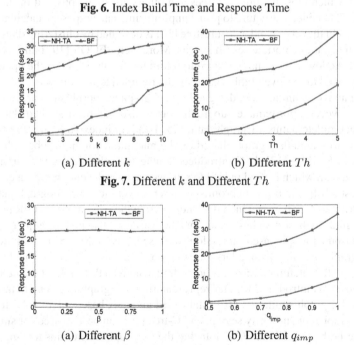

(a) Different k (b) Different Th

Fig. 7. Different k and Different Th

(a) Different β (b) Different q_{imp}

Fig. 8. Different β and Different q_{imp}

growth of data graph size for a query with certain size $|q| = 10$. It is because more matches need to be checked when the data graph size increases.

In Fig. 7, we reveal the impacts of the parameters k and Th. Both experiments are conducted on a $|V| = 10K$ data graph and the query size is set as $|q| = 10$. It is reasonable that more candidate results need to detect along with k or Th increasing.

Fig. 8(a) shows the response time on a $|V| = 10K$ graph under different β. The query size is $|q| = 10$. The increasing of β leads to attribute similarity becomes more important, which makes the TA algorithm halt in shorter time. On the other hand, it reveals the β has little effect on the response time. Fig. 8(b) reveals the impacts of q_{imp}. More nodes need to be approximately matched when q_{imp} increases.

6 Related Work

Subgraph search has been widely studied in recent years. For subgraph isomorphism problem, many algorithms such as Ullmann algorithm [2], VF algorithm [3], VF 2.0 algorithm [4] and QuickSI [5] have been proposed. Various techniques has been proposed for exact subgraph containment search on a graph database. GraphGrep [6] established an index structure on the feature set of frequent pathes. Both gIndex [8] and FG-Index [7] build the index structures on the subgraph feature set which captures more graph features compared to path feature. Tree+Delta [9] established an index based on the feature set in which most features are trees and others are subgraphs. All of the above techniques build an index structure to prune unpromising data graphs for calculating the candidate set. On the other hand, Closure-tree [10] is designed by clustering data graphs to establish tree-index to prune search spaces. Moreover, SEGOS [19] build two-level index, where the lower-level index is used to return top-k sub-units and the upper-level index is employed to retrieve graphs based on the returned top-k sub-units.

All of the above techniques are designed for subgraph isomorphism search on graph databases. However, approximate subgraph search also has been studied due to the noisy and incomplete nature of graph data. [25] build an index structure, the feature-graph matrix, to search data graphs based on the maximum common subgraph, which is also used in SIGMA [17]. [24] introduced connected substructure to search graph containing a query, which is used in [18] to retrieve data graph contained by a query. G-tree [28] is established to prune unpromising nodes to get the top-k subgraph matching based on attribute information which does not tolerate the edge missing. G-Ray [14] is proposed to find the best matches, exact-matches or near-matches, for a query pattern in a large attributed graph based on random walk score as TraM [15] to find the top-k similar subgraph matching from a large data graph. TreeSpan [26] is designed to compute similarity all-Matching for a query according to a given threshold of the number of missing edges. [21] introduced the star representation of a graph to calculate the graph edit distance for graph similarity. [16] established an index based on $\kappa - AT$ to index large sparse graphs for similarity search, and C-tree [10] can be also used for similarity search, since both $\kappa - AT$ and C-tree tackle the subgraph isomorphism using an approximate method called pseudo subgraph isomorphism, which uses level-n adjacent tree as an approximation of graph. TALE [23] is designed as a tool to retrieve a subset where each entry is similar to the query graph. [29] proposed some general subgraph matching kernels for graphs with labels or attributes which is however not involved in the problem of top-k.

7 Conclusion

In this paper, we introduced the problem of top-k similarity matching in a large graph with attributes. It is a NP-complete problem since subgraph isomorphism test is an extreme case, which is well known as NP-complete.

Different from other approximate matching problems, we take attribute similarity into consideration. For simplicity, we firstly employ NH-index in TALE to prune the unpromising data nodes. Then, TA strategy is applied to reduce the search space to find the top-k similarity matching based on attribute similarity with the constraints on structural similarity. Experiments performed on the DBLP collaboration graphs confirmed the effectiveness and the efficiency of our method.

In the future, we are plan to extend this work to a billion-node graph in distributed environment, which can improve the efficiency from another point of view. On the other hand, to protect the privacy of individuals in the published large graph, it is interesting to support effective top-k queries while guarantee differential privacy.

Acknowledgements. This work was supported by the National Natural Science Foundation of China under grant 61100060 and the Australia ARC discovery grant DP110103142, DP130104090.

References

1. Tassa, T., Cohen, D.: Anonymization of Centralized and Distributed Social Networks by Sequential Clustering. In: TKDE, pp. 1–14. IEEE Press, New York (2011)
2. Ullmann, J.R.: An Algorithm for Subgraph Isomorphism. J. ACM, 31–42 (1976)
3. Cordella, L.P., Foggia, P., Sansone, C., Vento, M.: Subgraph Transformations for the Inexact Matching of Attributed Relational Graphs, pp. 43–52. Springer, Vienna (1998)
4. Cordella, L.P., Foggia, P., Sansone, C., Vento, M.: An Improved Algorithm for Matching Large Graphs. In: 3rd IAPR-TC15 Workshop on Graph-based Representations in Pattern Recognition, pp. 149–159 (2001)
5. Shang, H., Zhang, Y., Lin, X., Yu, J.X.: Taming Verification Hardness: An Efficient Algorithm for Testing Subgraph Isomorphism. In: VLDB, pp. 364–375 (2008)
6. Shasha, D., Wang, J.T.L., Giugno, R.: Algorithmics and Applications of Tree and Graph Searching. In: PODS, pp. 39–52. ACM Press, New York (2002)
7. Cheng, J., Ke, Y., Ng, W., Lu, A.: FG-Index: Towards Verification-Free Query Processing on Graph Databases. In: SIGMOD, pp. 857–872. ACM Press, New York (2007)
8. Yan, X., Yu, P.S., Han, J.: Graph Indexing: A Frequent Structure-based Approach. In: SIGMOD, pp. 335–346. ACM Press, New York (2004)
9. Zhao, P., Yu, J.X., Yu, P.S.: Graph Iindexing: Tree+ Delta $>=$ Graph. In: VLDB, pp. 938–949 (2007)
10. He, H., Singh, A.K.: Closure-Tree: An Index Structure for Graph Queries. In: ICDE, pp. 38–49. IEEE Press, New York (2006)
11. Fagin, R., Lotem, A., Naor, M.: Optimal Aggregation Algorithms for Middleware. In: PODS, pp. 102–113 (2001)
12. Garey, M.R., Johnson, D.S.: Computers and Intractability: A Guide to the Theory of NP-Completeness. Freeman, San Francisco (1979)

13. Han, J., Kamber, M., Pei, J.: Data mining: Concepts and Techniques. Morgan Kaufmann (2006)
14. Tong, H., Gallagher, B., Faloutsos, C., Eliassi-Rad, T.: Fast Best-Effort Pattern Matching in Large Attributed Graphs. In: ACM KDD, New York, pp. 737–746 (2007)
15. Amin, M.S., Finley Jr., R.L., Jamil, H.M.: Top-k Similar Graph Matching Using TraM in Biological Networks. In: TCBB, New York, pp. 1790–1804 (2012)
16. Wang, G., Wang, B., Yang, X., Yu, G.: Efficiently Indexing Large Sparse Graphs for Similarity Search. In: TKDE, pp. 440–451. IEEE Press, New York (2012)
17. Mongiovi, M., Natale, R.D., Giugno, R., Pulvirenti, A., Ferro, A.: Sigma: A Set-Cover-Based Inexact Graph Matching Algorithm. Journal of Bioinformatics and Computational Biology, 199–218 (2010)
18. Shang, H., Zhu, K., Lin, X., Zhang, Y., Ichise, R.: Similarity Search on Supergraph Containment. In: ICDE, pp. 637–648. IEEE Press, New York (2004)
19. Wang, X., Ding, X., Tung, A.K.H., Ying, S., Jin, H.: An Efficient Graph Indexing Method. In: ICDE, pp. 210–221. IEEE Press, New York (2012)
20. Zhao, X., Xiao, C., Lin, X., Wang, W.: Efficient Graph Similarity Joins with Edit Distance Constraints. In: ICDE, pp. 834–845. IEEE Press, New York (2012)
21. Zeng, Z., Tung, A.K.H., Wang, J., Feng, J., Zhou, L.: Comparing Stars: On Approximating Graph Edit Distance. In: VLDB, pp. 25–36 (2009)
22. Khan, A., Li, N., Yan, X., Guan, Z., Chakraborty, S., Tao, S.: Neighborhood Based Fast Graph Search in Large Networks. In: SIGMOD, New York, pp. 901–912 (2011)
23. Tian, Y., Patel, J.M.: TALE: A Tool for Approximate Large Graph Matching. In: ICDE, pp. 963–972. IEEE Press, New York (2008)
24. Shang, H., Lin, X., Zhang, Y., Yu, J.X., Wang, W.: Connected Substructure Similarity Search. In: SIGMOD, pp. 903–914. ACM Press, New York (2010)
25. Yan, X., Yu, P.S., Han, J.: Substructure Similarity Search in Graph Databases. In: SIGMOD, pp. 766–777. ACM Press, New York (2005)
26. Zhu, G., Lin, X., Zhu, K., Zhang, W., Yu, J.X.: TreeSpan: Efficiently Computing Similarity All-Matching. In: SIGMOD, New York, pp. 529–540 (2012)
27. Sun, Z., Wang, H., Wang, H., Shao, B., Li, J.: Efficient Subgraph Matching on Billion Node Graphs. In: VLDB, pp. 788–799 (2012)
28. Zou, L., Chen, L., Lu, Y.: Top-K Subgraph Matching Query in A Large Graph. In: CIKM, pp. 139–146 (2007)
29. Kriege, N., Mutzel, P.: Subgraph Matching Kernels for Attributed Graphs. In: ICML, pp. 1–8 (2012)

On Perspective-Aware Top-k Similarity Search in Multi-relational Networks*

Yinglong Zhang[1,2], Cuiping Li[1], Hong Chen[1], and Likun Sheng[2]

[1] Key Lab of Data Engineering and Knowledge Engineering of MOE, and
Department of Computer Science, Renmin University of China, China
zhang_yinglong@126.com, cuiping_li@263.net
[2] JiangXi Agricultural University, China

Abstract. It is fundamental to compute the most *"similar"* k nodes w.r.t. a given query node in networks; it serves as primitive operator for tasks such as social recommendation, link prediction, and web searching. Existing approaches to this problem do not consider types of relationships (edges) between two nodes. However, in real networks there exist different kinds of relationships. These kinds of network are called multi-relational networks, in which, different relationships can be modeled by different graphs. From different perspectives, the relationships of the objects are reflected by these different graphs. Since the link-based similarity measure is determined by the structure of the corresponding graph, similarity scores among nodes of the same network are different w.r.t. different perspectives. In this paper, we propose a new type of query, *perspective-aware top-k similarity query*, to provide more insightful results for users. We efficiently obtain all top-k similar nodes to a given node simultaneously from all perspectives of the network. To accelerate the query processing, several optimization strategies are proposed. Our solutions are validated by performing extensive experiments.

Keywords: Random walk, Multi-relational network, Graph, Proximity.

1 Introduction

Recent years have seen an astounding growth of networks in a wide spectrum of application domains, ranging from sensor and communication networks to biological and social networks [1]. At the same time, a number of important real world applications (e.g. link prediction in social networks, collaborative filtering in recommender networks, fraud detection, and personalized graph search techniques) rely on querying the most "similar" k nodes to a given query node. The measure of "similarity" between two nodes is the proximity between two nodes

* This work was supported by National Basic Research Program of China (973 Program)(No. 2012CB316205), NSFC under the grant No.61272137, 61033010, 61202114, and NSSFC (No: 12&ZD220). It was partially done when the authors worked in SA Center for Big Data Research in RUC. This Center is funded by a Chinese National 111 Project Attracting.

S.S. Bhowmick et al. (Eds.): DASFAA 2014, Part II, LNCS 8422, pp. 171–187, 2014.

172   Y. Zhang et al.

   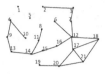

Fig. 1. A coauthor network G

Fig. 2. The graph G_{DB} from perspective of DB

Fig. 3. The graph G_{DM} from perspective of DM

Fig. 4. The graph G_{IR} from perspective of IR

based on the paths connecting them. For example, random walk with restart (RWR) [2], Personalized PageRank (PPR) [3], SimRank [4], and hitting time [5] are all such kinds of measures. These measures are computed based on the structure of graphs.

The question, computation of the most "similar" k nodes to a given query node, has been studied in these researches [2,6,7,8]. Although their works are excellent, they did not consider the query **under a specific viewpoint**. A query, top k similar authors w.r.t. *Jiawei Han* **in the database field**, is more interesting and useful than the query, that without the viewpoint, for people who are interested in the research of database.

Actually, as mentioned in [9,10,11,12,13], there may exist different kinds of relationships between any two nodes in real networks. For example, in a typical social network, there always exist various relationships between individuals, such as friendships, business relationships, and common interest relationships [10]. So different relationships can be modeled by different graphs in multi-relational networks. And these different graphs reflect relationships among objects from different perspectives. Correspondingly, the top k similarity query based on these graphs will return different answers.

Here an example is given:

Example 1. A network G of coauthor relationships is showed in Figure 1. In the figure, relationships are extracted based on the publish information from database (DB), data mining (DM), and information retrieval (IR) fields. Relationships of coauthor in different fields are denoted by different colors (DB, DM, and IR are denoted by red, green, and black edges respectively in Figure 1). The graph showed in figure 2 is modeled based on coauthor relationships of DB field. Similarly, Figure 3 and Figure 4 show the graphs from DM and IR perspective respectively. $G = G_{DB} \bigcup G_{DM} \bigcup G_{IR}$ is also considered as the graph from perspective of DB or DM or IR. Obviously, the corresponding structure from different perspective is different for G. The graph G' in figure 5 reflects the coauthor relationship among authors without considering the specific research field, on which the traditional top-k query is performed. G' is the corresponding simple graph of G.

Given a query node q, if we want to know the most "similar" nodes w.r.t. q from DB perspective, the result will be determined by the graph in figure 2 (rather than G or G'). Given the query node 4, Table 1 shows its top-3 similar nodes from different perspectives. The result of the traditional query based on G' (without considering a specific viewpoint) is (5, 10, 1), while the result is

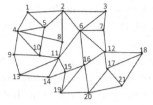

Fig. 5. The graph G' without considering perspectives in the network G

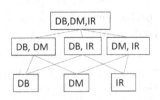

Fig. 6. All Perspectives

(1, 5, 8) from perspective DB, from perspective DM result is (10, 9, 11), and from perspective DB or DM or IR the result is (10, 9, 5).   □

From the example, the perspective-aware top-k search provides more insightful information to users. It is used in a lot of applications. For example, in e-commerce activities, if product a is frequently co-purchased with product b, then we construct a product co-purchasing network which contains an edge (a, b). For a young customer who bought a product a, recommending top-k most similar products w.r.t a from the perspective of young people to the customer is a targeted marketing effort in contrast with that without considering the viewpoint. Sometimes people desire to query the most "similar" k nodes w.r.t. a query node from different perspectives rather than under a specific viewpoint. For the network G of coauthor relationship in figure 1, its perspectives and the relationships among them are showed in figure 6. It is more interesting and useful how the result of the query varies as the perspectives change from bottom to top along the relationship showed in figure 6. For instance, the corresponding results of the query w.r.t. node 5 are showed in table 1 when the perspectives changed along DB→(DB DM),(DB IR)→(DB DM IR). The corresponding query results are almost same although perspectives are different. This information is interesting and it motivates us to think that the person (node 5) may be a pure database researcher. The assumption can be further verified by comparing G_{DB} (figure 2) with G (figure 1): the person (node 5) collaborates merely with other researchers in database field, and most of his coauthors also collaborate with other researchers in database field. Therefore, by computing the query from different perspectives, we can explore the relationship between query node and perspectives.

From the above discussion, the advantages of perspective-aware top-k search are the following:

Table 1. Top 3 query from different perspective on G using RWR measure

Query node	Perspective	Top-3 nodes	Query node	Perspective	Top-3 nodes
4	DB	1, 5, 8	5	DB	4, 1, 2
	DM	10, 9, 11		DB DM	4, 1, 2
	DB DM IR	10, 9, 5		DB IR	4, 1, 2
	G'	5,10,1		DB DM IR	4, 1, 2

- We can retrieve the most "similar" k nodes to a given query node from any specific viewpoint. Some results of the query can not be achieved by the traditional top-k query.
- We can discover the relationship between query node and perspectives. By exploring results of the query from different perspectives, we can find how the results change with perspectives. These are useful to comprehend both the query node and corresponding results for users.

In the paper we choose RWR as our proximity measure. RWR is a given node's personalized view of the importance of nodes on the graph. This is compliant with our problem: given a query node we want to find k most similar nodes based on the view of the query node.

To the best of our knowledge, our work is the first one to propose the perspective-aware top-k query in multi-relational networks. Due to the complexity for the computation of similarity and the huge size of the graphs, the challenge of the problem is whether we can traverse once to efficiently obtain all top-k nodes about all perspectives simultaneously to the query node. To address the challenge, we design a concise structure of graphs which contains information of all perspectives, then we accelerate speed of the query by merely searching the neighborhood of the query node.

The contributions of this paper are summarized as below:

1. We define a new type of query, perspective-aware top-k query, in multi-relational networks. The query can provide more meaningful and rich information than the traditional top-k query.

2. RWR is adopted as the measure of similarity. To accelerate the query processing, the corresponding bounding of proximity is given.

3. We propose an efficient query processing algorithm. By designing a concise data structure of graphs and with the help of boundings of the proximity, we merely traverse once the neighborhood of a query node to obtain all its top-k nodes simultaneously from all perspectives. Also, we can achieve top-k nodes from any specific perspective.

Related Work. Recently there are several works [2,6,7,8] based on link-based similarity measures to compute the most "similar" k nodes to a given query node. Theses algorithms are excellent but they did not consider the situations that perspective-aware top-k query.

Graph OLAP [14,15] provide a tool which can view and analyze graph data *from different perspectives*. The idea of Graph OLAP inspired our works. However Graph OLAP is fundamentally different from our problem. Vertex-specific attributes are considered as the dimensions of a network for Graph OLAP. We consider features of whole graph as perspectives. From different perspectives the structure of graph is different.

As dicussed in [11], the multi-relational network is not new. Some researches about multi-relational networks mainly focus on the community mining [9,10,12]). [13] gave the basis for multidimensional network analysis.

2   Problem Formulation

Multi-relational networks are modeled by multigraphs. For the sake of simplicity, we only consider undirected multigraphs and these can be easily extended to directed multigraphs. In the paper, all discussions are based on the following model and definitions.

A multigraph is denoted as $G = < V, E, \widetilde{F} >$ where V is a set of nodes; E is a set of labeled edges; $\widetilde{F}$ is a set of base perspectives: $\widetilde{F} = \{f_1, f_2, ..., f_m\}$. $(u, v, f) \in E$ ($u, v \in V$ and $f \in \widetilde{F}$) means there is a relationship between u and v from perspective f. Each pair of nodes in G is connected by at most $|\widetilde{F}|$ possible edges.

Definition 1. *From any perspective* $F (F \subseteq \widetilde{F}$ *and* $F \neq \varnothing)$, *the corresponding graph is an edge-induced subgraph* $G(S)$ *where* $S = \{(u, v, f)|f \in F \wedge (u, v, f) \in E\}$. *The subgraph* $G(S)$ *is called* **perspective graph** *of* F.

If $|F| = 1$, *the corresponding subgraph is a* **base perspective graph**. *The graph* G *is called* **top perspective graph**.

For a base perspective $f \in \widetilde{F}$, the corresponding base perspective graph is denoted by G_f. Based on definition 1 we conclude that $G = \bigcup_{f \in \widetilde{F}} G_f$.

Given a query node q and a number k, from perspective of F ($F \subseteq \widetilde{F}$ and $F \neq \varnothing$), the result of query, *the top-k similarity nodes of* q , is $T_k(q) = \{t_1, ..., t_k\}$ iif similarity score $P(q, t_i) \geq P(q, t)$ ($\forall t \in V(G(S))/T_k(q)$) on the graph $G(S)$ where $S = \{(u, v, f)|f \in F \wedge (u, v, f) \in E\}$.

Problem statement (On Perspective-Aware Top-k Similarity Search): Given a query node q and a number k, return all lists of top-k similar nodes of q from all the different perspectives.

From above statements and analyses, the number of corresponding perspective graphs is $2^m - 1$ where m is the number of base perspectives. So the size of results of the query is $2^m - 1$ from all different perspectives.

In practice, the set of base perspectives $\widetilde{F}$ is determined by domain experts.

3   Proximity Measure

A multigraph is consider as a weighted graph where weight A_{uv} is the number of edges (u, v). So similarity measures based on random walk can be defined in multigraphs which are represented by weighted graphs. RWR is same as PPR when the preference set of PPR contains merely one node q. According to the work [3], the RWR score between q and v, denoted by $r(q, v)$, is:

$$r(q, v) = \sum_{t: q \sim v} P(t)c(1 - c)^{l(t)} \tag{1}$$

where $c \in (0, 1)$ is called a constant decay factor, the summation is taken over all paths t (paths that may contain cycles) starting at q to **random walk** and ending at v, the term $P(t)$ is the probability of traveling t, and $l(t)$ is the length of path t.

Random Walks on Multigraphs: random walking on a multigraph is considered as random walking on the corresponding weighted graph where weight A_{ij} is the number of edges (i, j). Given a multigraph $G(V, E)$, A is its adjacency matrix, where A_{ij} is the number of edges (i, j) if edge $(i, j) \in E$ otherwise $A_{ij} = 0$. $d_i = \sum_i A_{ij}$ is the degree of node i on the multigraph.

Based on the work [16], the transition probability of from node i to node j is:

$$p'(i, j) = A_{ij}/d_i \quad . \tag{2}$$

So given any path $t : (w_1, w_2, ..., w_n)$, the probability of random surfer traveling t, $P(t)$, is $\prod_{i=1}^{n-1} \frac{A_{w_i w_{i+1}}}{d_{w_i}}$.

4   Naïve Method

As discussed in the previous section, each perspective graph is considered as a weighted graph. Given a query node q, the naïve method of perspective-aware top-k search consists in the computation of the similarity scores between query node and other nodes on each perspective graph respectively.

The naïve method is an inefficient method due to heavy overheads in both time and space. The time of fast executing a top-k query on a single graph is $O(n^2)$ and expensive [2]. Given a query node, we must execute the top-k query on each perspective graph and need to store the $2^m - 1$ perspective graphs adopting the method described in the work [2].

5   Top-k Algorithm

The naïve method is infeasible in practice because of heavy overheads in both time and space. So at this section we devise a concise data structure and give the bounding of the proximity to address the challenge. With aid of the data structure and bounding, we propose a new method to obtain all lists of top-k similarity nodes w.r.t a query node by merely searching the neighborhood of the query node.

5.1   Data Structure of Graph with All Perspectives Information

The goal of the data structure is: starting from a query node we traverse once the multigraph to compute RWR scores about all perspectives simultaneously, avoiding storing and traversing each perspective graph separately.

Given an edge we distinguish which perspective graph the edge belongs to. Since the size of base perspectives is m, we adopt m bits to denote the perspective graph the edge belongs to. Iff an edge (a, b) only belongs to a base perspective graph of G_{f_i}, ith bit of the bits is one and the rest is zero. If an edge (a, b) belongs to several base perspective graphs: $(a, b) \in \bigcap_{i=i_1}^{i_t} G_{f_i}$, each corresponding ith $(i = i_l, 1 \leq l \leq t)$ bit of the bits is one and the rest is zero. So edges (src, dst)

are denoted by a **triplet** $(src, dst, perspectiveFlag)$, where src and dst are nodes in the multigraph, and ***perspectiveFlag*** is the bits.

Analogously, we also adopt m bits to represent perspective graph $\bigcup_{i=i_1}^{i_t} G_{f_i}$, each corresponding ith ($i_1 \leq i \leq i_t$) bit of the bits is one and the rest value of the bits is zero. The bits is denoted by ***persIdent*** which represents corresponding perspective graph.

Therefore, given an edge (a, b, e), any perspective graph $\bigcup_{i=i_1}^{i_t} G_{f_i}$ and corresponding value of $persIdent$ is p we have:

$$
\begin{aligned}
(a, b, e) &\in \bigcup_{i=i_1}^{i_t} G_{f_i}, \text{ if } (e \text{ BITAND } p) \mathrel{!=} 0 \\
(a, b, e) &\notin \bigcup_{i=i_1}^{i_t} G_{f_i}, \text{ otherwise}
\end{aligned}
\tag{3}
$$

, where $BITAND$ is bitwise AND operator. The weight of the edge is the number of non-zero bit in $(e \text{ BITAND } p)$ on the perspective graph $\bigcup_{i=i_1}^{i_t} G_{f_i}$.

A graph G contains information of all perspective graphs when each edge of G is represented by the **triplet** format.

5.2   Bounding RWR

Using Eq.(1), we must traverse all paths which start from q and end at v to obtain the similarity. However to obtain all the paths is time consuming. At the same time, $P(t)(1 - c)^{l(t)}$ decreases exponentially with increasing of $l(t)$. This means when a random-walk path is more longer it contributes less to value of $r(p, v)$ in Eq.(1). Based on the observation, the following formula is utilized to approximate $r(q, v)$:

$$
r_d(q, v) = \sum_{\substack{t:q \sim v \\ l(t) \leq d}} P(t) c (1 - c)^{l(t)} \quad .
\tag{4}
$$

Obviously $r_d(q, v) \leq r(q, v)$ and $r(q, v) = \lim_{d \to \infty} r_d(q, v)$. It is unpractice to accurately compute $r(q, v)$. Therefor we compute $r_z(q, v)$ instead of $r(q, v)$:

$$
|r_z(q, v) - r(q, v)| < \varepsilon
\tag{5}
$$

where ε controls the accuracy of $r_z(q, v)$ in estimating $r(q, v)$, and z is the minimum value that satisfies the inequation.

We fast compute $r_{d+1}(q, v)$ from $r_d(q, v)$ by the following iteration :

$$
r_{d+1}(q, v) = r_d(q, v) + c(1 - c)^{d+1} \sum_{\substack{t:q \sim v \\ l(t) = d+1}} P(t) \quad .
\tag{6}
$$

Using Eq.(6), we efficiently compute $r_{d+1}(q, v)$ expanding one step from paths, whose length is d, when $r_d(q, v)$ has been obtained. The summation $\sum_{\substack{t:q \sim v \\ l(t) = d+1}} P(t)$ is computed by the algorithms 2 at section 5.

E.q. (6) is the lower bound of $r(q, v)$. It was shown in [17], that at dth iteration the upper bound of RWR is

$$
r_d(q, v) + \varepsilon_d
\tag{7}
$$

, where $\varepsilon_d = (1-c)^{d+1}$. The upper bound is very coarse because it is obtained in the extreme case that $\sum_{\substack{t:q\sim v \\ l(t)=i}} P(t)$, which is the probability that a surfer at q can reach v at the ith step, is 1. In most cases, $\sum_{\substack{t:q\sim v \\ l(t)=i}} P(t)$ is far less than 1. To attain more tight upper bound we assume that at the ith ($i \geq d+1$) step there is only one path along which a surfer at q can reach v and the probability of the path is estimated by a large value, which is less than 1.

Upper bound of RWR is introduced by the following proposition:

For all paths $t : (w_1, w_2, ..., w_n)$ which are obtained by breadth-first traversing from w_1, at dth iteration the maximum transition probability is $p_d = MAX\{p'(w_d, w_{d+1})\}$ where $p'(w_d, w_{d+1})$ is the transition probability from node w_d to node w_{d+1}.

Proposition 1. *At dth iteration the upper bound of RWR is:*

$$r_d(q,v) + \varepsilon_d \tag{8}$$

, where $\varepsilon_d = (1-c)^{d+1}\prod_{i=1}^{d} p_i$ and $p_i = MAX\{p'(w_i, w_{i+1})\}$.

Proof. According to Eq.(1): $r(q,v) = \sum_{t:q\sim v} P(t)c(1-c)^{l(t)} = \sum_{\substack{t:q\sim v \\ l(t)\leq d}} P(t)c(1-c)^{l(t)} + \sum_{\substack{t:q\sim v \\ l(t)\geq d+1}}^{\infty} P(t)c(1-c)^{l(t)} = r_d(q,v) + \sum_{\substack{t:q\sim v \\ l(t)\geq d+1}}^{\infty} P(t)c(1-c)^{l(t)}$. For any path $t = < w_1,...,w_{n-1}, w_n >$ which is obtained by breadth-first traversing from w_1 where $q = w_1$ and $v = w_n$, $P(t) = \prod_{i=1}^{n-1} p'(w_i, w_{i+1}) \leq \prod_{i=1}^{d} p_i \prod_{i=d+1}^{n-1} p'(w_i, w_{i+1}) \leq \prod_{i=1}^{d} p_i$ at dth iteration because the transition probability $p'(w_i, w_{i+1}) \leq 1$.

Thus at dth iteration: $\sum_{\substack{t:q\sim v \\ l(t)=d+1}}^{\infty} P(t)c(1-c)^{l(t)} \leq \sum_{\substack{t:q\sim v \\ l(t)=d+1}}^{\infty} ((\prod_{i=1}^{d} p_i)c(1-c)^{l(t)})$
$= c(\prod_{i=1}^{d} p_i)(\sum_{\substack{t:q\sim v \\ l(t)=d+1}}^{\infty} (1-c)^{l(t)}) = (1-c)^{d+1}\prod_{i=1}^{d} p_i = \varepsilon_d$ according to $\sum_{\substack{t:q\sim v \\ l(t)=d+1}}^{\infty} (1-c)^{l(t)} = \frac{(1-c)^{d+1}}{c}$. □

$r_d(q,v)$ and $r_d(q,v)+\varepsilon_d$ is lower bound and upper bound of $r(q,v)$ respectively at dth iteration. Given two nodes v and v', $r(q,v) < r(q,v')$ if $r_d(q,v) + \varepsilon_d < r_d(q,v')$. So using bounding of RWR we accelerate the top-k query in the paper.

5.3   On Perspective-Aware Top-k Similarity Search

Given a set of base perspectives $\widetilde{F} = \{f_1, f_2, ..., f_m\}$ and a multigraph $G = < V, E >$ which is represented by the data structure described in section 5.1, starting at a query node q, we do a breadth-first traverse to visit remaining nodes. At dth iteration, when a node v is visited, which perspective graphs the node belongs to is judged. Then we compute $r_d(q,v)$ and its upper value on each corresponding perspective graph. Each node v is associated with a list to store the values of $r_d(q,v)$ and its upper values.

After dth iteration, we then find a set of k nodes with the highest scores of lower bounds. Let T_k be the kth largest score on the corresponding perspective graph which the query node belongs to. We terminate the query and obtain the final result of the top-k query on the corresponding perspective graph based on following theorem:

Theorem 1. *At dth iteration, R is a set of k nodes with the highest scores r_d of lower bounds w.r.t. the query node q on any respective graph G', T_k is the kth largest r_d, and P is a set of nodes which already are visited by traversing on the G'. R is the exact theoretical top-k set w.r.t q on G' if one of following conditions is true:*

- *the value of its upper bound is less than T_k for any node $p \in P \setminus R$ and $T_k > \varepsilon_d$*
- *$\varepsilon_d \leq \varepsilon$.*

Proof. For any $p \in P \setminus R$, $r(q, p) < T_k$ and p is not in the set of the top-k nodes because its upper bound value is less than T_K. For any $v \in V(G') \setminus P$, v is still not visited so $r_d(q, v) = 0$. According to proposition 1, $r(q, v) \leq r_d(q, v) + \varepsilon_d = \varepsilon_d < T_k$ and v is not in the set of the top-k nodes. Therefore R is the result.

If $\varepsilon_d \leq \varepsilon$, according to inequality (5), $r_z(q, v)(\forall v \in P)$ is achieved and considered as final value of $r(q, v)$. While $r_z(q, v')$ ($\forall v' \in V(G') \setminus P$) is estimated as 0. So R is the result. □

Our method merely considers the neighborhood of the query node and avoids searching the whole graph.

5.4 Optimization Strategies

By relaxing terminating condition of traversing we improve the query speed at the expense of accuracy. In contrast to theorem 1, we terminate the query on the corresponding respective graph if $T_k > \varepsilon_d$ or $\varepsilon_d \leq \varepsilon$ is true. Although the new conditions do not guarantee accuracy in theory, in practice results of the query are almost accurate due to ε_d approaches 0 drastically as the increasing of d.

On the other hand, based on the general idea of the algorithm, we must sort the visited nodes for trying to obtain top-k nodes on each corresponding perspective graphs at each iteration. These would lead to overhead. So we adopt following strategy: at each iteration we merely try to obtain top-k nodes on **top perspective graph** until the corresponding T_k is greater than ε_d before we try to obtain top-k on other perspective graphs.

The bound of RWR accelerates the top-k query. We desire a more tight bound of RWR with the purpose of getting more faster response time. As discussed in subsection 5.2, for all paths $t : (w_1, w_2, ..., w_n)$ we can achieve transition probability $p'(w_i, w_{i+1})$ $(1 \leq i \leq d - 1)$ at i iteration, and we approximate the transition probability $p'(w_i, w_{i+1})$ $(d \leq i \leq n - 1)$ by average value, $\widetilde{p}$, of values $p'(w_i, w_{i+1})$ $(1 \leq i \leq d - 1)$. Then at dth iteration, the upper bound of RWR is:

$$r_d(q, v) + \varepsilon_d \tag{9}$$

, where $\varepsilon_d = \frac{c\widetilde{p}(1-c)^{d+1}}{1-\widetilde{p}(1-c)} \prod_{i=1}^{d} p_i$.

Algorithm 1. Top-k similarity queries from different perspectives

---

input : Graph g,c,v,k
output: $2^m - 1$ top-k ranking lists of v corresponding to each perspective

```
1   Set pathProb←{1.0,1.0,...,1.0};
2   push pair (v, pathProb) into queue que;
3   push (−1, pathPob) into queue que;
4   degree←g.getDegree(v);i←1;
5   for m ← 1 to sizeOfPerspectives do
6       if degree[m] ≠ 0 then
7           actualSize←actualSize + 1;
8           perspective[m]←1;

9   while obtained.size()<actualSize do
10      (currentNode, pathProb)←que.front();
11      que.pop();
12      if currentNode == −1 then
13          i←i + 1;
14          rwrScore ←calRWR(g,que,queTemp,c);
15          if obtained.has(top perspective) is false then
16              sort rwrScore[top perspective] to obtain top k nodes;
17              if its Tₖ > εᵢ or εᵢ < ε then
18                  obtained.insert(top perspective)

19          if obtained.has(top perspective) is true then
20              try to obtain top-k on other perspective graphs, and insert a identify of a graph into
                obtained if the top-k result obtained on corresponding graph ;

21          push (−1, pathProb) into que;
22          clear queTemp;
23          continue;
24      else
25          walkToNeighbors(g,currentNode,pathProb,
26          perspective,queTemp);                                              // update queTemp

27  return result;
```

5.5   The Details of the Algorithm

In this subsection we examine the details of the algorithm adopting optimization strategies.

Algorithm 1 describes the main framework of the algorithm. Starting at a query node v we do a breadth-first traversal to obtain perspective-aware top-k nodes w.r.t v on a graph G.

In the algorithm, the current visiting node is allocated a list $pathProb$ where each entry of the list is the probability of paths from the query node to the visiting node on corresponding perspective graph. In line 1, we first initialize each entry of $pathProb$ to be one. In lines 5~8 we judge which perspective graph the query node belongs to. $actualSize$ is the total size of perspective graphs the query node belongs to.

In line 12, if current node popped from queue is -1 : we call method $calRWR$ (line 16) to compute RWR scores of the nodes visited at ith iteration and update the queue, at each iteration we merely try to obtain top-k nodes on the **top perspective graph** until the corresponding T_k is greater than ε_i or $\varepsilon_i \leq \varepsilon$, then we try to obtain top-k nodes on the others perspective graphs (lines 15~20). If the result is achieved on a corresponding graph, then the identity of the graph is inserted into $obtained$.

Algorithm 2. walkToNeighbors

> **input** : Graph $g, currentNode, pathProb, perspective$
> **output**: $queTemp$
> // update $queTemp$

1 $i \leftarrow currentNode$; $degree \leftarrow g.getDegree(i)$;
2 **foreach** a neighbors j of i **do**
3   **for** $m' \leftarrow 1$ **to** $sizeOfPerspectives$ **do**
4   **if** $perspective[m'] \neq 0$ **and** $(eflag(j)$ & $m')$ **and** $pathProb[m'] \neq 0$ **then**
5              $probValue \leftarrow \frac{pathProb[m'] \times A_{ij}}{degree[m']}$;
6              $queTemp[j][m'] \leftarrow queTemp[j][m'] + probValue$;   // update $queTemp$

7 return;

If the current node is not -1 we call method *walkToNeighbors* (line 24) to visit its neighbors and calculate probability of paths from query node to the neighbors.

Algorithm 3. calRWR

> **input** : $g, que, queTemp, v, c$
> **output**: $que, rwrScore$
> // update $que, rwrScore$

1 **foreach** element i of $queTemp$ **do**
2   **foreach** element j of $queTemp[i]$ **do**
3          $rwrScore[j][i] \leftarrow rwrScore[j][i] + queTemp[i][j] \times c \times (1 - c)^{step}$;
4          $temp[j] \leftarrow queTemp[i][j]$;
5   push $(i, temp)$ into que;
6   clear temp;

7 return;

Algorithm 2 is the method *walkToNeighbors* mentioned above. The conditions (line 4) are **key factors** that we can traverse once on the graph to simultaneously compute RWR scores about all perspectives starting from the query node. The condition $perspective[m'] \neq 0$ is tested to judge whether or not the query node belongs to corresponding perspective graph whose identifier is m'. The condition $eflag(j)$ & m' refers to Eq.(3). The last condition, $pathProb[m'] \neq 0$, is true means there exits at least one path from query node to node i on the perspective graph of m'. We compute the probability of paths that start at the query node and via the current node end at its neighbors on perspective graph of m'. Then we accumulate the probability of the paths whose length is current iteration number (line 6). In a word the algorithm compute the summation $\sum_{\substack{t:q \sim v \\ l(t)=k+1}} P(t)$ of the Eq.(6) on each corresponding perspective graph the query node belongs to. And $queTemp$ contains all nodes visited at current iteration and the probability of paths from query node to those nodes.

Algorithm 3 (the method *calRWR* in the algorithm 1) compute RWR score between query node and each node of $queTemp$ on each corresponding respective graph based on Eq.(6), and then update the queue.

Time complexity of the algorithm is $max\{O(DNM), O(DN'log_2N')\}$ where D is maximum iterations, N is average number of visited nodes at each iteration, M is total number of perspective graphs which the query node belongs to, and N' is the average number of all visited nodes.

6   Experimental Study

In this section, we report our experimental studies to evaluate the effectiveness and efficiency of the proposed perspective-aware top-k query. We implemented all experiments on a PC with i3-550 CPU, 4G main memory, running windows 7 operating system. All algorithms are implemented in C++. The default values of our parameters are: c = 0.2, and $\varepsilon = 10^{-6}$. In the experiments the accurate method, which is described at section 5.3 and adopts ε_d in Eq.(7), is used to test effectiveness of our query, the method adopting ε_d in Eq.(8) and the method adopting the optimization strategies are denoted as RWR-approxity1 and RWR-approxity2 respectively .

6.1   Experimental Data Sets

Table 2. Major conferences chosen for constructing the co-authorship network

Area	Conferences
DB	SIGMOD, PVLDB, VLDB, PODS, ICDE, EDBT
DM	KDD, ICDM, SDM, PAKDD, PKDD
IR	SIGIR, WWW, CIKM, ECIR, WSDM
AI	IJCAI, AAAI, ICML, ML, CVPR, ECML

We conduct our experiments on two real-world data sets. The **DBLP**[1] Bibliography data is downloaded in September, 2012. Four research areas are considered as base perspectives: database (DB), data mining (DM), information retrieval (IR) and artificial intelligence (AI). We construct the network based on publication information from major conferences in the four research areas which are showed in table 2. The number of nodes and edges in the network is 38,412 and 110,486 respectively. The **IMDB**[2] data was extracted from the Internet Movies Data Base (IMDB). Movies contained in the data were released at the time between 1990 and 2000. We construct the network as following: we choose eight types of genres as base perspectives, including Action, Animation, Comedy, Drama, Documentary, Romance, Crime and Adventure. Assuming T is any one of the eight genres, one of the relationships of two movies is T if the two movies have same genre T and they also have a same actor/actress or a same writer or a same director at least. There are 51,532 vertices and 2,220,321 edges in the network.

---

[1] http://www.informatik.uni-trier.de/~ley/db/index.html
[2] http://www.imdb.com/interfaces

Table 3. Top-5 similar query from different perspectives on DBLP

Query author	Perspective	Top-5 authors
Jennifer Widom	IR	Robert Ikeda Semih Salihoglu,Glen Jeh Beverly Yang,Hector Garcia-Molina
	DM	Glen Jeh
	DB DM IR AI	Robert Ikeda Semih Salihoglu,Glen Jeh Hector Garcia-Molina,Jeffrey D. Ullman
	without perspectives	Hector Garcia-Molina Jeffrey D. Ullman,Shivnath Babu Robert Ikeda,Arvind Arasu
Jiawei Han	IR	Tim Weninger Xin Jin,Jiebo Luo Yizhou Sun,Ding Zhou
	DB DM IR AI	Zhenhui Li Xiaofei He,Deng Cai Jian Pei,Xifeng Yan
	without perspectives	Xifeng Yan Jian Pei,Yizhou Sun Hong Cheng,Philip S. Yu
Jim Gray	DB * * *	Alexander S. Szalay Peter Z. Kunszt,Ani Thakar Betty Salzberg,Michael Stonebraker

We also generate a series of synthetic data sets to evaluate the performance. All the top-k queries are repeated 200 times and the reported values are average values.

6.2 Effectiveness Evaluation

We evaluate the effectiveness of perspective-aware top-k query by comparing it with the traditional query. The difference between our query and the traditional query lies in structure of networks. Although the principle of their proximity measure is the same, our query contains viewpoints while traditional query does not.

In table 3 top-5 similar authors w.r.t given authors based on RWR from different perspectives are showed. And in the table the corresponding query results of *without perspectives* actually are the results obtained by the traditional query that without considering any specific viewpoint.

From perspective of DM, the similar authors w.r.t *Jennifer Widom* are *Glen Jeh*. However *Glen Jeh* is not in the top-5 candidates list that without considering any specific viewpoint (table 3). From perspective of (DB DM IR AI) the corresponding first three similar authors all collaborated with *Jennifer Widom* in two different fields. In contrast, the authors in the results of the traditional query merely collaborated with *Jennifer Widom* in DB field. The top-5 similar authors from perspective of IR are important for peoples interested in IR because the first three authors collaborate with *Jennifer Widom* in IR field whereas no one in the result of the traditional query collaborate with *Jennifer Widom* in IR field.

There is similar situation for querying *Jiawei Han* as showed in table 3. From perspective (DB DM IR AI) the corresponding first three similar authors collaborated with *Jiawei Han* in the all four field whereas the authors in the result of traditional query collaborated with *Jiawei Han* in three fields at most.

Let the order of basic perspectives is (DB DM IR AI), * means the corresponding base perspective exist or does not exist and 0 means the corresponding base perspective does not exist. From table 3 we conclude *Jim Gray* focus on the research of only one field DB because the top-5 candidate list is almost same from a group of perspectives (DB * * *).

Therefore our perspective-aware top-k query can provide more meaningful and insightful results in contrast to traditional query.

Examples mentioned above are based on only several authors, so we further evaluate the effectiveness of the perspective aware top-k query randomly choosing 200 query nodes. The two real data sets are used in this subsection. For IMDB, we choose first four types of genres as base perspectives. For the query node q, R is its top-k result on the top perspective graph.

Let $NP(q, p)$ denote the <u>N</u>umber of <u>P</u>espective relationships between q and p. For example, $NP(q, p) = 3$ means q and p co-published papers in 3 different area in DBLP. Given a metric $ANP = \frac{\sum_{p \in R} NP(q,p)}{|R|}$. We test whether results of our query reflect more perspectives information than the results of the traditional query by comparing ANP of our query on top perspective graph with ANP of the traditional query. The larger ANP is, the more perspective information our query can reflect. As illustrated in figure 7(a), our query considers more perspectives information than the traditional query does.

Given a query node q, R' $(R' \neq R)$ is its top-k result on any perspective graph. Then we evaluate whether the new query can provide rich and insightful results by the following metric: $\#N = |\{R' \mid (|R'| - |R \cap R'|) \geq \frac{1}{3}|R'|\}|$. $\#N$ is the number of perspective graphs, on which at least one third nodes in the results are different with the nodes in the results on top perspective graph for a given query node. For example showed in table 3, the nodes in query results from perspective IR almost are different with the nodes in the results on the top perspective graph for *Jiawei Han*, then we know who are most similar to *Jiawei Han* in IR field while these peoples are not contained in the results on top perspective graph. The experimental result based on 200 query nodes is showed in figure 7(b) and it verified the effectiveness of the new query.

6.3   Efficiency Evaluation

In this section, we evaluate the efficiency of our perspective-aware top-k query. First we assess the query time of our method in different situations. Then We use P@k (Precision at k) to measure the accuracy of top-k lists based on approximate mehtod by comparing it with the accurate top-k lists. At last we evaluate the efficiency of bounding of our proximities on synthetic data because bounding of RWR is adopted to accelerating speed of the query. Figure 8(a) shows the query time of the top-k query for different k values on the two real networks, where we choose the first four types of genres as base perspectives for IMDB. As illustrated in figure 8(a) approximate methods are much faster than the accuracy method,

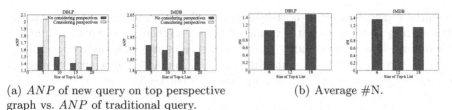

(a) *ANP* of new query on top perspective   (b) Average #N.
graph vs. *ANP* of traditional query.

Fig. 7. Effective of the new query

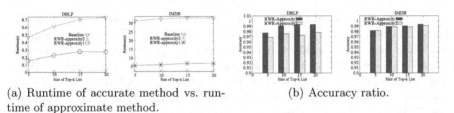

(a) Runtime of accurate method vs. run-   (b) Accuracy ratio.
time of approximate method.

Fig. 8. Efficiency of the new query

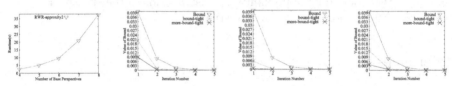

Fig. 9. Query time   **Fig. 10.** Bounds on   **Fig. 11.** Bounds on   **Fig. 12.** Bounds
vs. number of base   scale-free graph   Erdős Rényi graph   on random regular
perspectives on   graph
IMDB

while approximate method achieves a very high precision (>96.5%) as showed
in figure 8(b) .

We also test how the number of base perspective affects the runtime of the
RWR-approxity2 method on IMDB data set. As showed in figure 9 the runtime
becomes large as the number of base perspective increases. Our method is efficient for the data set with a small number of base perspectives (<8). Our future
work will focus on the top-k query when the number of base perspectives is large
(≥8).

ε_d in Eq.(8) and (9) are denoted as *bound-tight* and *more-bound-tight* respectively. The bound of PPR (Eq.(4) in [17]) is a baseline and denoted as *bound* to
compare with our bounds, where RWR is same as PPR because its preference
set of PPR is itself for a query node q.

For simplicity we evaluate the efficiency of bounding of our proximities on
simple graphs. Three type synthetic graphs that are used to test the bounds
are scale-free graph [18], Erdős Rényi graph [19] and random regular graph [19]
respectively. The average degree of the three graph is 6, 3 and 5 respectively.
And their number of nodes all is 1000. As analyzed in theorem 1, ε_d can quicken

the speed of the query if ε_d decreases drastically as iteration number increases. Figures from 10 to 12 show the results of bounds on the three synthetic graphs. The results show that bounds ε_d sharply decline as iteration number increases although the types of graphs are different. Our method is efficient because ε_d becomes very small after 3th iteration based on the results.

7 Conclusions

We have proposed a novel and practical perspective-aware top-k query in multi-relational networks. We not only achieve the most "similar" k nodes to a given query node from any specific viewpoint, but also can observe how the results change with perspectives to full understand query node and the results. With aid of the concise data structure of graphs and bounding of RWR, starting from a query node we can traverse once on the graphs and merely search the neighborhood of the query node to obtain all top-k nodes about all perspectives. Then we accelerated speed of the query by adopting several optimization strategies including tighter bounding of proximity. At last we showed the effectiveness and efficiency at the section of experimental study.

References

1. Zhao, P., Li, X., Xin, D., Han, J.: Graph cube: On warehousing and olap multidimensional networks. In: SIGMOD Conference, pp. 853–864 (2011)
2. Fujiwara, Y., Nakatsuji, M., Onizuka, M., Kitsuregawa, M.: Fast and exact top-k search for random walk with restart. PVLDB 5(5), 442–453 (2012)
3. Jeh, G., Widom, J.: Scaling personalized web search. In: WWW, pp. 271–279 (2003)
4. Jeh, G., Widom, J.: Simrank: A measure of structural-context similarity. In: KDD, pp. 538–543 (2002)
5. Sarkar, P., Moore, A.W., Prakash, A.: Fast incremental proximity search in large graphs. In: ICML, pp. 896–903 (2008)
6. Sarkar, P., Moore, A.W.: Fast nearest-neighbor search in disk-resident graphs. In: KDD, pp. 513–522 (2010)
7. Avrachenkov, K., Litvak, N., Nemirovsky, D.: Quick detection of top-k personalized pagerank lists. In: WAW, pp. 50–61 (2011)
8. Sun, Y., Han, J., Yan, X., Yu, P.S., Wu, T.: Pathsim: Meta path-based top-k similarity search in heterogeneous information networks. PVLDB 4(11), 992–1003 (2011)
9. Ahn, Y.Y., Bagrow, J.P., Lehmann, S.: Link communities reveal multiscale complexity in networks. Nature 466(7307), 761–764 (2010)
10. Cai, D., Shao, Z., He, X., Yan, X., Han, J.: Community mining from multi-relational networks. In: Jorge, A.M., Torgo, L., Brazdil, P.B., Camacho, R., Gama, J. (eds.) PKDD 2005. LNCS (LNAI), vol. 3721, pp. 445–452. Springer, Heidelberg (2005)
11. Rodriguez, M.A., Shinavier, J.: Exposing multi-relational networks to single-relational network analysis algorithms. J. Informetrics 4(1), 29–41 (2010)
12. Berlingerio, M., Coscia, M., Giannotti, F.: Finding and characterizing communities in multidimensional networks. In: ASONAM 2011, pp. 490–494 (2011)

13. Berlingerio, M., Coscia, M., Giannotti, F., Monreale, A., Pedreschi, D.: Foundations of multidimensional network analysis. In: ASONAM 2011, pp. 485–489 (2011)
14. Zhao, P., Han, J., Sun, Y.: P-rank: A comprehensive structural similarity measure over information networks. In: CIKM, pp. 553–562 (2009)
15. Chen, C., Yan, X., Zhu, F., Han, J., Yu, P.: Graph olap: A multi-dimensional framework for graph data analysis. Knowledge and Information Systems 21, 41–63 (2009)
16. Bollobás, B.: Modern Graph Theory. Springer (1998)
17. Sun, L., Cheng, R., Li, X., Cheung, D.W., Han, J.: On link-based similarity join. PVLDB 4(11), 714–725 (2011)
18. Barabasi, A.L., Albert, R.: Emergence of scaling in random networks. Science 286, 509–512 (1999)
19. Bollobás, B.: Random Graphs. Cambridge University Press (2001)

ρ-uncertainty Anonymization by Partial Suppression*

Xiao Jia[1], Chao Pan[1], Xinhui Xu[1], Kenny Q. Zhu[1], and Eric Lo[2]

[1] Shanghai Jiao Tong University, China
kzhu@cs.sjtu.edu.cn
[2] Hong Kong Polytechnic University, China

Abstract. We present a novel framework for set-valued data anonymization by partial suppression regardless of the amount of background knowledge the attacker possesses, and can be adapted to both space-time and quality-time trade-offs in a "pay-as-you-go" approach. While minimizing the number of item deletions, the framework attempts to either preserve the original data distribution or retain mineable useful association rules, which targets statistical analysis and association mining, two major data mining applications on set-valued data.

1 Introduction

Set-valued data sources are valuable in many data mining and data analysis tasks. For example, retail companies may want to know what items are top sellers (e.g., milk), or whether there is an association between the purchase of two or more items (e.g., people who buy flour also buy milk). According to our observation there are two main categories of set-valued data analysis: one is *statistical analysis* such as computing max, min and average values; the other is *mining of association rules* between items. In many cases, analysis tasks are *outsourced* to other external companies or individuals, or simply *published* to the general masses for scientific and public research purposes.

Publishing set-valued, and especially transactional data, can pose significant privacy risks. Set-valued transactions consist of one or more data items, which can be divided into two categories: *non-sensitive* and *sensitive*. Privacy is in general associated with the sensitive items. Table 1(a) shows an example of retail transactions in which each record (row) represents a set of items purchased in a single transaction by an individual. All the items are non-sensitive, except the *condom* which is sensitive. An individual's privacy is breached if he or she can be *re-identified*, or associated with a record in the data which contains one or more sensitive items. Past research has shown that such breach is possible through *linking attacks* [7], e.g. linking milk with condom in Table 1(a).

The privacy model we want to achieve is called ρ-uncertainty, where no sensitive association rules can be inferred with a confidence higher than ρ [3]. The

* Kenny Q. Zhu is the contact author and is supported by NSFC grants 61100050, 61033002, 61373031 and Google Faculty Research Award.

S.S. Bhowmick et al. (Eds.): DASFAA 2014, Part II, LNCS 8422, pp. 188–202, 2014.

Table 1. A Retail Dataset and Anonymization Results

(a) Original Dataset

TID	Transaction
1	bread, milk, *condom*
2	bread, milk
3	milk, *condom*
4	flour, fruits
5	flour, *condom*
6	bread, fruits
7	fruits, *condom*

(b) Global Suppression

TID	Transaction
1	bread, milk, ~~*condom*~~
2	bread, milk
3	milk, ~~*condom*~~
4	flour, fruits
5	flour, ~~*condom*~~
6	bread, fruits
7	fruits, ~~*condom*~~

(c) Our Approach 1

TID	Transaction
1	bread, milk, ~~*condom*~~
2	bread, milk
3	milk, *condom*
4	flour, fruits
5	~~flour~~, *condom*
6	bread, fruits
7	fruits, *condom*

(d) Our Approach 2

TID	Transaction
1	bread, milk, ~~*condom*~~
2	bread, milk
3	milk, *condom*
4	flour, fruits
5	flour, ~~*condom*~~
6	bread, fruits
7	fruits, *condom*

most popular approach to achieve ρ-uncertainty is called "global suppression" [3] in which once an occurrence of an item t is determined to be removed from one record, all occurrences of t are removed from the whole dataset. We instead opt to *partially* suppress the data set so only *some* occurrences of item t are deleted. Table 1 shows the example dataset and three anonymized datasets produced by global suppression and our approaches. The orginal dataset is not safe because sensitive rules such as $\{bread,\ milk\} \to condom$ can be inferred with confidence great than $1/3$ ($1/2$ in this case), which is our threshold. Table 1(b) is the anonymized dataset where all the occurrences of the sensitive item *condom* are deleted due to global suppression. Table 1(c) shows the anonymized dataset produced by our first approach, which is optimized to preserve data distribution, so different items (*condom* and flour) are deleted to make the dataset safe. Table 1(d) is the result of our second approach, which is optimized to preserve important data association in the original dataset, so only two occurrences of the item *condom* are deleted for safety. Even in such a small dataset, both our approaches outperform global suppression in the number of deletions (2 vs. 4) while retaining useful information at the same time.

To the best of our knowledge, the partial suppression technique has not been studied in the context of set-valued data anonymization before. We choose to solve the set-valued data anonymization problem by partial suppression because global suppression tends to delete more items than necessary, and the removal of all occurrences of the same item not only changes the data distribution significantly but also makes mining association rules about the deleted items impossible. The problem of anonymization by suppression (global or partial) is very challenging [1,18], exactly because, (i) the number of possible inferences from a given dataset is exponential, and (ii) the size of the search space, i.e. the number of ways to suppress the data is also exponential to the number of data items. We therefore propose two heuristics in this paper to anonymize input data in two different ways, giving rise to the two kinds of output in Table 1.

The main contributions of this paper are as follows.

1. To the best of our knowledge, we are the first to propose an effective *partial* suppression framework for anonymizing set-valued data (Section 2 and 3).
2. We adopt a "pay-as-you-go" approach based on divide-and-conquer, which can be adapted to achieve both space-time and quality-time trade-offs (Section 3 and 4). Our two heuristics can be adapted to either preserve data distribution or retain useful association in the data (Section 3).
3. Experiments show that our algorithm outperforms the peers in preserving the original data distribution (more than 100 times better than peers) or retaining mineable useful association rules while reducing the item deletions by large margins (Section 4).

2   Problem Definition

This section introduces the privacy model and data utility before formally describing the problem of partial suppression.

2.1   Privacy Model

$X \to Y$ is a *sensitive association rule* iff Y contains at least one sensitive item. The principle privacy model in this paper maintains that a table is *safe* iff no sensitive rules can be inferred with a confidence higher than threshold ρ [3]. It is easy to show that, if all sensitive association rules with one-item consequent from T are safe then all sensitive association rules are safe and hence T is safe.

Formally, we define quasi-identifier (also *qid*) to be any itemset (including sensitive items) drawn from any record in table T. A *qid* q is safe w.r.t. ρ iff $conf(q \to e) \le \rho$, for any sensitive item e in T. We say T is safe w.r.t. ρ iff q is safe w.r.t. ρ for any *qid* q in T. A *suppressor* is a function $S : T \mapsto T'$ where T' is a suppressed table which is safe w.r.t. ρ. There are many different ways to suppress a table. The goal is to find a suppressor that maximizes the *utility* of the suppressed table.

2.2   Data Utility

In this paper, we identify two major uses of an anonymized table: *statistical analysis* and *association rule mining*. In the first case, we want the anonymized table to have a distribution as close to the original table as possible; in the second case, we would like the anonymized data to retain all non-sensitive association rules while introducing few or no spurious rules. In both scenarios, the common goal is to minimize the *information loss*, i.e., the total number of items suppressed.

With these two scenarios in mind, we define two variants of an objective function $f(T, T')$ as:

$$f(T, T') = \begin{cases} NS(T, T') \cdot KL(T' \| T) & \text{(data distribution)} \\ \frac{NS(T, T')}{J(nr(T), nr(T'))} & \text{(rule mining)} \end{cases} \tag{1}$$

where

$$NS(T, T') = \frac{\sum_{e \in D(T)} (sup_T(e) - sup_{T'}(e))}{\sum_{e \in D(T)} sup_T(e)} \tag{2}$$

$$KL(P \| Q) = \sum_i Q(i) log \frac{Q(i)}{P(i)} \tag{3}$$

$$J(A, B) = \frac{|A \cap B|}{|A \cup B|}. \tag{4}$$

Here $D(T)$ denotes the domain of items in T, $sup_T(e)$ denotes the support of item e in T, $nr(T)$ denotes the set of all non-sensitive associations rules mineable from T with sufficient support and confidence, the functions NS, KL and J represent total number of suppressions (normalized to 1), K-L divergence[11] and Jaccard similarity[10], respectively. K-L divergence measures the distance between two probability distributions, while Jaccard similarity measures the similarity between two sets.

2.3   Optimal Partial Suppression Problem

The optimal partial supression problem is to find a *Partial Suppressor S* which anonymizes an input set-valued table T to minimize the objective function:

$$\min_S f(T, S(T))$$

such that $S(T)$ is *safe* w.r.t. to our privacy model.

3   Partial Suppression Algorithm

The Optimal Suppression Problem defined in Section 2 is an NP-hard problem. We therefore present the partial suppression algorithm as a heuristic solution to the Optimal Suppression Problem. To simplify the discussion of the algorithm, we make the following definitions.

Definition 1 (Number of Suppressions). *To disable an unsafe rule $q \to e$, the number of items of type $t \in q \cup \{e\}$ that need to be suppressed is*

$$N_s(t, q \to e) = \begin{cases} sup(q \cup \{e\}) - sup(q)\rho & t = e \\ \frac{sup(q \cup \{e\}) - sup(q)\rho}{1 - \rho} & t \in q \end{cases}$$

In other words, for each sensitive rule r, we need to delete $\min_t N_s(t, r)$ items of type t to make it safe. In this work, we select these items randomly for deletion.

Definition 2 (Leftover Items). *The leftover of item type t is defined as*

$$leftover(t) = sup_{T'}(\{t\})/sup_T(\{t\})$$

T is the original data and T' is the intermediate suppressing result. The ratio shows the percentage of remaining items of type t in the intermediate result T'.

The key intuition of our algorithm is that although the total number of "bad" sensitive association rules maybe, in the worst case, exponential in the original data, incremental "invalidation" of some of the rules through partial suppression of a small number of affected items can massively reduce the number of these bad rules, which leads to quick convergence to a solution, that is, a safe data set. Next we present the basic algorithm of this framework.

3.1   The Basic Algorithm

PARTIALSUPPRESSOR (Algorithm 1) presents the top-level algorithm. The partial suppressor iterates over the table T, and for each record $T[i]$, the algorithm first generates $qids$ from $T[i]$ and sanitizes the unsafe ones. The suppressor terminates when the whole table is scanned and there is no unsafe qid.

Algorithm 1. PARTIALSUPPRESSOR$(T, b_{\max})$

```
1:  T₀ ← T (original table)
2:  loop
3:      Initialize the sup of all qids to 0
4:      while |B| < b_max and i ≤ |T| do
5:          Fill B with qids generated by T[i]
6:          Update sup of all qids
7:          i ← i + 1
8:      end while
9:      if B contains an unsafe qids then
10:         SANITIZEBUFFER(T₀, T, B)
11:         safe ← false
12:     end if
13:     if i ≥ |T| and safe then
14:         break
15:     else if i ≥ |T| then
16:         i ← 1
17:         safe ← true
18:     end if
19: end loop
```

A qid is a combination of different items, and the number of distinct $qids$ to be enumerated is exponential. We therefore introduce a qid buffer of capacity $b_{\max}$ to balance the space consumption with the generation time. The value of

b_{max} is significant. Small b_{max} values cause repetitive generation of $qids$, while large b_{max} values cause useless generation of $qids$ which do not exist by the time to process them in the queue.

3.2   Buffer Sanitization

Each time qid buffer B is ready, SANITIZEBUFFER (Algorithm 2) is invoked to start processing $qids$ in B and make all of them safe. $D_S(T)$ denotes the domain of all sensitive items in T. We first partition $qids$ in B into two groups, *safe* and *unsafe*. Then in each iteration (Lines 2-18), SANITIZEBUFFER picks the "best" (according to heuristic functions H) unsafe sensitive association rule to sanitize (Lines 6 and 8). SUPPRESSIONPOLICY in SANITIZEBUFFER uses one of the the following two heuristic function.

Algorithm 2. SANITIZEBUFFER(T_0, T, B)

1: $\mathcal{P} \leftarrow$ SUPPRESSIONPOLICY$()$
2: **repeat**
3:     pick an unsafe qid q from B
4:     $E \leftarrow \{e \mid conf(q \rightarrow e) > \rho \wedge e \in D_S(T)\}$
5:     **if** $\mathcal{P} = Distribution$ **then**
6:         $(d, q, e) \leftarrow \underset{d \in q \cup E, q, e \in E}{\arg\max} H_{dist}(d, q, e, T_0, T)$
7:     **else if** $\mathcal{P} = Mine$ **then**
8:         $(d, q, e) \leftarrow \underset{d \in q \cup E, q, e \in E}{\arg\min} H_{mine}(d, q, e)$
9:     **end if**
10:    $X \leftarrow q \cup \{e\}$
11:    $k \leftarrow N_s(d, q \rightarrow e)$
12:    **while** $k > 0$ **do**
13:        pick a record R from T where $R \subseteq \mathcal{C}(X)$
14:        $R \leftarrow R - \{d\}$
15:        Update sup of $qids$ contained in R
16:        $k \leftarrow k - 1$
17:    **end while**
18: **until** there is no unsafe qid in B

Preservation of Data Distribution. Consider an unsafe sensitive association rule $q \rightarrow e$ where $conf(q \rightarrow e) > \rho$, and $q \in B$. To reduce $conf(q, e)$ below ρ, we suppress a number of items of type $t \in q \cup \{e\}$ from $\mathcal{C}(q \cup \{e\})$.[1] We hope to minimize $KL(T \parallel T_0)$ (see Equation (3)). From Equation (3), we observe that by suppressing some items of type t where $T(t) > T_0(t)$,[2] the KL divergence tends to decrease, thus we define the following heuristic function

$$H_{dist}(t, q, e, T_0, T) = \frac{T(t) log \frac{T(t)}{T_0(t)}}{N_s(t, q \rightarrow e)}. \tag{5}$$

---

[1] We define $\mathcal{C}(X) = \{T[i] \mid X \subseteq T[i], 1 \leq i \leq |T|\}$.
[2] We denote the probability of item type t in T as $T(t)$, which is computed by $\frac{sup_T(t)}{|T|}$.

The maximizing this function aims at suppressing item type t which maximally recovers the original data distribution and minimizes the number of deletions.

Preservation of Useful Rules. A spurious rule $(q \nrightarrow e)$ is introduced when the denominator of $conf(q \rightarrow e)$, $sup(q)$, is sufficiently small so that the confidence appears large enough. However, if $sup(q)$ is too small, the rule would not have enough support and can be ignored. Therefore, our objective is to suppress those items which have been suppressed before to minimize the support of the potential spurious rules. Therefore, we seek to minimize

$$H_{mine}(t, q, e) = leftover(t) \cdot N_s(t, q \rightarrow e)$$

3.3   Optimization with Divide-and-Conquer

When data is very large we can speed up by a divide-and-conquer (DnC) framework that partitions the input data dynamically, runs PARTIALSUPPRESSOR on them individually and combines the results in the end. This approach is correct in the sense that if each suppressed partition is safe, so is the combined data. This approach also gives rise to the parallel execution on multi-core or distributed environments which provides further speed-up (this will be shown in Section 4).

Algorithm 3. DNCSPLITDATA$(T, t_{\max})$

1: **if** $Cost(T) > t_{\max}$ **then**
2:     Split T equally into T_1 , T_2
3:     DNCSPLITDATA$(T_1, t_{\max})$
4:     DNCSPLITDATA$(T_2, t_{\max})$
5: **else**
6:     PARTIALSUPPRESSOR$(T, b_{\max})$
7: **end if**

Algorithm 3 splits the input table whenever the estimated cost of suppressing that table is greater than $t_{\max}$. Cost is estimated as:

$$Cost(T) = \frac{|T| \cdot 2^{\frac{N}{|T|}}}{|D(T)|} \tag{6}$$

where N is the total number of items in T.

4   Experimental Results

We conducted a series of experiments on 4 main datasets in Table 2. BMS-POS and BMS-WebView are introduced in [20] and are commonly used for

Table 2. Five Original Datasets

Dataset	Description	Recs	Dom. Size	Sensitive items	Non-Sens. items
BMS-POS (POS)	Point-of-sale data from a large electronics retailer	515597	1657	1183355	2183665
BMS-WebView (WV)	Click-stream data from e-commerce web site	77512	3340	137605	220673
Retail	Retail market basket data	88162	16470	340462	568114
Syn	Synthetic data with max record length = 50	493193	5000	828435	1242917

data mining. Retail is the retail market basket data [2]. Syn is the synthetic data in which each item is generated with equal probability and the max record length is 50. We randomly designate 40% of the item types in each dataset as sensitive items and the rest as non-sensitive. To evaluation the performance of rule mining, we produce four additional datasets by truncating all records in the original datasets to 5 items only, and denote such datasets as "cutoff = 5".

We compare our algorithm with the global suppression algorithm (named Global) and generalization algorithm (named TDControl) of Cao *et al.* [3].[3] Our algorithm has two variants, *Dist* and *Mine*, which optimize for data distribution and rule mining, respectively. Experiments that failed to complete in 2 hours is marked as "N/A" or an empty place in the bar charts. We run all the experiments on Linux 2.6.34) with an Intel 16-core 2.4GHz CPU and 8GB RAM.

In what follows, we first present results in data utility, then the performance of the algorithms, and then the effects of changing various parameters in our algorithm. Finally, we compare with a permutation method which utilizes a similar privacy model but with different optimization goals. Unless otherwise noted, we use the following default parameters: $b_{max} = 10^6$, $t_{max} = 500$.

4.1   Data Utility

We compare the algorithms in terms of information loss, data distribution, and association rule mining.

Figure 1 shows that *Mine* is uniformly better among the other four techniques. It suppresses only about 26% items in POS and WV and about 35% items in Retail, while the other four techniques incur on average 10% more losses than *Mine* and up to 75% losses in the worst case. We notice that *Dist* performs worse than *Global* even though it tries to minimize the information loss at each iteration. The reason is that it also tries to retain the data distribution. Further, we argue that for applications that require data statistics, the distribution, that is, summary information, is more useful than the details, hence losing some detailed information is acceptable. Note that Global and TDControl failed to complete in some datasets, because these methods don't scale very well.

---

[3] The source code of these algorithms was directly obtained from Cao.

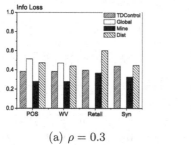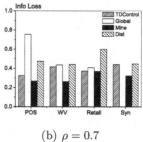

(a) $\rho = 0.3$ (b) $\rho = 0.7$

Fig. 1. Comparisons in Information Loss

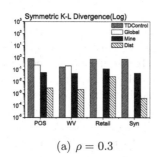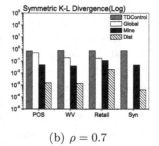

(a) $\rho = 0.3$ (b) $\rho = 0.7$

Fig. 2. Comparisons in Symmetric K-L Divergence

To determine the similarity between the item frequency distribution of original data and that of the anonymized data, we use the Kullback-Leibler divergence (also called relative entropy) as our standard. To prevent zero denominators, we modified Equation (4) to a symmetric form [6] defined as

$$\mathcal{S}(H_1 \| H_2) = \frac{1}{2} KL(H_1 \| H_1 \oplus H_2) + \frac{1}{2} KL(H_2 \| H_1 \oplus H_2)$$

where $H_1 \oplus H_2$ represents the union of distributions H_1 and H_2. Figure 2 shows that *Dist* outperforms the peers as its output has the highest resemblance to the original datasets. On the contrary, TDControl is the worst performer since generalization algorithm creates a lot of new items while suppressing too many item types globally. Since the symmetric relative entropy of *Dist* is very small, y-axis is in logrithmic scale to improve visibility. Therefore, the actual difference in K-L divergence is two or three orders of magnitude.

The most common criticism of partial suppression is that it changes the support of good rules in the data and introduces spurious rules in rule mining. In this experiment, we test the algorithms on data sets with the max record length=5 (cutoff=5), and check the rules mined from the anonymized data with support equals to 0.05% [4] and confidence equals to 70% and 30%. Figure 3 gives

---

[4] We choose this support level just to reflect a practical scenario.

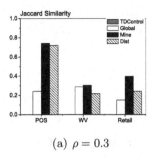

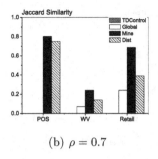

(a) $\rho = 0.3$ (b) $\rho = 0.7$

Fig. 3. Association Rules Mining with Support 0.05%

the results. Both TDControl and Global perform badly in this category, with negligible number of original rules remaining after anonymization. Conversely, all of the partial suppression algorithms manage to retain most of the rules and the Jaccard Similarity reaches 80% in some datasets which shows our heuristic works very well. Specifically, *Mine* performs the best among partial algorithms. The rules generated from TDControl are all in general form which is totally different from the original one. To enable comparison, we specialize the more general rules from the result of TDControl into rules of original level of abstraction in the generalization hierarchy. For example, we can specialize a rule {dairy product → grains} into: milk → wheat, milk → rice, yogurt → wheat, etc. Take WV as an example, there are 4 rules left in the result of TDControl when the ρ is 0.7 and the number becomes 28673 after specialization, which makes the results almost invisible.

4.2   Performance

Next we evaluate the time performance and scalability of our algorithms.

Table 3. Comparison in Time Performance ($\rho = 0.7$, $t_{max} = 300$)

Algorithm	POS	WV	Retail	Syn
TDControl	**183**	**30**	**156**	476
Global	1027	81	646	N/A
Dist	395	151	171	**130**
Mine	1554	478	256	132

From Table 3, TDControl is the clear winner for two of the four datasets. *Mine* does not perform well in BMS-POS. The reason is that *Mine* incurs the least information loss among all the competing methods. This means most of the original data remains unsuppressed. Given the large scale of BMS-POS, checking whether the dataset is safe in each iteration is therefore more time

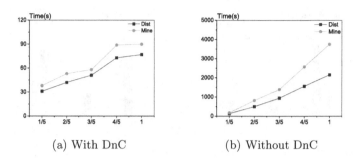

(a) With DnC   (b) Without DnC

Fig. 4. Scale-up with Input Data ($\rho = 0.7$)

consuming than other methods or in other datasets. Results for Global are not available for Syn because it runs out of memory.

Next experiment illustrates the scalability of our algorithm w.r.t. data size. We choose *Retail* as our target dataset here because *Retail* has the maximum average record length of 10.6. We run partial algorithms on 1/5, 2/5 through 5/5 of *Retail* respectively. Figure 4 shows the time cost of our algorithm increases reasonably with the input data size with or without DnC. Furthermore, increased level of partitioning causes the algorithm to witness superlinear speedup in Figure 4(a). In particular, the dataset is automatically divided into 4, 8, 16 and 32 parts at 1/5, 2/5, 3/5 and the whole of the data, respectively.

4.3   Effects of Parameters on Performance

In this section, we study the effects of t_{max}, b_{max} on the quality of solution (in terms of information loss) and time performance.

We choose *Retail* as the target dataset again since *Retail* is the most time-consuming dataset that can terminate within acceptable time without DnC strategy. The value of t_{max} determines the size of a partition in DnC. Here, we evaluate how partitioning helps with time performance and its possible effects on suppression quality. Figure 5(a) shows the relationship between partitions and information loss. The lines of *Dist* is flat, indicating that increasing t_{max} doesn't cost us the quality of the solution. *Mine* shows a slight descending tendency at first and then tends to be flat. We argue that a reasonable t_{max} will not cause our result quality to deteriorate. On the other hand, Figure 5(b) shows that time cost increases dramatically with the increase of t_{max}. The reason is that partitioning decreases the cost of enumerating qids which is the most time-consuming part in our algorithm. Moreover, parallel processing is also a major reason for the acceleration.

Next experiment (See Figure 6) illustrates the impact of varying b_{max} on performance. We choose WV as our target dataset since the number of distinct qids are relatively smaller than other datasets and our algorithm can terminate even when we set a small b_{max}.

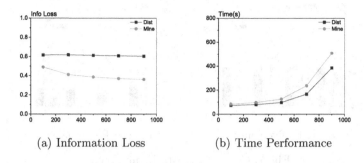

(a) Information Loss   (b) Time Performance

Fig. 5. Variation of t_{max} ($\rho = 0.7$)

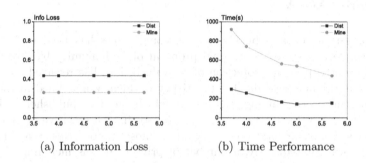

(a) Information Loss   (b) Time Performance

Fig. 6. Variation of Buffer Size b_{max} ($\rho = 0.7$)

Note first that varying b_{max} has no effect on the information loss which indicates that this parameter is purely for performance tuning. At lower values, increasing b_{max} gives almost exponential savings in running time. But as b_{max} reaches a certain point, the speedup saturates, which suggests that given the fixed size of the data, when B is large enough to accommodate all $qids$ at once after some iterations, further increase in b_{max} is not useful. The line for $Mine$ hasn't saturdated because $Mine$ suppresses fewer items and retains more $qids$, hence requires a much larger buffer.

4.4 A Comparison to Permutation Method

In this section, we compare our algorithms with a permutation method [8] which we call M. The privacy model of M states that the probability of associating any transaction $R \in T$ with any sensitive item $e \in D_S(T)$ is below $1/p$, where p is known as a privacy degree. This model is similar to ours when $\rho = 1/p$, which allows us to compare three variants of our algorithm against M where $p = 4, 6, 8, 10$ on dataset WV which was reported in [8]. Figure 7(a) shows the result on K-L divergence. All variants of our algorithm outperform M in preserving the data distribution. Figure 7(b) shows timing results. Even though M is faster, our algorithms terminate within acceptable time.

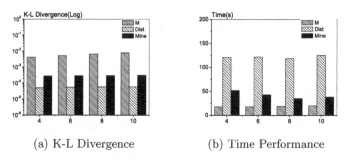

(a) K-L Divergence   (b) Time Performance

Fig. 7. Comparison with Permutation

5   Related Work

Privacy-preserving data publishing of relational tables has been well studied in the past decade since the original proposal of k-anonymity by Sweeney *et al.* [12]. Recently, privacy protection of set-valued data has received increasing interest. The original set-valued data privacy problem was defined in the context of association rule hiding [1,15,16], in which the data publisher wishes to "sanitize" the set-valued data (or *micro-data*) so that all sensitive or "bad" associate rules cannot be discovered while all (or most) "good" rules remain in the published data. Subsequently, a number of privacy models including (h, k, p)-coherence [18], k^m-anonymity [14], k-anonymity [9] and ρ-uncertainty [3] have been proposed. k^m-anonymity and k-anonymity are carried over directly from relational data privacy, while (h, k, p)-coherence and ρ-uncertainty protect the privacy by bounding the confidence and the support of any sensitive association rule inferrable from the data. This is also the privacy model this paper adopts.

A number of anonymization techniques were developed for these models. These generally fall in four categories: *global/local generalization* [14,9,3], *global suppression* [18,3], *permutation* [8] and *perturbation* [19,4]. Next we briefly discuss the pros and cons of these anonymization techniques.

Generalization replaces a specific value by a generalized value, e.g., "milk" by "dairy product", according to a generalization hierarchy [7]. While generalization preserves the correctness of the data, it compromises accuracy and preciseness. Worse still, association rule mining is impossible unless the data users have access to the same generalization taxonomy and they agree to the target level of generalization. For instance, if the users don't intend to mine rules involving "dairy products", then all generalizations to "dairy products" are useless.

Global suppression is a technique that deletes all items of some types so that the resulting dataset is safe. The advantage is that it preserves the support of existing rules that don't involve deleted items and hence retains these rules [18], and also it doesn't introduce additional/spurious association rules. The obvious disadvantage is that it can cause unnecessary information loss. In the past, partial suppression has not been attempted mainly due to its perceived side effects of changing the support of inference rules in the original data [18,3,15,16]. But our work shows that partial suppression introduces limited amount of new rules while preserving many more original ones than global suppression. Furthermore, it preserves the data distribution much better than other competing methods.

Permutation was introduced by Xiao *et al.* [17] for relational data and was extended by Ghinita *et al.* [8] for transactional data. Ghinita *et al.* propose two novel anonymization techniques for sparse high-dimensional data by introducing two representations for transactional data. However the limitation is that the quasi-identifier is restricted to contain only *non-sensitive items*, which means they only consider associations between quasi-identifier and sensitive items, and not *among* sensitive items. Manolis *et al.* [13] introduced "disassociation" which also severs the links between values attributed to the same entity but does not set a clear distinction between sensitive and non-sensitive attributes. In this paper, we consider all kinds of associations and try best to retain them.

Perturbation is developed for statistical disclosure control [7]. Common perturbation methods include *additive noise, data swapping*, and *synthetic data generation*. Their common criticism is that they damage the data integrity by adding noises and spurious values, which makes the results of downstream analysis unreliable. Perturbation, however, is useful in non-deterministic privacy model such as differential privacy [5], as attempted by Chen *et al.* [4] in a probabilistic top-down partitioning algorithm based on a context-free taxonomy.

The most relevant work to this paper is by Xu *et al.* [18] and Cao *et al.* [3]. The (h, k, p)-coherence model by Xu *et al.* requires that the attacker's prior knowledge to be no more than p public (non-sensitive) items, and any inferrable rule must be supported by at least k records while the confidence of such rules is at most $h\%$. They believe private items are essential for research and therefore only remove public items to satisfy the privacy model. They developed an efficient greedy algorithm using global suppression. In this paper, we do not restrict the size or the type of the background knowledge, and we use a partial suppression technique to achieve less information loss and also better retain the original data distribution.

Cao *et al.* [3] proposed a similar ρ-uncertainty model which is used in this paper. They developed a global suppression method and a top-down generalization-driven global suppression method (known as TDControl) to eliminate all sensitive inferences with confidence above a threshold ρ. Their methods suffer from same woes discussed earlier for generalization and global suppression. Furthermore, TD-Control assumes that data exhibits some monotonic property under a generalization hierarchy. This assumption is questionable. Experiments show that our algorithm significantly outperforms the two methods in preserving data distribution and useful inference rules, and in minimizing information losses.

6   Conclusion

We proposed a partial suppression framework including two heuristics which produce anonymized data that is highly useful to data analytics applications. Compared to previous approaches, this framework generally deletes fewer items to satisfy the the same privacy model. We showed that the first heuristic can effectively limit spurious rules while maximally preserving the useful rules mineable from the original data. The second heuristic which minimizes the K-L divergence between the anonymized data and the original data helps preserve the data distribution, which is a feature largely ignored by the privacy community in the past. Finally the divide-and-conquer strategy effectively controls the execution time with limited compromise in the solution quality.

References

1. Atallah, M., Bertino, E., Elmagarmid, A., Ibrahim, M., Verykios, V.: Disclosure limitation of sensitive rules. In: KDEX (1999)
2. Brijs, T., Swinnen, G., Vanhoof, K., Wets, G.: Using association rules for product assortment decisions: A case study. In: Knowledge Discovery and Data Mining, pp. 254–260 (1999)
3. Cao, J., Karras, P., Raïssi, C., Tan, K.-L.: ρ-uncertainty: inference-proof transaction anonymization. In: VLDB, pp. 1033–1044 (2010)
4. Chen, R., Mohammed, N., Fung, B.C.M., Desai, B.C., Xiong, L.: Publishing set-valued data via differential privacy. VLDB, 1087–1098 (2011)
5. Dwork, C.: Differential privacy: A survey of results. In: Agrawal, M., Du, D.-Z., Duan, Z., Li, A. (eds.) TAMC 2008. LNCS, vol. 4978, pp. 1–19. Springer, Heidelberg (2008)
6. Fisher, K., Walker, D., Zhu, K.Q., White, P.: From dirt to shovels: Fully automatic tool generation from ad hoc data. In: POPL, pp. 421–434 (2008)
7. Fung, B.C.M., Wang, K., Chen, R., Yu, P.S.: Privacy-preserving data publishing: A survey of recent developments. ACM Comput. Surv. (2010)
8. Ghinita, G., Kalnis, P., Tao, Y.: Anonymous publication of sensitive transactional data. TKDE, 161–174 (2011)
9. He, Y., Naughton, J.F.: Anonymization of set-valued data via top-down, local generalization. VLDB, 934–945 (2009)
10. Jaccard, P.: The distribution of the flora in the alphine zone. New Phytologist 11, 37–50 (1912)
11. Kullback, S., Leibler, R.A.: On information and sufficiency. Annals of Mathematical Statistics 21(1), 79–86 (1951)
12. Sweeney, L.: k-anonymity: A model for protecting privacy. Int. J. Uncertain. Fuzziness Knowl.-Based Syst., 557–570 (2002)
13. Terrovitis, M., Liagouris, J., Mamoulis, N., Skiadopoulos, S.: Privacy preservation by disassociation. PVLDB (2012)
14. Terrovitis, M., Mamoulis, N., Kalnis, P.: Privacy-preserving anonymization of set-valued data. VLDB, 115–125 (2008)
15. Verykios, V.S., Elmagarmid, A.K., Bertino, E., Saygin, Y., Dasseni, E.: Association rule hiding. TKDE, 434–447 (2004)
16. Wu, Y.-H., Chiang, C.-M., Chen, A.L.P.: Hiding sensitive association rules with limited side effects. TKDE, 29–42 (2007)
17. Xiao, X., Tao, Y.: Anatomy: Simple and effective privacy preservation. In: PVLDB, pp. 139–150 (2006)
18. Xu, Y., Wang, K., Fu, A.W.-C., Yu, P.S.: Anonymizing transaction databases for publication. In: KDD, pp. 767–775 (2008)
19. Zhang, Q., Koudas, N., Srivastava, D., Yu, T.: Aggregate query answering on anonymized tables. In: ICDE, pp. 116–125 (2007)
20. Zheng, Z., Kohavi, R., Mason, L.: Real world performance of association rule algorithms. In: KDD, pp. 401–406 (2001)

Access Control for Data Integration in Presence of Data Dependencies

Mehdi Haddad[1,*], Jovan Stevovic[2,3], Annamaria Chiasera[3],
Yannis Velegrakis[2], and Mohand-Saïd Hacid[4]

[1] INSA-Lyon, CNRS, LIRIS, UMR5205, F-69621, France
[2] Information Engineering and Computer Science, University of Trento, Italy
[3] Centro Ricerche GPI, Trento, Italy
[4] Université Lyon 1, CNRS, LIRIS, UMR5205, F-69622, France
{mehdi.haddad,mohand-said.hacid}@liris.cnrs.fr,
{stevovic,velgias}@disi.unitn.eu,
annamaria.chiasera@cr-gpi.it

Abstract. Defining access control policies in a data integration scenario is a challenging task. In such a scenario typically each source specifies its local access control policy and cannot anticipate data inferences that can arise when data is integrated at the mediator level. Inferences, e.g., using functional dependencies, can allow malicious users to obtain, at the mediator level, prohibited information by linking multiple queries and thus violating the local policies. In this paper, we propose a framework, *i.e.*, a methodology and a set of algorithms, to prevent such violations. First, we use a graph-based approach to identify sets of queries, called violating transactions, and then we propose an approach to forbid the execution of those transactions by identifying additional access control rules that should be added to the mediator. We also state the complexity of the algorithms and discuss a set of experiments we conducted by using both real and synthetic datasets. Tests also confirm the complexity and upper bounds in worst-case scenarios of the proposed algorithms.

1 Introduction

Data integration offers a convenient way to query different data sources while using a unique entry point that is typically called *mediator*. Although this ability to synthesize and combine information maximizes the answers provided to the user, some privacy issues could arise in such a scenario. The authorization policies governing the way data is accessed are defined by each source at local level without taking into consideration data of other sources. In relational and other systems, data constraints or hidden associations between attributes at the mediator level could be used by a malicious user to retrieve prohibited information. One type of such constraints are the *functional dependencies* (FDs). When FDs are combined with authorized information, they may allow the disclosure of

---

* This work is supported by both Rhône-Alpes Region Explora'Doc Framework and Rhhône-Alpes Region ARC6.

S.S. Bhowmick et al. (Eds.): DASFAA 2014, Part II, LNCS 8422, pp. 203–217, 2014.
© Springer International Publishing Switzerland 2014

some prohibited information. In these cases, there is a need for providing additional mechanisms at the mediator level to forbid the leakage of any prohibited information.

In this work we aim at assisting administrators in identifying such faults and defining additional access control rules at the mediator level to remedy the inference problem. Given a (relational) schema of the mediator, the sources' policies and a set of FDs, we propose a set of algorithms that are able to identify violating transactions. These transactions correspond to sets of queries that violate the sources' policies if used in conjunction with FDs. To avoid the completion of a transaction, and therefore the violation of any source's policy, we propose a query cancellation algorithm that identifies a *minimum set* of queries that need to be forbidden. The identified set of queries is then used to generate additional rules to be added to the existing set of rules of the mediator.

The reminder of the paper is organized as follows. Section 2 gives an overview of research effort in related areas. Section 3 provides definitions of the main (technical) concepts we use in the paper. Section 4 introduces a motivating scenario, the integration approach and challenges posed by functional dependencies. In Section 5 we describe our methodology. Section 6 describes the *detection phase* that identifies the policy violations. Section 7 describes the *reconfiguration phase* that deals with flaws identified in the detection phase. Section 8 describes the experiments. Finally, we conclude in Section 9.

2   Related Work

Access control in information integration is a challenging and fundamental task to be achieved [7]. In [3], the authors consider the use of metadata to model both purposes of access and user preferences. Our work does not consider these concepts explicitly, but they could be simulated by using predicates while defining the authorization rules. The authors of [14] analyzed different aspects related to access control in federated contexts [25]. They identified the role of administrators at mediator and local levels and proposed an access control model that accommodates both mediator and source policies. In [21], the authors propose an access control model based on both allow and deny rules and algorithms that check if a query containing joins can be authorized. This work, like the previous one, does not consider any association/correlation between attributes or objects that can arise at global level when joining different independent sources.

Sensitive associations happen when some attributes, when put together, lead to disclosure of prohibited information. Preventing the access to sensitive associations became crucial (see, e.g., [2,11]) in a distributed environment where each source could provide one part of it. In [2], the authors proposed a distributed architecture to ensure no association between attributes could be performed while in [11] fragmentation is used to ensure that each part of sensitive information is stored in a different fragment. In [12], the authors propose an approach to evaluate whether a query is allowed against all the authorization rules. It targets query evaluation phase while our goal is to *derive additional authorization* rules to be added to the mediator.

In [28], the authors provide an anonymization algorithm that considers FDs while identifying which portion of data needs to be anonymized. In our case, we focus on defining access control policies that should be used to avoid privacy breaches instead of applying privacy-preserving techniques [16].

Inferences allow indirect accesses that could compromise sensitive information (see [15] for a survey). A lot of work has been devoted to the inference channel in multilevel secure database. In [13] and [26] for each inference channel that is detected, either the schema of the database is modified or security level is increased. In [13], a conceptual graph based approach has been used to capture semantic relationships between entities. The authors show that this kind of relationships could lead to inference violations. In [26], the authors consider inference problem using FDs. This work does not consider authorization rules dealing with implicit association of attributes; instead, the authors assume that the user knows the mapping between the attributes of any FD. Other approaches such as [9,27] analyze queries at runtime and if a query creates an inference channel then it is rejected. In [27], both queries and authorization rules are specified using first order logic. While the inference engine considers the past queries, the functional dependencies are not taken into account. In [9], a history-based approach has been considered for the inference problem. The authors have considered two settings: the first one is related to the particular instance of the database. The second is only related to the schema of both relations and queries. In our work, we focus on inferences to identify additional access rules to be added to the mediator. In [17], we investigated how join queries could lead to authorization violations. In this current paper, we generalize the approach in [17] by considering the data inference problem.

3   Preliminaries

Before describing our approach, a number of introductory definitions are needed:

Definition 1 (Datalog rule). *[1] A (datalog) rule is an expression of the form $R_1(u_1):-R_2(u_2),...,R_n(u_n)$, where $n \geq 1$, $R_1,...,R_n$ are relation names and $u_1,...,u_n$ are free tuples of appropriate arities. Each variable occurring in u_1 must also occur in at least one of $u_2,...,u_n$.*

Definition 2 (Authorization policy). *An authorization policy is a set of authorization rules. An authorization rule is a view that describes the part of data that is prohibited to the user. An authorization rule will be expressed using an augmented datalog rule. This augmentation consists in adding a set of predicates characterizing the users to whom the authorization rule applies.*

Definition 3 (Violating Transaction). *A violating transaction T is a set of queries such that if they are executed and their results combined, they will lead to disclosure of sensitive information and thus violating the authorization policy.*

Definition 4 (Functional Dependency). *[23] A functional dependency over a schema R (or simply an FD) is a statement of the form:*

$R : X \rightarrow Y$ (or simply $X \rightarrow Y$ whenever R is understood from the context), where $X, Y \subseteq schema(R)$. We refer to X as the left hand side (LHS) and Y as the right hand side (RHS) of the functional dependency $X \rightarrow Y$.

A functional dependency $R : X \rightarrow Y$ is satisfied in a relation r over R, denoted by $r \models R : X \rightarrow Y$, iff $\forall\, t_1, t_2 \in r$ if $t_1[X] = t_2[X]$, then $t_1[Y] = t_2[Y]$.

Definition 5 (Pseudo transitivity rule). [23] The pseudo transitivity rule is an inference rule that could be derived from Armstrong rules [5]. This rule states that if $X \rightarrow Y$ and $YW \rightarrow Z$ then $XW \rightarrow Z$.

Without loss of generality we consider functional dependencies having only one attribute in their RHS. A functional dependency of the form $X \rightarrow YZ$ could always be replaced by $X \rightarrow Y$ and $X \rightarrow Z$ by using the decomposition rule [23] which is defined as follows: if $X \rightarrow YZ$, then $X \rightarrow Y$ and $X \rightarrow Z$.

4   Motivating Scenario

We consider a healthcare scenario inspired by one of our previous works while developing an Electronic Health Record (EHR) in Italy [4]. The EHR represents the mediator which provides mechanisms to share data and to enforce the appropriate authorizations [21]. From that scenario we extract an example that describes how FDs can impact access control and can be challenging to tackle at the mediator level.

Global as View Integration. We consider a data integration scenario where a Global As View (GAV) [22] approach is used to define a mediator over three sources. Particularly, we consider the sources S_1, S_2 and S_3 with the following local schemas: $S_1(SSN, Diagnosis, Doctor)$ contains the patient social security number (SSN) together with the diagnosis and the doctor in charge of her/him, $S_2(SSN, AdmissionT)$ provides the patient admission timestamp, $S_3(SSN, Service)$ provides the service to which a patient has been assigned.

The mediator virtual relation, according to the GAV integration approach, is defined by using relations of the sources. We consider a single virtual relation to simplify the scenario but the same reasoning applies for a mediator's schema composed by a set of virtual relations. In our example, the mediator will combine the data of the sources joined over the SSN attribute as shown by rule (1).

$$M(SSN, Diagnosis, Doctor, AdmissionT, Service) : - \atop S_1(SSN, Diagnosis, Doctor), S_2(SSN, AdmissionT), S_3(SSN, Service). \tag{1}$$

Authorization Policies are specified by each source on its local schema and propagated to the mediator. In our example, we assume two categories of users: doctors and nurses. For S_1, doctors can access SSN and $Diagnosis$ while nurses can access either SSN or $Diagnosis$ but not their association (i.e., simultaneously). The rule (2) expresses this policy in form of a prohibition.

$$R_1(SSN, Diagnosis) : -S_1(SSN, Diagnosis), role = nurse. \tag{2}$$

The other sources allow accessing to their content without restrictions both for doctors and nurses, therefore there are no more authorization rules to specify.

At the Mediator, authorization rules are propagated by the sources aiming at preserving their policies. The propagation can lead to policy inconsistencies and conflicts [14]. These issues are out of the scope of this paper. In our example there is only one rule defined by S_1 to be propagated at the mediator.

We then assume that at the mediator the following FDs are identified, either manually during the schema definition or by analyzing the data with algorithms such as TANE[20]:

$$(AdmissionT, Service) \rightarrow SSN \tag{F_1}$$
$$(AdmissionT, Doctor) \rightarrow Diagnosis \tag{F_2}$$

F_1 holds because at each service there is only one patient that is admitted at a given time $AdmissionT$. Note that $AdmissionT$ represents the admission timestamp including hours, minutes and seconds. F_2 holds because at a given timestamp, a doctor could make only one diagnosis.

Let see how $\mathcal{FD}$ could be used by a malicious user to violate the rule (2). Let us assume the following queries are issued by a nurse: $Q_1(SSN, AdmissionT, Service)$ and then $Q_2(Diagnosis, AdmissionT, Service)$. Combining the results of the two queries and using the functional dependency F_1, the nurse can obtain SSN and $Diagnosis$ simultaneously, which induces the violation of the authorization rule (2). To do so, the nurse could proceed as follows: (a) join the result of Q_1 with those of Q_2 on the attributes $AdmissionT$ and $Service$; (b) take advantage of F_1 to obtain the association between SSN and $Diagnosis$.

From now on, we refer to a query set like $\{Q_1, Q_2\}$ as a *violating transaction*. Indeed, both F_1 and F_2 do not hold in any source. They both use attributes provided by different sources. Thus, the semantic constraints expressed by these functional dependencies could not be considered by any source while defining its policy. This example highlights the limitation of the naïve propagation of the policies of the sources to the mediator. In the next section, we propose an intuitive approach for solving this problem.

5   Approach

We propose a methodology that aims at detecting all the possible violations that could occur at the mediator level by first identifying all the violating transactions and then disallowing completion of such violating transactions.

Our approach relies on the following settings: we consider the relational model as the reference model, both user queries and datalog expressions denoting authorization rules (see Section 3) are conjunctive queries and the mediator is defined following the GAV (Global As a View) data integration approach. This means that each virtual relation of the mediator is defined using a conjunctive query over some relations of the sources.

Currently we do not consider other types of inferences or background, external or adversarial knowledge that refer to the additional knowledge the user may

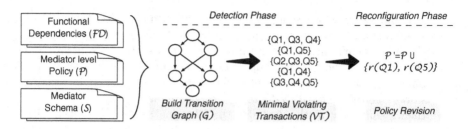

Fig. 1. The proposed methodology to identify violating transactions and define additional rules

have while querying a source of information [24,10]. These aspects are important but they are out of the scope of this paper.

The proposed methodology, as shown in Figure 1, consists of a sequence of phases and steps involving appropriate algorithms. It takes as input a set of functional dependencies ($\mathcal{FD}$), the policy ($\mathcal{P}$) and the schema ($\mathcal{S}$) of the mediator and applies the following phases:

1. **Detection phase:** aims at identifying all the violations that could occur using $\mathcal{FD}$. Each of the resulting transactions represents a potential violation. Indeed, as shown previously, the combination of all the queries of a single transaction induces an authorization violation. This phase is performed by the following steps:
   - *Construction of a transition graph ($\mathcal{G}$):* this is done for each authorization rule by using the set of provided functional dependencies ($\mathcal{FD}$).
   - *Identification of the set of Minimal[1] Violating Transactions ($\mathcal{VT}$):* it consists in identifying all the different paths between nodes in $\mathcal{G}$ to generate the set of minimal violating transactions.
2. **Reconfiguration phase:** it proposes an approach to forbid the completion of each transaction in $\mathcal{VT}$ identified in the previous phase. By completion of a transaction we mean issuing and evaluating all the queries of that transaction. A rule is violated only if the entire transaction is completed. This phase modifies/repairs the authorization policy in such a way that no $\mathcal{VT}$ could be completed.

6   Detection Phase

In the detection phase we enumerate all the violating transactions that could occur considering the authorization rules as queries that need to be forbidden. The idea is to find all the transactions (i.e., all sets of queries) that could match the query corresponding to the authorization rule.

---

[1] The concept of minimality is detailed in Section 6.2.

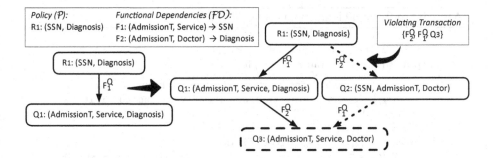

Fig. 2. Graph construction and violating transactions identification

6.1   Building the Transition Graph

The aim of the transition graph is to list all the queries that could be derived from an authorization rule using functional dependencies. For each authorization rule we use $\mathcal{FD}$ to derive a transition graph $(\mathcal{G})$ as shown in Figure 2. To build $\mathcal{G}$ we resort to Algorithm 1 as follows:

1. Consider the set of attributes of an authorization rule as the initial node.
2. For each FD in $\mathcal{FD}$ that has the RHS attribute inside the current node (starting from the root):
   (a) Create a new node by replacing the RHS attribute of the node with the set of attributes of the LHS of FD.
   (b) Create an edge between the two nodes and label it with F^Q (see Definition 6) corresponding to the FD that has been used.
3. Apply the same process for the new node.

6.2   Identifying Violating Transactions

The set of minimal violating transactions $(\mathcal{VT})$ is constructed as follows. First a path between the initial node (the node representing the authorization rule) and every other node is considered. As shown in Figure 2, from this path a transaction (*i.e.*, a set of queries) is constructed. Each query that is used as a label on this path is added to the transaction. Finally, the query of the final node of the path is also added to the transaction. This is done for all nodes and paths in $\mathcal{G}$. Before showing how minimality of the $\mathcal{VT}$ is ensured let us introduce the following definitions.

Definition 6 (Building a query from a functional dependency). *Let F be a functional dependency. We define F^Q as the query that projects on all the attributes that appear in F, either in the RHS or in the LHS. For example, let $R(A_1, A_2, A_3, A_4)$ be a relation and let F be the functional dependency $A_1, A_2 \rightarrow A_3$ that holds on R. In this case F^Q is the query that projects on all the attributes that appear in F. F^Q is the query $F^Q(A_1, A_2, A_3):-R(A_1, A_2, A_3, A_4)$.*

Algorithm 1. BuildTransitionGraph (BuildG)

input : r_i the rule $r_i \in P$,
        $\mathcal{FD}$ the set of functional dependencies.
output: $\mathcal{G}(V, E)$ the transition graph

1   $V := \{v(r_i)\};$ // create the root v with the attributes of r_i
2   $W := \{v(r_i)\};$ // add v also to a set W of vertexes to visit
3   **foreach** $w \in W$ **do**
4     $W := W - \{w\};$
5   **foreach** $FD(LHS \rightarrow RHS) \in \mathcal{FD}$ **do**
6   **if** $RHS \in w$ **then**   // RHS is one attribute
7         $x := w - \{RHS\} + LHS;$ // create new vertex
8   **if** $x \notin V$ **then**
9           $V := V + \{x\};$
10          $W := W + \{x\};$
11         $e := (w, x, LHS + \{RHS\});$ // e is a new edge from w to x
            with as transition the attributes $LHS + \{RHS\}$
12   **if** $e \notin E$ **then**   // if not already in E add it
13           $E := E + \{e\};$

14 **return** $\mathcal{G}(V, E)$;

Definition 7 (Minimal Query). *A query Q is minimal if all its attributes are relevant, that is $\forall Q' \subset Q : Q'$ cannot be used instead of Q in a violating transaction.*

Definition 8 (Minimal Violating Transaction). *A violating transaction T (see Section 3) is minimal if: (a) all its queries are minimal, and (b) all its queries are relevant i.e. $\forall Q \in T : T \setminus \{Q\}$ is not a violating transaction.*

To generate the minimal set of transactions ($\mathcal{VT}$) that is compliant with the definition 8, we use the recursive Algorithm 2. The initial call to the algorithm is: $\mathcal{VT} := FindVT(\mathcal{G}, root, \emptyset, \emptyset)$

The example in Figure 2 contains three nodes Q_1, Q_2 and Q_3 in addition to the initial node R_1. If we apply Algorithm 2, it will generate, for each node Q_i, a transaction containing each F^Q on the path between R_1 and Q_i, and Q_i itself. For example, to generate T_3 that represents the path between R_1 and Q_3, we start by adding each F_i on the path from R_1 to Q_3. Here, F_1 and F_2 are translated into F_1^Q and F_2^Q respectively. Finally, we add Q_3. Thus, we obtain $T_3 = \{F_1^Q, F_2^Q, Q_3\}$. In the example the returned $\mathcal{VT}$ is: $\mathcal{VT} = \{T_1 = \{Q_1, F_1^Q\}, T_2 = \{Q_2, F_2^Q\}, T_3 = \{Q_3, F_1^Q, F_2^Q\}\}$. At this stage we emphasized the fact that $\mathcal{FD}$ could be combined with authorized queries to obtain sensitive information. In our example, this issue is illustrated by the fact that if all the queries of any transaction T_i are issued then the authorization rule $R_1(SSN, Diagnosis)$ is violated. To cope with this problem and prohibit transaction completion, we propose an approach that

---

Algorithm 2. FindViolatingTransactions (FindVT)

---

input : $\mathcal{G}(V, E)$ the transition graph, v current vertex, c_t current path, $\mathcal{VT}$ current set of transactions.

output: $\mathcal{VT}$ the set of minimal violating transactions.

1 **foreach** $e \in$ *outgoing edges of* v **do**
2     $t := c_t + e.transition + e.to$;
    `// e.transition is the set of attributes of the transition`
      `while e.to is the destination node`
3   **if** $\nexists k \in VT \mid k \subseteq t$ **then** `//if` t `is minimal with respect to` $\forall k \in \mathcal{VT}$
4       $\mathcal{VT} := \mathcal{VT} + \{t\}$;
5   **foreach** $k \in \mathcal{VT}$ **do**
6   **if** $t \subseteq k$ **then**   `// if` k `is not minimal with respect to` t
7           $\mathcal{VT} := \mathcal{VT} - \{k\}$;
          `// reducing further` VT

8   **return** $FindVT(\mathcal{G}, e.to, c_t + e.transition, \mathcal{VT})$;
    `// recursive call with the` v `reached by` e `(e.to) by adding the`
      `e.transition to the current` VT

---

repairs the set of authorization rules with additional rules in such a way that no violation could occur.

7 Reconfiguration Phase

This phase aims at preventing a user from issuing all the queries of a violating transaction. If a user could not complete the execution of all the queries of any violating transaction then no violation could occur.

The reconfiguration phase revises the policy by adding new rules such that no violating transaction could be completed. A naïve approach could be to deny one query for each transaction. Although this naïve solution is safe from an access control point of view, it is not desired from an availability point of view. To achieve a trade off between authorization enforcement and availability, we investigate the problem of finding the minimal set of queries that denies at least one query for each violating transaction. We refer to this problem as *query cancellation problem*. We first *formalize* and characterize the *complexity* of the query cancellation problem for one rule. Then, we discuss the case of a policy (*i.e.*, a set of rules).

7.1 Problem Formalization

Let $\mathcal{VT} = \{T_1, \ldots, T_n\}$ be a set of minimal violating transactions and let $\mathcal{Q} = \{Q_1, \ldots, Q_m\}$ be a set of queries such that $\forall i \in \{1, \ldots, n\} : T_i \in \mathcal{P}(Q) \setminus \emptyset$. We define the following *Query Cancellation (QC)* recognition (decision) problem as follows:

- **Instance:** a set $\mathcal{VT}$, a set $\mathcal{Q}$ and a positive integer k.
- **Question:** is there a subset $Q \subseteq \mathcal{Q}$ with $|Q| \leq k$ such that $\forall i \in \{1, \ldots, n\}$: $T_i \smallsetminus Q \neq T_i$? Here, $|Q|$ denotes the cardinality of Q.

Algorithm 3. QueryCancellation

---

    input : $\mathcal{VT}$ is the set of minimal violating transactions.
             Q is the set of all the queries that appear in $\mathcal{VT}$
    output: S is the set of all solutions

1   **foreach** $q \in Q$ **do**
2   **if** $\forall t \in \mathcal{VT}, t \cap q \neq \emptyset$ **then**
3         $S := S \cup q$;

4   **return** S ;

Thus, the optimization problem, which consists in finding the *minimum number of queries* to be cancelled is called *Minimum Query Cancellation (MQC)*.

7.2 Problem Complexity

In this section, we show the NP-completeness of QC. We propose a reduction from the domination problem in split graphs [8]. In an undirected graph $\mathcal{G} = (V, E)$, where V is the node (vertex) set and E is the edge set, each node *dominates* all nodes joined to it by an edge (neighbors). Let $D \subseteq V$ be a subset of nodes. D is a dominating set of $\mathcal{G}$ if D dominates all nodes of $V \smallsetminus D$. The usual *Dominating Set (DS)*[8] decision problem is stated as follows:

- **Instance:** a graph G and a positive integer k.
- **Question:** does G admits a dominating set of size at most k ?

This problem has been proven to be NP-complete even for split graphs [8]. Recall that a split graph is a graph whose set of nodes is partitioned into a clique C and an independent set I. In other words, all nodes of C are joined by an edge and there is no edge between nodes of I. Edges between nodes of C and nodes of I could be arbitrary.

Theorem 1. *QC is NP-complete.*

Proof. QC belongs to NP since checking if the deletion of a subset of queries affects all transactions could be performed in polynomial time. Let G be a split graph such that C is the set of nodes forming the clique and I is the set of nodes forming the independent set. We construct an instance QC of query cancellation problem from $\mathcal{G}$ as follows: $\mathcal{Q} = C$, $\mathcal{VT} = I$ and each transaction T_i is the set of queries that are joined to it by an edge in G. We then prove that G admits a dominating set of size at most k if and only if QC admits a subset $Q \subseteq \mathcal{Q}$ of size at most k such that $\forall i \in \{1, \ldots, n\} : T_i \smallsetminus Q \neq T_i$.

Assume QC admits a subset $Q \subseteq \mathcal{Q}$ of size at most k such that $\forall i \in \{1, \ldots, n\} : T_i \smallsetminus Q \neq T_i$. Q is also a dominating set of $\mathcal{G}$. In fact, all nodes of I are dominated since all the transactions are affected by Q and all remaining nodes in the clique C are also dominated since they all are connected with nodes of Q. Assume $\mathcal{G}$ admits a dominating set D of size at most k. Observe that D could be transformed into a dominating set D' of same size and having all its nodes in C. To ensure this transformation it is sufficient to replace all nodes of D that are in I by any of their neighbors in C. Note that the obtained set D' is also a dominating set of $\mathcal{G}$. The subset of queries to be canceled is then computed by setting Q to D'.

Thus, we can deduce the following:

Corollary 1. *MQC is NP-hard.*

To generate the set of queries that need to be canceled we use Algorithm 3. It returns all the (candidate) sets of queries that have a non-empty intersection with each violating transaction. We can use different metrics to determine which set to choose. The first metric is the cardinality of the smallest set. Other metrics could be defined by the administrator. Indeed, some queries can be identified as more relevant to the application. In this case, the set of queries to be chosen could be the one that does not contain any relevant query. The minimal set of queries MQ is defined using one of the previous metrics. For each query Q in MQ a new authorization rule is added to prevent from the evaluation of Q.

In our example, the QC algorithm will return three different candidate sets of solutions to be added to $\mathcal{P}$. These sets are: $\{r(Q_1), r(F_2^Q)\}$, $\{r(Q_2), r(F_1^Q)\}$, $\{r(F_1^Q), r(F_2^Q)\}$. If we choose the first candidate set then we will have $\mathcal{P} = \{R_1(SSN, Diagnosis), R_2(AdmT, Service, Diag.), R_3(AdmT, Doctor, Diag.)\}$.

Algorithm 4. GenerelizationForPolicy

    input : $\mathcal{P}$ the set of authorization rules.
    output: $\mathcal{P}$ augmented with new rules.

1 **foreach** $r_i \in \mathcal{P}$ **do**
2     $\mathcal{G} := BuildG(r_i);$
3     $\mathcal{VT} := FindVT(\mathcal{G}, root, \emptyset, \emptyset);$
4     $\mathcal{S} := QueryCancellation(\mathcal{VT}, Q);$ // Q is obtained listing $\mathcal{VT}$
5     $N_R := \emptyset;$ // N_R is the set of new rules
6   **foreach** $q \in \mathcal{S}$ **do**
7        $N_R := N_R \cup \{r(q)\};$ // Generate a new authorization rule r
            from q
8   **if** N_R *is not empty* **then**
9        $N_R := GenerelizationForPolicy(N_R);$

10 $\mathcal{P} := \mathcal{P} \cup N_R;$
11 **return** $\mathcal{P}$;

7.3   Generalization for a Policy

Algorithm 4 deals with query cancellation for the whole policy. We denote by $\mathcal{P}$ the policy (i.e., the set of rules). We denote by N_R the set of new rules that has been generated. The new policy set $(\mathcal{P})$ will be the union of $\mathcal{P}$ and N_R $(\mathcal{P} = \mathcal{P} \cup N_R)$. A new rule could generate other new rules and so on until no rule is added. Let N_S be the set of attributes of the mediator schema. Since N_S is finite then the maximum number N_r of rules that could be defined is also finite. Let $N_{\mathcal{P}}$ be the number of rules in $\mathcal{P}$. Let n be the difference between N_r and $N_{\mathcal{P}}$. At each recursive call of the algorithm either no rule has been generated or n decreases since N_r increases. Thus, the algorithm terminates.

8   Experiments

We have conducted a number of experiments on real and synthetic datasets to validate each of the steps of our methodology. With synthetic datasets we generated particular configurations (e.g. worst-case scenarios) while with the real datasets (downloaded from the UCI ML Repository [6]) we first extracted $\mathcal{FD}$ by using a well-known algorithm called TANE [20] and then we run our algorithms with sets of rules having different number of attributes (from 2 to 10). We also tested the algorithms on specific subsets of $\mathcal{FD}$ (i.e., 100 and 200 extracted from the Bank dataset) that were not present in real datasets (Sub 1 and 2 in Table 1). The source code of the algorithms is released under GPL v3 free software licence and is available at the following address [18].

Table 1. Features data sets together with results of the experiments

Dataset desc.			Identified $\mathcal{FD}$		Performed experiments and results					
Name	$|S|$	$|\mathcal{FD}|$	$\mathcal{FD}_l$	$|\mathcal{G}(V)|$	$|\mathcal{G}(E)|$	BuildG	$|\mathcal{VT}|$	FindVT	$|P'|$
Yeast	8	10	3.88	6	10	5	5	4	7
Chess	20	22	9.14	21	20	3	20	14	21
Breast W.	11	37	4.13	41	165	26	37	65	20
Abalone	8	44	3.79	87	835	60	17	42	23
Sub 1	17	100	4.41	217	1312	193	130	197	54
Sub 2	17	200	4.92	453	8152	1502	1737	16596	263
Bank	17	433	6.47	14788	879241	3826	9137	335607	513

The reports about measures performed on each dataset shown in Table 1 are as follows:

1. **Detection Phase:** FD_l is the average number of attributes that appear in $\mathcal{FD}$, $|\mathcal{G}(V)|$ is the number of nodes and $|\mathcal{G}(E)|$ is the number of edges of the generated graph, $BuildG$ is the time in ms to build $\mathcal{G}$, $|\mathcal{VT}|$ is the number of generated $\mathcal{VT}$ and $FindVT$ is the time in ms to construct $\mathcal{VT}$.
2. **Reconfiguration Phase:** $|P'|$ is the number of rules that need to be added to the policy in order to forbid the completion of any transaction in $\mathcal{VT}$.

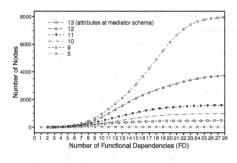

Fig. 3. Number of nodes with respect to number of FDs

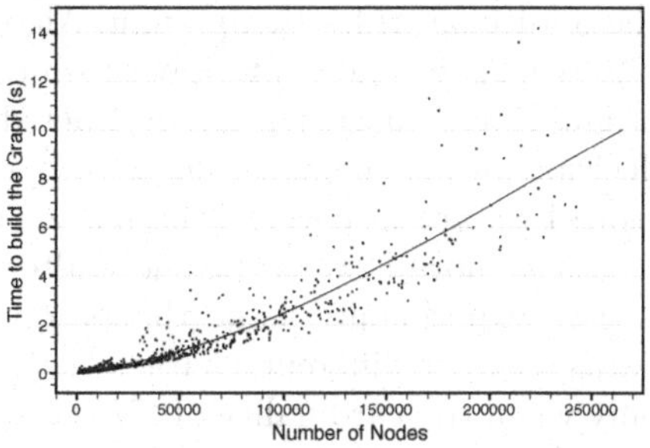

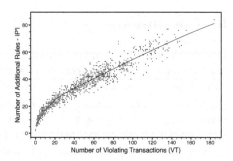

Fig. 4. Time to build the graph with respect to the number of nodes

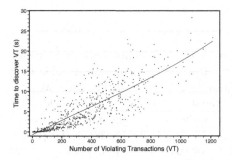

Fig. 5. Number of VTs and time to identify them

Fig. 6. Number of VTs and number of added rules

For each of the tests reported in Table 1 we calculated the mean value for 100 different executions generating rules with a number of attributes ranging from 2 to 10.

While Table 1 reports on the approach practicability on real datasets, the graphs in Figures 3, 4, 5 and 6 show tests performed on synthetic datasets. Also in this case we run multiple tests while varying parameters that are not subject to the evaluation. In particular, Figure 3 shows the relation between the number of nodes and the cardinality of randomly generated $\mathcal{FD}$. We report different tests while varying the number of attributes at the mediator schema. The tests show that by increasing the cardinality of $\mathcal{FD}$, the number of nodes increases very fast until, at a certain point, it starts slowing and approaching its upper bound as expected theoretically. Figure 4 shows the relation between the number of nodes and the time needed for building $\mathcal{G}$ with fixed attributes in the mediator schema. As we can see, the time to build $\mathcal{G}$ increases proportionally with respect to the number of nodes. This is mainly because we use binary trees to manage the nodes. The dots in figures represent single executions while the line has been generated using the Spline algorithm [19]. Figure 5 reports the performances on identifying $\mathcal{VT}$ from previously built graphs. The time grows proportionally with respect to the number of transactions. With the discovered $\mathcal{VT}$ we extract

the additional rules by applying Algorithm 4 to forbid transaction completion.
Figure 6 shows the relation between the number of transactions and the number
of additional rules that are extracted. In particular, at each cycle, we pick as
decision metric the new rule that appear more often in $\mathcal{VT}$. We observe that the
more FDs are discovered the more rules need to be added. This is due to the
fact that more FDs induce more alternatives to policy violations.

The experiments show the practicability of our methodology on different
datasets with different characteristics. The approach showed some limitation
only when the cardinality of $\mathcal{FD}$ becomes very large (e.g., greater than 1500 for
a single relation) being not able to discover transactions in an acceptable amount
of time. We believe that this amount of FDs does not represent a typical scenario.
Nevertheless, we will further investigate such situations.

9   Conclusions

In this work we have investigated the problem of illicit inferences that result
from combining semantic constraints with authorized information showing that
these inferences could lead to policy violations. To deal with this issue, we proposed an approach to detect the possible violating transactions. Each violating
transaction expresses one way to violate an authorization rule. Once the violating transactions are identified, we proposed an approach to repair the policy.
This approach aims at adding a minimal set of rules to the policy such that no
transaction could be completed.

As future work we are extending this approach to partial FDs (i.e., the FDs
that are not satisfied by all tuples but can lead to policy violations). We also plan
to investigate other kinds of semantic constraints such as inclusion dependencies and multivalued dependencies. Finally, we could consider other integration
approaches such as LAV and GLAV where same issues can arise.

References

1. Abiteboul, S., Hull, R., Vianu, V.: Foundations of databases, vol. 8. Addison-Wesley, Reading (1995)
2. Aggarwal, G., Bawa, M., Ganesan, P., Garcia-Molina, H., Kenthapadi, K., Motwani, R., Srivastava, U., Thomas, D., Xu, Y.: Two can keep a secret: A distributed architecture for secure database services. In: CIDR 2005 (2005)
3. Agrawal, R., Kiernan, J., Srikant, R., Xu, Y.: Hippocratic databases. In: VLDB 2002, pp. 143–154 (2002)
4. Armellin, G., Betti, D., Casati, F., Chiasera, A., Martinez, G., Stevovic, J.: Privacy preserving event driven integration for interoperating social and health systems. In: Jonker, W., Petković, M. (eds.) SDM 2010. LNCS, vol. 6358, pp. 54–69. Springer, Heidelberg (2010)
5. Armstrong, W.W.: Dependency structures of data base relationships. In: IFIP Congress 1974, pp. 580–583 (1974)
6. Bache, K., Lichman, M.: UCI machine learning repository (2013)
7. Bertino, E., Jajodia, S., Samarati, P.: Supporting multiple access control policies in database systems. In: IEEE Symp. on Security and Privacy, pp. 94–107 (1996)

8. Bertossi, A.A.: Dominating sets for split and bipartite graphs. Inf. Process. Lett. 19(1), 37–40 (1984)
9. Brodsky, A., Farkas, C., Jajodia, S.: Secure databases: Constraints, inference channels, and monitoring disclosures. TKDE 2000 12(6), 900–919 (2000)
10. Chen, B.-C., Le Fevre, K., Ramakrishnan, R.: Privacy skyline: Privacy with multidimensional adversarial knowledge. In: VLDB 2007, pp. 770–781 (2007)
11. Ciriani, V., De Capitani di Vimercati, S., Foresti, S., Jajodia, S., Paraboschi, S., Samarati, P.: Keep a few: Outsourcing data while maintaining confidentiality. In: Backes, M., Ning, P. (eds.) ESORICS 2009. LNCS, vol. 5789, pp. 440–455. Springer, Heidelberg (2009)
12. De di Capitani Vimercati, S., Foresti, S., Jajodia, S., Paraboschi, S., Samarati, P.: Assessing query privileges via safe and efficient permission composition. In: CCS 2008, pp. 311–322 (2008)
13. Delugach, H.S., Hinke, T.H.: Wizard: A database inference analysis and detection system. IEEE Trans. on Knowl. and Data Engineering 8(1), 56–66 (1996)
14. di Vimercati, S.D.C., Samarati, P.: Authorization specification and enforcement in federated database systems. J. Comput. Secur. 5(2), 155–188 (1997)
15. Farkas, C., Jajodia, S.: The inference problem: A survey. ACM SIGKDD Explorations Newsletter 4(2), 6–11 (2002)
16. Fung, B.C.M., Wang, K., Chen, R., Yu, P.S.: Privacy-preserving data publishing: A survey of recent developments. ACM Comput. Surv. 42(4), 14:1–14:53 (2010)
17. Haddad, M., Hacid, M.-S., Laurini, R.: Data integration in presence of authorization policies. In: IEEE 11th Int. Conference on Trust, Security and Privacy in Computing and Communications (TrustCom), pp. 92–99 (2012)
18. Haddad, M., Stevovic, J., Chiasera, A., Velegrakis, Y., Hacid, M.-S.: Access control for data integration project homepage (2013), http://disi.unitn.it/%7estevovic/acfordi.html
19. Hastie, T.J., Tibshirani, R.J.: Generalized additive models. Chapman & Hall, London (1990)
20. Huhtala, Y., Kärkkäinen, J., Porkka, P., Toivonen, H.: Tane: An efficient algorithm for discovering functional and approximate dependencies. The Computer Journal 42(2), 100–111 (1999)
21. Le, M., Kant, K., Jajodia, S.: Cooperative data access in multi-cloud environments. In: Li, Y. (ed.) DBSec. LNCS, vol. 6818, pp. 14–28. Springer, Heidelberg (2011)
22. Lenzerini, M.: Data integration: A theoretical perspective. In: Proc. of the Symp. on Principles of Database Systems, pp. 233–246 (2002)
23. Levene, M., Loizou, G.: A guided tour of relational databases and beyond. Springer (1999)
24. Martin, D.J., Kifer, D., Machanavajjhala, A., Gehrke, J., Halpern, J.Y.: Worst-case background knowledge in privacy. In: ICDE, pp. 126–135 (2007)
25. Sheth, A.P., Larson, J.A.: Federated database systems for managing distributed, heterogeneous, and autonomous databases. ACM Comput. Surv. 22(3), 183–236 (1990)
26. Su, T.-A., Özsoyoglu, G.: Data dependencies and inference control in multilevel relational database systems. In: IEEE S. on Sec. and Privacy, pp. 202–211 (1987)
27. Thuraisingham, M.: Security checking in relational database management systems augmented with inference engines. Computers & Security 6(6), 479–492 (1987)
28. Wang, H(W.), Liu, R.: Privacy-preserving publishing data with full functional dependencies. In: Kitagawa, H., Ishikawa, Y., Li, Q., Watanabe, C. (eds.) DASFAA 2010. LNCS, vol. 5982, pp. 176–183. Springer, Heidelberg (2010)

Thwarting Passive Privacy Attacks in Collaborative Filtering

Rui Chen[1], Min Xie[2], and Laks V.S. Lakshmanan[2]

[1] Department of Computer Science, Hong Kong Baptist University
ruichen@comp.hkbu.edu.hk
[2] Department of Computer Science, University of British Columbia
{minxie,laks}@cs.ubc.ca

Abstract. While recommender systems based on *collaborative filtering* have become an essential tool to help users access items of interest, it has been indicated that collaborative filtering enables an adversary to perform *passive privacy attacks*, a type of the most damaging and easy-to-perform privacy attacks. In a passive privacy attack, the dynamic nature of a recommender system allows an adversary with a moderate amount of background knowledge to infer a user's transaction through temporal changes in the *public* related-item lists (RILs). Unlike the traditional solutions that manipulate the underlying user-item rating matrix, in this paper, we respond to passive privacy attacks by directly anonymizing the RILs, which are the real outputs rendered to an adversary. This fundamental switch allows us to provide a novel rigorous inference-proof privacy guarantee, known as *δ-bound*, with desirable data utility and scalability. We propose anonymization algorithms based on suppression and a novel mechanism, *permutation*, tailored to our problem. Experiments on real-life data demonstrate that our solutions are both effective and efficient.

1 Introduction

In recent years, recommender systems have been increasingly deployed in diverse applications as an effective tool to cope with information overload. Among various approaches developed for recommender systems, *collaborative filtering* (CF) [1] is probably the most successful technique that has been widely adopted. As a standard practice, many CF systems release *related-item lists* (RILs) as a means of engaging users. For example, e-commerce service providers like *Amazon* and *Netflix* have incorporated CF as an essential component to help users find items of interest. Amazon provides RILs as the "Customers who bought this item also bought" feature, while Netflix presents RILs as the "More like" feature. These RILs serve the role of explanations of sorts, which can motivate users to take recommendations seriously.

Though successful as a means of boosting user engagement, it has been recently shown by Calandrino et al. [2] that release of RILs brings substantial privacy risks w.r.t. a fairly simple attack model, known as *passive privacy attack*. In a passive privacy attack, an adversary possesses a moderate amount of *background knowledge* in the form of a subset of items that a *target user* has bought/rated and aims to infer whether a *target item* exists in the target user's transaction. In the sequel, we use the terms *buy* and *rate* interchangeably. The adversary monitors the *public* RIL of each background

S.S. Bhowmick et al. (Eds.): DASFAA 2014, Part II, LNCS 8422, pp. 218–233, 2014.

	i1	i2	i3	i4	i5	i6	i7	i8
u1	-	2	-	5	1	-	-	-
u2	3	-	4	-	-	-	-	1
u3	1	-	-	1	3	-	-	-
u4	-	1	-	-	-	2	-	3
u5	-	3	4	-	2	5	5	5
u6	2	2	1	-	2	1	3	3
u7	-	2	-	-	2	-	-	1
u8	-	1	5	-	-	3	-	-

☐ Ratings given before time T_1
▨ Ratings given during time $(T_1, T_2]$

(a) A sample user-item rating matrix

Related-item list release at time T_1							
i1	i2	i3	i4	i5	i6	i7	i8
i3	i7	i8	i2	i8	i3	i8	i7
i5	i8	i2	i5	i7	i2	i2	i2
i8	i3	i6	i1	i2	i1	i5	i5

Related-item list release at time T_2							
i1	i2	i3	i4	i5	i6	i7	i8
i3	i8	i6	i2	i2	i8	i8	i7
i5	i7	i8	i5	i7	i7	i6	i6
i8	i6	i2	i1	i8	i3	i2	i2

(b) Public RILs at different timestamps

Fig. 1. A sample user-item rating matrix and its public RILs

item (i.e., items in the background knowledge) over a period of time. If the target item appears afresh and/or moves up in the RILs of a sufficiently large subset of the background items, the adversary infers that the target item has been added to the target user's transaction. Here is an example that illustrates the idea of passive privacy attacks.

Example 1. Consider a recommender system associated with the user-item rating matrix in Fig. 1(a). Suppose at time T_1 an attacker knows that Alice (user 5) has bought items i_2, i_3, i_7 and i_8 from their daily conversation, and is interested to learn if Alice has bought a sensitive item i_6. The adversary then monitors the temporal changes of the public RILs of i_2, i_3, i_7 and i_8. Let the new ratings made during $(T_1, T_2]$ be the shaded ones in Fig. 1(a). At time T_2, by comparing the RILs with those at T_1, the attacker observes that i_6 appears or moves up in the RILs of i_2, i_3, i_7 and i_8, and consequently infers that Alice has bought i_6. ■

Example 1 demonstrates the possibility of a passive privacy attack. In a real-world recommender system, each change in an RIL is the effect of thousands of transactions. The move-up or appearance of a target item in some background items' RILs may not even be caused by the target user. Thus, one natural question to ask is *"how likely will a passive privacy attack succeed in a real-world recommender system?"*. Calandrino et al. [2] perform a comprehensive experimental study on four real-world systems, including *Amazon, Hunch, LibraryThing* and *Last.fm*, and show that it is possible to infer a target user's unknown transaction with over 90% accuracy on Amazon, Hunch and LibraryThing and 70% accuracy on Last.fm. In particular, passive privacy attacks are able to successfully infer a third of the test users' transactions with *no error* on Hunch. This finding is astonishing as it suggests that the simple passive privacy attack model is surprisingly effective in real-world recommender systems. Therefore there is an urgent need to develop techniques for preventing passive privacy attacks.

Privacy issues in CF have been studied before. With the exception of very few works [3, 4], most proposed solutions [5–11] resort to a distributed paradigm in which user information is kept on local machines and recommendations are generated through the collaboration between a central server and client machines. While this paradigm provides promising privacy guarantees by shielding individual data from the server, it is *not* the current practice of real-world recommender systems. The distributed solution requires substantial architectural changes to existing recommender systems. Worse, the distributed setting does *not* prevent passive privacy attacks because the attacks do not require access to individual user data, but instead rely on aggregate outputs.

Unfortunately, the only works [3, 4] in the centralized setting do not address passive privacy attacks either. Polat and Du [3] suggest to add uniform noise to the user-item rating matrix. However, no formal privacy analysis is provided. In fact, we show in Section 6 that adding uniform noise does not really prevent passive privacy attacks and cannot achieve meaningful utility for RILs. McSherry and Mironov [4] ground their work on *differential privacy* [12], which is known for its rigorous privacy guarantee. They study how to construct a differentially private item covariance matrix, however they do not consider updates to the matrix, an intrinsic characteristic of recommender systems. Furthermore, we argue that differential privacy is *not* suitable for our problem because recent research [13] indicates that differential privacy does *not* provide inferential privacy, which is vital to thwart passive privacy attacks.

Our Contributions. To our best knowledge, *ours is the first remedy to passive privacy attacks in CF, a type of the most damaging and easy-to-perform privacy attacks.* Our contributions are summarized as follows.

First, we analyze the cause of passive privacy attacks, and accordingly propose a novel inference-proof privacy model called δ-*bound* to limit the probability of a successful passive privacy attack. We establish the critical condition for a user-item rating matrix to satisfy δ-bound, which enables effective algorithms for achieving δ-bound.

Second, deviating from the direction of existing studies that manipulates the underlying user-item rating matrix, we address the problem by directly anonymizing RILs. This departure is supported by the fact that, in real-life recommender systems, an adversary does *not* have access to the underlying matrix, and is critical to both data utility and scalability. We propose two anonymization algorithms, one based on suppression and the other based on a novel anonymization mechanism, *permutation*, tailored to our problem. We show that permutation provides better utility.

Third, our anonymization algorithms take into consideration the inherent dynamics of a recommender system. We propose the concept of *attack window* to model a real-world adversary. Our algorithms ensure that the released RILs are private within any attack window in that they satisfy δ-bound w.r.t. passive privacy attacks.

Finally, through an extensive empirical study on real data, we demonstrate that our approach can be seamlessly incorporated into existing recommender systems to provide formal protection against passive privacy attacks while incurring slight utility loss.

2   Related Work

Centralized Private Recommender Systems. There are very few studies on providing privacy protection in centralized recommender systems [3, 4]. Polat and Du [3] suggest users to add uniform noise to their ratings and then send the perturbed ratings to a central recommender system. However, this approach neither provides a formal privacy guarantee and nor prevents passive privacy attacks. McSherry and Mironov [4] show how to generate differentially private item covariance matrices that could be used by the leading algorithms for the Netflix Prize. However, it is *not* known how to apply their approach to a dynamic setting. In contrast, our method aims to support a dynamic recommender system. With a different goal, Machanavajjhala et al. [14] study the privacy-utility trade-offs in personalized social recommendations. The paper shows that, under differential privacy, it is *not* possible to obtain accurate social recommendations without disclosing sensitive links in a social graph in many real-world settings. These findings motivate us to define a customized privacy model for recommender systems.

Distributed Private Recommender Systems. A large body of research [5–11] resorts to distributed storage and computation of user ratings to protect individual privacy. Canny [5] addresses privacy issues in CF by cryptographic techniques. Users first construct an aggregate model of the user-item rating matrix and then use local computation to get personalized recommendations. Individual privacy is protected by multi-party secure computation. In a later paper [6], Canny proposes a new method based on a probabilistic factor analysis model to achieve better accuracy. Zhang et al. [7] indicate that adding noise with the same perturbation variance allows an adversary to derive significant amount of original information. They propose a two-way communication privacy-preserving scheme, where users perturb their ratings based on the server's guidance. Berkvosky et al. [8] assume that users are connected in a pure decentralized P2P platform and autonomously keep and maintain their ratings in a pure decentralized manner. Users have full control of when and how to expose their data using three general data obfuscation policies. Ahn and Amatriain [10] consider a variant of the traditional CF, known as *expert CF*, in which recommendations are drawn from a pool of domain experts. Li et al. [11] motivate their approach by an active privacy attack model. They propose to identify item-user interest groups and separate users' private interests from their public interests. While this method reduces the chance of privacy attacks, it fails to provide a formal privacy guarantee.

A related research area is *privacy-preserving transaction data publishing* [15] whose goal is to release anonymized transaction databases that satisfy certain privacy models. However, anonymizing the underlying database (e.g., the rating matrix) leads to undesirable data utility and scalability in our problem.

3 Preliminaries

3.1 Item-to-Item Recommendation

A common recommendation model followed by many popular websites is to provide, for every item, a list of its top-N related items, known as *item-to-item recommendation* [2]. Item-to-item recommendations take as input a user-item rating matrix M in which rows correspond to users and columns correspond to items. The set of all users form the user universe, denoted by U; the set of all items form the item universe, denoted by I. Each cell in this matrix represents a user's stated preference (e.g., ratings for movies or historical purchasing information) on an item, and its value is usually within a given range (e.g., $\{1, \cdots, 5\}$) or a special symbol "-", indicating that the preference is unknown. A sample user-item rating matrix is illustrated in Fig. 1(a).

To generate a list of related items for an item i, we calculate *item similarity scores* between i and other items. The similarity scores can be calculated based on some popular approaches, such as *Pearson correlation* and *vector cosine similarity* [16]. The *related item list* (RIL) of an item i is then generated by taking the top-N items that have the largest similarity scores. We call all RILs for all items published at a timestamp T_k an *RIL release*, denoted by $\mathcal{R}_k$. We denote a single RIL of an item j at timestamp T_k by R_k^j. Two sample RIL releases are given in Fig. 1(b).

3.2 Attack Model

In this section, we briefly review passive privacy attacks [2] in CF. In the setting of passive privacy attacks, an adversary possesses some *background knowledge* in the form of a subset of items that have been rated by a *target user*, and seeks to infer whether some other item, called a *target item*, has been rated/bought by the user, from the *public* RIL releases published by the recommender system.

As mentioned in Section 3.1, in item-to-item recommendations, for each item, the recommender system provides an RIL according to item similarity scores. Let an adversary's background knowledge on a target user u_t be B and the target item be $i_t \notin B$. The adversary monitors the changes of the RIL of each *background item* in B over time. If i_t appears afresh and/or moves up in the RILs of a sufficiently large number of background items, indicating the increased similarities between background items and i_t, the adversary might infer that i_t has been added to u_t's transaction, i.e., u_t has bought i_t, with high accuracy.

In reality, an adversary could launch passive privacy attacks by observing the temporal changes between any two RIL releases. However, it is unrealistic to assume that an adversary will perform privacy attacks over an unreasonably long timeframe (e.g., several months or even several years). Therefore, we propose the concept of *attack window* to model a real-world adversary. Without loss of generality, we assume that the RIL releases are generated at consecutive discrete timestamps and an adversary performs attacks at a particular timestamp. We note that this reflects the behavior of real-world recommender systems as RILs are indeed periodically updated. At time T_k, an adversary's attack window $\mathcal{W}_{T_k}$ contains the RIL releases generated at timestamps $T_k, T_{k-1}, \cdots, T_{k-|\mathcal{W}_{T_k}|+1}$, where $|\mathcal{W}_{T_k}|$ is the size of $\mathcal{W}_{T_k}$, namely the number of RIL releases within $\mathcal{W}_{T_k}$. The adversary performs privacy attacks by comparing any two RIL releases within his attack window.

4   Our Privacy Model

To thwart passive privacy attacks in CF, a formal notion of privacy is needed. In the context of *privacy-preserving data publishing*, where an anonymized relational database is published, a plethora of privacy models have been proposed [17]. In contrast, in our problem, recommender systems never publish anonymized rating matrices but only aggregate RILs. In this paper, we propose a novel *inference-proof* privacy notion, known as δ-*bound*, tailored for passive privacy attacks in CF. Let $\mathsf{Tran}(u)$ denote the transaction of user u, i.e., the set of items bought by u.

Definition 1. *(δ-**bound**) Let B be the background knowledge on user u in the form of a subset of items drawn from* $\mathsf{Tran}(u)$, *i.e.,* $B \subset \mathsf{Tran}(u)$. *A recommender system satisfies* δ-bound *with respect to a given attack window* $\mathcal{W}$ *if by comparing any two RIL releases* $\mathcal{R}_1$ *and* $\mathcal{R}_2$ *within* $\mathcal{W}$, $\max_{u \in U, i \in (I-B)} Pr(i \in \mathsf{Tran}(u) \mid B, \mathcal{R}_1, \mathcal{R}_2) \leq \delta$, *where*

$i \in (I - B)$ is any item that either appears afresh or moves up in $\mathcal{R}_2$, and $0 \le \delta \le 1$ is the given privacy requirement. ∎

Intuitively, the definition of δ-bound thwarts passive privacy attacks in item-to-item CF by limiting the probability of a successful attack on *any* user with *any* background items to at most δ. A smaller δ value provides more stringent privacy protection, but may lead to worse data utility. This unveils the fundamental trade-off between privacy and data utility in our problem. We will explore this trade-off in designing our anonymization algorithms in Section 5.

We now analyze the cause of passive privacy attacks and consequently derive the critical condition under which a recommender system enjoys δ-bound. The fundamental cause of passive privacy attacks is that the target user u_t's rating a target item i_t will increase the similarity scores between i_t and the background items B, which might lead to its move-up or appearance in some background items' RILs. So essentially B acts as the *quasi-identifier*, which could potentially be leveraged to identify u_t. u_t's privacy is at risk if B is possessed by only very few users. Consider an extreme example, where u_t is the only user who previously rated B. Suppose that no user who previously rated just part of B rated i_t during the time period $(T_1, T_2]$. Then observing the appearance or move-up of i_t in the RILs of B at T_2 allows the adversary to infer that u_t has rated i_t with 100% probability.

Based on this intuition, one possible way to alleviate passive privacy attacks is to require every piece of background knowledge to be shared by a sufficient number of users. However, this criterion alone is still *not* adequate to ensure δ-bound. Consider an example where, besides u_t, there are another 9 users who also rated B. Suppose, during $(T_1, T_2]$, *all* of them rated i_t. By observing the appearance or move-up of i_t in B's RILs, an adversary's probability of success is still 100%. So, to guarantee δ-bound, it is critical to limit the portion of users who are associated with the background knowledge B and also rated i_t. Let $\mathsf{Sup}(B) \ge 1$ be the number of users u associated with B (i.e., $B \subset \mathsf{Tran}(u)$) at time T_2, $\mathsf{Sup}(B \cup \{i_t\})$ be the number of users who are associated with both B and i_t at T_2. We establish the theorem below.

Theorem 1. *Consider an adversary with background knowledge B on any target user u_t. The adversary aims to infer the existence of the target item $i_t \in (I - B)$ in $\mathsf{Tran}(u_t)$ by comparing two RIL releases $\mathcal{R}_1$ and $\mathcal{R}_2$ published at time T_1 and T_2, respectively. If $\frac{\mathsf{Sup}(B \cup i_t)}{\mathsf{Sup}(B)} \le \delta$, then $Pr(i_t \in \mathsf{Tran}(u_t) \mid B, \mathcal{R}_1, \mathcal{R}_2) \le \delta$.* ∎

Theorem 1 bridges the gap between an attacker's probability of success and the underlying user-item rating matrix, and enables us to guarantee δ-bound by examining the supports in the matrix.

5 Anonymization Algorithms

Achieving δ-bound deals with privacy guarantee. Another equally important aspect of our problem is preserving utility of the RILs. For simplicity of exposition, in this paper we consider the standard utility metric *recall* [18] to measure the quality of anonymized RILs. Essentially, an anonymization algorithm results in better recall if the original RILs and the anonymized RILs contain more common items. It is straightforward to extend our algorithms to other utility metrics.

Our solution employs two *anonymization mechanisms*: *suppression*, a popular mechanism used in privacy-preserving data publishing [17], and *permutation*, a novel mechanism tailored to our problem. Suppression refers to the operation of suppressing an item from an RIL, while permutation refers to the operation of permuting an item that has moved up in an RIL to a position equal to or lower than its original position.

Before elaborating on our algorithms, we give the terminology and notations used in our solution. Recall that an RIL release at timestamp T_k is the set of RILs of all items published at T_k, denoted by $\mathcal{R}_k$. The RIL of an item j at T_k is denoted by R_k^j. Given two timestamps T_1 and T_2 with $T_1 < T_2$ (i.e., T_1 is before T_2), we say that an item i *distinguishes* between R_1^j and R_2^j if one of the following holds: 1) i appears in R_2^j but not in R_1^j, *or* 2) i appears in both R_1^j and R_2^j but its position in R_2^j is higher than its position in R_1^j (i.e., i moves up in R_2^j).

5.1 Suppression-Based Anonymization

Static Release. We start by presenting a simple case, where we are concerned with only two RIL releases (i.e., the attacker's attack window is of size 2). We refer to such a scenario as *static release*. Our goal is to make the second RIL release satisfy δ-bound w.r.t. the first release. We provide an overview of our approach in Algorithm 1.

Algorithm 1. Suppression-based anonymization algorithm for static release

---

Input: User-item rating matrix M, RIL release $\mathcal{R}_1$ at time T_1, privacy parameter δ
Output: Anonymized RIL release $\mathcal{R}_2$ at time T_2
1: Generate $\mathcal{R}_2$ from M;
2: **for** each item $i \in I$ **do**
3:     Generate the set of items S_i whose RILs are distinguished by i;
4: **for** each item $i \in I$ with $S_i \neq \emptyset$ **do**
5:     $V_i = \mathsf{GenerateViolatingBorder}(S_i, \delta, M)$
6:     $L_i = \mathsf{IdentifySuppressionLocation}(V_i)$;
7:     **for** each location $l \in L_i$ **do**
8:         $\mathsf{Suppress}(i, l, M)$;
9: **return** Suppressed $\mathcal{R}_2$;

Identify Potential Privacy Threats (Lines 2-3). Since an adversary leverages the temporal changes of the RILs to make inference attacks, the first task is to identify, for each item i, the set of items whose successive RILs at time T_1 and T_2 are distinguished by i. This set of items are referred to as *potential violating items* of i, denoted by S_i. For example, for the two RIL releases in Fig. 1(b), the set of potential violating items of i_6 is $S_{i_6} = \{i_2, i_3, i_7, i_8\}$. An adversary could use any subset of S_i as his background knowledge to infer the existence of i in a target user's transaction.

Determine Suppression Locations (Lines 5-6). Not all these potential violating items (or their combinations, i.e., itemsets) will cause actual privacy threats. Among potential violating items, we identify the itemsets where real privacy threats arise, that is, when an adversary uses the itemsets as his background knowledge, he is able to infer the target item with probability $> \delta$. We eliminate the threats by suppressing the target item from some RILs while achieving minimum utility loss. There are two major technical challenges in doing this, which Algorithm 1 addresses: 1) how to calculate a set of suppression locations s.t. the resultant utility loss is minimized (Lines 5-6); 2) how to suppress an item from an RIL without incurring new privacy threats (Line 8).

For the first challenge, we show that the problem is NP-hard (see Theorem 3 below) and provide an approximation algorithm. For a target item i_t, an adversary's background knowledge could be *any* subset of S_{i_t}. Therefore, we have to guarantee that the probability of inferring the presence of i_t in *any* target user's transaction from *any* itemset $B \subseteq S_{i_t}$, viewed as background knowledge on the target user, is $\leq \delta$. We refer to this probability as the *breach probability* associated with the background knowledge (i.e., itemset) B. We point out that the problem structure does not satisfy any natural monotonicity: indeed, the breach probability associated with an itemset may be more or less than that of its superset. Thus, in the worst case, we must check the breach probability for every itemset (except the empty set) of S_{i_t}, which has exponential complexity. Note that *every* item $i \in I$ could be a target item.

To help tame the complexity of checking all subsets of S_{i_t}, where i_t is *any* candidate target item, we develop a pruning scheme. Define an itemset $s \subset S_{i_t}$ to be a *minimal violating itemset* provided s has a breach probability $> \delta$ and every proper subset of s has a breach probability $\leq \delta$. Let V_{i_t} be the *violating border* of i_t, consisting of *all* minimal violating itemsets of i_t. By definition of minimality, to thwart the privacy attacks on V_{i_t}, it is enough to suppress i_t from the RIL of one item in v, for every minimal violating itemset $v \in V_{i_t}$. The reason is that, for any $v \in V_{i_t}$, no proper subset of v can be used to succeed in an attack. We next show that it is sufficient to guarantee δ-bound on all itemsets in S_{i_t} by ensuring δ-bound on V_{i_t}.

Theorem 2. *For two RIL releases $\mathcal{R}_1$ and $\mathcal{R}_2$, a target user u_t and a target item i_t, $\forall v \in V_{i_t}$, suppressing i_t from the RIL of one item in v ensures $\forall s \subseteq S_{i_t}, Pr(i_t \in \mathsf{Tran}(u_t)|s, \mathcal{R}_1, \mathcal{R}_2) \leq \delta$.* ∎

The general idea of GenerateViolatingBorder (Line 5) is that if an itemset violates δ-bound, then there is no need to further examine its supersets. We impose an arbitrary total order on the items in S_i to ensure that each itemset will be checked exactly once. We iteratively process the itemsets with increasing sizes. The minimal violating itemsets with size k come from a candidate set generated by *joining* non-violating itemsets of size $k-1$. Two non-violating itemsets, $c_1 = \{i_1^1, i_2^1, \cdots, i_l^1\}$ and $c_2 = \{i_1^2, i_2^2, \cdots, i_l^2\}$, can be joined if for all $1 \leq m \leq l-1, i_m^1 = i_m^2$ and $\mathsf{Order}(i_l^1) > \mathsf{Order}(i_l^2)$. The joined result is $c_1 \bowtie c_2 = \{i_1^1, i_2^1, \cdots, i_l^1, i_l^2\}$.

For a target item i_t whose potential violating items do not cause any privacy threat, we still need to consider all $2^{|S_{i_t}|} - 1$ itemsets before concluding that there is no threat. To alleviate the computational cost of these items, we make use of a simple pruning strategy. Let the number of users who rated i_t at time T_2 be $\mathsf{Sup}(i_t)$, the number of users rated S_{i_t} at T_2 be $\mathsf{Sup}(S_{i_t})$, and the number of users who rated both S_{i_t} and i_t at T_2 be $\mathsf{Sup}(S_{i_t} \cup \{i_t\})$. Since $\frac{\mathsf{Sup}(S_{i_t} \cup \{i_t\})}{\mathsf{Sup}(S_{i_t})} \leq \frac{\mathsf{Sup}(i_t)}{\mathsf{Sup}(S_{i_t})}$, to guarantee that the breach probability $\frac{\mathsf{Sup}(S_{i_t} \cup \{i_t\})}{\mathsf{Sup}(S_{i_t})} \leq \delta$, it is enough to ensure that $\mathsf{Sup}(S_{i_t}) \geq \frac{\mathsf{Sup}(i_t)}{\delta}$. Notice that, for any subset $s \subset S_{i_t}$, $\frac{\mathsf{Sup}(s \cup \{i_t\})}{\mathsf{Sup}(s)} \leq \frac{\mathsf{Sup}(i_t)}{\mathsf{Sup}(s)} \leq \frac{\mathsf{Sup}(i_t)}{\mathsf{Sup}(S_{i_t})} \leq \delta$. Thus, there is no need to make any checks for subsets of S_{i_t}.

Achieving δ-bound on V_{i_t} requires to find a set of items from whose RILs we suppress the target item i_t, so that after the suppression, for each *potential* background knowledge (i.e., itemset) B, either the breach probability associated with i_t is $\leq \delta$ or i_t does not distinguish the successive RILs of at least one item in B. From a *recall* point of view, we would like to minimize the number of items to be suppressed, since each item suppression leads to a utility loss of 1. This problem can be defined as follows.

Definition 2. *(IdentifySuppressionLocation) Given the violating border V_{i_t}, select a set of items L_{i_t} such that $\forall v \in V_{i_t}(\exists l \in L_{i_t}(l \in v))$ and $|L_{i_t}|$ is minimized.* ■

The problem is identical to the *minimal hitting set* (MHS) problem [19], and therefore we have the following theorem.

Theorem 3. IdentifySuppressionLocation *is NP-hard. There is an $O(\ln |V_{i_t}|)$-approximation algorithm to the optimal solution, which runs in $O(|V_{i_t}||I|)$ time.* ■

Algorithm 2 shows a simple greedy algorithm, which repeatedly picks the item that belongs to the maximum number of "uncovered" itemsets in V_{i_t}, where an itemset is said to be "covered" if one of the items in the current hitting set belongs to it.

Algorithm 2. IdentifySuppressionLocation

Input: The violating border V_i of item i
Output: A set of locations (items) to suppress L_i
1: $L_i = \emptyset$;
2: $C \leftarrow$ the set of items in V_i;
3: **while** $V_i \neq \emptyset$ **do**
4:     **for** each item $j \in C$ **do**
5:         $n_j = |\{v \in V_i : j \in v\}|$;
6:     Add the item j with the maximum n_j to L_i;
7:     $V_i = V_i - \{v \in V_i : j \in v\}$
8:     $C \leftarrow$ the set of items in V_i;
9: **return** L_i;

Perform Suppression (Algorithm 1, Line 8). To thwart privacy attacks, we suppress the target item i_t from the RILs of the items identified by Algorithm 2. Suppressing i_t from a RIL will make items with a position lower than $\mathsf{Pos}(i_t)$ (i.e., the position of i_t in the RIL) move up one position and introduce a new item into the RIL. Note that the move-up or appearance of these items might cause many *new* privacy threats, resulting in both higher complexity and lower utility. To alleviate this problem, instead of changing the positions of all items below $\mathsf{Pos}(i_t)$ and introducing a new item to the RIL, we directly insert a new item at $\mathsf{Pos}(i_t)$ and check its breach probability.

Even inserting a new item i directly at $\mathsf{Pos}(i_t)$ in j's RIL might lead to substantial computational cost, because, in the worst case, it demands to examine every possible combination of the itemsets derived from S_i with j, which is of $O(2^{|S_i|})$ complexity. So we are only interested in items with $S_i = \emptyset$. In this case, we can perform the check in constant time. More specifically, we iteratively consider the items not in the RIL in the descending order of their similarity scores until we find an item to be inserted at $\text{Pos}(i_t)$ without incurring new privacy threats. If an item i not in the RIL has $S_i \neq \emptyset$, we skip i and consider the next item. This process terminates when a qualified item is found. When i is inserted into j's RIL, S_i is accordingly updated: $S_i = \{j\}$.

Multiple Release. We next deal with the case of multiple releases. As discussed in Section 3.2, at any time T_k, an adversary performs passive privacy attacks by comparing any two RIL releases within the attack window $\mathcal{W}_{T_k}$. Hence, whenever a recommender system generates a new RIL release, it has to be secured with respect to all previous $|\mathcal{W}_{T_k}| - 1$ releases. We assume that the attack window size of an adversary is fixed at different timestamps. In reality, this assumption can be satisfied by setting a large enough window size.

We explain the key idea of extending Algorithm 1 for this case. Anonymizing the RIL release $\mathcal{R}_k$ at time T_k works as follows. First, we should generate the potential violating items of *every* item i in $\mathcal{R}_k$ with respect to each of $\mathcal{R}_{k-1}, \mathcal{R}_{k-2}, \cdots, \mathcal{R}_{k-|\mathcal{W}_{T_k}|+1}$. Let $S_i^{\mathcal{R}_j}$ be the potential violating items of i generated by comparing $\mathcal{R}_k$ and $\mathcal{R}_j$, where $k - |\mathcal{W}_{T_k}| + 1 \leq j \leq k - 1$. We calculate the violating border over each $S_i^{\mathcal{R}_j}$, denoted by $V_i^{\mathcal{R}_j}$. To make $\mathcal{R}_k$ private for the entire attack window, we need to eliminate all itemsets from these $|\mathcal{W}_{T_k}| - 1$ borders. We take the union of all the borders $V_i = V_i^{\mathcal{R}_{k-1}} \cup \cdots \cup V_i^{\mathcal{R}_{k-|\mathcal{W}|+1}}$. We prune all itemsets that are the supersets of an itemset in V_i, i.e., retain only minimal sets in V_i. The rationale of this pruning step is similar to that of Theorem 2. Second, when we bring in a new item to an RIL, its breach probability needs to be checked with respect to *each* of the previous $|\mathcal{W}_{T_k}| - 1$ releases.

5.2   Permutation-Based Anonymization

In the suppression-based solution, we do not distinguish between an item's appearance and move-up. For items that newly appear in an RIL, we have to suppress them. However, for items that move up in an RIL, we do not really need to suppress them from the RIL to thwart passive privacy attacks. To further improve data utility, we introduce a novel anonymization mechanism tailored to our problem, namely *permutation*. The general idea of permutation is to permute the target item to a lower position in the RIL so that it cannot be used by an adversary to perform a passive privacy attack. If we cannot find a position to permute without generating new privacy threats, we suppress the target item from the RIL. So our permutation-based anonymization algorithm employs both permutation and suppression, but prefers permutation whenever a privacy threat can be eliminated by permutation.

Static Release. Similarly, we first generate the potential violating items S_i for each item i. Unlike in the suppression-based method, we label each item in S_i with either *suppress* or *permute*. If an item gets into S_i due to its appearance in an RIL, it is labeled *suppress*; otherwise it is labeled *permute*. For example, in Fig. 1, we label the occurrences of i_6 in the RILs of i_2, i_7 and i_8 with *suppress* and its occurrence in i_3's RIL with *permute*.

The violating border of S_i can be calculated by the GenerateViolatingBorder procedure described in Section 5.1. For *recall*, it can be observed that permutation does not incur any utility loss. For this reason, we take into consideration the fact that *suppress* and *permute* are associated with different utility loss when identifying items to anonymize. We call this new procedure IdentifyAnonymizationLocation. We model IdentifyAnonymizationLocation as a *weighted minimum hitting set* (WMHS) problem. IdentifyAnonymizationLocation chooses at every step the item that maximizes the score, namely the ratio between the number of uncovered itemsets containing it and its weight. The weight of an item is calculated based on its utility loss. For an item labeled *suppress*, its weight is 1. For an item labeled *permute*, it does not result in any utility loss and should receive a weight value 0. However, this leads to a divide-by-zero problem. Instead, we assign the item a sufficiently small weight value $\frac{1}{|V_i|+1}$. This is sufficient to guarantee that items labeled *permute* are always preferred over items labeled *suppress*, because the maximum score of an item labeled *suppress* is $|V_i|$ while the minimum score of an item labeled *permute* is $|V_i| + 1$.

To tackle the anonymization locations identified by IdentifyAnonymizationLocation, we start by suppressing items labeled *suppress* because these privacy threats cannot be solved by permutation. Similar to the Suppress procedure described in Section 5.1, we look for the first item i outside an RIL with $S_i = \emptyset$, which does not incur any new privacy threat, as a candidate to replace the suppressed item. One exception is that in the permutation-based solution we can stop searching once we reach the first item that was in the previous RIL (for this type of items there is *no* need to check their breach probability because our following steps make sure that they cannot be used in passive privacy attacks, as is shown later). For the moment, we do not assign a particular position for i and wait for the permutation step. After suppressing all items labeled *suppress*, we perform permutation on the RILs that contain locations returned by IdentifyAnonymizationLocation. In an RIL, for all items that were also in the RIL at the previous timestamp T_1, we assign them the same positions as those at T_1; for all items that were not in the RIL at T_1, we randomly assign them to one of the remaining positions.

We next show the correctness of our permutation-based solution. For an item that needs to be suppressed, it is replaced by a new item, whose appearance is examined to be free of privacy threats, and thus randomly assigning a position does not violate the privacy requirement. For an item that needs to be permuted, we freeze its position to be the same as before, i.e., as in the previous RIL release, and therefore it cannot be used by the adversary to perform passive privacy attacks. So the anonymized RILs are resistant to passive privacy attacks.

Multiple Release. Finally, we explain our permutation-based algorithm for the *multiple release* scenario. Algorithm 3 presents our idea in detail. We compare the true $\mathcal{R}_k$ at time T_k with each of the previous $|\mathcal{W}_{T_k}| - 1$ RIL releases within the attack window $\mathcal{W}_{T_k}$ to generate the corresponding potential violating items for each item i, denoted by $S_i^{\mathcal{R}_j}$ (Lines 2-4). In addition to labeling each potential violating item by *suppress* or *permute*, for an item i labeled *permute*, we record its position in the RIL in which it moves up (Line 5).

For each $S_i^{\mathcal{R}_j}$, we calculate its violating border $V_i^{\mathcal{R}_j}$ (Line 6). Since we have to make $\mathcal{R}_k$ private with respect to all previous $|\mathcal{W}_{T_k}| - 1$ releases, we perform a union

---

Algorithm 3. Permutation-based anonymization algorithm for multiple release

Input: User-item rating matrix M, the attack window $\mathcal{W}_{T_k}$ at time T_k, previous $|\mathcal{W}_{T_k}| - 1$ RIL releases, privacy parameter δ

Output: Anonymized RIL release $\mathcal{R}_k$ at time T_k

  1: Generate $\mathcal{R}_k$ from M;
  2: **for** each previous RIL release $\mathcal{R}_j$ **do**
  3:     **for** each item $i \in I$ **do**
  4:         Generate the set of items $S_i^{\mathcal{R}_j}$ whose RILs are distinguished by i between $\mathcal{R}_k$ and $\mathcal{R}_j$;
  5:         Label items in $S_i^{\mathcal{R}_j}$ by $suppress$ or $permute$ and record $permute$ position;
  6:         $V_i^{\mathcal{R}_j} = \mathsf{GenerateViolatingBorder}(S_i^{\mathcal{R}_j}, \delta, M)$;
  7: **for** each item $i \in I$ **do**
  8:     $V_i = V_i^{\mathcal{R}_{k-1}} \cup \cdots \cup V_i^{\mathcal{R}_{k-|\mathcal{W}_{T_k}|+1}}$;
  9:     $V_i = \mathsf{Label}(V_i)$;
10:     $V_i = \mathsf{Prune}(V_i)$;
11:     **for** each item $i \in I$ with $V_i \neq \emptyset$ **do**
12:         $\langle L_i, C_i \rangle = \mathsf{IdentifyAnonymizationLocation}(V_i)$;
13:         **for** each location-code pair $\langle l, c \rangle \in \langle L_i, C_i \rangle$ **do**
14:             **if** $c = suppress$ **then**
15:                 $\mathsf{SuppressMR}(i, l, M)$;
16:             **else**
17:                 $\mathsf{PermuteMR}(i, l, M)$;
18: **return** Anonymized $\mathcal{R}_k$;

---

over all $V_i^{\mathcal{R}_j}$ (Line 8). In the case of multiple release, the same item might be labeled both *suppress* and *permute* in different $V_i^{\mathcal{R}_j}$ and by different positions. To resolve this inconsistency, we let *suppress* take precedence over *permutation*. That is, if an item i is labeled *suppress* in any $V_i^{\mathcal{R}_j}$, it will be labeled *suppress* in V_i (Line 9), because a new item's entering in an RIL cannot be hidden by permuting its position. Also, the position associated with an item labeled *permute* is updated to the lowest position of all its positions in different $V_i^{\mathcal{R}_j}$. We call this lowest position the *safe position*. It is not hard to see that only if the item is permuted to a position lower than or equal to its safe position, it can be immune to passive privacy attacks within the entire attack window. A similar pruning strategy can be applied on V_i, which removes all supersets of an itemset in V_i (Line 10).

V_i is then fed into IdentifyAnonymizationLocation (Line 12). The outputs are a set of items (i.e., locations) in whose RIL i should be anonymized, their corresponding anonymization codes (either *suppress* or *permute*), and safe positions for items labeled *permute*. For items labeled *suppress*, they are processed with the same procedure as the suppression-based solution for the multiple release scenario (SuppressMR). Here we focus on PermuteMR (Line 17). In static release, we can restore the items labeled *permute* to their previous positions to thwart privacy attacks. However, this is not sufficient for multiple release, because changes of the underlying user-item rating matrix are different in different time periods.

The key observation is that we have to permute the target item i_t to a position lower than or equal to its safe position. We iteratively switch i_t with the items in the RIL with position lower than or equal to i_t's safe position and check if the switch incurs any new privacy threat with respect to all previous $|\mathcal{W}_{T_k}| - 1$ releases. If we cannot find a permutation without violating the privacy requirement, we suppress i_t instead.

6 Experiments

In this section, we study the performance of the proposed anonymization algorithms over the public real-life datasets *MovieLens* and *Flixster*. We compare our suppression-based anonymization algorithm (*SUPP*) and permutation-based anonymization algorithm (*PERM*) with the randomized perturbation approach (*RP*) [3]. Due to the reason explained in Section 2, we cannot perform a meaningful comparison with the approach in [4]. All implementations were done in Python, and all experiments were run on a Linux machine with a 4 Core Intel Xeon CPU and 16GB of RAM.

The objectives of our experiments are: 1) evaluate the utility of various anonymization algorithms under different parameters; 2) examine the probability of successful passive privacy attacks after performing different anonymization algorithms; and 3) demonstrate the efficiency of our proposed algorithms.

The first dataset *MovieLens* (http://www.movielens.org) is a popular recommendation benchmark. It contains 1 million ratings over 4K movies and 6K users. The second dataset *Flixster* was crawled from the Flixster website [20], and contains 8.4 million ratings over 49K movies and 1 million users. Both datasets are time-stamped, and in all experiments, we follow the classical item-based recommendation framework studied in [18] to calculate item similarity scores. For *RP*, we use zero-meaned uniform noise with small variances. Experimental results obtained under different variances exhibit similar trends. Due to space limit, we only report the results with the variance equal to 1.

For all experiments, we select the initial timestamp such that the initial RIL release is generated based on approximately 10% of all ratings in the dataset. For the time gap between two consecutive RIL releases, we consider it to be a time period for generating a multiple of 1% of total ratings, e.g., if the time gap is 5, then the number of ratings generated between the two consecutive RIL releases will be approximately 5% of all ratings. Results obtained from other settings of these two parameters are very similar, and hence omitted here. In all experiments, we consider the effect of four tunable parameters: the attack window size, the time gap between two consecutive RIL releases, the privacy requirement δ, and the number of items in an RIL N. The following default values are used unless otherwise specified: 4 for the attack window size, 5 for the time gap, 0.1 for δ, and 5 for N.

Utility Study. As discussed before, *SUPP* and *PERM* only anonymize a few RILs in which real privacy risks for passive privacy attacks arise. Thus, they will leave most of the RILs intact. This is confirmed by *overall recall*, which is defined as the percentage of items in *all* original RILs that are retained after the anonymization. We show in Fig. 2 the overall recall of different algorithms on both datasets by varying the four parameters, namely attack window size, time gap between two consecutive RIL releases, δ, and N. It can be observed that both *SUPP* and *PERM* consistently achieve high overall recall, while *RP* cannot provide desirable utility in terms of RILs.

To further examine the utility loss just on the anonymized RILs (by ignoring RILs which are intact after the anonymization), we also consider *targeted recall*, which is

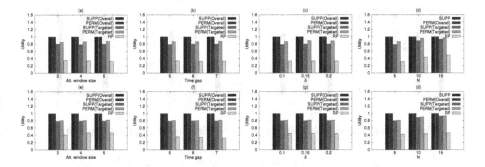

Fig. 2. Utility results on: MovieLens (a)–(d); Flixster (e)–(h)

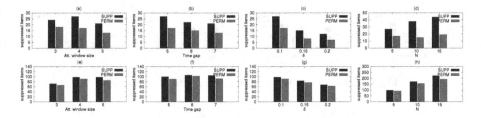

Fig. 3. Number of items suppressed by different algorithms: MovieLens (a)–(d); Flixster (e)–(h)

defined as the percentage of items retained in the anonymized RILs (i.e., the RILs in which suppression and/or permutation are performed). This utility metric is of importance because we do not want to have anonymized RILs that are substantially different from the original ones. The experimental results on both datasets, as shown in Fig. 2, suggest that our algorithms do not significantly destroy the usefulness of any RIL. We can also observe that *PERM* achieves better utility than *SUPP*. We present the numbers of suppressed items by both *PERM* and *SUPP* under varying parameters in Fig. 3. The results confirm that *PERM* is more preferable than *SUPP* in all cases.

Privacy Study. In Fig. 4, for both datasets, we demonstrate that *RP* cannot prevent passive privacy attacks: the worst case breach probability over the RILs generated from the perturbed user-item rating matrix is still extremely high (e.g., 100% for some target user). In contrast, our algorithms ensure that the breach probability over anonymized RILs is always less than the given privacy parameter δ.

Efficiency Study. Finally, we show the run-time of our proposed anonymization algorithms under various settings over both datasets in Fig. 5. As can be observed, both proposed algorithms are efficient, and in most situations, *PERM* is at least twice as fast as *SUPP*. The reason is that the cost of permutation is often much smaller than suppression, since the latter may need to explore many items beyond an RIL before finding a qualified replacement. Therefore, we conclude that empirically *PERM* is a better choice than *SUPP* in terms of both utility and efficiency.

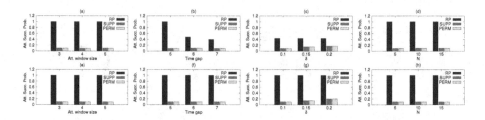

Fig. 4. Attack success probability results on: MovieLens (a)–(d); Flixster (e)–(h)

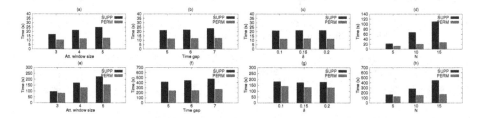

Fig. 5. Efficiency results on: MovieLens (a)–(d); Flixster (e)–(h)

7   Conclusion

The recent discovery of passive privacy attacks in item-to-item CF has exposed many real-life recommender systems to a serious compromise of privacy. In this paper, we propose a novel inference-proof privacy notion called δ-*bound* for thwarting passive privacy attacks. We develop anonymization algorithms to achieve δ-*bound* by means of a novel anonymization mechanism called *permutation*. Our solution can be seamlessly incorporated into existing recommender systems as a post-processing step over the RILs generated using traditional CF algorithms. Experimental results demonstrate that our solution maintains high utility and scales to large real-life data.

References

1. Goldberg, D., Nichols, D., Oki, B.M., Terry, D.: Using collaborative filtering to weave an information tapestry. Communications of ACM 35(12), 61–70 (1992)
2. Calandrino, J.A., Kilzer, A., Narayanan, A., Felten, E.W., Shmatikov, V.: "You might also like": Privacy risks of collaborative filtering. In: S&P (2011)
3. Polat, H., Du, W.: Privacy-preserving collaborative filtering using randomized perturbation techniques. In: ICDM (2003)
4. McSherry, F., Mironov, I.: Differentially private recommender systems: building privacy into the Netflix prize contenders. In: SIGKDD (2009)
5. Canny, J.: Collaborative filtering with privacy. In: S&P (2002)
6. Canny, J.: Collaborative filtering with privacy via factor analysis. In: SIGIR (2002)
7. Zhang, S., Ford, J., Makedon, F.: A privacy-preserving collaborative filtering scheme with two-way communication. In: EC (2006)

8. Berkvosky, S., Eytani, Y., Kuflik, T., Ricci, F.: Enhancing privacy and preserving accuracy of a distributed collaborative filtering. In: RecSys (2007)
9. Aimeur, E., Brassard, G., Fernandez, J.M., Onana, F.S.M.: ALAMBIC: A privacy-preserving recommender system for electronic commerce. International Journal of Information Security 7(5), 307–334 (2008)
10. Ahn, J.W., Amatriain, X.: Towards fully distributed and privacy-preserving recommendations via expert collaborative filtering and restful linked data. In: WI-IAT (2010)
11. Li, D., Lv, Q., Xia, H., Shang, L., Lu, T., Gu, N.: Pistis: A privacy-preserving content recommender system for online social communities. In: WI-IAT (2011)
12. Dwork, C., McSherry, F., Nissim, K., Smith, A.: Calibrating noise to sensitivity in private data analysis. In: Halevi, S., Rabin, T. (eds.) TCC 2006. LNCS, vol. 3876, pp. 265–284. Springer, Heidelberg (2006)
13. Cormode, G.: Personal privacy vs population privacy: Learning to attack anonymization. In: SIGKDD (2011)
14. Machanavajjhala, A., Korolova, A., Sarma, A.D.: Personalized social recommendations: Accurate or private. PVLDB 4(7), 440–450 (2011)
15. Ghinita, G., Tao, Y., Kalnis, P.: On the anonymization of sparse high-dimensional data. In: ICDE (2008)
16. Su, X., Khoshgoftaar, T.M.: A survey of collaborative filtering techniques. Advances in Artificial Intelligence (2009)
17. Fung, B.C.M., Wang, K., Chen, R., Yu, P.S.: Privacy-preserving data publishing: A survey of recent developments. ACM Computing Surveys 42(4), 14:1–14:53 (2010)
18. Karypis, G.: Evaluation of item-based top-n recommendation algorithms. In: CIKM (2001)
19. Ausiello, G., D'Atri, A., Protasi, M.: Structure preserving reductions among convex optimization problems. Journal of Computer and System Sciences 21(1), 61–70 (1980)
20. Jamali, M., Ester, M.: A matrix factorization technique with trust propagation for recommendation in social networks. In: RecSys (2010)

Privacy-Preserving Schema Reuse

Nguyen Quoc Viet Hung, Do Son Thanh, Nguyen Thanh Tam, and Karl Aberer

École Polytechnique Fédérale de Lausanne
{quocviethung.nguyen,sonthanh.do,tam.nguyenthanh,karl.aberer}@epfl.ch

Abstract. As the number of schema repositories grows rapidly and several web-based platforms exist to support publishing schemas, *schema reuse* becomes a new trend. Schema reuse is a methodology that allows users to create new schemas by copying and adapting existing ones. This methodology supports to reduce not only the effort of designing new schemas but also the heterogeneity between them. One of the biggest barriers of schema reuse is about privacy concerns that discourage schema owners from contributing their schemas. Addressing this problem, we develop a framework that enables privacy-preserving schema reuse. Our framework supports the contributors to define their own protection policies in the form of *privacy constraints*. Instead of showing original schemas, the framework returns an *anonymized schema* with maximal *utility* while satisfying these privacy constraints. To validate our approach, we empirically show the efficiency of different heuristics, the correctness of the proposed utility function, the computation time, as well as the trade-off between utility and privacy.

1   Introduction

Schema reuse is a new trend in creating schemas by allowing users to copy and adapt existing ones. The key driving forces behind schema reuse are the slight differences between schemas in the same domain; thus making reuse more realistic. Reusing existing schemas supports to reduce not only the effort of creating a new schema but also the heterogeneity between schemas. Moreover, as the number of publicly available schema repositories (e.g. schema.org[2], Factual[3]) grows rapidly and several web-based platforms (e.g. Freebase [14], Google Fusion Tables [27]) exists to support publishing schemas, reusing them becomes a great interest in both academic and industrial worlds.

One of the biggest barriers of reuse is about privacy concerns that discourage contributors from contributing their schemas [11]. In traditional approaches, all original schemas and their own attributes are presented to users [17]. However, in practical scenarios, the linking of attributes to their containing schemas, namely *attribute provenance*, is dangerous because of two reasons. First, providing the whole schemas (and all of their attributes) leads to privacy risks and potential attacks on the owner database. Second, since some attributes are the source of revenue and business strategy, the schema contributors want to protect their sensitive information to maintain the competitiveness. As a result, there is a need of developing new techniques to cope with these requirements.

In this paper, we develop a privacy-preserving schema reuse framework that protects the attribute provenance from being disclosed. To this end, our framework enables schema owners to define their own protection policies in terms of privacy constraints. Unlike previous works [17,20,36], we do not focus on finding and ranking schemas

S.S. Bhowmick et al. (Eds.): DASFAA 2014, Part II, LNCS 8422, pp. 234–250, 2014.

relevant to a user search query. Instead, our framework takes as input these relevant schemas and visualizes them in a unified view, namely *anonymized schema*, which satisfies pre-defined privacy constraints. Constructing such an anonymized schema is challenging because of three reasons. First, defining the representation for an anonymized schema is non-trivial. The anonymized schema should be concise enough to avoid overwhelming but also generic enough to provide comprehensive understanding. Second, for the purpose of comparing different anonymized schemas, we need to define a utility function to measure the amount of information they carry. The utility value must reflect the conciseness and the completeness of an anonymized schema. Third, finding an anonymized schema that maximizes the utility function and satisfies privacy constraints is NP-complete.

The main goal of this paper is to construct an anonymized schema with maximal utility while preserving the privacy constraints, which are defined to prevent an adversary from linking the shown attributes back to the original schemas (and to the schema owners). Our key contributions are summarized as follows.

- We model the setting of schema reuse with privacy constraints by introducing the concept of *affinity matrix* (represents a group of relevant schemas) and *presence constraint* (is a privacy constraint that translates human-understandable policies into mathematical standards).
- We develop a quantitative metric for assessing the utility of an anonymized schema by capturing two important aspects: (i) *attribute importance*—which reflects the popularity of an attribute—and (ii) *completeness*—which reflects the diversity of attributes in the anonymized schema.
- We show the intractability result for the problem of finding an anonymized schema with maximal utility, given a set of privacy constraints. We propose a heuristic-based algorithm for this problem. Through experiments on real and synthetic data, we show the effectiveness of our algorithm.

The paper is organized as follows. Section 2 gives an overview of our approach. Section 3 formally introduces the notion of schema group, anonymized schema and privacy constraint. Section 4 demonstrates the intractability result and the heuristic-based algorithm to the *maximizing anonymized schema* problem. Section 5 empirically shows the efficiency of our framework. Section 6 and 7 present related work and conclusions.

2 Overview

System Overview. Fig. 1 illustrates a schema reuse system, in which there is a repository of schemas that are willingly contributed by many participants in the same domain. We focus on the scenario in which end-users want to design a new database and reuse the existing schemas as hints, by exploring the repository through search queries [17,20,36] to find schemas of relevance to their applications. In traditional approaches, all the relevant schemas and their whole attributes are shown to users. Nevertheless, it is important to support the contributors to preserve the privacy of their schemas [18] because of several privacy issues. One possible issue is the threats of being attacked and unprivileged accesses to the owner database systems. For example, knowing the schema information (e.g. schema name, attribute names), an adversary can use SQL injection [28] to extract the data without sufficient privileges (details are described in the report [41]). Another possible issue is the policies of schema owners that require hiding

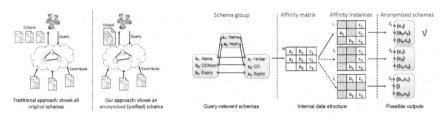

Fig. 1. System Overview   **Fig. 2.** Solution Overview

a part of schemas under some circumstances. For instance, a hospital needs to hide the personal information of its patient records due to legal concerns [49], or an enterprise makes an agreement with its clients about the privacy of business practices [7]. Since schema reuse only shares schema information, the scope of this paper is to preserve the privacy of schema only.

Solution Overview. To encourage schema reuse while preserving privacy, we propose a novel approach that shows a unified view of original schemas, namely *anonymized schema*, instead of revealing all of the schemas. In our approach, schema owners are given the rights to protect sensitive attributes by defining privacy constraints. The resulting anonymized schema has to be representative to cover original ones but opaque enough to prevent leaking the provenance of sensitive attributes (i.e. the linking to containing schemas). Towards our approach, we propose a framework that enables privacy-preserving schema reuse. The input of our framework is a group of schemas (which is relevant to user query as returned by a search engine) and a set of privacy constraints. Respecting these constraints, the framework returns an anonymized schema with maximal utility, which reflects the amount of information it carries. The problem of constructing such an anonymized schema is challenging and its details will be described in Section 4. Fig. 2 depicts the simplified process of our solution for this problem, starting with a schema group and generated correspondences (solid lines) that indicates the semantic similarity between attributes. First, we represent a schema group by an *affinity matrix*. In that, attributes in the same column belongs to the same schema, while attributes in the same row have similar meanings. Next, we derive various *affinity instances* by eliminating some attributes from the affinity matrix. For each affinity instance, we construct an anonymized schema with many *abstract attributes*. Each abstract attribute is a set of original attributes in the same row. Among constructed anonymized schemas, we select the one that maximizes the utility function and then present this "best" anonymized schema to user as final output. Table 1 summarizes important notations, which will be described in the next section.

3   Model

In this section, we describe three important elements of our schema reuse approach: (i) schema group – a set of schemas relevant to a user search query, (ii) anonymized schema – a unified view of all relevant schemas in the group, and (iii) privacy constraint – a mean to represent human-understandable policies by mathematical standards. All these elements are fundamental primitives for potential applications built on top of our framework. Right after defining the anonymized schema, we also propose a single comprehensive notion to quantify its utility in Section 3.3.

3.1 Schema Group

We model a schema as a finite set of attributes $A_s = \{a_1, ..., a_n\}$, which is generic enough for various types of schemas [43]. Let $S = \{s_1, ..., s_n\}$ be a schema group that is a set of schemas relevant to a specific search query. Let A_S denote the set of attributes in S, i.e. $A_S = \bigcup_i A_{s_i}$. Two schemas can share a same-name attribute; i.e. $A_{s_i} \cap A_{s_j} \neq \emptyset$ for some i, j. Given a pair of schemas $s_1, s_2 \in S$, an attribute correspondence is an attribute pair $(a \in A_{s_1}, b \in A_{s_2})$ that reflects the similarity of the two attributes' semantic meanings [12]. The union of attribute correspondences for all pairs of schemas is denoted as C, each of which can be generated by state-of-the-art schema matching techniques [8,45]. Note that the quality of generated correspondences is not our concern, since there is a large body of research on reconciling and improving schema matching results [29,44,42]. In the context of schema reuse, the generated correspondences are useful for users to see possible variations of attributes among schemas.

To model a schema group as well as generated correspondences between attributes of its schemas, we define an affinity matrix $M_{m \times n}$. Each of n columns represents an original schema and its attributes. Each of m rows represents the set of equivalent attributes, each pair of which have correspondences between them. An entry being null means that the schema-column of that entry does not have an equivalent attribute against the attributes of other schema-columns in the same row.

Table 1. Summary of Notations

Symbol	Description
$S = \{s_i\}$	a group of schemas
A_s	a set of attributes of the schema s
C	a set of attribute correspondences
M, I	affinity matrix, affinity instance
$\hat{S}, \hat{A}$	anonymized schema, abstract attribute
$u(\hat{S})$	utility of an anonymized schema
$\Gamma = \{\gamma_i\}$	a set of privacy constraints

Definition 1 *Affinity Matrix. Given a schema group S and a set of correspondences C, an affinity matrix is a matrix $M_{m \times n}$ such that: (i) $\forall i, j : M_{ij} \in A \cup \{null\}$, (ii) $\forall i, j : M_{ij} \in A_{s_j}$, (iii) $\forall i, j_1 \neq j_2, M_{ij_1} \neq null, M_{ij_2} \neq null : (M_{ij_1}, M_{ij_2}) \in C$, and (iv) $\nexists i_1 \neq i_2, j_1, j_2 : (M_{i_1j_1}, M_{i_2j_2}) \in C$.*

In order to construct a unique affinity matrix, we define an order relation on A_S. We consider each attribute $a \in A_S$ as a character. The order relation between any two attributes is the alphabetical order between two characters. Consequently, a row $M_{i.}$ of the affinity matrix M is a string of n characters. An affinity matrix is *valid* if the string of row r_1 is less than the string of row r_2 ($r_1 < r_2$) w.r.t. the lexicographical order. Since the alphabetical and lexicographical order are unique, a valid affinity matrix is unique.

Example 1. Consider a group of 3 schemas $S = \{s_1, s_2, s_3\}$ and their attributes $A_{s_1} = \{a_1, a_2\}$, $A_{s_2} = \{b_1, b_2, b_3\}$, $A_{s_3} = \{c_1, c_2, c_3\}$. We have 8 correspondences $C = \{(a_1, b_1), (a_1, c_1), (b_1, c_1), (a_2, b_2), (a_2, c_2), (b_2, c_2), (b_3, c_3)\}$. Assume that the lexicographical order of all attributes is $a_1 < a_2 < b_1 < b_2 < b_3 < c_1 < c_2 < c_3$. Then we can construct a unique affinity matrix M as in Fig. 2. The string of the first, second, third row of M is "$a_1b_1c_1$", "$a_2b_2c_2$", "b_3c_3" respectively. M is valid because "$a_1b_1c_1$" < "$a_2b_2c_2$" < "b_3c_3" w.r.t. the assumed lexicographical order.

3.2 Anonymized Schema

An anonymized schema should meet the following desirable properties: (i) representation, and (ii) anonymization. First, the anonymized schema has to be representative to cover the attributes of a schema group. It should show to end-user not only all possible attributes but also the variations of each attribute. Second, the anonymized schema should be opaque enough to prevent adversaries from reconstructing a part or the whole of an original schema.

Our approach is inspired by the generalization technique [10], with some modifications, to protect the owners of original schemas. The core idea is that by not showing all the attributes but only a subset of them (e.g. remove one or many cells of the affinity matrix), we preserve the privacy of important attributes. However, hiding too many attributes would lead to poor information presented to end-users. Thus, the problem of selecting which attributes and how many of them is challenging and will be formulated in Section 4. Here we provide the basic notion of an anonymized schema by representing it as an *affinity instance*. An affinity instance $I_{m \times n}$ is constructed from an affinity matrix $M_{m \times n}$ by removing one or many cells; i.e. $\forall i,j : I_{ij} = M_{ij} \vee I_{ij} = null$. Formally, an anonymized schema $\hat{S}$ is transformed from an affinity instance $I_{m \times n}$ as follows.

$$\hat{S} = \{\hat{A}_1, \ldots, \hat{A}_m \mid \hat{A}_i = \bigcup_{1 \le j \le n} I_{ij}\} \tag{1}$$

where $\hat{A}$ is an abstract attribute (a set of attributes at the same row). In other words, an abstract attribute of an anonymized schema is a set of equivalent attributes (i.e. any pair of attributes has a correspondence). Following the running example in Fig. 2, a possible affinity instance is I_1 and the corresponding anonymized schema is $\hat{S}_1 = \{\hat{A}_1, \hat{A}_2, \hat{A}_3\}$ with $\hat{A}_1 = \{c_1\}$, $\hat{A}_2 = \{a_2, c_2\}$, and $\hat{A}_3 = \{b_3, c_3\}$.

With this definition, the anonymized schema meets two aforementioned properties—representation and anonymization. First, the representation property is satisfied because by showing abstract attributes, users can observe all possible variations to design their own schemas. Second, the anonymization property is satisfied because it is non-trivial to recover the original affinity matrix from a given anonymized schema. By knowing only the anonymized schema (as a set of abstract attributes), brute-force might be the only way for an adversary to disclose the ownerships of contributors, their schemas, and schema attributes. In Section 3.4, we will illustrate this property in terms of probability theory.

3.3 Anonymized Schema Utility

The utility of an anonymized schema $\hat{S}$ is measured by a function $u : \overline{S} \rightarrow \mathbb{R}$, where $\overline{S}$ is the set of all possible anonymized schemas. The utility value reflects the amount of information $\hat{S}$ carries. To define this utility function, we first introduce two main properties of an anonymized schema: *importance* and *completeness*.

Importance. The importance of an anonymized schema is defined as a function $\sigma : \overline{S} \rightarrow \mathbb{R}$. We propose to measure $\sigma(\hat{S})$ by summing over the popularity of attributes it contains; i.e. $\sigma(\hat{S}) = \sum_{a \in \hat{S}} t(a)$, where $t(a)$ is the popularity of attribute a. The higher popularity of these attributes, the higher is the importance of the anonymized schema. Intuitively, an attribute is more popular than another if it appears in more original schemas. At the same time, an original schema is more significant if it contains more popular attributes. To capture this relationship between attributes and their containing schemas, we attempt to develop a model that propagates the popularity along attribute-schema links. We notice some similarities to the world-wide-web setting, in which the significance of a web page is determined by both the goodness of the match of search terms on the page itself and the links connecting to it from other important pages. Adapting this idea, we apply the hub-authority model [32] to quantify the popularity of an attribute and the significance of related schemas, whose details are described in our report [41] for the sake of brevity.

Completeness. The completeness reflects how well an anonymized schema collectively covers the attributes of original schemas, where each abstract attribute is responsible for original attributes it represents. This coverage is necessary to penalize trivial anonymized schemas that contain only one attribute appearing in all original schemas. To motivate the use of this property, we illustrate some observations in Fig. 2. In this running example, $\hat{S}_2$ has more abstract attributes than $\hat{S}_3$ ($\hat{S}_2$ has three while $\hat{S}_3$ has two). As $\hat{S}_3$ lacks information about *CCNum* or *CC* attributes, users would prefer $\hat{S}_2$ in order to have a better overview of necessary attributes for their schema design. Intuitively, an anonymized schema is more preferred if it covers more information of original schemas. In this paper, we measure the completeness of an anonymized schema by the number of abstract attributes it contains. The more abstract attributes, the higher is the completeness. Technically, the completeness is defined as a function $\lambda : \bar{S} \to \mathbb{R}$ and $\lambda(\hat{S}) = |\hat{S}|$.

Put It All Together. As mentioned above, there are two properties (importance and completeness) that should be considered for the purpose of comparing different anonymized schemas. We attempt to combine these properties into a single comprehensive notion of *utility* of an anonymized schema. Formally, we have:

$$u(\hat{S}) = w \cdot \sigma(\hat{S}) + (1 - w) \cdot \lambda(\hat{S}) \tag{2}$$

where $w \in [0, 1]$ is a regularization parameter that defines the trade-off between completeness and importance. In terms of comparison, an anonymized schema is more preferred than another anonymized schema if its utility is higher. Back to the running example in Fig. 2, consider three anonymized schemas $\hat{S}_1$, $\hat{S}_2$ and $\hat{S}_3$. By definitions, we have $\sigma(\hat{S}_1) = 5$, $\sigma(\hat{S}_2) = \sigma(\hat{S}_3) = 4$ and $\lambda(\hat{S}_1) = \lambda(\hat{S}_2) = 3$, $\lambda(\hat{S}_3) = 2$. Assuming $w = 0.5$, we have $u(\hat{S}_1) > u(\hat{S}_2) > u(\hat{S}_3)$. $\hat{S}_1$ is more preferred than $\hat{S}_2$ since $\hat{S}_1$ contains more original attributes (importance property). While, $\hat{S}_2$ is more preferred than $\hat{S}_3$ since $\hat{S}_2$ contains more abstract attributes (completeness property).

3.4 Privacy Constraint

Privacy constraint is a mean to represent human-understandable policies by mathematical standards. In order to define a privacy constraint, we have to identify two elements: *sensitive information* and *privacy requirement*. In our setting, the sensitive information is attributes of a given schema. Each attribute has different levels of sensitivity (e.g. credit card number is more sensitive than phone number) and some of them are the source of revenue and business strategy and can only be shared under specific conditions. Whereas, the privacy requirement is to prevent adversaries from linking sensitive attributes back to original schemas. To capture these elements, we employ the concept of *presence constraint* that defines an upper bound of the probability $Pr(D \in s|\hat{S})$ that an adversary can disclose the presence of a set of attributes D in a particular schema s, if he sees an anonymized schema $\hat{S}$. Formally, we have:

Definition 2 *Presence Constraint.* A presence constraint γ is a triple $\langle s, D, \theta \rangle$, where s is a schema, D is a set of attributes, and θ is a pre-specified threshold. An anonymized schema $\hat{S}$ satisfies the presence constraint γ if $Pr(D \in s|\hat{S}) \leq \theta$.

Intuitively, the presence constraint protects a given schema from being known what attributes it has. The more chances of knowing the exact attribute(s) a schema contains, the riskier an adversary can attack and exploit the owner database of that schema. Hence, the presence threshold θ is given to limit the possibilities of linking the attributes back to original schemas by hiding the attributes whose provenances (i.e. attribute-schema links) can be inferred with probabilities higher than θ. Choosing an appropriate value for θ depends on the domain applications and beyond the scope of this work.

Checking Presence Constraint. Given an anonymized schema $\hat{S}$ constructed from the affinity instance I and a presence constraint $\gamma = \langle s, D, \theta \rangle$, we can check whether $\hat{S}$ satisfies γ by computing the probability $Pr(D \in s \mid \hat{S})$ as follows:

$$Pr(D \in s \mid \hat{S}) = \begin{cases} \frac{1}{\prod_{i=1}^{k} |\hat{A}_i|}, & \text{if } \hat{A}_1, \ldots, \hat{A}_k \neq \emptyset \\ 0, & \text{otherwise} \end{cases} \tag{3}$$

where $D = \{a_1, \ldots, a_k\}$ is a subset of attributes of s and $\hat{A}_i \in \hat{S}$ is the abstract attribute (a group of attributes at the same row of I) that contains a_i. In case there is no abstract attribute which contains a_i, we consider $\hat{A}_i = \emptyset$. Eq. 3 comes from the assumption that to check an attribute a_i belongs to s, an adversary has to exhaustively try every attribute of $\hat{A}_i$, thus the probability of a successful disclosure is $Pr(a_i \in s|\hat{S}) = 1/|\hat{A}_i|$. Consequently, the probability of checking a subset of attributes is simply a product of elemental probabilities, since the anonymized schema generated by our model does not imply any dependence between abstract attributes.

Example 2. Continuing our running example in Fig. 2, we consider an anonymized schema $\hat{S} = \{\hat{A}_1, \hat{A}_2\}$, where $\hat{A}_1 = \{a_1, c_1\}$ and $\hat{A}_2 = \{a_2, b_2\}$. An adversary wants to check if two attributes a_1 and a_2 belongs to s_1 but he can only see $\hat{S}$. Based on Eq. 3, we have $Pr(\{a_1, a_2\} \in s_1 \mid \hat{S}) = \frac{1}{|\hat{A}_1| \cdot |\hat{A}_2|} = 0.25$. If we consider a presence constraint $\gamma = \langle s_1, \{a_1, a_2\}, 0.6 \rangle$, the anonymized schema $\hat{S}$ can be accepted as system output.

Privacy vs. Utility Trade-Off. It is worth noting that there is a trade-off between privacy and utility [35]. In general, anonymization techniques reduce the utility of querying results when trying to provide privacy protection [15]. On one hand, the utility of an anonymized schema is maximal when there is no presence constraint (i.e. $\theta = 1$). On the other hand, when a contributor set the threshold θ of a presence constraint $\langle s, D, \theta \rangle$ very small (closed to 0), D will not appear in the anonymized schema. Consequently, the utility of this anonymized schema is low since we loss the information of these attributes. As a result, privacy should be seen as a spectrum on which contributors can choose their place. From now on, we use two terms—*privacy constraint* and *presence constraint*—interchangeably to represent the privacy spectrum defined by contributors.

4   Maximizing Anonymized Schema

Maximizing anonymized schema is the selection of an optimal (w.r.t. utility function) anonymized schema among potential ones that satisfy pre-defined privacy constraints. Formally, the maximization problem of our schema reuse setting is defined as follows.

Problem 1. *(Maximizing Anonymized Schema) Given a schema group S and a set of privacy constraints Γ, construct an anonymized schema $\hat{S}$ such that $\hat{S}$ satisfies all constraints Γ, i.e. $\hat{S} \models \Gamma$, and the utility value $u(\hat{S})$ is maximized.*

Theorem 1. *Let S be a schema group and Γ be a set of privacy constraints. Then, for a constant U, the problem of determining if there exists an anonymized schema $\hat{S} \models \Gamma$, whose utility value $u(\hat{S})$ is at most U, is NP-complete.*

The key of the proof is to transform the problem to *Maximum Independent Set* problem, which is known to be NP-complete [31]. For brevity sake, further details are referred to our report [41]. In this section, we consider the heuristic-based approach to relax optimization constraints and find the approximate solution in polynomial time.

4.1 Algorithm

In light of Theorem 1, we necessarily consider heuristic approaches to the maximizing anonymized schema problem. Our heuristic-based algorithm takes two parameters as input: a schema group modeled by an affinity matrix M and a set of presence constraints Γ. It will efficiently return an approximate solution—an anonymized schema $\hat{S}$ which satisfies Γ—with the trade-off that the utility value $u(\hat{S})$ is not necessarily maximal. The key difficulty is that during the optimization process, repairing the approximation solution to satisfy one constraint can break another. At a high level, our algorithm makes use of greedy and local-search heuristics to search the optimal anonymized schema in the space of conflict-free candidates (i.e. anonymized schemas without violations). The details are illustrated in Algorithm 1 and described in what follows.

Technically, we begin by constructing a valid affinity instance from the original affinity matrix (line 2 to line 7). The core idea is we generate an initial affinity instance $I = M$ and then continuously remove the attributes of I until all constraints are satisfied ($\Gamma_s = \Gamma$). The challenge then becomes how to choose which attributes for deletion such that the resulting instance has maximal utility. To overcome this challenge, we use a repair routine (line 3 to line 7) which greedily removes the attributes out of I to eliminate all violations. The greedy choice is that at each stage, we remove the attribute $\hat{a}$ with lowest utility reduction (line 5); i.e. choose an attribute a such that the utility of I after removing a is maximized. However, this greedy strategy could make unnecessary deletions since some constraints might share common attributes, thus making the utility not maximal. For example in Fig. 2, consider the anonymized schema $\hat{S} = \{\{a_1, c_1\}, \{a_2, b_2\}\}$ with two predefined constraints $\gamma_1 = Pr(a_1 \in s_1 \mid \hat{S}) \le 0.4$ and $\gamma_2 = Pr(\{a_1, a_2\} \in s_1 \mid \hat{S}) \le 0.1$. The best way to satisfy these constraints is removing only attribute a_2, but the algorithm will delete a_1 first and then a_2. Hence, we need to insert a_2 back into $\hat{S}$ to increase the utility. For this purpose, A_{del} stores all deleted attributes (line 6) for the following step.

The above observation motivates the next step in which we will explore neighbor instances using local search and keep track of the instance with highest utility (line 9 to line 20). Starting with the lower-bound instance from the previous step ($I_{op} = I$), the local search is repeated until the termination condition Δ_{stop} is satisfied (e.g. k iterations or time-out). In each iteration, we incrementally increase the utility of the current instance I by reverting unnecessary deletions from the previous step. The raising issue is how to avoid local optima if we just add an attribute with highest utility gain; i.e. choose an attribute a such that the utility of I is maximized after adding a. To tackle this issue, we perform two routines. (1) *Insertion:* Each attribute $a \in A_{del}$ is a possible candidate to insert into I (line 11). We use Roulette-wheel selection [26] as a non-deterministic strategy, where an attribute with greater utility has higher probability to add first (line 12). (2) *Repair:* When a particular attribute is inserted, it might make some constraints

---

Algorithm 1. Heuristics-based Algorithm

---

    input : $\langle M, \Gamma \rangle$ // An affinity matrix M and a set of predefined constraints Γ
    output : $\mathcal{I}_{op}$ // Maximal Affinity Instance
    denote : $u(I^{-a})$: utility of I after removing attribute a out of I
                   $u(I^{+a})$: utility of I after adding attribute a into I
1   $A_{del} = \emptyset$ // Set of deleted attributes; $\Gamma_s = \emptyset$ // Set of satisfied constraints
2   $I = M$
3   **while** $\Gamma_s \neq \Gamma$ **do**
4   /* **Violation Repair: greedy deletion until satisfying constraints** */
5       $\hat{a} = \text{argmax}_{a \in \{I_{ij}\}} \, u(I^{-a})$
6       $I.delete(\hat{a}); A_{del} = A_{del} \cup \{\hat{a}\}$
7       $\Gamma_s = I.checkConstraints()$
8   $I_{op} = I; T = Queue[k]$ // a queue with fixed size k
9   **while** Δ_{stop} **do**
10   /* **Local search by non-deterministic insertion** */
11      $\Omega = \{\langle a, u(I^{+a}) \rangle : a \in A_{del}\}$
12      $\hat{a} = \Omega.rouletteSelection()$
13      $I.add(\hat{a})$; $T.add(\hat{a})$
14      $\Gamma_s = I.checkConstraints()$
15   **while** $\Gamma_s \neq \Gamma$ **do**
16   /* **Violation Repair: greedy deletion until satisfying constraints** */
17         $\hat{a} = \text{argmax}_{a \in \{I_{ij}\} \setminus T} \, u(I^{-a})$
18         $I.delete(\hat{a})$
19         $\Gamma_s = I.checkConstraints()$
20   **if** $u(I) > u(I_{op})$ **then** $I_{op} = I$
21 **return** I_{op}

---

unsatisfied ($\Gamma_s \neq \Gamma$). Thus, the repair routine is invoked again to eliminate new violations by removing the potential attributes out of I (line 15 to line 19).

In an iteration of local-search, an attribute could be inserted into an affinity instance and then removed immediately by the repair routine, making this instance unchanged and leading to being trapped in local optima. For this reason, we employ the Tabu search method [25] that uses a fixed-size "tabu" (forbidden) list T of attributes so that the algorithm does not consider these attributes repeatedly. In the end, the maximal affinity instance I_{op} is returned by selecting the one with the highest utility among explored instances (line 20).

4.2 Algorithm Analysis

Our algorithm is terminated and correct. Termination follows from the fact that the stopping condition Δ_{stop} can be defined to execute at most k iterations (k is a tuning parameter). The correctness follows from the following points. (1) When a new attribute a added into I (line 13) violates some constraints, I is repaired immediately. This repair routine takes at most $O(|M|)$, where $|M|$ is the number of attributes of the schema group represented by M. (2) The newly added attributes are not removed in the repair routine since they are kept in the "tabu" list T. The generation of Ω takes at most $O(|M|)$. (3) I_{op} always maintains the instance with maximal utility (line 20). Sum it up, the worse-case running time is $O(k \times |M|^2)$—which is reasonably tractable for real datasets.

Looking Ahead for Repair Cost. We have a heuristic to improve the repair technique. The idea of this heuristic is to avoid bad repairs (that cause many deletions) by adding some degree of lookahead to the repair cost (i.e. number of deletions). In Algorithm 1, the utility value $u(I^{-a})$ is used as an indicator to select an attribute for deletion. Now we modify this utility value to include a look-ahead approximation of the cost of deleting one further attribute. More precisely, when an attribute a is considered for deletion (first step), we will consider a next attribute a' given that a is already deleted (second step). Our look-ahead function will be the utility of the resulting instance after these two steps. In other words, we prevent a bad repair by computing the utility of the resulting instances one step further. Due to space limit, further heuristics for improve the algorithm performance are omitted and can be found in the report version of this work [41].

5 Experiments

To verify the effectiveness of the proposed framework in designing and constructing the anonymized schema, following experiments are performed: (i) effects of different heuristics for the maximizing anonymized schema problem, (ii) the correctness of utility function, (iii) the computation time of our heuristic-based algorithm in various settings, and (iv) the tradeoff between utility and privacy. We proceed to report the results on real datasets and synthetic data.

5.1 Experimental Settings

Datasets. Our experiments are conducted on two types of data: real data and synthetic data. While the real data provides a pragmatic view on real-world scenarios, the synthetic data helps to evaluate the performance with flexible control of parameters.

- *Real Data.* We use a repository of 117 schemas that is available at [1]. To generate attribute correspondences, we used the well-known schema matcher COMA++ [8].
- *Synthetic Data.* We want to simulate practical cases in which one attribute is used repeatedly in different schemas (e.g. 'username'). To generate a set of n synthetic schemas, two steps are performed. In the first step, we generate k core schemas, each of which has m attributes, such that all attributes within single schema as well as across multiple schemas are completely different. Without loss of generality, for any two schemas a and b, we create all 1-1 correspondences between their attributes; i.e. $(a_i, b_i)_{1 \leq i \leq m}$. In the second step, $n-k$ remaining schemas are permuted from k previous ones. For each schema, we generate m attributes, each of which is copied randomly from one of k attributes which have all correspondences between them; i.e. attributes at the same row of the affinity matrix.

Privacy Constraints. To generate a set of privacy constraints Γ for simulation, we need to consider two aspects:

- *The number of attributes per constraint.* To mix the constraints with different size, we set a parameter α_i as the number of constraints with i attributes. Since the chance for an adversary to find the correct provenance of a group of four or more attributes is often small with large abstract attributes (Eq. 3), we only consider constraints with less than four attributes (i.e. $\alpha_{i \geq 4} = 0$). For all experiments, we set $\alpha_1 = 50\%$, $\alpha_2 = 30\%$, and $\alpha_3 = 20\%$.
- *Privacy threshold.* We generate the privacy threshold θ of all constraints according to uniform distribution in the range $[a, b]$, $0 \leq a, b \leq 1$. In all experiments, we set $a = (1/n)^i$ and $b = (2/n)^i$, where n is the number of schemas and i is the number of attributes in the associated constraint.

In general, the number of constraints is proportional to the percentage of the total number of attributes; i.e. $|\Gamma| \propto m \cdot n$). In each experiment, we will vary the percentage value differently.

Evaluation Metrics. Beside measuring the computation time to evaluate proposed heuristics, we also consider two important metrics:

- *Privacy Loss:* measures the amount of disagreement between the two probability distributions: actual privacy and expected privacy. Technically, given an anonymized schema $\hat{S}$ and a set of presence constraints $\Gamma = \{\gamma_1, \ldots, \gamma_n\}$ where $\gamma_i = \langle s_i, D_i, \theta_i \rangle$, we denote $p_i = Pr(D_i \in s_i | \hat{S})$. Privacy loss is computed by the Kullback-Leiber divergence between two distributions $P = \{p_i\}$ (actual) and $\Theta = \{\theta_i\}$ (expected):

$$\Delta p = KL(P \parallel \Theta) = \sum_i p_i \log \frac{p_i}{\theta_i} \tag{4}$$

- *Utility Loss:* measures the amount of utility reduction with regard to the existence of privacy constraints. Technically, denote u_0 is the utility of an anonymized schema without privacy constraints and u_Γ is the utility of an anonymized schema with privacy constraints Γ, utility loss is calculated as:

$$\Delta u = \frac{u_0 - u_\Gamma}{u_0} \tag{5}$$

We implement our schema reuse framework in Java. All the experiments ran on an Intel Core i7 processor 2.8 GHz system with 4 GB of RAM.

5.2 Effects of Heuristics

This experiment will compare the utility loss of the anonymized schema returned by different heuristics. In this experiment, we study the proposed algorithm with two heuristics: without lookahead (Greedy) and with lookahead (Lookahead). We use a synthetic data with a random number of schemas $n \in [50, 100]$ and each schema has a random number of attributes $m \in [50, 100]$. In that, the copy percentage is varied from 10% to 20%; i.e. $k/n \in [0.1, 0.2]$. The number of constraints is varied from 5% to 50% of total number of attributes. The results for each configuration are averaged over 100 runs. Regarding utility function, we set $w = 0.5$ which means the completeness is equally preferred as the importance.

Fig. 3 depicts the result of two described heuristics. The X-axis shows the number of constraints, while the Y-axis shows the utility loss (the lower, the better). A noticeable observation is that the Lookahead heuristic performs better than the Greedy heuristic. Moreover, when there are more constraints, this difference is more significant (up to 30% when the number of constraints is 50%). This is because the Lookahead heuristic enables our algorithm to foresee and avoid bad repairs, which cause new violations when eliminating the old ones, thus improves the quality of the result.

5.3 Correctness of Utility Function

In this experiment, we would like to validate the correctness of our utility function by comparing the results of the proposed algorithm with the choices of users. Technically, we expect that the ranking of anonymized schemas by utility function is equivalent to their ranking by users. Regarding the setting, we construct 20 different schema groups from the real data by extracting a random number of schemas and attributes. For each group, we present five of constructed anonymized schemas (ranked #1, #2, #3, #4, #5 by the decreasing order of utility) to 10 users (users do not know the order). These presented anonymized schemas are not top-5, but taken from the highest such that the

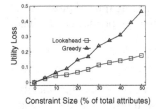

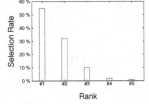

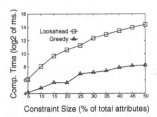

Fig. 3. Utility Loss (the lower, the better)

Fig. 4. Correctness of Utility Function

Fig. 5. Computation Time vs. #Constraints

utility difference between two consecutive ranks is at least 10% to avoid insignificant differences. Each user is asked to choose the best anonymized schema in his opinion (200 votes in total). We fix the number of constraints at 30% of total attributes. Since the Lookahead strategy is better than the Greedy strategy as presented in the previous experiment (Section 5.2), we use it to construct anonymized schemas.

The results are depicted in Fig. 4. X-axis shows top-5 anonymized schemas returned by our algorithm. Y-axis presents the selection rate (percentage of users voting for an anonymized schema), averaged over all schema groups. A key finding is that the order of selection rate corresponds to the ranking of solutions by the utility function. That is, the first-rank anonymized schema (#1) has highest selection rate (55%) and the last-rank anonymized schema (#5) has lowest selection rate (1%). Moreover, most of users opt for the first-rank solution (55%) and second-rank solution (33%), indicating that our algorithm is able to produce anonymized schemas similar to what users want in real-world datasets.

5.4 Computation Time

In this experiment, we design various settings to study the computation time with two above-mentioned strategies (Greedy vs. Lookahead) using synthetic data. In each setting, we measure the average computation time over 100 runs.

Effects of Input Size. In this setting, the computation time is recorded by varying the input size of the affinity matrix; i.e. the number of schemas and the number of attributes per schema. A significant observation in Table 2 is that the Greedy strategy outperforms the Lookahead strategy. This is because the cost of looking-ahead utility change is expensive with the trade-off that the utility loss is smaller as presented in Section 5.2.

Table 2. Computation Time (log2 of msec.)

Size of M^*	Greedy	LookAhead
10×20	2.58	5.04
10×50	5.13	10.29
20×50	7.30	13.36
20×100	9.95	17.60
50×100	12.62	21.58

* $m \times n$: m attributes (per schema) and n schemas

Effects of Constraints Size. In this setting, the computation time is recorded by varying #constraints from 5% to 50% of total number of attributes. Based on results of the previous setting, we fix the input size at 20×50 for a low starting point. Fig. 5 illustrates the output, in which the X-axis is the constraint size and the Y-axis is the computation time (ms.) in logarithmic scale of base 2. A noticeable observation is that the Lookahead strategy is much slower than the Greedy strategy. In fact, adding degrees of look-ahead into utility function actually degrades the performance. The running time of those extra computations increases due to the large number of combinations of privacy constraints.

5.5 Trade-Off between Privacy and Utility

In this experiment, we validate the trade-off between privacy and utility. Technically, we relax each presence constraint $\gamma \in \Gamma$ by increasing the threshold θ to $(1 + r)\theta$, where r is called relaxing ratio and varied according to normal distribution $\mathcal{N}(0.3, 0.2)$. Based on previous experiments, we use the Lookahead strategy on the synthetic data with different input size and fix the number of constraints at 30% of total attributes. For a resulting anonymized schema, privacy is quantified by privacy loss (Eq. 4), while utility is quantified by utility loss (Eq. 5). For comparison purposes, both utility loss and privacy loss values are normalized to [0, 1] by the thresholding technique: $\Delta u = \frac{\Delta u - min_{\Delta u}}{max_{\Delta u} - min_{\Delta u}}$ and $\Delta p = \frac{\Delta p - min_{\Delta p}}{max_{\Delta p} - min_{\Delta p}}$, where $max_{\Delta u}$, $min_{\Delta u}$, $max_{\Delta p}$, $min_{\Delta p}$ are the maximum and minimum values of utility loss and privacy loss in this experiment.

Fig. 6 shows the distribution of all the points whose x and y values are privacy loss and utility loss, respectively. Each point represents the resulting anonymized schema in a specific configuration of the relaxing ratio and the input size. One can easily observe a trade-off between privacy and utility: the utility loss decreases when the privacy loss increases and vice-versa. For example, the maximal privacy (privacy loss = 0) is achieved when the utility is minimal (utility loss = 1). This is because the decrease of presence threshold θ will reduce the number of presented attributes, which contributes to the utility of resulting anonymized schemas.

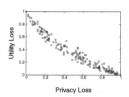

Fig. 6. Trade-off between privacy and utility

6   Related Work

We now review salient work in schema reuse, privacy models, anonymization techniques, and utility measures that is related to our research.

Schema Reuse. There is a large body of research on schema-reuse. Some of the works focus on building large-scale schema repositories [38,17,47,27], designing management models to maintain them [24,13], and establishing semantic interoperability between collected schemas in those repositories [12,19,8]. While, some of the others support to find relevant schemas according to a user query [20,16,36], techniques to speed-up the querying process [39], and visualization aspects [51,46]. Unlike these works, we study the problem of preserving privacy of schema information when presenting the schema attributes as hints for end-users to design new schemas.

Privacy Models. Privacy model is a mean to prevent information disclosure by representing human-defined policies under mathematical standards. Numerous types of information disclosure have been studied in the literature [33], such as attribute disclosure [22] and identity disclosure [23], and membership disclosure [40]. In the context of schema reuse, the linking of attribute to its containing schema can be considered as an information disclosure. There is a wide body of work have proposed privacy models at data level, including differential entropy [5], k-anonymity [48], l-diversity [37], and t-closeness [34]. Different from these works, the model proposed in this paper employs the concept of presence constraint defined at schema-level.

Anonymization Techniques. To implement privacy models, there are a wide range of techniques have been proposed in literature such as generalization, bucketization, randomization, suppression, and differential privacy. In terms of schema reuse, these techniques can be interpreted as follows. The generalization technique [10] replaces the sensitive attribute with a common attribute, which still preserves the utility of attributes shown to users (no noises). The bucketization technique [50] extends the generalization technique for multi-dimensional data through partitioning. The randomization technique [4,6] adds a "noisy" attribute (e.g. 'aaa'). The suppression technique [30] replaces a sensitive attribute by a special value (e.g. '*'). The differential privacy technique [23] enhances both generalization and suppression techniques by balancing between preserving the utility of attributes and adding noises to protect them. Similar to most of other works, we adopt the generalization technique (which does not add noises) to preserve the utility of attributes shown to users. This utility is important in schema design since adding noisy attributes could lead to data inconsistencies.

Quality Measures. To quantify the quality of a schema unification solution, researchers have proposed different quantitative metrics, the most important ones among which are completeness and minimality [9,21]. While the former represents the percentage of original attributes covered by the unified schema, the latter ensures that no redundant attribute appears in the unified schema. Regarding the quality of a schema attribute, [52] also proposed a metric to determine whether it should appear in schema unification. Combining these works and applying their insights in the schema-reuse context, this paper proposed a comprehensive notion of *utility* that measures the quality of an anonymized schema by capturing how many important attributes it contains and how well it covers a wide range of attribute variations.

7   Conclusions

In this work, we addressed the challenge of preserving privacy when integrating and reusing schemas in schema-reuse systems. To overcome this challenge, we introduced a framework for enabling schema reuse with privacy constraints. Starting with a generic model, we introduce the concepts of schema group, anonymized schema and privacy constraint. Based on this model, we described utility function to measure the amount of information of an anonymized schema. After that, we formulated privacy-preserving schema reuse as a maximization problem and proposed a heuristic-based approach to find the maximal anonymized schema efficiently. In that, we investigated various heuristics (greedy repair, look-ahead, constraint decomposition) to improve the computation time of the proposed algorithm as well as the utility of the resulting anonymized schema. Finally, we presented an empirical study that corroborates the efficiency of our framework with main results: effects of heuristics, the correctness of utility function, the guideline of computation time, and the trade-off between privacy and utility.

Our work opens up several future research directions. One pragmatic direction is to support a "pay-as-you-use" fashion in schema-reuse systems. In that, reuse should be defined at a level finer than complete schemas, as schemas are often composed of meaningful building blocks. Another promising direction is the scalability issues of browsing, searching, and visualizing the anonymized schema when the number of original schemas becomes very large. Moreover, our techniques can be applied to other domains such as business process and component-based design.

Acknowledgment. The research has received funding from the EU-FP7 EINS project (grant number 288021) and the EU-FP7 PlanetData project (grant number 257641).

References

1. http://lsirwww.epfl.ch/schema_matching/
2. http://schema.org/
3. http://www.factual.com/
4. Adam, N.R.: Security-control methods for statistical databases: a comparative study. In: CSUR, 515–556 (1989)
5. Agrawal, D.: On the design and quantification of privacy preserving data mining algorithms. In: PODS 2001, pp. 247–255 (2001)
6. Agrawal, R., Srikant, R.: Privacy-preserving data mining. SIGMOD Rec., 439–450 (2000)
7. Antón, A.I., Bertino, E., Li, N., Yu, T.: A roadmap for comprehensive online privacy policy management. Communications of the ACM 50(7), 109–116 (2007)
8. Aumueller, D., Do, H.-H., Massmann, S., Rahm, E.: Schema and ontology matching with coma++. In: SIGMOD, pp. 906–908 (2005)
9. Batista, M.C.M., Salgado, A.C.: Information quality measurement in data integration schemas. In: QDB, pp. 61–72 (2007)
10. Bayardo, R.J., Agrawal, R.: Data privacy through optimal k-anonymization. In: ICDE, pp. 217–228 (2005)
11. Bentounsi, M., Benbernou, S., Deme, C.S., Atallah, M.J.: Anonyfrag: an anonymization-based approach for privacy-preserving bpaas. In: Cloud-I, pp. 9:1–9:8 (2012)
12. Bernstein, P.A., Madhavan, J., Rahm, E.: Generic Schema Matching, Ten Years Later. In: VLDB, pp. 695–701 (2011)
13. Bernstein, P.A., Melnik, S.: Model management 2.0: manipulating richer mappings. In: SIGMOD, pp. 1–12 (2007)
14. Bollacker, K., Evans, C., Paritosh, P., Sturge, T., Taylor, J.: Freebase: a collaboratively created graph database for structuring human knowledge. In: SIGMOD, pp. 1247–1250 (2008)
15. Brickell, J., Shmatikov, V.: The cost of privacy: destruction of data-mining utility in anonymized data publishing. In: KDD, pp. 70–78 (2008)
16. Cafarella, M.J., Halevy, A., Wang, D.Z., Wu, E., Zhang, Y.: Webtables: exploring the power of tables on the web. In: VLDB, pp. 538–549 (2008)
17. Chen, K., Kannan, A., Madhavan, J., Halevy, A.: Exploring schema repositories with schemr. SIGMOD Rec., 11–16 (2011)
18. Clifton, C., Kantarcioglu, M., Doan, A., Schadow, G., Vaidya, J., Elmagarmid, A., Suciu, D.: Privacy-preserving data integration and sharing. In: Proceedings of the 9th ACM SIGMOD Workshop on Research Issues in Data Mining and Knowledge Discovery, pp. 19–26. ACM (2004)
19. Sarma, A.D., Dong, X., Halevy, A.: Bootstrapping Pay-As-You-Go Data Integration Systems. In: SIGMOD, pp. 861–874 (2008)
20. Sarma, A.D., Fang, L., Gupta, N., Halevy, A., Lee, H., Wu, F., Xin, R., Yu, C.: Finding related tables. In: SIGMOD, pp. 817–828 (2012)
21. Duchateau, F., Bellahsene, Z.: Measuring the quality of an integrated schema. In: Parsons, J., Saeki, M., Shoval, P., Woo, C., Wand, Y. (eds.) ER 2010. LNCS, vol. 6412, pp. 261–273. Springer, Heidelberg (2010)

22. Duncan, G.T., Lambert, D.: Disclosure-limited data dissemination. In: JASA, pp. 10–18 (1986)
23. Dwork, C.: Differential privacy. In: Bugliesi, M., Preneel, B., Sassone, V., Wegener, I. (eds.) ICALP 2006. LNCS, vol. 4052, pp. 1–12. Springer, Heidelberg (2006)
24. Franklin, M., Halevy, A., Maier, D.: From databases to dataspaces: a new abstraction for information management. SIGMOD Rec., 27–33 (2005)
25. Glover, F., McMillan, C.: The general employee scheduling problem: an integration of ms and ai. COR, 563–573 (1986)
26. Goldberg, D.E.: Genetic algorithms in search, optimization and machine learning. Addison-Wesley (1989)
27. Gonzalez, H., Halevy, A.Y., Jensen, C.S., Langen, A., Madhavan, J., Shapley, R., Shen, W., Goldberg-Kidon, J.: Google fusion tables: web-centered data management and collaboration. In: SIGMOD, pp. 1061–1066 (2010)
28. Halfond, W., Viegas, J., Orso, A.: A classification of sql-injection attacks and countermeasures, pp. 65–81. IEEE (2006)
29. Hung, N.Q.V., Tam, N.T., Miklós, Z., Aberer, K.: On leveraging crowdsourcing techniques for schema matching networks. In: Meng, W., Feng, L., Bressan, S., Winiwarter, W., Song, W. (eds.) DASFAA 2013, Part II. LNCS, vol. 7826, pp. 139–154. Springer, Heidelberg (2013)
30. Iyengar, V.S.: Transforming data to satisfy privacy constraints. In: SIGKDD, pp. 279–288 (2002)
31. Karp, R.M.: Reducibility Among Combinatorial Problems. In: CCC, pp. 85–103 (1972)
32. Kleinberg, J.M.: Authoritative sources in a hyperlinked environment. J. ACM, 604–632 (1999)
33. Lambert, D.: Measures of disclosure risk and harm. In: JOS, p. 313 (1993)
34. Li, N., Li, T., Venkatasubramanian, S.: t-closeness: Privacy beyond k-anonymity and l-diversity. In: ICDE, pp. 106–115 (2007)
35. Li, T., Li, N.: On the tradeoff between privacy and utility in data publishing. In: SIGKDD, pp. 517–526 (2009)
36. Limaye, G., Sarawagi, S., Chakrabarti, S.: Annotating and searching web tables using entities, types and relationships. In: VLDB, pp. 1338–1347 (2010)
37. Machanavajjhala, A., Kifer, D., Gehrke, J., Venkitasubramaniam, M.: L-diversity: Privacy beyond k-anonymity. TKDD, 24 (2007)
38. Madhavan, J., Bernstein, P.A., Doan, A.-H., Halevy, A.Y.: Corpus-based schema matching. In: ICDE, pp. 57–68 (2005)
39. Mahmoud, H.A., Aboulnaga, A.: Schema clustering and retrieval for multi-domain pay-as-you-go data integration systems. In: SIGMOD, pp. 411–422 (2010)
40. Nergiz, M.E., Atzori, M., Clifton, C.: Hiding the presence of individuals from shared databases. In: SIGMOD, pp. 665–676 (2007)
41. Viet, Q., Nguyen, H., Do, S.T., Nguyen, T.T., Aberer, K.: Towards enabling schema reuse with privacy constraints, EPFL-REPORT-189971 (2013)
42. Nguyen, Q.V.H., Luong, H.X., Miklós, Z., Quan, T.T., Aberer, K.: Collaborative Schema Matching Reconciliation. In: CoopIS (2013)
43. Nguyen, Q.V.H., Thanh, T.N., Miklos, Z., Aberer, K., Gal, A., Weidlich, M.: Pay-as-you-go Reconciliation in Schema Matching Networks. In: ICDE (2014)
44. Quoc Viet Nguyen, H., Wijaya, T.K., Miklós, Z., Aberer, K., Levy, E., Shafran, V., Gal, A., Weidlich, M.: Minimizing human effort in reconciling match networks. In: Ng, W., Storey, V.C., Trujillo, J.C. (eds.) ER 2013. LNCS, vol. 8217, pp. 212–226. Springer, Heidelberg (2013)
45. Peukert, E., Eberius, J., Rahm, E.: Amc - a framework for modelling and comparing matching systems as matching processes. In: ICDE, pp. 1304–1307 (2011)
46. Smith, K., Bonaceto, C., Wolf, C., Yost, B., Morse, M., Mork, P., Burdick, D.: Exploring schema similarity at multiple resolutions. In: SIGMOD, pp. 1179–1182 (2010)

47. Smith, K.P., Mork, P., Seligman, L., Leveille, P.S., Yost, B., Li, M.H., Wolf, C.: Unity: Speeding the creation of community vocabularies for information integration and reuse. In: IRI, pp. 129–135 (2011)
48. Sweeney, L.: k-anonymity: a model for protecting privacy. IJUFKS, 557–570 (2002)
49. Tsui, F.-C., Espino, J.U., Dato, V.M., Gesteland, P.H., Hutman, J., Wagner, M.M.: Technical description of rods: a real-time public health surveillance system. Journal of the American Medical Informatics Association 10(5), 399–408 (2003)
50. Xiao, X., Tao, Y.: Anatomy: Simple and effective privacy preservation. In: VLDB, pp. 139–150 (2006)
51. Yost, B., Bonaceto, C., Morse, M., Wolf, C., Smith, K.: Visualizing Schema Clusters for Agile Information Sharing. In: InfoVis, pp. 5–6 (2009)
52. Yu, C., Jagadish, H.V.: Schema summarization. In: VLDB, pp. 319–330 (2006)

Any Suggestions? Active Schema Support for Structuring Web Information

Silviu Homoceanu, Felix Geilert, Christian Pek, and Wolf-Tilo Balke

IFIS TU Braunschweig, Mühlenpfordstraße 23, 38106 Braunschweig, Germany
{silviu,balke}@ifis.cs.tu-bs.de,
{f.geilert,c.pek}@tu-bs.de

Abstract. Backed up by major Web players schema.org is the latest broad initiative for structuring Web information. Unfortunately, a representative analysis on a corpus of 733 million Web documents shows that, a year after its introduction, only 1.56% of documents featured any schema.org annotations. A probable reason is that providing annotations is quite tiresome, hindering wide-spread adoption. Here even state-of-the-art tools like Google's Structured Data Markup Helper offer only limited support. In this paper we propose SASS, a system for automatically finding high quality schema suggestions for page content, to ease the annotation process. SASS intelligently blends supervised machine learning techniques with simple user feedback. Moreover, additional support features for binding attributes to values even further reduces the necessary effort. We show that SASS is superior to current tools for schema.org annotations.

Keywords: Schema.org, semantic annotation, metadata, structuring unstructured data.

1 Introduction

The Web is a vast source of information and continues to grow at a very fast pace. Unfortunately most of the information is in the form of unstructured text, making it hard to query. Recognizing the importance of structured data for enabling complex queries, the problem of algorithmically structuring information on the Web has been extensively researched, see e.g., [3, 4, 15]. However, current *automatic approaches* still face quality problems and require significant effort for extracting, transforming and loading data. Thus, from a practical perspective they are not yet mature enough to keep up with the volume and velocity, at which new data is published on the Web.

In contrast, the Linked Open Data (LOD) [1, 2] initiative tried a *manual approach*. It offers technology for information providers to directly publish data online in structured form and interlinked with other data. LOD is very flexible since it allows for each data publisher to define its own structure. But this flexibility comes at a price [7]: Although data stores may overlap in terms of the data stored, the vocabulary used for structuring (and thus querying) may seriously differ. Ontology alignment has been proposed as a remedy, but the quality of results is still not convincing [11, 12].

S.S. Bhowmick et al. (Eds.): DASFAA 2014, Part II, LNCS 8422, pp. 251–265, 2014.

To avoid all these problems while improving their query capabilities, major Web search engine providers went a slightly different way. Their *managed approach* builds on a collection of ready-made schemas accessible on schema.org, which are centrally managed by Bing, Google, Yahoo! and Yandex. These schemas are used as a vocabulary to be embedded in the HTML source code of a page using microdata. The main incentive for page owners to use schema.org is that once a Web page features content annotated with schema.org's vocabulary, any search engine can present it as a rich snippet. Furthermore, the Web page has a higher chance of being found by users interested in that very specific content, too. Indeed, motivating page owners to annotate their data with schema.org vocabulary has multiple advantages:

- The effort is spread over many shoulders reducing the effects of volume and velocity at which new data comes to the Web;
- annotations are of high quality – the one creating the data should understand its semantic meaning best;
- the structure is centrally managed and data can be queried globally without complicated alignment operations like in the case of LOD;
- complex queries with Web data are enables, ultimately fostering semantic search for the next generation Web.

But is schema.org being adopted by page owners? An in depth analysis on the acceptance of schema.org reveals, that the number of annotations is in fact very small. The main reason is that the annotation process is quite demanding. Annotators have to repeatedly switch between the page to annotate and schema.org, while browsing through more than 500 schemas with numerous attributes each to find the best matches. Furthermore, adding the actual markup can be tiresome, especially for publishers using "What You See Is What You Get" content management systems.

Sharing the confidence that schema.org will empower complex queries with Web data, we propose *SASS* (Schema.org Annotation Support System), a two stage approach offering support for annotating with schema.org. Analyzing any web page, in the first stage the system finds suggestions for schemas matching page content. For this purpose, simple models are trained with common *machine learning techniques*. But to gain high precision SASS then relies on *user feedback* to validate and fine tune proper schema annotations. The second stage specifically focuses on aiding users in associating schema attributes to values from the page content: *typical attributes* for items of a certain schema have a higher chance of being mentioned in the item data. SASS thus encourages users to consider attributes in order of their typicality. Finally, the system directly generates the HTML code enriched with schema.org annotations. Through the semi-automatic schema matching process and the benefits of considering typical attributes first, our system is superior to solutions like Google's Structured Data Markup Helper (www.google.com/webmasters/markup-helper) or the method recently presented in [13] offering only graphical interface support.

The contribution of this paper can be summarized as follows: we first perform an extensive analysis on the acceptance of semantic annotation technologies with an in-depth focus on schema.org. We then present the design of our annotation support system SASS that relies on lessons learned from the analysis; and finally we present and evaluate machine learning methods for supporting semi-automatic annotations.

2   Web-Scale Analysis on Data Annotation Technologies

Semantic annotation technologies for Web content started to gain importance about a decade ago. Introduced in 2005, microformats are the first important semantic annotation technology. But microformats cover only few entity types annotated through generic HTML "class" attribute. This complicates both the process of annotation and of finding annotations. Attempting to tackle the problems that microformats have, the W3C proposed RDFa as a standard in 2008. It introduces special data annotation attributes for HTML. However it has no centralized vocabulary, leading to heavy fragmentation: Different sources may use different vocabulary to describe the same data. This hinders the process of automatically interpreting or querying data as a whole.

To tackle this problem, search engine providers proposed microdata, a semantic annotation technique relying on few global attributes and a standard vocabulary. The first such vocabulary was "data-vocabulary" proposed by Google in 2009. In mid-2011 Bing, Yahoo and Yandex joined Google's initiative. The "data-vocabulary" was extended to a collection of schemas. Made available on *schema.org*, this collection is evolving continuously to reflect data being published on the Web. To provide for a high level of quality, new schema proposals are reviewed for approval by a standardization committee. An example of a Web page for the movie "Iron Man 3" annotated with schema.org is presented in Figure 1.

a)

```
<div>
    <h1>Iron Man 3</h1>

    <div>
    When Tony Stark's world is torn apart by a formidable terrorist called the
    Mandarin, Stark starts an odyssey of rebuilding and retribution.
    </div>
    <div>
        Director: <span>Shane Black</span>
    </div>
    <div>
        <span>10</span> stars from <span>1337</span> users.
        Reviews: <span>42</span>.
    </div>
</div>
```

b)

```
<div itemscope itemtype="http://schema.org/Movie">
    <h1 itemprop="name">Iron Man 3</h1>

    <div itemprop="description">
        When Tony Stark's world is torn apart by a formidable terrorist called the
        Mandarin, Stark starts an odyssey of rebuilding and retribution.
    </div>

    <div itemtype="http://schema.org/Person" itemscope itemprop="director">
        Director: <span itemprop="name">Shane Black</span>
    </div>

    <div itemtype="http://schema.org/AggregateRating" itemscope itemprop="aggregateRating">
        <span itemprop="ratingValue">10</span> stars from <span itemprop="ratingCount">1337</span> users.
        Reviews: <span itemprop="reviewCount">42</span>.
    </div>
</div>
```

Fig. 1. Web Page section presenting information about movie Iron Man 3. (a) before and b) after annotating it with schema.org.

Currently, there are 529 schemas on schema.org, organized in a 5 level hierarchical structure. Each level introduces a higher degree of specificity. At the root of the hierarchy, there is an all-encompassing schema called "Thing" with 6 general attributes suitable for describing all kinds of entities. Attributes are inherited from parent to child schemas. They may be of basic types e.g., text, boolean, number, etc. or they may in turn represent schemas. For example, for schema Movie, the attribute actor is of type Person which is itself a schema on schema.org. From a structural stance schema.org is similar to DBpedia, the central data repository in LOD. Overall DBpedia comprises 458 schemas. DBpedia maps its structural information to schema.org through the "owl:equivalentClass" attribute. However, the mapping is relatively small: Only 45 links are provided in the current version of DBpedia despite far more semantic similarities easy to spot on manual samples.

To assess the acceptance of schema.org we analyzed ClueWeb12, a publicly available corpus comprising English sites only. The corpus has about 733 million pages, crawled between February and May 2012. It comprises pages of broad interest: The initial seeds for the crawl consisted of 3 million websites with the highest PageRank from a previous Web scale crawl. Our analysis shows that about a year after its introduction, only 1.56% of the websites from ClueWeb12 used schema.org to annotate data. The numbers of pages annotated with mainstream data annotation techniques found in the ClueWeb12 documents are presented in Table 1. The use of different standards reflects the chronology of their adoption: Microformats are the most spread followed by RDFa, microdata and schema.org. It's interesting to notice that while microdata was introduced just a year after RDFa, there is a noticeable difference between their usage rates. The reason for this behavior is that when it was introduced, RDFa was presented as the prime technology for semantic annotation. Many content providers adopted it. Further developments brought by Google in 2009 have been regarded as yet another annotation method. It was only in mid-2011 when microdata became the main annotation technology for the newly proposed schema.org that microdata started gaining momentum. In fact, out of 15 million documents annotated with microdata, 12 million (80%) represent schema.org annotations.

Table 1. Distribution of Annotations in ClueWeb12

Data	Found URLs
Microformats	97,240,541 (12.44%)
RDFa	59,234,836 (7.58%)
Microdata	15,210,614 (1.95)
schema.org	12,166,333 (1.56%)

Out of the 296 schemas available in mid-2012 when ClueWeb12 was crawled, only 244 schemas have been used. To retrieve the state of schema.org at that time we used the Internet Archive (web.archive.org/web/20120519231229/www.schema.org/docs/full.html). The number of annotations per schemas (Table 2) follows a power law distribution with just 10 highest ranking schemas being used for 80% of the annotations and 17 schemas making for already 90% of all annotations. From the low occurring schemas in the long tail, 127 schemas occur less than 1000 times and 96 schemas occur even less than 100 times.

Schemas on schema.org are quite extensive. They include on average 34 attributes. "Thing" is with 6 attributes the smallest schema while "ExercisePlan" with 71 attributes is the most extensive schema on schema.org. The annotations however are by far not as extensive as the structure allows. On average over all annotations, only 4.7 attributes were used. This accounts for about 10% of the attributes available in the corresponding schemas despite remaining data and existing matching attributes. It seems users are satisfied with just annotating some of the attributes. Most probably, this behavior is driven by the fact that rich snippets can only present a few attributes. In consequence users annotate only those few attributes that they consider should be included in the rich snippet. This way, from a user perspective, both the effort of annotating additional information and the risk that the rich snippet would present a random selection out of a broader number of annotated attributes are minimized.

Table 2. Top-20 Schema.org Annotations on the ClueWeb12 Corpus

Schemas	Occurrences	Average Nr. of Attributes	Percentage (Schema.org)
http://schema.org/Blog	5,536,592	5.56	19.57%
http://schema.org/PostalAddress	3,486,397	3.62	12.32%
http://schema.org/Product	2,983,587	2.28	10.54%
http://schema.org/LocalBusiness	2,720,790	3.29	9.62%
http://schema.org/Person	2,246,303	4.97	7.94%
http://schema.org/MusicRecording	1,580,764	2.77	5.59%
http://schema.org/Offer	1,564,257	1.32	5.53%
http://schema.org/Article	1,127,413	1.04	3.99%
http://schema.org/NewsArticle	823,572	3.81	2.91%
http://schema.org/BlogPosting	767,382	3.32	2.71%
http://schema.org/WebPage	659,964	4.11	2.33%
http://schema.org/Review	470,343	3.20	1.66%
http://schema.org/Organization	407,557	1.35	1.44%
http://schema.org/Event	400,721	2.69	1.42%
http://schema.org/VideoObject	396,993	0.47	1.40%
http://schema.org/Place	380,055	2.50	1.34%
http://schema.org/AggregateRating	342,864	1.66	1.21%
http://schema.org/CreativeWork	232,585	2.30	0.82%
http://schema.org/MusicGroup	223,363	1.15	0.78%
http://schema.org/JobPosting	168,542	4.38	0.60%

3   Learning to Annotate Unstructured Data with Schema.org

The benefits of building a structured Web are obvious and the approach followed by schema.org seems promising. Most of the data annotated with schema.org vocabulary represents e-shopping entities like products, restaurants or hotels. Indeed economic factors may have driven the adoption of schema.org for e-shopping relevant data. For the rest, the benefit of having pages presented as rich snippets seems rather small when compared to the effort of annotating data.

Unfortunately, as we have seen, schema.org annotations are not yet used broadly. The main reason invoked insistently on technology blogs on the Web is that the actual process of annotating Web data with schema.org is quite demanding, see e.g., http://readwrite.com/2011/06/07/is_schemaorg_really_a_google_land_grab. In particular, the structure is centrally managed by schema.org and not at the liberty of annotators like in the case of RDFa. This means that when annotating pages one has to:

a)   repeatedly switch between the Web page to annotate and schema.org,

b)   to browse through hundreds of schemas with tens of attributes each trying to find those schemas and attributes that best match the data on the Web page,

c)   and finally to write the microdata annotation with corresponding schema.org URL resources into the HTML code of the page.

With such a complicated process it's no wonder that 1.1% of all found annotations are erroneous. Most frequent errors were bad resource identifiers caused by misspelled schemas or attributes or by schemas and attribute names incorrectly referred through synonyms.

We believe that providing support for the annotation process will make using schema.org much more attractive for all kinds of data. For this purpose, we propose SASS, a system to assist page owners in:

1.   matching schemas from schema.org to the content of a web page,

2.   linking the attributes of the matched schemas to the corresponding values from the page,

3.   and automatically generating the updated HTML page to include the schema.org annotations.

In contrast to simple tool s like Google's Structured Data Markup Helper or the system presented in [13], SASS goes beyond a mere user interface and actively analyzes page content, to find and propose the best matching schemas, to the user.

Let us take a closer look at SASS's basic interactive annotation workflow (cf. also Algorithm 1): Once a page has been created and before publishing it on the Web, the owner loads the HTML source file into SASS's Web-based annotation support system. First the system finds matches between schemas and pieces of page content, using models that have been trained with machine learning techniques on data annotated in ClueWeb12. Theoretically, any selection comprising consecutive words from the page content is a possible candidate for the matching. But considering all possible selections of page content is not feasible. Fortunately, the layout expressed through HTML elements says much about how information is semantically connected. With the help of the Document Object Model (DOM) API the HTML page is represented as a logical structure that connects HTML elements to page content in a hierarchical DOM tree node structure. These nodes envelop the pieces of content that are matched to the schemas. Starting from the most fine-granular nodes (nodes are processed in the reversed order of the depth-first search) the content of each node is checked for possible match with all schemas from schema.org. Once a match is found, the user is requested to provide feedback. If the match is accepted by the user, the system goes in the second stage of linking schema attributes to values.

If the proposed match is not correct the user may browse through the set of next best matching schemas. If there is no suitable match, the user can always refuse the match recommendation. In this case, the system proceeds to the next node. The process continues until all nodes have been considered. In Figure 2.a. we present a snapshot of SASS proposing "Movie" as best matching schema for the Web page content framed in red. Other schemas showing weaker but still relevant match to the same content area are also presented on the left hand side (listed in the descending order of matching strength). In some cases, fine-tuning may be necessary to adjust the size of the selection marked by the red square. If the user considers that a proposed schema matches the marked content, but the selection square is either too broad or too small, the size control elements enable moving up and down the DOM tree node structure to adjust selection size. Of course, allowing users to fine-tune the selection region may also lead to conflicting assignments: for instance if the new selection corresponds to a node that has already been matched to another schema. Since the content in each node may be associated to just one schema, in such cases the user is asked to confirm which assignment is valid.

Once a match has been confirmed, the system proceeds to the second phase of linking the attributes of the chosen schema values from the selected page content. For schemas, the annotations found in ClueWeb12 are enough for learning schema models to support the schema-to-page-content matching process. However, our analysis presented in Section 2 shows that just about 10% of the attributes have actually been used for annotations. Data for attribute annotations is thus very sparse and learning

---

Algorithm 1. Content to schemas matching workflow.

Input: *HTML* – the HTML page content, *SORG* – the set of schemas from schema.org

Output: *R* – *result* set comprising matching relations between nodes and schemas, and corresponding list of attributes and values

```
1:  doc ← DOMDocument(HTML)
2:  N ← DFS(doc).reverse
3:  R ← ∅;
4:  foreach n in N do
5:      S ← ∅; s ← null; A ← ∅
6:      if n in R then
7:          continue        // skip n as it was already added probably by the
                            user's adjusting the selection for some other node
8:      end if
9:      S ← match(n, SORG)
            // returns all schemas from schema.org matching the content from n
10:     if S ≠ ∅ then
11:         s = USER_FEEDBACK(S)
12:         if s ≠ null then
13:             A ← bind(n, s)  // A contains attribute value pairs obtained
                                from the attributes to page content associations made by the
                                user with drag&drop functionality
14:             R.add(n, s, A)   // if n already exists in R the tuple will be
                                updated with the new matching
15:         end if
16:     end if
17: end for
```

algorithms didn't prove as successful as for the schemas matching. The lack of proper attribute annotations shows that this phase would benefit the most from any level of support the system can offer. We assume that attributes showing higher degree of typicality for some item-type have a higher chance of appearing in the respective item-data than other attributes. For the confirmed schema, our system displays the attributes in the descending order of their typicality value and encourages users to consider at least those attributes showing highly typicality for the chosen schema. Those attributes are made available on the right hand side of the screen and can be dragged-and-dropped on the corresponding values if provided in the page content (Figure 2.b.). Start-end selection sliders allow for fine content selection.

With the obtained schema matching and attribute value pairs, the system then enriches the original HTML with the corresponding microdata. The generated HTML source is finally validated with MicrodataJS (github.com/foolip/microdatajs).

Let us now provide a detailed description of all core functionalities of the SASS approach within the workflow described above. In Section 3.1 we present the process of matching schemas to unstructured data - page content enclosed in DOM nodes. In Section 3.2 the process of computing attribute typicality is discussed.

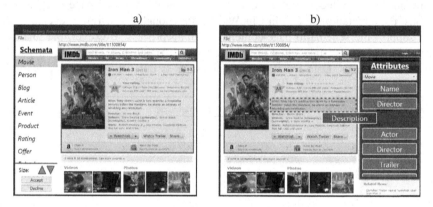

Fig. 2. Schema.org Annotation Support System: a) Matching schema to unstructured data; b) Mapping schema attributes to HTML elements with corresponding values

3.1   Matching Schemas to Unstructured Data

From a sequence of words (the content of a DOM node) and the list of schemas from schema.org, the matching process finds those schemas that "best" match the content. More formally, given $W_n=\{w_1, w_2, ..., w_k\}$ the sequence of words representing the content of node n, and S, the set of URIs for schemas from schema.org, the schemas that best match W_n are:

$$S_{W_n} = \{s_i \mid s_i \in S \wedge match(W_n, s_i) \geq \theta\} \tag{1}$$

where θ is a quality regulating parameter (for our experiments θ was set to 0.5), and $match:\{\text{Words} \times \text{URIs}\} \rightarrow [-1,1]$ is the function for computing the confidence that a certain schema matches the given set of words. The expression of this function depends on the method that is chosen to perform the matching. There are various such methods. For instance, given that schemas published on schema.org describe various types of entities, one of the first approaches that come to one's mind for binding these schemas to unstructured data is entity recognition and named entity recognition. This has proven to work well for some entity types like products, persons, organizations or diseases [18]. However, considering the popular entities annotated on the ClueWeb12 corpus (Table 2), most of them describe more abstract entities e.g. "Blog", "Review", "Offer", "Article", "BlogPosting", etc. In fact, out of the top-20 entity types, entity recognitions systems like OpenNLP (opennlp.apache.org) or StandfordNER [5] recognize less than half of them. Given an observation W_n, and the annotations extracted from ClueWeb12 as a training set comprising a large number of observations whose category of membership is known (the annotated schema) this becomes a problem of identifying the class for observation W_n. Machine learning methods like Naïve Bayes classification or Support Vector Machines have proven successful for text classification tasks even for more abstract entity types ([8, 9]).

Naïve Bayes classifiers rely on probabilities to estimate the class for a given observation. It compares the "positive" probability that some word sequence is the observation for some schema to the "negative" probability that the same word sequence is an observation for other schemas. In this case the matching function is:

$$match_{Bayes}(W_n, s) = P(s|W_n) - P(\bar{s}|W_n) \qquad (2)$$

But neither of the two probabilities can be computed directly from the training set. With the help of Bayes's Theorem $P(s|W_n)$ can be rewritten in computable form as $P(s|W_n) = \frac{P(W_n|s)*P(s)}{P(W_n)}$. Since W_n is a sequence of words that may get pretty long $(W_n=\{w_1, w_2, ..., w_n\})$, and this exact same sequence may occur rarely in the training corpus, to achieve statistically significant data samples "naive" statistical independence between the words of W_n is assumed. The probability of W_n being an observation for schema s becomes: $P(s|W_n) = \frac{\prod_{j=1} P(w_j|s)*P(s)}{\prod_{j=1} P(w_j)}$, and all elements of this formula can be computed based on the training set: $P(s)$ can be computed as the relative number of annotations for schema s, $P(w_j|s)$ the number of annotations for schema s that include w_j relative to the total number of annotations for s, and $P(w_j)$ as the relative number of annotations including w_j. The negative probability $P(\bar{s}|W_n)$ is computed analogously and the matching function on the Bayes classifier can be rewritten as:

$$match_{Bayes}(W_n, s) = \prod_{j=1} P(w_j|s) * P(s) - \prod_{j=1} P(w_j|\bar{s}) * P(\bar{s}) \qquad (3)$$

Being common to all matching involving W_n, $\prod_{j=1} P(w_j)$ can safely be reduced without negative influence on the result. Probabilities for all words from the training set comprising annotations from ClueWeb12 (excluding stop words) build the statistical language models for all schemas, which are of course efficiently precomputed before performing the actual Web site annotations.

Support Vector Machines use a different approach for classification. For each schema, a training set is built. It comprises annotations of the schema ("positive annotations") and annotations of other schemas ("negative annotations") in equal proportions. Each training set is represented in a multidimensional space (the Vector Space Model) with terms from all annotations as the space axes and annotations as points in space. In this representation, SVM finds the hyperplane that best separates the positive from the negative annotations for each schema. In the classification process, given observation W_n, and a schema s, SVM represents W_n in the multidimensional term space and determines the side W_n is positioned in with respect to the hyperplane of s. If it's the positive side then there is a match. The normalized distance from W_n to the hyperplane reveals the confidence of the assignment. The closer W_n is to the hyperplane of s, the less reliable the assignment. In this case, the *match* function is:

$$match_{SVM}(W_n, s) = distance(W_n, H_s) \tag{4}$$

3.2   The Typicality of Attributes for a Chosen Schema

Schemas on schema.org on average have 34 attributes. But our analysis on annotated content shows that on average only 4 attributes are actually being annotated. On manual inspection over annotations for multiple schemas we observed that some of the prominent attributes were left un-annotated although there was matching content available. For instance, for movie data, the 'title' was always annotated along with maybe the 'description' or 'director' attributes. But the 'genre' or 'actors' were often left un-annotated. In fact only 38% of the movie annotations also include 'genre' although simply by performing keyword search with a list of genres we found that the information was available in more than 60% of the cases. Clearly, attributes that are typically associated with the concept of movie will most probably also appear in content about movies. Unfortunately many of those attributes had less than a hundred annotations, not enough for building reliable classification models. To support users in providing more extensive annotations we make it easy for them to find those attributes having a high chance to appear in the content. For this purpose we ask users to consider the attributes in the order of their *typicality* w.r.t. to the chosen schema.

Following on the concept of *typicality* from the field of cognitive psychology in [10] we define *attribute typicality* and present a novel and practical rule for actually calculating it. This method doesn't require that attributes themselves be annotated, schema annotations are enough. It's built on top of open information extraction tools and works directly with unstructured data. Starting from content that has been annotated with a certain schema the method is able to compute typicality values for the schema attributes that it finds, even if no annotation is provided for the attributes. To be self-contained, in the following we briefly describe the core of attribute typicality.

The Concept of Typicality. It has been often shown that some instances of a semantic domain are more suitable than others to represent that domain: For instance Jimmy Carter is a better example of an American president than William Henry Harrison. In her quest for defining the psychological concept of typicality, Eleanor Rosch showed empirically that the more similar an item was to all other items in a domain, the more typical the item was for that domain. In fact, the experiments show that typicality strongly correlates (Spearman rhos from 0.84 to 0.95 for six domains) with *family* *resemblance* a philosophical idea made popular by Ludwig Wittgenstein in [19]. For family resemblance Wittgenstein postulates that the way in which family members resemble each other is not defined by a (finite set of) specific property(-ies), but through a variety of properties that are shared by some, but not necessarily all members of a family. Based on this insight, Wittgenstein defines a simple family-member similarity measure based on property sharing:

$$S(X_1, X_2) = |X_1 \cap X_2| \tag{5}$$

where X_1 and X_2 are the property sets of two members of the same family. But this simple measure of family resemblance assumes a larger number of common properties to increase the perceived typicality, while larger numbers of distinct properties do not decrease it. In [16] Tversky suggests that typicality increases with the number of shared properties, but to some degree is negatively affected by distinctive properties:

$$S(X_1, X_2) = \frac{|X_1 \cap X_2|}{|X_1 \cap X_2| + \alpha|X_1 - X_2| + \beta|X_2 - X_1|} \tag{6}$$

where α and $\beta \geq 0$ are parameters regulating the negative influence of distinctive properties. For $\alpha = \beta = 1$ this measure becomes the well-known Jaccard coefficient.

Following on the theory introduced by Wittgenstein and extended by Tversky, properties that an entity shares with its family are more typical for the entity than properties that are shared with other entities. Also in the context of Web data, similar entities, in our case items that share the same schema, can be considered to form families. When talking about factual information extracted from the Web we have to restrict the notion of family resemblance based on generic properties (like characteristics, capabilities, etc.) to clear cut attributes. Attributes are in this case given by predicates extracted from (subject, predicate, object) triple-relations extracted from text. Given some family F consisting of n entities $E_1, \dots, E_n$ all being annotated with the same schema, let's further assume a total of k distinct attributes given by predicates $p_1, \dots, p_k$ are observed for family F in the corresponding entities' textual annotations. Let X_i and X_j represent the attribute sets for two members E_i and E_j, then:

$$|X_i \cap X_j| = 1_{X_i \cap X_j}(p_1) + 1_{X_i \cap X_j}(p_2) + \dots + 1_{X_i \cap X_j}(p_k) \tag{7}$$

where $1_X(p) = \begin{cases} 1 \text{ if } p \in X \\ 0 \text{ if } p \notin X \end{cases}$ is a simple indicator function.

Now we can rewrite Tversky's similarity measure to make all attributes explicit:

$$S(X_i, X_j) = \frac{\sum_{l=1}^{k} 1_{X_i \cap X_j}(p_l)}{|X_i \cap X_j| + \alpha|X_i - X_j| + \beta|X_j - X_i|} \tag{8}$$

According to Tversky, each attribute shared by X_i and X_j contributes evenly to the similarity score between X_i and X_j. This allows us to calculate the *contribution score* of each attribute of any member of the family (in our case schema) to the similarity of each pair of members: Let p be an attribute of a member from F. The *contribution score* of p to the similarity of any two attribute sets X_i and X_j, denoted $C_{X_i, X_j}(p)$, is:

$$C_{X_i, X_j}(p) = \frac{1_{X_i \cap X_j}(p)}{|X_i \cap X_j| + \alpha|X_i - X_j| + \beta|X_j - X_i|} \tag{9}$$

Attribute Typicality. Let F be a set of n entities $E_1, \dots, E_n$ having the same schema type, represented by their respective attribute sets $X_1, \dots, X_n$. Let U be the set of all distinct attributes of all entities from F. The typicality $T_F(p)$ of an attribute $p \in U$ w.r.t. F is the average contribution of p to the pairwise similarity of all entities in F:

$$T_F(p) = \frac{1}{C_2^n} \cdot \sum_{i=1}^{n-1} \sum_{j=i+1}^{n} c_{X_i, X_j}(p) \tag{10}$$

where C_2^n represents the number of possible combinations of entities from F.

Typicality values change slowly over time and are influenced only by significant data evolution. For the purpose of this application, the typicality values for each attribute of each schema can be computed as an offline process, on a monthly basis.

4   Evaluation

The approach introduced in this paper requires two stages for performing a full annotation: the first is for matching a piece of content to a schema and the second stage is for associating attributes proposed by the system in the descending order of their typicality, to page content. The main merits of the system are to suggest best matching schemas for the first, and to compute attribute typicality for the second stage.

For evaluating the schema matching functionality we prepared two data sets. Each has about 60,000 annotated web pages randomly harvested from ClueWeb12 comprising annotations with about 110 different schemas each. One of them is used as a training corpus for the classification methods. The other one is used as a test set. The test set is stripped of all annotations and provided to our system. We disabled user feedback at this stage for evaluation purposes. Instead, each match proposed by the system is accepted as correct. We compare the pages annotated by the system for both Naïve Bayes and SVM, to the pages from the original test set and measure the schema matching effectiveness in terms of precision and recall.

On inspection over the results, about 5% of the schemas were not detected at all. The reason for this behavior is the fact that these schemas are present in the test set but they have no or almost no occurrences (up to 10) in the training set. Increasing the size of the training set helps reducing the number of undetected schemas. In fact, initial experiments with 10,000 and 30,000 web pages as training sets, with smaller schema annotation coverage, showed higher numbers of undetected schemas.

Overall, on the 110 schema annotations the system achieves on average 0.59 precision and 0.51 recall for Naïve Bayes and 0.74 precision and 0.76 recall for SVM, Simply matching schemas at random, as a comparison method, results in precision and recall lower than 0.01. The result values vary strongly from schema to schema. For brevity reasons, in Table 3 we show the results for 15 schemas. The system is counting on user feedback to even out the so called "false alarms" emphasized by precision. But the "false dismissals" emphasized by recall are much harder to even out by the user. The system proposes a list of alternatives (formula 1), but if the matching schema is missing from this list, the corresponding annotation will most probably fail. For this reason, the 15 schemas presented in Table 3 are chosen to cover the whole spectrum of F_2-measure values, given that the F_2-measure weights recall twice as much as precision.

No correlation between the number of occurrences in the training set and results could be observed. Having hundreds of schema annotations seems to lead to results similar to having tens of thousands of annotations. A few schemas, especially in the case of Naïve Bayes, have catastrophic precision and recall values (less than 0.01), despite occurring more than 4,000 times in the training set. These are schemas with very broad meaning e.g. "WebPage" or "Thing". Overall, SVM does better than Naïve Bayes. But it is interesting to notice that for many schemas the two approaches seem to complement each other: schemas where the Bayes achieves bad results are handled much better by SVM and vice versa. This finding encourages us to believe that approaches relying boosting meta-algorithms like the well know AdaBoost [6] will provide even better results.

For the second stage, correctness of the typicality value of an attribute can only be assessed through broad user studies. Our experiments, presented in detail in [10] show that with the attribute typicality method, the top-10 most typical attributes are selected with average precision and recall values of 0.78 and 0.6 respectively. The lower recall value is explained by the fact that attributes are extracted from text, and don't always exist in the corresponding schema.org schemas.

Table 3. Precision and Recall values for the matching of schemas with Bayes and SVM

		Naïve Bayes				SVM				
Nr.	Schemata	Prec.	Rec.	#	F_2	F_2	Prec.	Rec.	#	Schemata Nr.
1	JobPosting	0.80	0.91	4,129	0.89	0.95	0.94	0.95	10,622	Blog 10
2	Event	0.63	0.78	15,856	0.74	0.85	0.79	0.87	4,129	JobPosting 1
3	LocalBusiness	0.86	0.71	65,691	0.74	0.82	0.83	0.82	126,220	Product 9
4	Offer	0.49	0.82	73,907	0.72	0.81	0.72	0.84	65,691	LocalBusiness 3
5	MusicRecording	0.42	0.77	12,848	0.66	0.77	0.69	0.79	34	EducationEvent 14
6	Movie	0.38	0.38	3,407	0.38	0.68	0.63	0.69	10,395	Place 8
7	Review	0.53	0.32	4,551	0.35	0.68	0.59	0.70	4,551	Review 7
8	Place	0.47	0.30	10,395	0.32	0.58	0.54	0.60	1,153	MobileApplication 13
9	Product	0.92	0.26	126,220	0.30	0.57	0.52	0.59	14,398	Article 12
10	Blog	0.11	0.16	10,622	0.15	0.46	0.44	0.46	3,407	Movie 6
11	Organization	0.52	0.10	10,558	0.12	0.46	0.41	0.47	12,848	MusicRecording 5
12	Article	0.06	0.05	14,398	0.05	0.45	0.43	0.46	4,725	WebPage 15
13	MobileApplication	0.01	0.04	1,153	0.02	0.41	0.41	0.41	15,856	Event 2
14	EducationEvent	0.00	0.25	34	0.01	0.39	0.38	0.40	10,558	Organization 11
15	WebPage	0.00	0.00	4,725	0.00	0.16	0.17	0.16	73,907	Offer 4

On average the system matches schemas correctly even without user feedback in 2 out of 3 cases. But the overall quality of the results doesn't encourage us to believe in the feasibility of a fully automatic annotation system. User input is especially important for the attribute annotations. The attributes being presented first, show high typicality values and have a higher chance of appearing in the content to be annotated. This reduces the effort needed for broader annotations and has the benefit of controlling that at least the important attributes are included in the annotation.

5   Related Work

Acknowledging the difficulties users have when annotating data with schema.org, a variety of tools have recently been proposed (schema.rdfs.org/tools.html). Schema-Creator (schema-creator.org) and microDATAGenerator (microdatagenerator.com) are two important solutions for form based interface-focused tools for schema.org annotations. Google's Structured Data Markup Helper offers more elaborate GUI support for more comfortable manual content annotation. An in-depth analysis regarding the visualization of un-structured semantic content along with a formal description of visualization concepts is presented in [13]. Building on these concepts the authors propose and evaluate a graphical user interface on two application scenarios for annotating data with schema.org. In contrast, our system goes beyond a simple GUI and makes high quality suggestions for the annotations.

In [17] the authors present MaDaME, a system that infers mappings between content highlighted by the user and schemas from schema.org. For this purpose the system relies on WordNet. But the system is only appropriate for annotating named entities and nouns known to WordNet. Everything else is not supported. Atomic content like nouns or names also have to be highlighted by the user for the annotation process as the system doesn't process pages as a whole. In [14], a tool for adding schema.org types automatically is presented. It relies on domain knowledge and NER to extract key terms and to generate structured microdata markup. It requires a high quality knowledge base with metadata represented in RDF for each schema to be annotated and has been show to work only on patent data. This will not scale for all schemas. Furthermore, NER alone cannot deal with all types of schemas from schema.org.

6   Conclusions and Future Work

The Web is an abundant source of information – however, mostly in unstructured form. Querying such data is difficult and the quality of results obtained by keyword-based approaches is far behind the quality offered by querying structured data. Sustained by major Web players and relying on a controlled set of schemas, schema.org has the potential to change this situation once and for all. Unfortunately, annotating data with schema.org still is a tiresome process, hindering its wide-spread adoption. Since state-of-the-art tools like Google's Markup Helper offer only limited benefits for annotators, we inspected the feasibility of providing better annotation support.

Driven by insights into schema.org annotations obtained by analyzing a large corpus of 733 million web documents, we derived a set of desirable design goals. A successful support has to always maintain high annotation suggestion quality, while using supervised machine learning techniques to cater for the highest possible amount of automation. Our innovative SASS approach shows that given the right blend of techniques integrated in an intelligent workflow, existing annotations can indeed be effectively used for training high quality models for schema matching. Relying on the concept of attribute typicality SASS first offers those attributes having higher chance of appearing in the page content. As our evaluations show this indeed essentially reduces the effort of searching through attribute lists. Currently, these features make our system superior to any other tool offering support for schema.org annotations.

Schema matching approaches have different strengths. In future work, we plan to evaluate the performance of boosting meta-algorithms employing multiple matching techniques. User feedback generated through the system usage can also be used to improve the quality of the suggestions, a subject we leave to future work.

References

1. Berners-Lee, T.: Linked Data. Design issues for the World Wide Web Consortium (2006), http://www.w3.org/DesignIssues/LinkedData.html
2. Bizer, C., et al.: Linked Data - The Story So Far. Int. J. Semant. Web Inf. Syst. (2009)
3. Cafarella, M.J., et al.: WebTables: Exploring the Power of Tables on the Web. PVLDB (2008)
4. Cafarella, M.J., Etzioni, O.: Navigating Extracted Data with Schema Discovery. Proc. of the 10th Int. Workshop on Web and Databases, WebDB (2007)
5. Finkel, J.R., et al.: Incorporating Non-local Information into Information Extraction Systems by Gibbs Sampling. In: Proc. of Annual Meeting of the Assoc. for Comp. Linguistics, ACL (2005)
6. Freund, Y., Schapire, R.E.: A Decision-Theoretic Generalization of On-Line Learning and an Application to Boosting. J. Comput. Syst. Sci. 55, 1 (1997)
7. Homoceanu, S., Wille, P., Balke, W.-T.: ProSWIP: Property-based Data Access for Semantic Web Interactive Programming. In: Alani, H., et al. (eds.) ISWC 2013, Part I. LNCS, vol. 8218, pp. 184–199. Springer, Heidelberg (2013)
8. Homoceanu, S., et al.: Review Driven Customer Segmentation for Improved E-Shopping Experience. In: Int. Conf. on Web Science, WebSci (2011)
9. Homoceanu, S., et al.: Will I Like It? Providing Product Overviews Based on Opinion Excerpts. IEEE (2011)
10. Homoceanu, S., Balke, W.-T.: A Chip Off the Old Block – Extracting Typical Attributes for Entities based on Family Resemblance (2013) (Under submission), http://www.ifis.cs.tu-bs.de/node/2859
11. Jain, P., et al.: Contextual ontology alignment of LOD with an upper ontology: A case study with proton. The Semantic Web: Research and Applications (2011)
12. Jain, P., et al.: Ontology Alignment for Linked Open Data. Information. Retrieval. Boston (2010)
13. Khalili, A., Auer, S.: WYSIWYM – Integrated Visualization, Exploration and Authoring of Un-structured and Semantic Content. In: WISE (2013)
14. Norbaitiah, A., Lukose, D.: Enriching Webpages with Semantic Information. In: Proc. Dublin Core and Metadata Applications (2012)
15. Suchanek, F.M., Weikum, G.: YAGO: A Core of Semantic Knowledge Unifying WordNet and Wikipedia. In: WWW (2007)
16. Tversky, A.: Features of similarity. Psychol. Rev. 84, 4 (1977)
17. Veres, C., Elseth, E.: Schema. org for the Semantic Web with MaDaME. In: Proc. of I-SEMANTICS (2013)
18. Whitelaw, C., Kehlenbeck, A., Petrovic, N., Ungar, L.: Web-scale named entity recognition. In: CIKM (2008)
19. Wittgenstein, L.: Philosophical investigations. The MacMillan Company, New York (1953)

ADI: Towards a Framework of App Developer Inspection

Kai Xing, Di Jiang, Wilfred Ng, and Xiaotian Hao

Department of Computer Science and Engineering
Hong Kong University of Science and Technology, Hong Kong
{kxing,dijiang,wilfred,xhao}@cse.ust.hk

Abstract. With the popularity of smart mobile devices, the amount of mobile applications (or simply called apps) has been increasing dramatically in recent years. However, due to low threshold to enter app industry, app developers vary significantly with respect to their expertise and reputation in the production of apps. Currently, there is no well-recognized objective and effective means to profile app developers. As the mobile market grows, it already gives rise to the problem of finding appropriate apps from the user point of view. In this paper, we propose a framework called *App Developer Inspector* (ADI), which aims to effectively profile app developers in aspects of their expertise and reputation in developing apps. ADI is essentially founded on two underlying models: the *App Developer Expertise* (ADE) model and the *App Developer Reputation* (ADR) model. In a nutshell, ADE is a generative model that derives the latent expertise for each developer and ADR is a model that exploits multiple features to evaluate app developers' reputation. Using the app developer profiles generated in ADI, we study two new applications which respectively facilitate app search and app development outsourcing. We conduct extensive experiments on a large real world dataset to evaluate the performance of ADI. The results of experiments demonstrate the effectiveness of ADI in profiling app developers as well as its boosting impact on the new applications.

Keywords: App Developer, Profiling, App Searching, App Development Outsourcing.

1 Introduction

Nowadays, more and more people use smart phones as their primary communication tools. This trend leads to the booming of the mobile app market, which is estimated to reach 25 billion US$ by 2015 [1]. In the face of the lucrativeness of the app market, many people plunge themselves into developing mobile apps. However, app development has relatively low threshold to enter and thus either individual amateurs or well-organized studios can release their apps to the

---

[1] http://www.prnewswire.com/news-releases/marketsandmarkets-world-mobile-applications-market-worth-us25-billion-by-2015-114087839.html

S.S. Bhowmick et al. (Eds.): DASFAA 2014, Part II, LNCS 8422, pp. 266–280, 2014.
© Springer International Publishing Switzerland 2014

market. In light of the huge discrepancy among app developers, we propose a formal framework that can effectively evaluate and comprehensively analyze app developers. The study and development of this framework can not only facilitate better app quality control but also promote new app development in the app ecosystem.

In this paper, we propose a framework named *App Developer Inspector* (ADI) for profiling app developers. ADI mainly consists of two underlying models, the *App Developer Expertise* (ADE) model and the *App Developer Reputation* (ADR) model, respectively profiling app developers' expertise and reputation. ADE is a novel generative model to derive app developers' expertise. ADR, which is a RankSVM-based model, undertakes the function of evaluating app developers' reputation. The expertise reveals which kind of app functionalities (such as sports games, communication applications and so on) an app developer is expert in. The reputation indicates a developer's overall trustworthiness and proficiency in developing high quality apps.

In order to demonstrate the use of app developer profiles obtained from ADI, we study the following two applications.

- **Facilitate App Search.** App developers' reputation and expertise are both significant factors in developing effective app search engine. There exist many "Look-Alike" apps having similar names and descriptions with the high quality ones[5]. When users search for a popular high quality app, app search engines sometimes rank the "Look-Alike" apps high in the searching results without taking app developers' reputation into consideration.
- **App Development Outsourcing.** App developer profiles are also useful in app development outsourcing. With the information of app developers' expertise, employers[2] are able to find proper developers to undertake app development.

We conduct exhaustive experiments on a large real world dataset to show the effectiveness of ADE and ADR. The results of experiments demonstrate that ADE performs well in generating app developers' expertise and ADR shows good performance in evaluating app developers' reputation. We also demonstrate the boosting impact of app developer profiles in facilitating app search and app development outsourcing by detailed experimentation.

In summary, the main contributions of this paper are as follows:

- To the best of our knowledge, this work is the first to systematically study how to profile app developers. We develop a new framework, *App Developer Inspector* (ADI), which seamlessly integrates multiple information sources to derive app developers' expertise and reputation.
- We conduct comprehensive experiments to verify the effectiveness of the ADI framework. The experimental results show that ADE effectively generates app developers' expertise in fine granularity and ADR objectively evaluates app developers' reputation.

---

[2] In this work, we refer companies or individuals who want to outsource app development as the employers.

– We show how to apply app developers' reputation in facilitating app search and verify the improvement by exhaustive experiments. For app development outsourcing, we propose a new method BMr-BR to recommend quality developers.

The rest of the paper is organized as follows. In Section 2, we review the related literature. In Section 3, we provide an overview of the *App Developer Inspector* framework. In Section 4, we propose ADE model to obtain app developers' expertise. In Section 5, we evaluate app developers' reputation with the ADR model. In Section 6, we introduce the applications of app developer profiles in facilitating app search and app development outsourcing. In Section 7, we present the experimental results. Finally, the paper is concluded in Section 8.

2   Related Work

Our framework involves a spectrum of techniques that are briefly discussed as follows.

Smart Phone App. There are some works on smart phone apps but most of them focused on app security. In [25], a systematic study was presented to detect malicious apps on both official and unofficial Android Markets. Alazab *et al.* [1] used the Android application sandbox Droidbox to generate behavioral graphs for each sample and these provided the basis of development of patterns to aid in identifying malicious apps. X.Wei *et al.* [24] described the nature and sources of sensitive data, what malicious apps can do to the data, and possible enterprise solution to secure the data and mitigate the security risks. Di *et al.* [11] proposed a framework of utilizing semantic information for app search.

Topic Modeling. In recent years, topic modelling is gaining momentum in data mining. Griffiths *et al.* [6] applied Latent Dirichlet Allocation (LDA) to scientific articles and studied its effectiveness in finding scientific topics. There follow more topic models that are proposed to handle the problems of document analysis that exist in specific domains. Kang *et al.* [13] proposed a topic-concept cube which supports online multidimensional mining of query log. Mei *et al.* [17] proposed a novel probabilistic approach to model the subtopic themes and spatiotemporal theme patterns simultaneously. Some recent work on query log analysis also studied the impact of temporal and spatial issues. Ha-Thuc *et al.* [7] proposed an approach for event tracking with emphasis on scalability and selectivity. Di *et al.* [9],[10] also studied the spatial issues in web search data with topic modeling. To the best of our knowledge, our work is the first one to systemically study how to utilize topic modeling to profile app developers' expertise.

Learn to Rank. Our work is related to learning to rank techniques, which is a intensively studied area in information retrieval and machine learning. RankSVM was proposed in [12] to optimize search engines via clickthrough data. In [22] and[8], RankSVM had been successfully applied to identify quality tweets on the social network *Twitter* and used social network user profile to personalize web search result.

Expert Finding. The application of app development outsourcing has some similarities with expert finding. Maarten de Rijke *et al.* [2] proposed two models to search experts on a given topic from an organization's document repositories. Although there are some similarities between expert finding and app development outsourcing, e.g. they all aim to find "expert", the difference between them is fundamental. Two models in [2] focused on mining experts from corporation documents while our app development outsourcing is intended to recommend proper and quality app developers.

3   Overview of ADI

In this section, we provide a general overview of the ADI framework, including its architecture and a glance of app developer profiles.

3.1   Architecture of ADI

As Figure 1 shows, there are two models, ADE and ADR, to generate app developer profiles. Before applying ADR, We first extract some features, such as popularity, good ratio, web site quality and so on, from multiple information sources, i.e., information from apps, users and developers. Then ADR model generates app developers' reputation on basis of the extracted features. Meanwhile, a novel generative model ADE is employed to compute app developers' expertise from app descriptions and categories. Finally, an app developer profile that consists of app developers' expertise and reputation is generated.

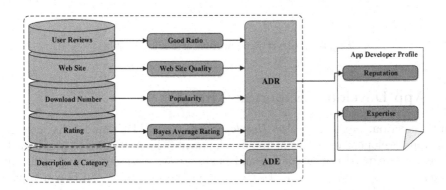

Fig. 1. Simplified Architecture of App Developer Inspector (ADI)

3.2 App Developer Profiles

In ADI, app developer profiles consists of two components, which are respectively app developers' reputation and app developers' expertise.

The first component, app developers' reputation, which indicates app developer's proficiency and trustworthiness in developing apps, is represented by a real value. We apply ADR to compute this value and higher the value is, higher reputation a developer achieves.

The other component, app developers' expertise, depicts functionality-based expertise of developers in app development. In app developers' expertise are also two subcomponents. One is a series of expertises, each of which is represented by a set of keywords. These keywords are generated by our ADE model to characterize a certain expertise. For each expertise, a proficiency is given to indicate how proficient a developer is in the expertise. Take Rovio (a famous game studio) as an example, in Figure 2 expertise 1 is represented by "game, birds, war..." and the corresponding proficiency is 0.7. The underlying reason of including this expertise part is that an app developer generally has more than one set of expertises and they may be in different proficiency level as well.

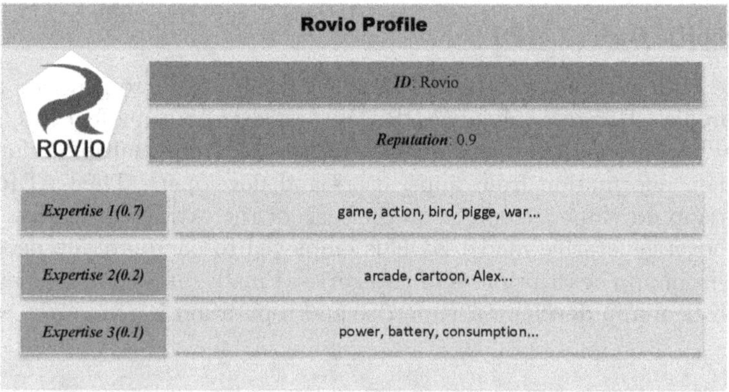

Fig. 2. Abstract Presentation of Rovio's profile

4 App Developer Expertise Model

In this section, we describe *App Developer Expertise* (ADE) model that derives the app developers' expertise in app development. The ADE aims to profile each developer's expertise in a concise and flexible way. We assume that each app developer has a Multinomial *expertise* (or in the metaphor of LDA, a topic) distribution. We first group the app descriptions of the same developer as a document. Then, we filter out the non-informative words according to a stop-word list provided in [14]. An interesting phenomenon in the app corpus is that developers are actually have implicit *links*, which can be obtained by analyzing the download records of the users. For example, if app a_1 developed by developer d_1 and app a_2 developed by developer d_2 are both downloaded by the same user, we create a link between d_1 and d_2. We utilize a $D \times D$ matrix M to store the link information and the entry $M[i,j]$ is computed by the number of times that i's apps have been downloaded with j's apps.

The generative process of this model is illustrated in Algorithm 1 and the notation used is summarized in Table 1.

Algorithm 1. Generative Process of ADE

---

1. **for** each topic $k \in 1, ..., K$ **do**
2.   draw a word distribution $\phi_k \sim \text{Dirichlet}(\beta)$;
3.   draw a category distribution $\phi'_k \sim \text{Dirichlet}(\delta)$;
4. **end for**
5. **for** each document $d \in 1, ..., D$ **do**
6.   draw d's topic distribution $\theta_d \sim \text{Dirichlet}(\alpha)$
7.   sample a linked developer d_i with proportion to link weight of $l(d, d_i)$, then draw a document specific distribution θ_{d_i};
8.   combine θ_d and θ_{d_i} by tuning parameter λ to generate a document distribution θ;
9.   **for** each sentence $s \in d$ **do**
10.   choose a topic $z \sim \text{Multinomial}(\theta)$;
11.   generate words $w \sim \text{Multinomial}(\phi_z)$;
12.   generate the category $c \sim \text{Multinomial}(\phi'_z)$;
13.   **end for**
14. **end for**

We aim to find an efficient way to compute the joint likelihood of the observed variables with hyperparameters:

$$P(\mathbf{w}, \mathbf{z}|\alpha, \beta, \delta, \lambda, \mathbf{l}) = P(\mathbf{w}|\mathbf{z}, \beta)P(\mathbf{c}|\mathbf{z}, \delta)P(\mathbf{z}|\alpha.\lambda, \mathbf{l}). \tag{1}$$

The probability of generating the words is given as follows:

$$P(\mathbf{w}|\mathbf{z}, \beta) = \int \prod_{d=1}^{D} \prod_{s=1}^{S_d} \prod_{i=1}^{W_{ds}} P(w_{dsi}|\phi_{z_{ds}})^{N_{dsw_{dsi}}} \prod_{z=1}^{K} P(\phi_z|\beta)d\Phi. \tag{2}$$

The probability of generating the categories is given as follows:

$$P(\mathbf{c}|\mathbf{z}, \beta) = \int \prod_{d=1}^{D} \prod_{i=1}^{S_d} P(c_{di}|\phi'_{z_{di}}) \prod_{z=1}^{K} P(\phi'_z|\beta)d\Phi. \tag{3}$$

After combining the formula terms, we apply Bayes rule and fold terms into the proportionality constant, the conditional probability of the kth topic for the ith sentence is defined as follows:

$$P(z_i = k|\mathbf{z}_{-i}, \mathbf{w}, \mathbf{l}, \alpha, \beta, \lambda) \propto$$

$$\frac{(1-\lambda)C_{dk}^{DK} + \lambda C_{lk}^{DK} + \alpha_k}{\sum_{k'=1}^{K}((1-\lambda)C_{dk'}^{DK} + \lambda C_{lk'}^{DK} + \alpha_{k'})} \frac{C_{kc}^{KC} + \delta_c}{\sum_{c'=1}^{C}(C_{kc}^{KC} + \delta_{c'})}$$

$$\frac{\Gamma(\sum_{w=1}^{W}(C_{kw}^{KW} + \beta_w))}{\Gamma(\sum_{w=1}^{W}(C_{kw}^{KW} + \beta_w + N_{iw}))} \prod_{w=1}^{W} \frac{\Gamma(C_{kw}^{KW} + \beta_w + N_{iw})}{\Gamma(C_{kw}^{KW} + \beta_w)} \qquad (4)$$

where C_{lk}^{DK} is the number of sentences that are assigned topic k in document l, which is a randomly sampled document that is linked by the document d.

After processing the app developers by the proposed model, the ith developer's profile is represented by a search topic vector $(\theta_{i1}, \theta_{i2}, ..., \theta_{in})$, where θ_{ik} is a real number that indicates the ith user's endorsement for the kth search topic. The value of θ_{ik} is computed as follows:

$$\theta_{ik} = \frac{C_{dk}^{DK} + \alpha_k}{\sum_{k'=1}^{K}(C_{dk'}^{DK} + \alpha_{k'})}. \qquad (5)$$

Table 1. Notations Used in the ADI Framework

Parameters	Meaning	Parameters	Meaning
D	the number of documents	λ	a parameter controlling the influence of the linked document
K	the number of topics	z_i	the topic of word i
z	a topic	$\mathbf{z}_{-i}$	the topic assignments for all words except word i
w	a word	$\mathbf{w}$	word list representation of the corpus
θ	multinomial distribution over topics	C_{kc}^{KC}	the number of times that c is assigned topic k
ϕ	multinomial distribution over words	C_{lk}^{DK}	the number of words assigned to topic k in the linked document l
ϕ'	multinomial distribution over categories	δ	Dirichlet prior vector for ϕ'
α	Dirichlet prior vector for θ	C_{dk}^{DK}	the number of sentences that are assigned topic k in document d
β	Dirichlet prior vector for ϕ	C_{kw}^{KW}	the number of times that w is assigned topic k
$\mathbf{l}$	link list		

We utilize the generative model to mine an individual developer's expertise of a specific field. Compared with the category information, the topics whose amount can be determined by the users represent the developer characteristics with finer granularity. Note that the topic amount K can be customized and thus, it strikes a good balance between efficiency and granularity. Note that the method proposed here is potentially scalable to very large datasets. For example, the Gibbs sampling is scaled to run very large sized datasets in [18].

5   App Developer Reputation Model

In this section, we introduce *App Developer Reputation* (ADR) model to evaluate app developers' reputation. Before elaborating how ADR works, we define app developers' reputation as the overall trustworthiness and proficiency in developing apps. ADR is model based on RankSVM[12] and works as follows. It first ranks each developer using a series of extracted features, i.e., popularity, rating and so forth. Then reputation of each developer can be calculated according to the generated ranking.

5.1   App Developer Reputation Generation

The generation process of ADR is carried out in three steps. We first input some pairwise training instances into RankSVM and get a generated rank model. Next, we utilize this rank model to rank a set of developers. Finally, the reputation of the developer i is given by:

$$R_i = 1 - \frac{rank_i}{N},$$

(6)

where $rank_i$ is the ranking of developer i, N is the total number of developers.

5.2   Extracted Features

There are four features, popularity, rating, web site quality, and good ratio, used in RankSVM. We now present the construction of these features from multiple information sources.

App Popularity. The popularity of an app is evaluated by the number of times that the app has been downloaded[5]. To compute a developer's *app popularity*, we first compute the popularity of an app as follows:

$$p_j = log(N_j).$$

(7)

Where N_j denotes the downloaded number of app j. Then the *app popularity* feature of a developer i is defined as follows:

$$Pop_i = \frac{\sum_{j \in A(i)} p_j}{|A(i)|},$$

(8)

where $A(i)$ is the collection of apps developed by the developer i.

Bayes Average Rating. The more ratings are given to an app, the more reliable the average of ratings is to reflect the app quality. Otherwise, app rating may be a misleading indicator. Here we use *Bayes average rating* to represent this intuition:

$$Br_j = \frac{N_{av} \cdot r_{av} + r_j \cdot N_j}{N_j + N_{av}}, \tag{9}$$

where N_{av} and r_{av} are respectively the average number of rating from users and average rating over all apps, r_j and N_j respectively denote the raw rating and the number of rating for app j.

If number of rating for an app is much smaller than the average number of rating over all apps, *Bayes average rating* is close to the average rating over all apps. Otherwise, *Bayes average rating* approximately equals app's raw rating. This property makes *Bayes average rating* a more reliable indicator.

Considering a developer may produce more than one apps, we use the average *Bayes average rating* of all the apps developed by the developer as the rating feature. Let $A(i)$ contain all the apps developed by developer i, rating feature is computed as follows:

$$\overline{Br_i} = \frac{1}{|A(i)|} \sum_{j \in A(i)} Br_j. \tag{10}$$

Web Site Quality. Good developers usually have their own web sites where they post information about their app products. In this case, these web sites can be an important auxiliary information for evaluating app developer reputation. Here we define *web site quality* feature of a developer as the content relevance between the developers' web sites and app development or app products. The process of extracting *web site quality* feature is done in two steps. First, we collect a corpus of words related to app development and app products. Then we compute cosine similarity in vector space model [21] between developers' web sites and the corpus collected in first step. This cosine similarity is considered as *web site quality*. For developers not having their own web sites, we simply set their *web site quality* to 0.

Good Ratio. In app marketplace such as Google Play, every app is open for app users to comment. Therefore, we can obtain a general user opinion of apps by mining the user reviews. Here we present *good ratio* feature, which is the proportion of positive reviews among all reviews. Considering that app user reviews are very short texts similar to tweets, we adapt a two-step SVM classifier model proposed by [3] to conduct sentiment analysis on app user reviews. The first step aims to distinguish subjective reviews from non-subjective reviews through a subjectivity classifier. Then we further classify the subjective reviews into positive and negative reviews, namely, polarity detection. The features used in these two SVM are word *meta features* in app user reviews.

Meta Features. For a word in app user reviews, we map it to its part-of-speech using a part-of-speech dictionary[3]. In addition to POS tags, we also map the word to its prior subjectivity and polarity. The prior polarity is switched from positive to negative or from negative to positive when a negative expression, e.g., "don't", "never" precedes the word.

---

[3] The POS dictionary is available at: `http://wordlist.sourceforge.net/pos-readme`

6   Applications

In this section, we show the use of app developer profiles in facilitating app search and app development outsourcing.

6.1   Facilitate App Search

In this application, rather than exploring the background rank schema in existing app search engines, we turn to rank aggregation which deals with problem of combining the result lists returned by multiple search engines. The rank aggregation method we utilize here is Borda Count[19], which is a rank based aggregation method.

We apply app developer profiles in facilitating app search in the following way. We first rank apps by their developers' reputation and get a ranking list, which we call reputation ranking list here. Then we utilize Border's method to aggregate the ranking list returned by existing app search engine and the reputation ranking list. The aggregated ranking list is considered as the ranking result after applying app developer profiles in app search. The experiment in Section 7.3 shows that this strategy improves the search quality of existing app search engines.

6.2   App Development Outsourcing

We consider the following scenario in app development outsourcing. Given a detailed description of desired app, which may contain the functionalities and the interface design, we aim to identify and recommend proper app developers to users. Let us call this app development outsourcing problem. Candidate Model and the Document Model proposed in [2] can be applied to app development outsourcing after some adaptation and are used as two baselines in our experiments. We now explain the details of the adaptation of the two models as follows: in the app development outsourcing scenario, each app description is considered as a document and the description of wanted app is referred to as query.

However, candidate model and document model merely take relevance between the given query and existing app descriptions into consideration, which can only recommend developers who have developed similar apps with the wanted one but can't guarantee that the recommended developers are all proficient. To tackle this defect, we propose a BM25 and reputation based recommendation (*BMr-BR*) methods to recommend excellent and experienced developers.

BMr-BR. *BMr-BR* not only considers the relevance between the given query and existing app descriptions but also takes into app developer reputation. In this way, the recommended developers can be guaranteed to be experienced and proficient. The ranking function to recommend developrs in *BMr-BR* is as follows:

$$R \cdot \sum_i score(T_i, q), \tag{11}$$

where T_i is the key words set of Expertise i generated by ADE, q is the given query, R is developer's reputation.

The value $score(T_i, q)$ is defined as:

$$score(T_i, q) = \sum_{j=1}^{n} IDF(t_j)\alpha_i \frac{n(t_j, T_i)(k_1 + 1)}{n(t_j, T_i) + k_1(1 - b + b\frac{|T_i|}{avgdl})}, \tag{12}$$

where t_j is a term in q, α_i is developer's proficiency in expertise i, $n(t_j, T_i)$ denotes the frequency of t_j in T_i, k_1 and b are free parameters. Usually, $k_1 \in [1.2, 2.0]$ and $b = 0.75$ according to [15].

7   Experiments

7.1   Experimental Setting

We select the official Android marketplace, Google Play, as the core information source of apps and developers.

We collected a total of 533,740 apps, which accounts for 82% of the whole apps on Google Play. From the webpage of an app on Google Play, we can obtain detailed information about this app such as the description, rating, download number and user reviews. Besides, the *URLs* of developer's web sites (if any) can be also accessed from Goog Play. We use these *URLs* to collect app developer's web sites.

7.2   Evaluation of ADE and ADR

Evaluation of ADE. An informal but important measure of the success of probabilistic topic models is the plausibility of the discovered search topics. For simplicity, we use the fixed symmetric Dirichlet distribution like [6], which demonstrates good performance in our experiments. Hyperparameter setting is well studied in probabilistic topic modeling and the discussion is beyond the scope of this paper. Interested readers are invited to refer [23] for further details.

Currently, very few probabilistic topic models are proposed to analyze app developers, thus it is hard to find counterparts for the proposed one. Thus, we select Latent Dirichlet Allocation (LDA) [4] and Pink-LDA as baselines, since they are general enough to be applied in the task. We use a held-out dataset to evaluate the proposed model's capability of predicting unseen data. Perplexity is a standard measure of evaluating the generalization performance of a probabilistic model [20]. It is monotonically decreasing in the likelihood of the held-out data. Therefore, a lower perplexity indicates better generalization performance. Specifically, perplexity is calculated according to the following equation:

$$Perplexity_{held-out}(\mathcal{M}) = (\prod_{d=1}^{D} \prod_{i=1}^{N_d} p(w_i|\mathcal{M}))^{\frac{-1}{\sum_{d=1}^{D}(N_d)}}, \tag{13}$$

where $\mathcal{M}$ is the model learned from the training process. The result of perplexity comparison is presented in Figure 3(a), from which we observe that the proposed models demonstrate much better capability in predicting unseen data comparing with the LDA baselines. For example, when the number of search topics set to 300, the perplexity of LDA is 420.65, the perplexity of PLink-LDA is 419.23 and that of the proposed topic model is 299.12. The result verifies that our proposed model provides better fit for the underlying structure of the information of each app developer. Thus, the proposed model has better performance to derive different facets of the expertness for app developers.

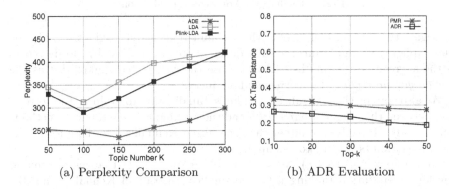

(a) Perplexity Comparison   (b) ADR Evaluation

Fig. 3. ADE and ADR Evaluations

Evaluation of ADR. To evaluate ADR, we manually label 400 pairwise comparison among about 800 developers which are used to train RankSVM. The comparison between two developers is performed according to their web sites and the average rating, average installations, user reviews of their apps. We first consider the average rating and average installations. If two app developers are very close in above two aspects, we then refer to their web sites and reviews of their apps. Apart from these pairwise comparison, we also rank 50 developers manually according to 10 volunteers' judgements as the ground truth to evaluate effectiveness of ADR and baselines.

We propose an intuitive baseline (PMR) which ranks developers according to the arithmetic product of popularity and average rating. The evaluation metric we use here is the well-known generalized Kendall's Tau distance[16].

The experiment result is showed in Figure 3(b) where ADR is about 0.1 lower than PMR in terms of Kendall's Tau Distance, which indicates that ADR gives much better reputation ranking.

7.3   Evaluation of Applications

Evaluation of Facilitating App Search. We conduct some experiments to show that app developer profiles make app search more effective. We first prepare the *pseudo ground truth* by rank aggregation. We use Borda Count[19] to

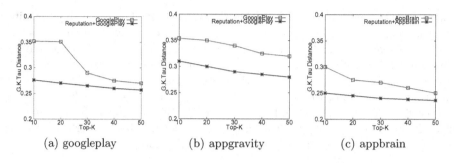

Fig. 4. Facilitating App Search Evaluation

aggregate the app ranking lists of three commercial search engines, i.e. Google Play, appgravity, appbrain, for each query. Apps are ranked by Borda scores in the aggregated ranking list. Then we use the aggregated ranking list as the *pseudo ground truth* of the corresponding query.

We also employ the generalized Kendall's Tau distance to evaluate the correlation between the *pseudo ground truth* and the ranking list generated by methods under study. Roughly, the larger the generalized Kendall's Tau distance is, the less correlation between two ranking lists have.

From Figure 4(a) to 4(c), we see that all ranking results that aggregate app developer reputation ranking list have smaller generalized Kendall's Tau distance than these not taking app developer reputation into consideration. After combined with app developer reputation ranking, the generalized Kendall's Tau distance of Google Play, appgravity, appbrain all decrease in studied top-k ranking results, which show app developer profiles effectively improves app search quality.

Evaluation of App Development Outsourcing. We implement *BMr-BR* and two baselines, candidate model and document model, in this section. 5000 developers are selected as the candidates set, in which some big famous studios are excluded. We first fabricate 10 queries which are descriptions of ten wanted apps. Then BMr-BR, candidate model and document model are applied to recommend developers for these 10 queries. In order to evaluate the recommended results, we manually check the top-10 and top-20 recommended developers for each query. The checking standard is that, if a developer has developed a similar[4] apps to the given query we consider it is a hit developer. Besides, a hit developer is more proper if its apps that are similar to the given query have higher rating as well as more good user reviews.

When implementing *BMr-BR*, we set the topic amount K=50 and the average length of expertise key words sets is 802. To evaluate the performance of these three methods, we compute the mean precision at top-10 and top-20 as well as Mean Reciprocal Rank (MRR) at top-20. As Figure 5(a) shows, *BMr-BR* has the

---

[4] "similar" means the two app have similar functions, game rules or undertake similar tasks.

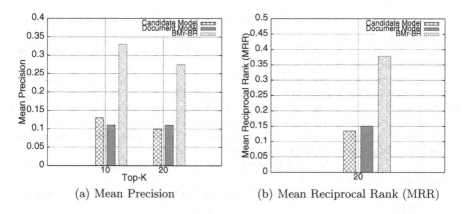

(a) Mean Precision   (b) Mean Reciprocal Rank (MRR)

Fig. 5. App Development Outsourcing Evaluation

highest mean precision both at top-10 and top-20, which are respectively 0.33 and 0.275. Besides, *BMr-BR* also outperforms candidate model and document model in MRR at top-20 respectively by 0.24 and 0.22.

8 Conclusions

In this paper, we present a new ADI framework which is founded on two underlying models of ADE and ADR to profile app developers. Within the framework, these two models take into account multiple information sources, such as app descriptions, ratings and user reviews, in order to effectively and comprehensively profile app developers. The ADE model uses app's description and category to generate app developer's functionality-based expertise. The ADR model utilizes information from apps, users and developers to evaluate the reputation of an app developer. The extensive experiments show that the two models are effective. In addition, we demonstrate the use of app developer profiles by two applications, facilitating app search and app development outsourcing. All the empirical results show that app developer profiling is extremely useful for both the users and developers. Our modeling approach paves the way to establish more sophisticated profiling, for example taking into account social information in a mobile platform to extend our ADI framework.

Acknowledgements. This work is partially supported by GRF under grant numbers HKUST 617610.

References

1. Alazab, M., Monsamy, V., Batten, L., Lantz, P., Tian, R.: Analysis of malicious and benign android applications. In: ICDCSW (2012)
2. Balog, K., Azzopardi, L., De Rijke, M.: Formal models for expert finding in enterprise corpora. In: SIGIR (2006)

3. Barbosa, L., Feng, J.: Robust sentiment detection on twitter from biased and noisy data. In: ACL (2010)
4. Blei, D.M., Ng, A.Y., Jordan, M.I.: Latent dirichlet allocation. The Journal of Machine Learning Research (2003)
5. Chia, P., Yamamoto, Y., Asokan, N.: Is this app safe?: A large scale study on application permissions and risk signals. In: WWW, pp. 311–320 (2012)
6. Griffiths, T.L., Steyvers, M.: Finding scientific topics. Proc. of the National Academy of Sciences of the United States of America (2004)
7. Ha-Thuc, V., Mejova, Y., Harris, C., Srinivasan, P.: A relevance-based topic model for news event tracking. In: SIGIR (2009)
8. Jiang, D., Leung, K., Ng, W.: Context-aware search personalization with concept preference. In: CIKM (2011)
9. Jiang, D., Ng, W.: Mining web search topics with diverse spatiotemporal patterns. In: SIGIR (2013)
10. Jiang, D., Vosecky, J., Leung, K.W.-T., Ng, W.: G-wstd: A framework for geographic web search topic discovery. In: CIKM (2012)
11. Jiang, D., Vosecky, J., Leung, K.W.-T., Ng, W.: Panorama: A semantic-aware application search framework. In: EDBT (2013)
12. Joachims, T.: Optimizing search engines using clickthrough data. In: SIGKDD (2002)
13. Kang, D., Jiang, D., Pei, J., Liao, Z., Sun, X., Choi, H.J.: Multidimensional mining of large-scale search logs: A topic-concept cube approach. In: WSDM (2011)
14. Manning, C.D., Raghavan, P., Schutze, H.: Introduction to information retrieval. Cambridge University Press, Cambridge (2008)
15. Manning, C.D., Raghavan, P., Schütze, H.: Introduction to information retrieval, vol. 1. Cambridge University Press, Cambridge (2008)
16. Mazurek, J.: Evaluation of ranking similarity in ordinal ranking problems. Acta academica karviniensia
17. Mei, Q., Liu, C., Su, H., Zhai, C.X.: A probabilistic approach to spatiotemporal theme pattern mining on weblogs. In: WWW (2006)
18. Newman, D., Asuncion, A., Smyth, P., Welling, M.: Distributed algorithms for topic models. The Journal of Machine Learning Research 10, 1801–1828 (2009)
19. Renda, M., Straccia, U.: Web metasearch: Rank vs. score based rank aggregation methods. In: Proceedings of the 2003 ACM symposium on Applied Computing (2003)
20. Rosen-Zvi, M., Griffiths, T., Steyvers, M., Smyth, P.: The author-topic model for authors and documents. In: Proceedings of the UAI Conference (2004)
21. Salton, G., Wong, A., Yang, C.: A vector space model for automatic indexing. Communications of the ACM 18(11), 613–620 (1975)
22. Vosecky, J., Leung, K.W.-T., Ng, W.: Searching for quality microblog posts: Filtering and ranking based on content analysis and implicit links. In: Lee, S.-G., Peng, Z., Zhou, X., Moon, Y.-S., Unland, R., Yoo, J. (eds.) DASFAA 2012, Part I. LNCS, vol. 7238, pp. 397–413. Springer, Heidelberg (2012)
23. Wallach, H.M.: Structured topic models for language. Unpublished doctoral dissertation. Univ.of Cambridge (2008)
24. Wei, X., Gomez, L., Neamtiu, I., Faloutsos, M.: Malicious android applications in the enterprise: What do they do and how do we fix it? In: ICDEW, pp. 251–254. IEEE (2012)
25. Zhou, Y., Wang, Z., Zhou, W., Jiang, X.: Hey, you, get off of my market: Detecting malicious apps in official and alternative android markets. In: NDSS (2012)

Novel Community Recommendation Based on a User-Community Total Relation

Qian Yu[1], Zhiyong Peng[1], Liang Hong[2,*], Bin Liu[1], and Haiping Peng[1]

[1] School of Computer, Wuhan University,
Wuhan, Hubei, 430072, China
hong@whu.edu.cn
[2] School of Information Management, Wuhan University,
Wuhan, Hubei, 430072, China
hong@whu.edu.cn

Abstract. With the exponential increase of Web communities, community recommendation has become increasingly important in sifting valuable and interesting communities for users. In this paper, we study the problem of novel community recommendation, and propose a method based on a user-community total relation (i.e. user-user, community-community, and user-community interactions). Our novel recommendation method suggests communities that the target user has not seen but is potentially interested in, in order to broaden the user's horizon. Specifically, a **W**eighted **L**atent **D**irichlet **A**llocation (WLDA) algorithm improves recommendation accuracy utilizing social relations. A definition of community novelty together with an algorithm for novelty computation are further proposed based on the total relation. Finally, a multi-objective optimization strategy improves the overall recommendation quality by combining accuracy and novelty scores. Experimental results on a real dataset show that our proposed method outperforms state-of-the-art recommendation methods on both accuracy and novelty.

Keywords: Novel community recommendation, user-community total relation, weighted Latent Dirichlet Allocation, novelty.

1 Introduction

Recommender systems have played a vital role in E-commerce sites such as Amazon and Netflix by sifting valuable information for users. Based on known item ratings, recommender systems predict ratings on non-rated items for the target user and use the predicted ratings for recommendation. Web communities are groups of users who interact through social media to pursue common goals or interests, which have various themes and topics. It is impossible for a user to browse all communities and choose interesting and valuable ones to join. To broaden users' horizon and promote development of communities, it is very important to recommend communities that are *novel* to users; novel communities are the ones users have not seen but are potentially interested in [1,2].

---

* Corresponding Author.

S.S. Bhowmick et al. (Eds.): DASFAA 2014, Part II, LNCS 8422, pp. 281–295, 2014.

Typical recommendation techniques (e.g. [3,4,5,6,7]) belong to **accurate recommendation**, which refers to the type of techniques purely pursuing the recommendation accuracy, and these techniques are designed to only improve the precision in the proximity to users' preference [8]. That is, the more similar an item to the target user's preference, the more accurate the item.

Accurate recommendation is successful evaluated by metrics for accuracy, but accuracy can hurt recommender systems [9]. Accurate recommendation suggests items close to users' preference and tends to recommend popular ones [10], so users may already know the recommended items or be bored with the popular ones. In addition, recommending excessive popular items can lead to profit decline of enterprises, because Matthew Effect[1] [11] can be caused and less accessed niche products can become a large share of total sales [12]. That is, techniques focusing on pure accuracy are limited by the lack of **novelty** [1,10,13].

In contrast to accurate recommendation which can hurt user experience and development of communities, novel recommendation tries to suggest novel items to users. Figure 1(a) shows the difference between accuracy and novelty in recommending communities. "Big Bang Theory" is close to "Friends" because they both are situation comedies, and the overlapped users can be similar users of u_q, so the community can be accurate. The "Gossip Coming" community is probably unknown to u_q, because it is distant from u_q in social relations and distant from "Friends" in semantics; but u_q can be interested in entertainment gossip about "Friends", and reach the community by social relations. So "Gossip Coming" can be novel to u_q.

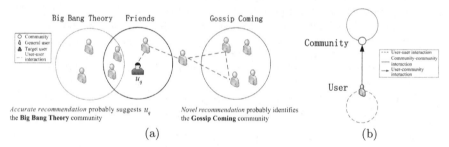

(a)   (b)

Fig. 1. (a) The example is excerpted from the real Douban dataset. There are three communities above, and the overlap between two communities represents their common users. The target user u_q joins the "Friends" community, and social relations of u_q are displayed by dashed lines. (b) A sketch map of the user-community total relation, which indicates the existence of three types of interactions.

Several novel recommendation methods have been proposed in recent years [8,10,13,15,16,17] to remedy the deficiency of pure accuracy. The proposed methods are effective in identifying novel items by digging deep into the user-item interactions, or more precisely, the user-item rating matrix. However, we argue that there are three drawbacks in applying the methods to novel community recommendation. (1) The methods put forward vague definitions of novelty in spite

---

[1] The "the rich get richer and the poor get poorer" phenomenon.

of designing subtle algorithms to identify novel items, which makes the methods less rational. (2) User-user interactions and their underlying social relations extensively exist in Web communities. The proposed methods ignore user-user interactions, although they can influence users' behaviors due to the homophily in social networks [14], and further impact on recommendation. (3) There are semantical relations between communities because they have distinct themes, and semantics can provide powerful help in identifying novel communities as illustrated in Figure 1(a). Nonetheless, the methods cannot make full use of the semantical relations, because they neglect item-item interactions.

In order to overcome the aforementioned drawbacks, we propose a novel community recommendation method based on a user-community total relation. The total relation, which is formally defined in subsection 3.1, refers to "interact with" over the set $\{user, community\}$ as shown in Figure 1(b). The total relation indicates the existence of user-user, user-community and community-community interactions, and our method integrates the three types of interactions.

Specifically, we first propose a **W**eighted **L**atent **D**irichlet **A**llocation (WLDA) method to further enhance accuracy, taking advantage of user-community interactions. Then, we define the community novelty and design an algorithm to compute novelty, using the total relation. Finally, a multi-objective optimization strategy is adopted to achieve a balance between accuracy and novelty, in order to improve the overall quality of recommendation. The main contributions of our work are as follows.

- We propose WLDA to utilize strength on social relations in Web communities, and the method outperforms the competitors [5,17] on accuracy.
- We define the community novelty based on the user-community total relation, and design an algorithm to compute novelty. First, We observe that user-user interactions constitute a social network and there exists clusters[2] in the network. Then we determine candidates, which are communities that the target user u_q is potentially interested in, of u_q by user-user interactions within the cluster l_q that u_q belongs to. Finally we compute the novelty of community c_i using the popularity of c_i, the semantical distance from c_i to u_q extracted from community-community interactions, and user-community interactions between c_i and other users in l_q.
- We conduct experiments to evaluate our proposed method as well as other recommendation techniques, using a real dataset from Douban[3]. The experimental results show that our method outperforms the competitors on accuracy and novelty.

The rest of the paper is organized as follows. We review related works in Section 2. Section 3 elaborates our proposed method. The experimental results are presented in Section 4, and we make the conclusion in Section 5.

---

[2] In the rest of this paper, "community" refers to a Web community, and "cluster" refers to the cluster of social network users.

[3] http://www.douban.com/

2   Related Works

In this section, we briefly review related works in two aspects: accurate recommendation and novel recommendation.

2.1   Accurate Recommendation

Accurate recommendation refers to the type of techniques purely pursuing accuracy, and covers most of the existing recommendation techniques.

Collaborative Filtering. Memory-base and model-based CF (Collaborative Filtering) are the most common CF techniques. Memory-based CF can be further divided into item-based [6] and user-based [7] CF. Item-based CF assumes that users tend to choose items similar to their preferences; user-based CF assumes that users tend to choose items favored by their similar users. Wang et al. [18] unify item-base and user-based CF.

Several model-based CF methods have been proposed in recent years. Chen et al. [5] experimentally prove that LDA [19] can generate more accurate results than association rule mining [20] in community recommendation. An original topic model which views a community as both a bag of users and a bag of words, is proposed in [21] for community recommendation.

Social Recommendation. With the development of social networking sites, social recommendation utilizing social information (e.g. friendship) to improve accuracy has emerged. Ma et al. [3] extend the basic matrix factorization [22] to combine social network structure and the user-item rating matrix. Jamali et al. [4] run a modified random walk [23] algorithm in a trust network, in order to recommend items by trust relationships. Explicit and implicit social recommendation are experimentally compared on their abilities of improving accuracy in [24].

2.2   Novel Recommendation

The conception of novel recommendation, that novel items are unknown but attractive to users, is first raised in [1] to the best of our knowledge. The method in [15] is an early work on novel recommendation; it reveals unexpected interests of users by detecting user clusters in a similarity network, in which users are nodes and similarities between users are edges.

Onuma et al. [8] treat the user-item rating matrix as a graph, and use the basic random walk algorithm to calculate TANGENT scores of items, and items with higher scores are more novel. Nakatsuji et al. [13] define the item novelty based on the item taxonomy, construct a similarity network like [15], and compute novelty of candidates identified in the network. Oh et al. [10] propose a conception called Personal Popularity Tendency (PPT) to explore patterns of user-item interactions, and design a PPT matching algorithm to achieve a balance between accuracy and novelty.

The following so-called serendipitous recommendation methods are classified into novel recommendation according to [9]: "*serendipity = novelty + user feedback*", because they are essentially novel recommendation and collect no user feedback. Kawamae et al. [16] add timestamps to the user-item rating matrix, calculate the probability that a user purchases an item under influences from other users, and take the probability as serendipity of the item. Zhang et al. [17] propose a reversed LDA to compute item similarities, build a similarity network taking items as nodes and item similarities as edges, and compute clustering coefficients of nodes in the network as serendipity.

We summarize the above novel recommendation methods from three aspects. Firstly, there are no formal definitions of item novelty in the methods except [13], and the drawback of the definition is detailed in subsection 3.4. Secondly, none of the methods take advantage of user-user interactions. Thirdly, only [13] utilizes item-item interactions in a quite direct manner.

3   The UCTR Method

In this section, we describe details of the proposed **User-Community Total Relation (UCTR)** method for novel community recommendation.

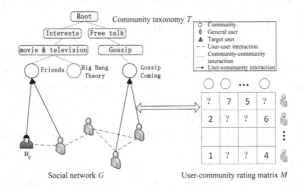

Fig. 2. An illustration of R that evolves from the example in Figure 1. User-user interactions constitute a social network G. User-community interactions correspond to the rating matrix M. Community-community interactions exist in the taxonomy T.

3.1   Problem Definition

We first elaborate the user-community total relation in this subsection.

Definition 1. (The User-Community Total Relation) Given a set $X = \{user, community\}$, the user-community total relation R is defined as: R=(X,X,G), where $G = X \times X$, and R refers to the binary relation "interact with".

The total relation R represents the three types of interactions as illustrated in Figure 2, by an example excerpted from the Douban dataset. The social network $G(V, E)$ is a weighted and undirected graph, where nodes denoted by V are users and edges denoted by E are user-user interactions. User-community interactions correspond to non-zero entries in user-community rating matrix M. Community-community interactions are indirect interactions between communities through category nodes and edges in community taxonomy T.

The problem to be addressed in this paper is defined as follows: Given user-community rating matrix M, social network G and community taxonomy T, for the target user u_q, find a list of communities highly novel and accurate to u_q.

3.2 Framework of the UCTR Method

In this subsection, we describe the framework of the proposed method. Our method aims to determine the vector $\overrightarrow{t_q} = (t_{q,i})_{c_i \in C}$. The element $t_{q,i}$ represents the recommendation degree of community c_i to u_q, C is the set of communities. In order to recommend communities that are unknown but attractive to u_q, the UCTR method consists of the following three parts.

1. Compute $\overrightarrow{a_q} = (a_{q,i})_{c_i \in C}$, accuracy scores of communities for u_q.
2. Compute $\overrightarrow{n_q} = (n_{q,i})_{c_i \in C}$, novelty scores of communities for u_q.
3. Compute UCTR scores $\overrightarrow{t_q}$ by merging two criteria above.

Computing $\overrightarrow{n_q}$ reuses a part of the intermediate result of computing $\overrightarrow{a_q}$, which reduces the computation. The next subsections detail the framework.

3.3 Computing Community Accuracy

We propose a WLDA (**Weighted LDA**) algorithm to compute community accuracy. The method utilizes LDA to dig into user-community interactions, i.e., matrix M, and emphasizes weights on user-community interactions to further enhance accuracy. Compared to traditional accurate recommendation methods such as memory-based CF methods[6,7] which utilize explicit ratings, LDA can be less affected by the sparseness [3] of matrix M, in the way of exploring implicit relations.

WLDA first transforms matrix M to text corpora to model M with LDA. The transformation takes a user as a text and takes communities the user joins as words in the text. Weights on user-community interactions correspond to entries in M, and play a significant role in WLDA as deciding how many times a community appears in the text representing the user. For example, if $m_{q,i} = 3$, then c_i appears three times in the text of u_q. That is, the weights determine the input, so they greatly impact on the output of WLDA.

WLDA then computes implicit relations, i.e., $\overrightarrow{\theta_q} = (\theta_{q,1}, \theta_{q,2}, \cdots, \theta_{q,K})$ and $\overrightarrow{\phi_i} = (\phi_{i,1}, \phi_{i,2}, \cdots, \phi_{i,K})$, where $\theta_{q,k}$ represents the relatedness between u_q and the latent topic k, $\phi_{i,k}$ represents the relatedness between c_i and topic k, K is the number of latent topics, and U is the set of users as shown in Algorithm 1. After a random assignment of latent topics, the four arrays are initialized: $N1[i,k]$ is the number of times c_i is assigned to topic k; $N2[k]$ is is the number of times topic k appears; $N3[q,k]$ is the number of times topic k is assigned to u_q; $N4[q]$ is the number of topics assigned to u_q. In each iteration, we re-assign a topic to each community in each user in step 7-12; the topic assignment will converge after enough iterations, which means that elements in $p[]$ become nearly constant in step 10. Implicit relations can be computed by the output of Algorithm 1. The algorithm is based on Gibbs Sampling [28].

Algorithm 1. Computing implicit relations

Input: text corpora transformed from matrix M
Output: $\overrightarrow{\theta_q}, u_q \in U$; $\overrightarrow{\phi_i}, c_i \in C$
1 Randomly assign a latent topic to each community in each user;
2 Initialize four arrays $N1,N2,N3,N4$;
3 **for** *each iteration* **do**
4   **for** *each user u_q* **do**
5   **for** *each community c_i in u_q* **do**
6   //ct refers to the current topic assigned to c_i
7           $N1[i, ct] - -, N2[ct] - -, N3[q, ct] - -, N4[q] - -$;
8   //K is the total number of latent topics
9   **for** *each topic $k = 1$ to K* **do**
10             $p[k] = \frac{N1[i,k]+\beta}{N2[k]+|C|\cdot\beta}\frac{N3[q,k]+\alpha}{N4[q]+K\cdot\alpha}$
11   // $p[nt]$ is the largest in $p[]$
12          $N1[i, nt] + +, N2[nt] + +, N3[q, nt] + +, N4[q] + +$;
13 **return** $\phi_{i,k} = \frac{N1[i,k]+\beta}{N2[k]+|C|\cdot\beta}$; $\theta_{q,k} = \frac{N3[q,k]+\alpha}{N4[q]+K\cdot\alpha}$

Definition 2. (Accuracy Score) Given $\overrightarrow{\theta_q}$ and $\overrightarrow{\phi_i}$, $a_{q,i}$ as the accuracy score of community c_i to u_q is defined as: $a_{q,i} = \overrightarrow{\theta_q} \cdot \overrightarrow{\phi_i}^T = \sum_{k=1}^{K} \theta_{q,k}\phi_{i,k}$.

WLDA finally computes $\overrightarrow{a_q} = (a_{q,i})_{c_i \in C}$, accuracy scores of communities for u_q, according to Definition 2. The definition utilizes the Bayes' rule, and our idea of computing community accuracy is that, the closer c_i and u_q on latent topics, the more accurate c_i. For example, if c_i and u_q are highly related to the same latent topic k, then $\theta_{q,k}$ and $\phi_{i,k}$ are simultaneously large, $a_{q,i}$ is large, and c_i is highly accurate to u_q.

3.4 Computing Community Novelty

We formally define community novelty and design an algorithm to compute $\overrightarrow{n_q}$ in this subsection, based on the user-community total relation R.

Community-community interactions existing in community taxonomy T are used to generate semantical relations between communities.

Definition 3. (Semantical Distance) Let $dis(i, j)$ be the distance from the category of c_i to the category of c_j in taxonomy T, then we define the semantical distance between u_q and c_i as follows:

$$d(q,i) = \arg \min_{c_j \in C_q} dis(i, j) \qquad (1)$$

where we set $dis(i, j) = 1$ if c_i and c_j belong to the same category, and C_q is the set of communities that u_q joins.

Referring to Figure 2, we use c_f to represent the "Friends" community, c_b the "Big Bang Theory" community and c_g the "Gossip Coming" community, then $C_q = \{c_f\}$. According to Equation 1, $d(q, b) = 1$ because $dis(f, b) = 1$, and $d(q, g) = 4$ because $dis(f, g) = 4$. So the "Gossip Coming" community is more distant from u_q than the "Big Bang Theory" community.

User-community interactions directly reflect users' preference, and implicit relations which are computed from these interactions by WLDA are reused to measure similarity between users, as mentioned in subsection 3.2.

Definition 4. (User Similarity) Given vectors $\overrightarrow{\theta_q}$ and $\overrightarrow{\theta_o}$, we define $s(q,o)$, the similarity between user u_q and u_o as follows:

$$s(q,o) = \frac{\sum_{k=1}^{K} \theta_{q,k}\theta_{o,k}}{\sqrt{\sum_{k=1}^{K}(\theta_{q,k})^2}\sqrt{\sum_{k=1}^{K}(\theta_{o,k})^2}} \tag{2}$$

where $\theta_{q,k}$ represents the relatedness between u_q and the latent topic k, $\theta_{o,k}$ the relatedness between u_o and topic k. Equation 2 is based on cosine similarity, which is a measure of similarity between two vectors. Notice that other similarity measures can also be applied to computing the user similarity.

User-user interactions establish a social network G and there exists clusters in the network, which we observe from the real dataset and illustrated in Figure 3. Without loss of generality, we assume that u_q and the other four users belong to one cluster l_q in G, as shown in Figure 2. So u_q joins tangible Web communities, and simultaneously belongs to a latent cluster l_q. Cluster l_q and social relations in l_q are used to compute community novelty as described next.

Definition 5. (Novelty Score) Given semantical distance $d(q,i)$, user similarity $s(q,o)$, user-community rating matrix M, and cluster l_q that u_q belongs to, we define $n_{q,i}$, the novelty score of c_i to u_q as follows:

$$n_{q,i} = -\log_2\left(\frac{|c_i|}{\arg\max_{c_j \in C}|c_j|}\frac{1}{d(q,i)}\sum_{u_o \in l_q, o \neq q} s(q,o)m_{o,i}\right) \tag{3}$$

where $|c_i|$ represents the number of members in c_i, $m_{o,i}$ the rating on c_i from u_o in M, $d(q,i)$ the semantical distance between c_i and u_q, and $s(q,o)$ the similarity between u_q and u_o.

Definition 5 starts from the intuition that novel items are unknown but attractive to users [1]. Further we assume that given c_i is attractive to u_q, the more unknown c_i to u_q, the more novel c_i. And we determine a community as a novel candidate if u_q is potentially interested in the community (other users in l_q joins it), and evaluate the extent of "unknown" of the candidate.

We identify novel candidates by social relations of u_q within cluster l_q. If $\forall u_o \in l_q \wedge o \neq q, m_{o,i} = 0$, then $n_{q,i} = 0$ in Equation 3. That is, c_i must be joined by other users in l_q to be attractive and a novel candidate to u_q. We use social relations within l_q for the identification according to the homophily in social networks [14], the finding that people who interact with each other are more likely to share interests [25], as well as the generally accepted assumption that users of the same cluster have densely intrinsic links [27].

We evaluate the extent of "unknown" from three aspects as shown in Equation 3. (1) Large $|c_i|$, which means that c_i is popular, leads to low novelty [2], because popular communities are less likely to be unknown for u_q. (2) Small $d(q,i)$ leads to low novelty, because u_q can know c_i if c_i is close to u_q through community-community interactions. (3) large $\sum s(q,o)m_{o,i}$ leads to low novelty, because it is more likely for u_q to know c_i through user-user interactions, if other uses in l_q frequently interact with c_i; user similarity is treated as the weight on user-user interactions. And the three points above are achieved by the entropy form, $-\log_2 X$, of Equation 3.

Our novelty definition can overcome drawbacks of existing definitions. The definition that novel items are unknown but attractive to users [1] can not be used for quantitative measurement. The definition that items with lower popularity are more novel [2] cannot be perfectly applied to Web communities, because interactions are a more intrinsic feature than popularity for communities. The definition, that novelty of an item equals the distance from it to interests of the target user according to the item taxonomy [13], can lead to confusion, because items belonging to the same category have the same novelty score under this definition. For example, in Figure 2 communities that belong to the category "Gossip" have the same novelty score "4" to u_q.

We design the following Algorithm 2 to compute $\vec{n_q}$. We choose the *SHRINK* algorithm [27] to detect clusters in social network $G(V, E)$ for two reasons. (1) *SHRINK* is efficient with the overall time complexity of $O(|E| \log |V|)$. (2) The algorithm overcomes the resolution limit so that clusters of relatively small size can be detected.

Algorithm 2. Computing community novelty

   Input: rating matrix M, social network G, taxonomy T, and the target user u_q
   Output: $\vec{n_q}$
1 Run *SHRINK* on G to detect l_q; $S_q = \varnothing$;
2 **for** *each* $u_o \in l_q, o \neq q$ **do**
3 | $S_q = S_q \cup (C_o - C_q)$;
4 **for** *each* $c_j \in S_q$ **do**
5 | compute $n_{q,j}$ using Equation 3;
6 **return** $\vec{n_q} = \left(n_{q,1}, \cdots, n_{q,|S_q|} \right)$;

In step 1, l_q is the cluster u_q belongs to, and S_q the set of novel candidates for u_q. Step 2-3 shows that S_q consists of communities joined by other users in l_q excluding C_q, where C_o is the set of communities u_o joins. The algorithm returns novelty scores of candidates for u_q.

3.5 Computing UCTR Scores

Novel recommendation needs multi-objective optimization techniques to achieve a balance between accuracy and novelty to generate recommendation results, because accuracy and novelty have different essence and are conflicting goals to some extent. For example, numerical multiplication is used in [8], probability multiplication is used in [16], and linear combination is used in [17].

And we use the following technique to achieve the balance.

$$t_{q,i} = \frac{n_{q,i}}{a_{q,i}} \frac{\arg\max\limits_{c_j \in S_q} a_{q,j}}{\arg\max\limits_{c_j \in S_q} n_{q,j}} \tag{4}$$

We choose the technique of numerical multiplication, because it is parameter-free and of high efficiency. Notice that in Equation 4 both novelty and accuracy scores are normalized to $[0, 1]$, and the form of inverse proportion between novelty and accuracy makes that higher UCTR score reflect higher novelty, and vice versa.

Take computing $\overrightarrow{t_q}$ as an example. The UCTR method first transforms user-community rating matrix M to text corpora and run WLDA, then accuracy scores of all users in U can be computed including $\overrightarrow{a_q}$, and $\overrightarrow{\theta}$ is kept for computing $\overrightarrow{n_q}$. The method next identifies cluster l_q by *SHRINK*. Assume that u_q and the other four users constitute l_q as shown in Figure 2, then S_q which consists of communities joined by the four users excluding C_q can be confirmed, and $|\overrightarrow{n_q}| = |S_q|$. Further take c_i ($c_i \in S_q$) as an example, $d(q, i)$ can be computed from taxonomy T, similarity between u_q and the four users can be computed using $\overrightarrow{\theta}$, interactions between the four users and c_i are stored in M, then $\overrightarrow{n_q}$ and $\overrightarrow{t_q}$ can be ultimately computed.

4   Experiments

We divide our experiments into three parts, which are conducted on the same dataset. The first part evaluates accuracy of Basic LDA [5], WLDA and Basic Auralist [17]. The second part evaluates WLDA, Our Novelty computed according to Definition 5, and B-Aware Auralist [17]. The third part compares the UCTR to Auralist [17] on both accuracy and novelty.

4.1   Data Description

We collect the experimental data from Douban (www.douban.com), which is a highly ranked social networking site and contains more than 70,000,000 users and 320,000 communities currently. To safeguard user privacy, all user and community data are anonymized.

In the user-community interactions part, we have 252,993 users, 229 communities. In the community-community part, we have 67 bottom-level category nodes and 8 high-level category nodes in taxonomy T. In the user-user interactions part, we extract a latent network containing 41,633 nodes (users) and 104,423 edges. We give the statistical graph of user degree distribution of the network in Figure 3. The degree distribution of the extracted network obeys the power-law distribution, and the *SHRINK* algorithm detects 3476 clusters in the network.

4.2   Accuracy Evaluation

Evaluation metric and protocol. The goal of this part is to measure the effectiveness of recommending top-ranked items, and we employ the top-k recommendations metric NDCG [29], which is also used in [5].

Firstly, for each user u we randomly choose a community c from communities u joins to form the training set. Secondly, for each user u, we randomly choose $(k - 1)$ communities that u does not join in. The objective is to compare the

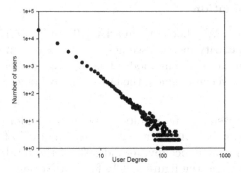

Fig. 3. User degree distribution in G under the log-log scale

rankings for c generated by the three methods: the higher the ranking, the more accurate the method.

Results. We fix the parameters of Basic LDA and WLDA to the same value for exact comparison: the number of latent topics is 67 (number of bottom-level category nodes in T), α is 0.28, β is 0.1, the number of iterations is set to 10000 to make sure that results of Gibbs Sampling are converged. And we set k to 180.

We compare the three methods based on the top-k recommendation metric. Figure 4 shows the cumulative distributions of ranks for withheld communities: the 0% rank indicates that the withheld community is ranked 1, while 100% indicates that it is ranked last in the list. WLDA consistently outperforms Basic Auralist and Basic LDA mainly because WLDA utilizes strength of social relaitons. Basic Auralist outperforms Basic LDA because the former method introduces the strategy of item-based CF.

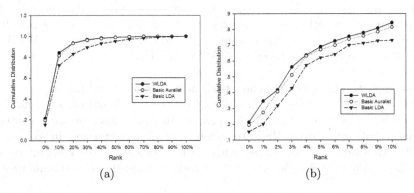

Fig. 4. We plot the cumulative distributions of ranks for withheld communities. The x-axis represents the percentile rank. (a) The macro view (0% - 100%) of top-k performance. (b)The micro view (0% - 10%) of top-k performance.

4.3 Novelty Evaluation

Evaluation metrics. We choose popularity [10] and coverage [13] as metrics for novelty. The popularity metric assumes that the more unpopular items recommended, the more novel the methods. The coverage metric assumes that the more individual items recommended, the larger the coverage, and the more novel the methods.

Results. Figure 5(a) shows the results on popularity. More than 95% of the communities have less than 10,000 members; less than 5% of the communities, which have high popularity, occupy around 60% of the total recommendations generated by WLDA, and the figure is 30% for Our Novelty.

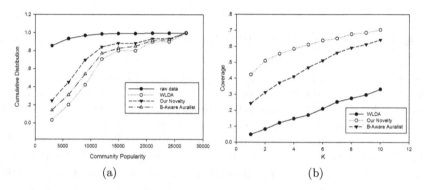

(a)   (b)

Fig. 5. We again plot the cumulative distributions of the recommendation results. (a) Performance on popularity. The distribution of the raw data is displayed, i.e., popularity of communities. And we consider top-10 recommendation for convenient comparison. (b) Performance on coverage.

The results based on the coverage metric are more obvious than popularity, as shown in Figure 5(b). When making top-1 recommendation, Our Novelty can cover about 45% of all the communities and the figure is 5% for WLDA. When making top-10 recommendation, WLDA still only covers 30% of the 229 communities.

In summary, WLDA recommends much more popular communities than the other two methods and is poor in coverage due to the drawbacks of accuracy. Our Novelty outperforms B-Aware Auralist because the former utilizes the user-community total relation but the latter ignores abundant interactions in Web communities; B-Aware Auralist outperforms WLDA because it takes advantage of clustering coefficients of nodes in social networks to identify novel communities.

4.4 Combinational Evaluation

Evaluation protocol. The ultimate recommendation results based on our UCTR scores and Auralist are compared as shown in Figure 6. We adopt the top-k recommendations metric and the coverage metric. The linear combination parameter of Auralist λ is set to 0.2 as claimed in [17], and the UCTR is parameter-free.

Results. UCTR outperforms Auralist on the novelty metric coverage. However, Auralist seems better under the metric of accuracy. The reason could be that UCTR pays more attention to novelty than Auralist and Auralist pays more attention to accuracy (λ is set to a fairly small number 0.2, and the larger λ, the more novel Auralist).

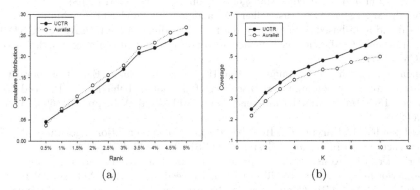

(a)   (b)

Fig. 6. (a) Comparison on micro view of top-k performance. (b) On coverage.

5   Conclusion

Features of Web communities drive us to make use of the user-community total relation. We emphasize strength of social relations in communities in WLDA to further enhance the recommendation accuracy. We define the community novelty and identify novel candidates for the target user according to a social network constituted by user-user interactions. And we evaluate novelty using popularity of communities, semantical distances extracted from community-community interactions, and user-community interactions of other users in the network. Finally we compute UCTR scores for candidates to generate recommendation results.

Our WLDA method outperforms the competitors on accuracy, our novelty method outperforms the competitors on novelty, and UCTR can achieve reasonable tradeoff between the two criteria. Future work could be focused on designing a cluster detection algorithm which is specifically targeted at the scenario of recommendation, and a parallel implementation of the proposed UCTR method.

Acknowledgments. This work is supported by the National Natural Science Foundation of China under Grant (No.61070011, 61303025, 61202034 and 61272275), Natural Science Foundation of Hubei Province, and the Doctoral Fund of Ministry of Education of China No.20100141120050.

References

1. Herlocker, J.L., Konstan, J.A., Terveen, L.G., Riedl, J.T.: Evaluating collaborative filtering recommender systems. ACM Transactions on Information Systems (TOIS) 22(1), 5–53 (2004)
2. Vargas, S., Castells, P.: Rank and relevance in novelty and diversity metrics for recommender systems. In: Proceedings of the Fifth ACM Conference on Recommender Systems, pp. 109–116. ACM (2011)
3. Ma, H., Yang, H., Lyu, M.R., King, I.: Sorec: Social recommendation using probabilistic matrix factorization. In: Proceedings of the 17th ACM Conference on Information and Knowledge Management, pp. 931–940. ACM (2008)
4. Jamali, M., Ester, M.: Trustwalker: A random walk model for combining trust-based and item-based recommendation. In: Proceedings of the 15th ACM SIGKDD International Conference on Knowledge Discovery and Data Mining, pp. 397–406. ACM (2009)
5. Chen, W.-Y., Chu, J.-C., Luan, J., Bai, H., Wang, Y., Chang, E.Y.: Collaborative filtering for orkut communities: Discovery of user latent behavior. In: Proceedings of the 18th International Conference on World Wide Web, pp. 681–690. ACM (2009)
6. Deshpande, M., Karypis, G.: Item-based top-n recommendation algorithms. ACM Transactions on Information Systems (TOIS) 22(1), 143–177 (2004)
7. Herlocker, J.L., Konstan, J.A., Borchers, A., Riedl, J.: An algorithmic framework for performing collaborative filtering. In: Proceedings of the 22nd Annual International ACM SIGIR Conference on Research and Development in Information Retrieval, pp. 230–237. ACM (1999)
8. Onuma, K., Tong, H., Faloutsos, C.: Tangent: A novel, surprise me, recommendation algorithm. In: Proceedings of the 15th ACM SIGKDD International Conference on Knowledge Discovery and Data Mining, pp. 657–666. ACM (2009)
9. McNee, S.M., Riedl, J., Konstan, J.A.: Being accurate is not enough: how accuracy metrics have hurt recommender systems. In: CHI 2006 Extended Abstracts on Human Factors in Computing Systems, pp. 1097–1101. ACM (2006)
10. Oh, J., Park, S., Yu, H., Song, M., Park, S.-T.: Novel recommendation based on personal popularity tendency. In: 2011 IEEE 11th International Conference on Data Mining (ICDM), pp. 507–516. IEEE (2011)
11. Merton, R.K.: The matthew effect in science. Science 159(3810), 56–63 (1968)
12. Anderson, C.: The long tail: Why the future of business is selling less of more. Hyperion Books (2008)
13. Nakatsuji, M., Fujiwara, Y., Tanaka, A., Uchiyama, T., Fujimura, K., Ishida, T.: Classical music for rock fans?: Novel recommendations for expanding user interests. In: Proceedings of the 19th ACM International Conference on Information and Knowledge Management, pp. 949–958. ACM (2010)
14. McPherson, M., Smith-Lovin, L., Cook, J.M.: Birds of a feather: Homophily in social networks. Annual Review of Sociology, 415–444 (2001)
15. Kamahara, J., Asakawa, T., Shimojo, S., Miyahara, H.: A community-based recommendation system to reveal unexpected interests. In: Proceedings of the 11th International Multimedia Modelling Conference, MMM 2005, pp. 433–438. IEEE (2005)
16. Kawamae, N.: Serendipitous recommendations via innovators. In: Proceedings of the 33rd International ACM SIGIR Conference on Research and Development in Information Retrieval, pp. 218–225. ACM (2010)

17. Zhang, Y.C., Saghdha, D., Quercia, D., Jambor, T.: Auralist: Introducing serendipity into music recommendation. In: Proceedings of the Fifth ACM International Conference on Web Search and Data Mining, pp. 13–22. ACM (2012)
18. Wang, J., De Vries, A.P., Reinders, M.J.: Unifying userbased and item-based collaborative filtering approaches by similarity fusion. In: Proceedings of the 29th Annual International ACM SIGIR Conference on Research and Development in Information Retrieval, pp. 501–508. ACM (2006)
19. Blei, D.M., Ng, A.Y., Jordan, M.I.: Latent dirichlet allocation. The Journal of Machine Learning Research 3, 993–1022 (2003)
20. Agrawal, R., Imielinski, T., Swami, A.: Mining association rules between sets of items in large databases. In: ACM SIGMOD Record, vol. 22, pp. 207–216. ACM (1993)
21. Chen, W.-Y., Zhang, D., Chang, E.Y.: Combinational collaborative filtering for personalized community recommendation. In: Proceedings of the 14th ACM SIGKDD International Conference on Knowledge Discovery and Data Mining, pp. 115–123. ACM (2008)
22. Koren, Y., Bell, R., Volinsky, C.: Matrix factorization techniques for recommender systems. Computer 42(8), 30–37 (2009)
23. Fouss, F., Pirotte, A., Renders, J.-M., Saerens, M.: Random-walk computation of similarities between nodes of a graph with application to collaborative recommendation. IEEE Transactions on Knowledge and Data Engineering 19(3), 355–369 (2007)
24. Ma, H.: An experimental study on implicit social recommendation. In: Proceedings of the 36th International ACM SIGIR Conference on Research and Development in Information Retrieval, pp. 73–82. ACM (2013)
25. Singla, P., Richardson, M.: Yes, there is a correlation:-from social networks to personal behavior on the web. In: Proceedings of the 17th International Conference on World Wide Web, pp. 655–664. ACM (2008)
26. Ziegler, C.-N., Lausen, G., Schmidt-Thieme, L.: Taxonomy-driven computation of product recommendations. In: Proceedings of the Thirteenth ACM International Conference on Information and Knowledge Management, pp. 406–415. ACM (2004)
27. Huang, J., Sun, H., Han, J., Deng, H., Sun, Y., Liu, Y.: Shrink: A structural clustering algorithm for detecting hierarchical communities in networks. In: Proceedings of the 19th ACM International Conference on Information and Knowledge Management, pp. 219–228. ACM (2010)
28. Griffiths, T.L., Steyvers, M.: Finding scientific topics. Proceedings of the National academy of Sciences of the United States of America 101(suppl. 1), 5228–5235 (2004)
29. Järvelin, K., Kekäläinen, J.: Cumulated gain-based evaluation of ir techniques. ACM Transactions on Information Systems (TOIS) 20(4), 422–446 (2002)

User Interaction Based Community Detection in Online Social Networks

Himel Dev, Mohammed Eunus Ali, and Tanzima Hashem

Department of Computer Science and Engineering
Bangladesh University of Engineering and Technology, Dhaka, Bangladesh
himeldev@gmail.com, {eunus,tanzimahashem}@cse.buet.ac.bd

Abstract. Discovering meaningful communities based on the interactions of different people in online social networks (OSNs) is an active research topic in recent years. However, existing interaction based community detection techniques either rely on the content analysis or only consider underlying structure of the social network graph, while identifying communities in OSNs. As a result, these approaches fail to identify *active communities*, i.e., communities based on actual interactions rather than mere friendship. To alleviate the limitations of existing approaches, we propose a novel solution of community detection in OSNs. The key idea of our approach comes from the following observations: (i) the degree of interaction between each pair of users can widely vary, which we term as *the strength of ties*, and (ii) for each pair of users, the interactions with mutual friends, which we term the *group behavior*, play an important role to determine their belongingness to the same community. Based on these two observations, we propose an efficient solution to detect communities in OSNs. The detailed experimental study shows that our proposed algorithm significantly outperforms state-of-the-art techniques for both real and synthetic datasets.

1 Introduction

Community detection in social networks has gained a huge momentum in recent years due to its wide range of applications. These applications include online marketing, friend/news recommendations, load balancing, and influence analysis. The community detection involves grouping of similar users into clusters, where users in a group are strongly bonded with each other than the other members in the network. Most of the existing community detection algorithms assume a social network as a graph, where each user in the social network is represented as a vertex, and the connection or interaction between two users is depicted as an edge connecting two vertices. One can then apply graph clustering algorithms [1] to find different communities in OSNs.

An earlier body of research in this domain focuses on identifying communities based on the underlying structure (e.g., number of possible paths or the common neighbors between two vertices) of the social network graph. Popular approaches include maximal clique [2], minimum cut [3], modularity [4], and edge betweenness [5]. These methods use only the structural information (e.g., connectivity) of the network for community prediction and clustering. However, they do not consider the level/degree of interactions among users and thus fail to identify communities based on actual interactions rather than mere friendship, which we call *active communities* in this paper.

S.S. Bhowmick et al. (Eds.): DASFAA 2014, Part II, LNCS 8422, pp. 296–310, 2014.

With the explosion of user generated contents in the social network, some recent approaches [6,7,8,9] focus on exploiting the rich sets of information of the contents to detect meaningful communities in the network. Among these works, [6] considers the contents associated with vertices, [8] considers the contents associated with edges, and [9] considers both vertex and edge contents for detecting communities. These works are mainly based on the content analysis that considers the semantic meaning of the contents while grouping users into different communities. Since these approaches analyze the contents to determine the communities, they are not suitable for real-time applications such as load balancing, which require online analysis. Moreover, none of these works considers the various interaction types available in OSNs and the degree of interactions among users, i.e., does not distinguish between the weak and strong links between the users.

We observe that the degree of user interactions can be vital in community detection in many applications that include load balancing and recommendation systems. For example, the astounding growth of social networks and highly interconnected nature among the end users make it a difficult problem to partition the load of managing a gigantic social network data among different commodity cheap servers in the cloud and make it scalable [10]. A random partitioning, which is the defacto standard in OSNs [11], will generate huge inter-server traffic for resolving many queries, especially, when two users with a high degree of interaction belong to two separate servers. However, if we are able to identify communities that have a high degree of interactions among the users within the community and low degree of interactions among inter-community users, place all users in a community in the same server, we can eliminate the inter-server communication to a great extent and at the same time can improve the query response time. Similarly, identifying communities based on the interactions of mutual friends can be an effective technique for predicting missing links between a pair of users and recommending friends based on those identified missing links.

In this paper, we propose a novel community detection technique that considers the structure of the social network and interactions among the users while detecting the communities. The key idea of our approach comes from the following observations: (i) the degree of interaction between each pair of users can widely vary, which we term as *the strength of ties*, and (ii) for each pair of users, the degree of interactions with common neighbors (e.g., mutual friends in Facebook), which we term the *group behavior*, play an important role to determine their belongingness to the same community. Based on these observations, we propose a community detection algorithm that identifies *active* communities into four phases. First, we model and quantify the interactions between every connected pair of users as the strength of ties, which allows us to differentiate strong and weak links in OSNs while detecting communities. Based on these interactions, in the second phase, we quantify the group behavior for every pair of users who are connected via common neighbors. Third, based on the interactions and group behavior, we build a probability graph, where each edge is assigned a probability denoting the likelihood of two users belonging to the same community. Finally, we apply hierarchical clustering algorithm on the computed probability graph to identify communities. Experimental results show that our approach outperforms state-of-the art community detection algorithms in a wide range of evaluation metrics.

In summary, our contributions are as follows:

– We model and quantify the degree of interactions between users and the group behaviors among users.
– We propose a novel community detection algorithm for OSNs based on user interaction and group behavior to identify active communities with a high accuracy.
– We conduct extensive experiments using both real and synthetic datasets to show the effectiveness and efficiency of our approach.

2   Related Work

With the recent advances in information networks, the problem of community detection has been widely studied in recent years. Many approaches formulated the problem in terms of dense region identification such as dense clique detection [2] [12] [13] and structural density [14] [15]. Further, there have been approaches formulating the problem in terms of minimum cut [3] and label propagation [16]. However, commonly used methods of community detection involve modularity [4] [17] [18] and edge betweenness [4].

Among the aforementioned approaches, SCAN (Structural Clustering Algorithm for Networks) [14] and Truss [15] use neighborhood concept of common neighbors, which have similarity with our group interaction concept for un-weighted graphs. However, the limitation associated with the aforementioned approaches is, most of these methods are pure link-based methods based on topological structures. They focus only on the information regarding the linkage behavior (connection) for the purposes of community prediction and clustering. They do not utilize different attributes present in networks and as a result their performance degrades in networks with rich contents, e.g., OSNs. In case of OSNs, these methods do not consider the various interaction types and degree of interactions among users, and hence fail to identify *active communities*.

The limitations of traditional methods along with the availability of rich contents in OSNs have led to the emergence of content based community detection methods [9] [8] [19] [6]. There is a growing body of literature addressing the issue of utilizing the rich contents available in OSNs to determine communities. Some recent works [19] [6] have shown the use of vertex content in improving the quality of the communities. Zhou et al. [6] combined vertex attributes and structural property to cluster graphs. Yang et al. [19] proposed a discriminative model of combining link and content analysis for community detection. However, these methods have been criticized [9] as certain characteristics of communities can not be modeled by vertex content. Again, there have been some recent proposals involving the use of edge content [8] [7]. Sachan et al. [8] used content and interaction to discover communities. These methods focus on utilizing edge content in the form of texts/images to discover communities. One major bottleneck of such content based community detection methods is their scalability, which makes them unsuitable for real-time community detection from large-scale data.

3   Community Detection

In this section, we propose a community detection algorithm for OSNs based on in-teractions among users that range from simple messaging to participating together in different types of applications like games, location-based services and video chat. More specifically, to identify whether two social network users fall into the same community, our algorithm considers both the impact of interaction between them and the impact of the group behavior that the users show based on the interaction with their common friends (i.e., common neighbors in the social graph). The key idea of our algorithm is to combine these impacts in a weighted manner and vary the weights to identify the active communities with a high accuracy.

The proposed community detection algorithm has four phases. In the first phase, the algorithm quantifies the degree of interaction between every connected pair of users in the OSNs and based on these interactions, in the second phase, the algorithm quantifies the group behavior for every pair of users who are connected via common neighbors. In the third phase, the algorithm determines the probability of two users belonging to the same community using the impact of interaction between them and their group behavior. Finally, in the fourth phase, the algorithm applies hierarchical clustering to detect communities based on the computed probabilistic measure.

3.1   Quantifying the Interaction

In this phase, the algorithm computes the degree of interaction between every connected pair of users in a social network. Given a social graph $G(V,E)$ and user interaction data, the algorithm constructs an *Interaction Graph*, $G_I(V,E,W)$, where each weight $w_{uv} \in W$ quantifies the degree of interaction between two users u and v. The higher the value of w_{uv}, the higher is the strength of tie between u and v. We have considered the following three factors to quantify interaction between two users: interaction type, average number of interactions for a particular interaction type, and relative interaction.

Interaction Type: Now-a-days OSNs involve interactions of different types. For ex-ample, Facebook users can interact via personal messages, wall posts, photo tags, page likes etc. To quantify the degree of interaction between two users, it is necessary to consider all interaction types. In addition, we observe that, some of these interaction types indicate stronger bonding than the others. Thus, it is important to prioritize the interaction types in an order. Prioritizing the interaction types in terms of bonding is es-pecially useful for applications such as friend recommendation, and influence analysis. The prioritization can also be done considering the data transfer overhead associated with each interaction type, which is useful for applications like load balancing. We prioritize different interaction types using weights.

In addition, in an OSN, there could be both active and passive users. Passive users establish friendship with others, but hardly interact with them. Sometimes there is no interaction involved in a link established by a passive user. This can also happen for newly joined users of an OSN. To determine the community in which a passive/new user belongs to, we consider the user's connection/link with others. More specifically, we consider the establishment of friendship between two users as a special type of interaction and to incorporate this special type of interaction, our communication detection algorithm provides a threshold value for each established friendship link.

Average Number of Interactions for a Particular Interaction Type: Another important factor to consider is the average number of interactions for a particular interaction type, which is not same for all interaction types. For example, average number of personal messages in Facebook is higher compared to average number of wall posts. To address this issue, we normalize the number of interactions for each type using the average value corresponding to the type. Otherwise, interaction types with higher average values eliminate the effect of type prioritization.

Relative Interaction: We observe that the importance of an interaction can vary among users based on their activities. To incorporate this issue in quantifying the interaction between users for our community detection algorithm, we take relative interaction into account. For ease of understanding, first consider an example in an unweighted graph shown in Fig. 1(left): Both users B and E are connected to user A and the number of connections of users B, E, and A are 2, 4, and 4, respectively. Thus, we observe that the friendship link AB has high importance to user B than user A as it represents 50% and 25% of activities of B and A, respectively. Further, the link AB is more significant than the link AE as the number of links associated with both A and E are higher than that of B.

Fig. 1. Relative Interaction

The relativity of interaction also applies to weighted graphs. In Figure 1(right), we find that E is a highly active user with total interaction weight 21 and B is the least active user with total interaction weight 4. User A shows moderate level of activities with total interaction weight 15. We observe that AB involves 13% and 50% of total interactions of A and B, respectively and AE involves 27% and 19% of total interactions of A and E, respectively. Thus, interactions associated with AB seem more significant compared to interactions associated with AE.

To quantify the impact of relative interaction, we normalize each interaction in terms of total interaction of the involved users.

Next, we formally quantify the interaction between two users based on the above three factors.

Quantification: Let $\{I^1, I^2, ..., I^t\}$ represent t interaction types in an OSN. To prioritize the interaction types, we associate a weight with every interaction type. Assume that weights $W^1, W^2, ..., W^t$ are associated with $I^1, I^2, ..., I^t$, respectively, where $W^i > W^j$, if I^i represents stronger bonding between users than that of I^j. Weights can be predefined by experts or even by social network users. In our experiments, we conduct a survey among social network users to determine the weights. Note that for a particular type of social network, weights can be set once and used multiple times to identify communities in that social network. We first quantify interaction between users based on

$\{I^1, I^2, ..., I^t\}$ and then to incorporate the impact of links without interaction, we add an additional threshold value, ε, to the quantified interaction.

Let i_{uv}^t represent the number of interactions of type t between users u and v, and n is the number of users in the social graph $G(V,E)$. The average number of interactions for a particular type t, $\bar{I}^t$, is computed as $\sum_{u \in V} \sum_{v \in V \wedge u \neq v} i_{uv}^t / n$. The normalized number of interactions of type t between u and v, I_{uv}^t, is computed as $i_{uv}^t / \bar{I}^t$.

Considering prioritized interaction types, links without interaction, and the average number of interaction for each interaction type, the algorithm quantifies interaction between two users u and v as $\hat{w}_{uv}$ using the following equations:

$$\hat{w}_{uv} = I_{uv}^1 \times W^1 + I_{uv}^2 \times W^2 + ... + I_{uv}^t \times W^t + \varepsilon \qquad (1)$$

Finally, to incorporate the impact of the relative importance of an interaction between users u and v, the algorithm considers the quantified interactions of u and v with their neighbors $N(u)$ and $N(v)$, respectively, using Equation 1 and modifies $\hat{w}_{uv}$ as w_{uv} using the following equation:

$$w_{uv} = \frac{1}{2}\left(\frac{\hat{w}_{uv}}{\sum_{x \in N(u)} \hat{w}_{ux}} + \frac{\hat{w}_{uv}}{\sum_{y \in N(v)} \hat{w}_{yv}} \right) \qquad (2)$$

Without loss of generalization, consider a scenario, where Facebook users A, B, C, D interact using the following public interactions:

Table 1. Interaction among A, B, C, D

Interaction Type	A,B	A,C	A,D	B,C	B,D
Wall Posts	12	15	12	9	9
Photo Tags	9	27	27	45	9
Page Likes	0	0	76	38	76

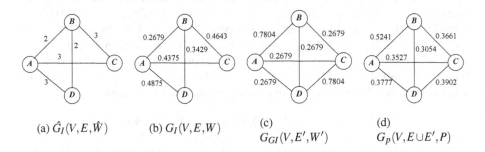

(a) $\hat{G}_I(V,E,\hat{W})$ (b) $G_I(V,E,W)$ (c) $G_{GI}(V,E',W')$ (d) $G_P(V,E \cup E',P)$

Fig. 2. Probability Graph Construction

Let the weights for interaction type wall posts, photo tags and page likes be $0.4, 0.3, 0.2$, respectively and $\varepsilon = 0.1$. The average number of wall posts, photo tags and page likes for all users are 3, 9, and 38, respectively. Then the weight w_{AB} between A and B in the interaction graph is computed as follows:

$$w_{\widehat{AB}} = \frac{12}{3} * 0.4 + \frac{9}{9} * 0.3 + \frac{0}{38} * 0.2 + 0.1 = 2, w_{AB} = \frac{\frac{2}{2+3+3} + \frac{2}{2+3+2}}{2} = 0.2679$$

By computing edge weights in the aforementioned way, we generate the interaction graph G_I as shown in Fig. 2(b). Note that, the weight corresponding to every edge of interaction graph, w_{uv} lies between 0 and 1.

3.2 Quantifying Group Behavior

In this phase, the algorithm computes the group behavior that every pair of users, who are connected via one or more common neighbors, show in a social network. Any two users of a social community usually have common neighbors (mutual friends). Most of these common neighbors are other members of that community who interact with both of them. Thus, the algorithm quantifies the group behavior of pair of users based on the number of common neighbors and their interactions with common neighbors.

Given an interaction graph, $G_I(V,E,W)$, the algorithm constructs a *Group Interaction Graph*, $G_{GI}(V,E',W')$, where there is an edge, $e'_{uv} \in E'$ between two users u and v, if they have at least a common neighbor. W' is a set of weights, where each weight quantifies the group behavior between two users with respect to common neighbors. Let $M_{uv} = \{m_1, m_2, ..., m_h\}$ represent h common neighbors (mutual friends) of u and v, where $m_1, m_2, ..., m_h, u, v \in V$. The interaction with a common neighbor $m_i \in M_{uv}$, $w_{uv}^{m_i}$ is quantified as $w_{uv}^{m_i} = min(w_{um_i}, w_{vm_i})$.

The proposed algorithm computes the interaction of every pair of users who are connected via one or more common neighbors with respect to each of their common neighbors and then quantifies the group behavior for pair of users, $w_{M_{uv}} \in W'$ as follows:

$$w_{M_{uv}} = \sum_{i=1}^{h} w_{uv}^{m_i} \tag{3}$$

Consider the interaction graph G_I in Fig. 2(b): A and B have two common neighbors C and D. The weights representing interactions of C with A and B are 0.4375 and 0.4643 respectively. Hence, the group behavior corresponding to A and B involving C is $w_{AB}^{C} = min(0.4375, 0.4643) = 0.4375$. Again, the interaction of D with A and B weights 0.4875 and 0.3429, respectively. Hence, the group behavior corresponding to A and B involving D is $w_{AB}^{D} = min(0.4875, 0.3429) = 0.3429$. Thus, the group behavior corresponding to A and B involving all common neighbors, $w_{M_{AB}}$ is computed as follows:

$$w_{M_{AB}} = w_{AB}^{C} + w_{AB}^{D} = 0.4375 + 0.3429 = 0.7804$$

By computing edge weights in the aforementioned way, we construct the group interaction graph G_{GI} as shown in Fig. 2(c). Note that, the weight corresponding to every edge of group interaction graph, $w_{M_{uv}}$ lies between 0 and $(1 - w_{uv})$.

3.3 Computing the Probabilities

Given an interaction graph, $G_I(V,E,W)$, and a group interaction graph, $G_{GI}(V,E',W')$, the algorithm constructs a *Probability Graph*, $G_p(V,E \cup E',P)$, where each weight

$p_{uv} \in P$ represents a probability between two vertices u and v to belong in the same community. The probability of two users sharing the same community depends on: the interaction between them and their group behavior. The group behavior between two users is determined in terms of their number of common neighbors and the interaction with them. If two social network users interact with a large number of common neighbors, it is very likely that they belong to the same community. The weights in W and W' represent user interaction and group behavior, respectively.

Formally, we define *probability* p_{uv} of two users u and v belonging to the same community as follows:

Definition 1. *(Probability of belonging to the same community) Let w_{uv} and $w_{M_{uv}}$ represent the weight of interaction between u and v and their group behavior, respectively, and α be a parameter used to combine the impact of w_{uv} and $w_{M_{uv}}$, where $0 \le \alpha \le 1$. The probability of u and v to belong to the same community, p_{uv}, is defined as follows:*

$$p_{uv} = \alpha * w_{uv} + (1 - \alpha) * w_{M_{uv}} \qquad (4)$$

Here, both α and w_{uv} lie between 0 and 1. Further, $w_{M_{uv}}$ lies between 0 and $(1 - w_{uv})$. Hence, the p_{uv} values generated by equation (4) lie between 0 and 1. We use the parameter α to control the impact of w_{uv} and $w_{M_{uv}}$ on probability for users u and v. We experimentally find the appropriate value of α that identifies the communities with a high accuracy.

Consider the interaction graph G_I in Fig. 2(b) and the group interaction graph G_{GI} in Fig. 2(c), where the interaction and group behavior corresponding to A,B,C,D are quantified. From G_I, we find that the quantified interaction between A and B, w_{AB}, is 0.2679 and from G_{GI}, we find that the quantified group behavior between A and B, $w_{M_{AB}}$, is 0.7804. The probability p_{AB} of A and B to belong to the same community is computed as follows:

$$p_{AB} = \alpha * w_{AB} + (1 - \alpha) * w_{M_{AB}} = \alpha * 0.2679 + (1 - \alpha) * 0.7804$$

For $\alpha = 0.5$, the probability of A and B belonging to the same community is 0.5*0.2679+0.5*0.7804, i.e., 0.5241.

3.4 Hierarchical Clustering on Probability Graph

The final phase of the proposed algorithm involves identifying communities by applying hierarchical clustering in the probability graph $G_p(V, E \cup E', P)$. We use agglomerative hierarchical clustering to build a hierarchy of clusters. By horizontally cutting the agglomerative hierarchical cluster tree at an appropriate height, we will get the required clusters (communities) $\{C_1, C_2 \ldots\}$. We use the following distance measure and linkage criterion for our hierarchical algorithm.

Distance Measure: In our approach, the probability $p_{uv} \in P$ of u and v to belong to the same community serves as the similarity measure. By subtracting p_{uv} from 1, i.e., $1 - p_{uv}$, we can have the distance between u and v.

Linkage Criterion: To estimate the distance between two clusters C_i and C_j, we have used 'Un-weighted Pair Group Method with Arithmetic Mean (UPGMA)'as the linkage criterion. According to UPGMA, the distance is computed between every pair of users $u \in C_i$ and $v \in C_j$. Then the distance between any two clusters C_i and C_j is computed as the average of all computed distances, i.e., $\frac{1}{|C_i| \cdot |C_j|} \sum_{u \in C_i} \sum_{v \in C_j} (1 - p_{uv})$.

3.5   Algorithm

Algorithm 1, *Detect_Communities*, shows the pseudocode for our community detection algorithm. The input to the algorithm is a social graph $G(V,E)$ and the user interaction data in the form of interaction matrices $A^1, A^2, \ldots, A^t$, where A^t represents interaction matrix for the interaction of type t. The size of each interaction matrix is $|V| \times |V|$ and each cell of an interaction matrix A^t denotes the number of interactions of type t between a pair of users in V. The output of the algorithm is a set of clusters, where each cluster represents an active community.

Algorithm 1. DETECT_COMMUNITIES$(G, A^1, A^2, \ldots, A^t)$

> **Input** : A social graph $G(V,E)$, Interaction matrices $A^1, A^2, \ldots, A^t$
> **Output**: A set of clusters $C = \{C_1, C_2, \ldots\}$

1.1 $G_I \leftarrow Construct\_Interaction\_Graph(G, A^1, A^2, \ldots, A^t)$;
1.2 $G_{GI} \leftarrow Construct\_Group\_Interaction\_Graph(G_I)$;
1.3 $currentOptimal \leftarrow 0$;
1.4 **for** $\alpha \leftarrow 0$ *to* 1 **do**
1.5 $\quad$ $G_P \leftarrow Construct\_Probability\_Graph(G_I, G_{GI}, \alpha)$;
1.6 $\quad$ $C^T \leftarrow Hierarchical\_Clustering(G_p)$;
1.7 $\quad$ **for** $clusterNum \leftarrow 1$ *to* $|V|$ **do**
1.8 $\quad\quad$ $C' \leftarrow Get(C^T, clusterNum)$;
1.9 $\quad\quad$ **if** $f(C') > currentOptimal$ **then**
1.10 $\quad\quad\quad$ $C \leftarrow C'$;
1.11 $\quad\quad\quad$ $currentOptimal \leftarrow f(C')$;
1.12 $\quad\quad$ $clusterNum \leftarrow clusterNum + 1$;
1.13 $\quad$ $\alpha \leftarrow \alpha + stepsize$;

The algorithm first constructs G_I and G_{GI} using functions *Construct_Interaction_Graph* and *Construct_Group_Interaction_Graph*, respectively (Lines 1.1-1.2). Then, the algorithm varies the value of α from 0 to 1, and for each α, the algorithm computes probability graph G_p (Line 1.5). Since hierarchical clustering is applied on the probability graph, the clustering tree C^T also varies with different values of α (Line 1.6). For every α, the algorithm varies the number of clusters, *clusterNum* from 1 to $|V|$, and for every *clusterNum*, the algorithm computes the set of clusters as C' from C^T (Line 1.8). The variable *clusterNum* is used to select the height at which the cluster tree C^T is cut and generate the clusters accordingly. The goal of our algorithm is to determine the pair $< \alpha, clusterNum >$ that provides us with the best *quality of clusters*. In our algorithm, we measure the quality of the computed set of clusters C' using a function $f(C')$ that evaluates the quality of clusters based on different standards such as modularity (Please see Section 4 for details). For every pair $< \alpha, clusterNum >$, the algorithm checks whether the quality of the computed set of clusters C' is better than the current best set of clusters C. If yes, then the algorithm updates C with C', otherwise C remains unchanged (Lines 1.9-1.10).

3.6 Time Complexity

Function *Construct_Interaction_Graph* in Line 1.1 needs $O(m)$ time, where m is the total number of edges in network $G(V,E)$. Again, function *Construct_Group_Interaction_Graph* in Line 1.2 needs $O(n\Delta^2)$ time, where the average vertex degree of the social network $G(V,E)$ is Δ. Similarly, the complexity of function *Construct_Probability_Graph* in Line 1.5 is $O(m + n\Delta^2)$ which is iterated for each value of α. However, the corresponding loop is iterated a constant number of times (typically 100). Further, time complexity corresponding to the efficient implementation of hierarchical clustering (UP-GMA) in Line 1.6 is $O(n^2)$ [20] and time complexity corresponding to the loop in Line 1.7 is $O(mn)$. So, the time complexity of the complete algorithm is $O(n\Delta^2 + n^2 + mn)$. Now, in a typical social graph $m > n$, the time complexity of the complete algorithm can be considered as $O((m + \Delta^2)n)$.

4   Experimental Evaluation

In this section, we evaluate the performance of our proposed community detection algorithm using both real and synthetic datasets. We compare our approach with the state-of-the-art community detection algorithms that include Girvan Newman [5], walktrap [21], fast greedy [17], leading eigenvector [22], infomap [23], label propagation [16] and multi-level [24] method using a number of evaluation metrics that include normalized mutual information (NMI) [19], pairwise F-measure (PWF) [19], and modularity [4].

4.1   Datasets

We have used four categories of data sets in our experiments.

Facebook User Interactions Labeled Dataset: In the first category, we have collected 2421 interactions from 221 Facebook users. These interaction types involve common page likes, photo tags, and wall posts. The average number of interactions corresponding to the aforementioned interaction types are 38, 9, and 3, respectively.[1]

Since, users may want to put more weight on a certain type of interactions than the others, we have conducted a survey among users (whose data have been collected) to assign a weight for each type of interactions that signifies the relative importance of that particular interaction type. Upon conducting the survey, we have found that user preferred relative weights are 0.15, 0.35 and 0.5 for common page likes, tagged photos and wall posts, respectively. These collected data are later labeled into different classes. These labels are used as ground-truth values to measure the qualities of our proposed community detection algorithm.

---

[1] Dataset: https://sites.google.com/site/mohammedeunusali/publications

Facebook-like Forums Weighted Un-labeled Dataset: In the second category, we have used publicly available data set of a Facebook-like Forum [25]. The forum represents an interesting two-mode network among 899 users with 7089 edges.

Classical Un-weighted Datasets: To show the wide range applicability of our proposed approach, in the third category, we have used publicly available data set of Facebook ego network [26] along with three popular classical data sets: Zachary Karate Club [27], Dolphins [28], Jazz Musicians by Gleiser and Danon [29].

Computer Generated Benchmark Datasets: In the fourth category, we have used computer-generated benchmark graphs [30] that account for the heterogeneity in terms of external/internal degree. The generated graphs vary in terms of the mixing parameter that corresponds to the average ratio of external degree/total degree. Thus, by varying the mixing parameter as 0.1, 0.2, 0.3, and 0.4, we have generated four sets of social graphs. Each graph contains 128 vertices and they form four groups (i.e., vertices are labeled) each having 32 vertices. The average degree of vertices in each graph is 16 and the maximum degree is 25.

4.2   Performance Metrics

Our datasets consist of both labeled data and un-labeled data. Since the labeled dataset has the ground-truth values, we use supervised metrics: normalized mutual information (NMI) and pairwise F-measure (PWF) to evaluate the performance of our community detection algorithm. On the other hand, for both labeled and un-labeled datasets we use the most common measurement metric, modularity to compare the performance of our approach with the other state-of-the-art techniques.

Normalized Mutual Information (NMI): Let, $C = \{C_1, C_2, ..., C_k\}$ be the true community structure, where C_i corresponds to the community containing all vertices labeled with $1 <= i <= k$. Let $C' = \{C_1', C_2', ..., C_k'\}$ be the community structure returned by a community detection algorithm, where C_i' corresponds to the community containing all vertices labeled with $1 <= i <= k$ by the algorithm. Then the mutual information between the two community structures is defined as follows.

$$MI(C,C') = \Sigma_{C_i, C_j'} p(C_i, C_j') log(\frac{p(C_i, C_j')}{p(C_i)p(C_j')})$$

By using the above defined mutual information, we can define the normalized mutual information as follows.

$$NMI(C,C') = \frac{MI(C,C')}{max(H(C),H(C'))}$$

Here, $H(C)$ and $H(C')$ are entropies of community structures C and C'.

Pairwise F-measure: Let T denote the set of vertex pairs that have the same label, S denote the set of vertex pairs that are assigned to the same community by the algorithm, $|T|$ and $|S|$ denote the cardinality of T and S, respectively. The pairwise F-measure (PWF) is computed from the pairwise precision (PWP) and pairwise recall (PWR), and can be defined as follows.

$$PWP = \frac{|S \cap T|}{|S|}, PWR = \frac{|S \cap T|}{|T|}, PWF = \frac{2*PWP*PWR}{PWP+PWR}$$

Modularity: The most popular measure of evaluating communities of a un-labeled social graph is the modularity proposed by Newman [4]. Modularity is the fraction of the edges that fall within the given groups minus the expected such fraction if edges were distributed at random. Modularity, Q, can be defined as follows.

$$Q = \frac{1}{2m} \Sigma_{ij} [A_{ij} - \frac{d_i d_j}{2m}] \delta(s_i, s_j)$$

Here, m denotes the number of edges corresponding to adjacency matrix A, d_i denotes the degree corresponding to vertex v_i, s_i denotes the community membership of vertex v_i and $\delta(s_i, s_j) = 1$ if $s_i = s_j$.

4.3   Experimental Results

Facebook User Interactions Labeled Dataset: In this set of experiments, we compare our approach with five existing state-of-the-art community detection techniques. The results are summarized in Table 2. We observe that our algorithm achieves the *highest* NMI (0.7946) and PWF (0.7903) values which are significantly higher compared to other algorithms. The modularity, which is the most common unsupervised evaluation metric, of our algorithm is 0.484, which is higher than other competitive algorithms. So, our proposed algorithm outperforms traditional algorithms in terms of both supervised and unsupervised metrics for the real Facebook dataset.

Facebook-like Forums Weighted Un-labeled Dataset: This dataset is weighted, where the weight of an edge represents number of messages between two vertices. However, the data set is not labeled, and thus no ground truth about existing communities is available for this dataset. For this un-labeled dataset, we measure the modularity achieved by each community detection algorithm. We summarize the results in Table 2. As the network is weighted and interactions in the network are like OSNs, our algorithm performs better in this dataset. We can see that our algorithm achieves modularity value of 0.315 which is better than the values of most competitive algorithms.

Table 2. Comparison of Different Algorithms on Weighted Datasets

Dataset →	Facebook Interactions Labeled Dataset					Forum
Algorithm ↓	NMI	PWP	PWR	PWF	Modularity	Modularity
Girvan-Newman	0.4860	**0.9947**	0.2347	0.3798	0.043	0.0488
Walktrap	0.7754	0.9339	0.6608	0.7739	0.4414	0.2031
Eigenvector	0.7279	0.9137	0.6382	0.7515	0.4254	0.1696
Infomap	0.3535	0.4266	0.2297	0.2986	-0.03	0.1372
Multi-level	0.6274	0.601	0.5229	0.5592	0.448	**0.3458**
Proposed Algorithm	**0.7946**	0.9738	**0.6650**	**0.7903**	**0.484**	0.315

Classical Un-weighted Datasets: These graphs are un-weighted and do not allow us to construct *interaction graph* as per our model. Hence, we assume equal weights (i.e., equal to one) for all connecting edges, normalize the edge weights in terms of user degree as per equation (2), and run our algorithms for subsequent phases. The results obtained for these datasets are summarized in Table 3. In Karate and Dolphins

Table 3. Comparison of Different Algorithms on Un-weighted Datasets

Algorithm ↓ Dataset →	Karate	Dolphin	Jazz	Ego	Dataset-1	Dataset-2	Dataset-3	Dataset-4
Girvan Newman	0.401	0.519	0.405	0.271	0.6533	0.5498	0.4541	0.341
Walktrap	0.353	0.489	0.438	0.398	0.6533	0.5498	0.4541	0.341
Fast Greedy	0.381	0.495	0.439	**0.442**	0.6533	0.5498	0.4541	0.2555
Leading Eigenvector	0.393	0.491	0.394	0.433	0.6533	0.4404	0.4541	0.3234
Infomap	0.402	0.523	0.28	0.418	0.6533	0.5498	0.4541	0.341
Label Propagation	0.36	0.379	0.282	0.343	0.6533	0.5498	0.4541	0.1797
Multi-level	**0.419**	0.519	**0.44**	0.427	0.6533	0.5498	0.4541	0.341
Proposed Algorithm	**0.419**	**0.524**	0.42	0.433	0.6533	0.5498	0.4541	**0.3427**

network, our algorithm performs better than other algorithms and achieves the *highest* modularity value of 0.419 and 0.524 respectively. However, in Jazz network, our algorithm performs moderate and achieves the modularity value 0.42 which is close to the maximum modularity value 0.44. Again, in Ego network our algorithm achieves the modularity value of 0.433 which is close to the maximum modularity value 0.442.

Computer Generated Benchmark Datasets: In the last set of experiments, we use computer-generated benchmark graphs to compare the effectiveness of our algorithm with other competitive methods. We have generated four graphs, namely, Dataset-1, Dataset-2, Dataset-3, and Dataset-4 for four different mixing parameter values 01, 0.2, 0.3, and 0.4, respectively. We have measured modularity performance of all algorithms for these graphs. The obtained results are summarized in Table 3. We can see that, in graphs with low connectivity outside clusters (mixing parameter values 0.1, 0.2 and 0.3), the modularity performance is almost identical for all algorithms. However, with an increased ratio of external degree/total degree (mixing parameter value 0.4), the modularity values of the most established algorithm degrades. We observe that our algorithm performs well under any external degree/total degree ratio.

4.4 Value of α for Different Networks

As we have discussed in Section 3, we can vary α to tune our algorithm to achieve a higher modularity value, which in turn means good quality communities. We vary α and see the effect of α in modularity (Q) for different datasets. For networks which have relatively sparse connections within the group, i.e., the number of common neighbors for any two members of the group is relatively low, the algorithm achieves better modularity values for higher values of α i.e, one-to-one interaction gets priority. Karate network (Fig. 3(i)) can be considered in this regard. It is to be noted that, in some cases

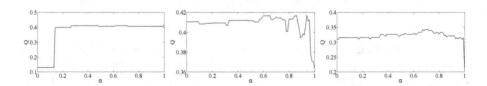

Fig. 3. α vs Q for Different Networks: (i) Karate Network, (ii) Jazz Network, (iii) Dataset-4

completely eliminating group interaction degrades modularity performance significantly. For example, Fig. 3(ii) shows that modularity value for Jazz Musicians dataset sharply falls, when α reaches to 1. Another interesting case appears for Dataset-4 (Fig. 3(iii)). We observe that when the network has almost balanced external/total degree ratio, we achieve the highest modularity value when α is near 0.5. These results validate our claim made in Section 3 that for different social network graphs the value of appropriate α can be different.

5 Conclusion

In this paper, we have proposed a community detection algorithm based on user interactions in online social networks. To identify active communities, we have considered both the impact of interaction between users and the impact of the group behavior that the users show based on the interaction with their common friends. We have combined both impacts in a weighted manner to identify the active communities with a high accuracy. Our detailed experimental results show the superiority of our approach over other state-of-the-art techniques for a wide range of datasets. We have observed that for a real Facebook user interaction dataset, our algorithm achieves the *highest* normalized mutual information (NMI) and pairwise F-measure (PWF) values, which are significantly higher than the NMI and PWF values achieved by other competitive methods. Moreover, our algorithm performs reasonably well for different benchmark datasets, for both weighted and un-weighted social graphs.

Acknowledgements. We are thankful to Samsung Innovation Lab - BUET, and Code-Crafters International and Investortools, Inc. for supporting this research. We would also like to thank Tanmoy Sen, Rajshakhar Paul, Ishat E Rabban and Madhusudan Basak for helping us in data collection.

References

1. Aggarwal, C.C., Wang, H. (eds.): Managing and Mining Graph Data. Advances in Database Systems, vol. 40. Springer (2010)
2. Abello, J., Resende, M.G.C., Sudarsky, S.: Massive quasi-clique detection. In: Rajsbaum, S. (ed.) LATIN 2002. LNCS, vol. 2286, pp. 598–612. Springer, Heidelberg (2002)
3. Newman, M.E.J.: Detecting community structure in networks. The European Physical Journal B - Condensed Matter and Complex Systems 38 (2004)
4. Newman, M.E.J., Girvan, M.: Finding and evaluating community structure in networks. Physical Review E 69 (2004)
5. Girvan, M., Newman, M.E.J.: Community structure in social and biological networks. Proceedings of the National Academy of Sciences 99(12), 7821–7826 (2002)
6. Zhou, Y., Cheng, H., Yu, J.X.: Graph clustering based on structural/attribute similarities. Proc. VLDB Endow. 2(1), 718–729 (2009)
7. Zhou, W., Jin, H., Liu, Y.: Community discovery and profiling with social messages. In: Proceedings of the 18th ACM SIGKDD International Conference on Knowledge Discovery and Data Mining, pp. 388–396 (2012)

8. Sachan, M., Contractor, D., Faruquie, T.A., Subramaniam, L.V.: Using content and interactions for discovering communities in social networks. In: Proceedings of the 21st International Conference on World Wide Web, pp. 331–340 (2012)
9. Qi, G.J., Aggarwal, C.C., Huang, T.S.: Community detection with edge content in social media networks. In: ICDE, pp. 534–545 (2012)
10. Pujol, J.M., Erramilli, V., Siganos, G., Yang, X., Laoutaris, N., Chhabra, P., Rodriguez, P.: The little engine(s) that could: Scaling online social networks. IEEE/ACM Trans. Netw. 20(4), 1162–1175 (2012)
11. Rothschild, J.: High performance at massive scale - lessons learned at facebook, http://cns.ucsd.edu/lecturearchive09.shtml
12. Zeng, Z., Wang, J., Zhou, L., Karypis, G.: Out-of-core coherent closed quasi-clique mining from large dense graph databases. ACM Trans. Database Syst. 32(2) (June 2007)
13. Pei, J., Jiang, D., Zhang, A.: On mining cross-graph quasi-cliques. In: Proceedings of the 11th ACM SIGKDD International Conference on Knowledge Discovery in Data Mining (2005)
14. Xu, X., Yuruk, N., Feng, Z., Schweiger, T.A.J.: Scan: A structural clustering algorithm for networks. In: Proceedings of the 13th ACM SIGKDD International Conference on Knowledge Discovery and Data Mining, pp. 824–833 (2007)
15. Cohen, J.: Trusses: Cohesive subgraphs for social network analysis (2008)
16. Raghavan, U.N., Albert, R., Kumara, S.: Near linear time algorithm to detect community structures in large-scale networks (September 2007)
17. Clauset, A., Newman, M.E.J., Moore, C.: Finding community structure in very large networks 70(6) (December 2004)
18. Meo, P.D., Ferrara, E., Fiumara, G., Provetti, A.: Generalized louvain method for community detection in large networks. In: ISDA, pp. 88–93. IEEE (2011)
19. Yang, T., Jin, R., Chi, Y., Zhu, S.: Combining link and content for community detection: A discriminative approach. In: Proceedings of the 15th ACM SIGKDD International Conference on Knowledge Discovery and Data Mining, pp. 927–936 (2009)
20. Mllner, D.: http://math.stanford.edu/muellner/fastcluster.html
21. Pons, P., Latapy, M.: Computing communities in large networks using random walks. J. of Graph Alg. and App. bf 10, 284–293 (2004)
22. Newman, M.E.J.: Finding community structure in networks using the eigenvectors of matrices. Phys. Rev. E 036104
23. Rosvall, M., Bergstrom, C.T.: Maps of random walks on complex networks reveal community structure. Proc. Natl. Acad. Sci., 1118 (2008)
24. Blondel, V., Guillaume, J., Lambiotte, R., Mech, E.: Fast unfolding of communities in large networks. J. Stat. Mech., P10008 (2008)
25. Opsahl, T.: Triadic closure in two-mode networks: Redefining the global and local clustering coefficients. Social Networks (August 2011)
26. SNAP: Stanford network analysis project
27. Zachary, W.: An information flow model for conflict and fission in small groups. Journal of Anthropological Research 33, 452–473 (1977)
28. Lusseau, D.: The emergent properties of a dolphin social network. Proceedings of the Royal Society of London. Series B: Biological Sciences 270(suppl. 2) (November 2003)
29. Gleiser, P.M., Danon, L.: Community structure in jazz. Advances in Complex Systems 6(4), 565–574 (2003)
30. Lancichinetti, A., Fortunato, S., Radicchi, F.: Benchmark graphs for testing community detection algorithms. Phys. Rev. E 78(4), 46110 (2008)

Object Semantics for XML Keyword Search

Thuy Ngoc Le[1], Tok Wang Ling[1], H.V. Jagadish[2], and Jiaheng Lu[3]

[1] National University of Singapore
[2] University of Michigan
[3] Renmin University of China
{ltngoc,lingtw}@comp.nus.edu.sg, jag@umich.edu,
jiahenglu@ruc.edu.cn

Abstract. It is well known that some XML elements correspond to objects (in the sense of object-orientation) and others do not. The question we consider in this paper is what benefits we can derive from paying attention to such object semantics, particularly for the problem of keyword queries. Keyword queries against XML data have been studied extensively in recent years, with several lowest-common-ancestor based schemes proposed for this purpose, including SLCA, MLCA, VLCA, and ELCA. It can be seen that identifying objects can help these techniques return more meaningful answers than just the LCA node (or subtree) by returning objects instead of nodes. It is more interesting to see that object semantics can also be used to benefit the search itself. For this purpose, we introduce a novel Nearest Common Object Node semantics (NCON), which includes not just common object ancestors but also common object descendants. We have developed XRich, a system for our NCON-based approach, and used it in our extensive experimental evaluation. The experimental results show that our proposed approach outperforms the state-of-the-art approaches in terms of both effectiveness and efficiency.

1 Introduction

XML has become a widely accepted standard for data storage and data exchange in many applications, such as electronic business[1], science[2], and text databases[3]. In addition, keyword search provides a simple and user-friendly query interface to access XML data in most applications. Therefore, keyword search on data-centric XML documents has attracted great interest. One of the most successful approaches to XML keyword search is the LCA semantics [5], which was inspired by the hierarchical structure of XML. Following this, many extensions of the LCA semantics such as SLCA [20], MLCA [13], ELCA [22] and VLCA [10] have been proposed to improve the effectiveness of the search.

1.1 Limitations with the LCA Semantics

While the LCA semantics and its variants work well for many types of XML documents, unfortunately, in several scenarios, they still suffer from two limitations: they

---

[1] http://www.ebxml.org
[2] http://www.biodas.org/documents/spec-1.53.html
[3] http://www-connex.lip6.fr/~denoyer/wikipediaXML/

S.S. Bhowmick et al. (Eds.): DASFAA 2014, Part II, LNCS 8422, pp. 311–327, 2014.

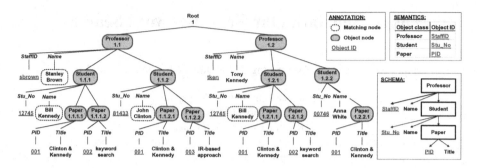

Fig. 1. An XML document with the corresponding schema and the discovered semantics

may return *meaningless* answers and *incomplete* sets of answers. Meaningless answers are returned when the LCA node (or its variants) just simply matches query keywords and does not provide any other additional information. More importantly, LCA-based approaches return incomplete sets of answers because they only search up from the matching nodes, i.e., the nodes containing keywords, for common ancestors, and never search down to find common information appearing as descendants. From now on, we use the term *common descendant* to refer to such common information. Example of these drawbacks can be seen in the context of the XML data tree in Fig. 1.

EXAMPLE 1 (**Meaninglessness**). *For query* {Stanley, Brown}, *an answer such as the value node* Stanley Brown *(in the left most) is meaningless since it does not provide any additional information about* Stanley Brown. *A meaningful answer should contain additional information about keywords, i.e., it should be the subtree rooted at node* Professor(1.1).

EXAMPLE 2 (**Incompleteness**). *For query* {Bill, John}, *the keywords match two students:* Student(1.1.1) *and* Student(1.1.2) *respectively. Their common ancestor* Professor(1.1) *is an answer returned by LCA-based approaches. However, this is not complete. Object* <Paper:001>[4], *which is represented by groups of nodes started at* Paper(1.1.1.1) *and* Paper(1.1.2.1), *should also be returned as an answer. This paper is a common descendant of these student nodes. Intuitively, these students are not only supervised by the same professor* <Stanley:Brown>, *but also co-authors of the same paper* <Paper:001>.

The problem of incompleteness happens when the relationship between object classes is many-to-many ($m : n$). Then, the child object is duplicated each time it occurs in the relationship, and two nodes may have the same object as the child. In practice, $m : n$ relationships occur in many real XML datasets, including IMDb[5] and NBA[6] (used in experiments of prevalent XML research works such as [17,16]). In IMDb, an actor or actress can play in many movies, and a company can produce several movies. In NBA, a player can play for several teams in different years. Moreover, due to the flexibility

---
[4] <Paper:001> denotes an object which belongs to object class Paper and has OID 001.
[5] http://www.imdb.com/interfaces
[6] http://www.nba.com

and exchangeability of XML, many relational datasets with $m : n$ relationships can be transformed to XML [8,4] with duplication. Thus, it is very likely that LCA-based approaches will return an incomplete answer set for such databases.

1.2 Our Novel Semantics Based on Object

To address limitations of the LCA semantics, we propose to use the concept of *object* in XML keyword search. In XML data, an object may be represented as different object *instances*, each of which corresponds to a group of nodes, rooted at a tag indicating object class, followed by a set of attributes and their associated values. We refer to this root node as *object node* and the others as *non-object nodes*. For example, object <Paper:001> has four instances starting from four object nodes Paper(1.1.1.1), Paper (1.1.2.1), Paper(1.2.1.1) and Paper(1.2.2.1) respectively. Other nodes such as PID, 001, Title, Clinton& Kennedy are non-object nodes. An object is identified by its *object class* and *OID (object identifier)*. Thus, *two instances represent the same object if they have the same object class and the same OID*. This is all the object orientation we rely upon. Using just this much, we show in this paper the benefits that accrue to keyword search.

Based on these object orientation concepts, we introduce a novel semantics, called *Nearest Common Object Node* (NCON) for XML keyword search. The NCON semantics has two key features. First, an NCON must be an *object node* rather than an arbitrary node. This reduces the number of meaningless answers. Second, an NCON can be either an LCOA *(lowest common object ancestor)* or an HCOD *(highest common object descendant)*. Although LCOA is similar to LCA [5,10,20], the important difference is that an LCOA must be an object node. An HCOD (1) is a common object descendant of a set of keywords, and (2) has no ancestor that is also a common object descendant of that set of keywords. The second feature includes *common descendants* into the answer set. Let us revisit the motivating examples introduced above and see how our proposed NCON semantics helps.

EXAMPLE 3 **(Example 1 Reprise).** *LCOA* Professor(1.1) *is the object node of the non-object node* Stanley Brown. *Returning the former is meaningful, whereas returning the latter is meaningless. LCOA semantics will return the object node, rather than the non-object node.*

Several works such as XSeek [14], XReal[1], and [19] have attempted to solve the problem of meaningless answers by identifying entity (object), and they can obtain more meaningful answers in several cases. However, these works do not use OIDs as ours, and therefore do not always distinguish an object from an aggregation node, a composite attribute, and a multi-valued attribute.

EXAMPLE 4 **(Example 2 Reprise).** *We discover that* Paper(1.1.1.1) *and* Paper (1.1.2.1) *refer to the same object: paper* <PID:001>, *because they belong to the same object class* Paper *and have the same OID value* 001. *HCOD will find this paper and return it as an answer. In contrast, LCA based approaches cannot detect this common paper because it appears as a descendant, not as an ancestor.*

For an XML document with ID/IDREF, graph-based approaches such as [2,11,7] can provide common descendants. However, to maintain the tree structure, XML designers may duplicate information instead of using ID/IDREF. Moreover, those graph-based approaches can find common descendants only if XML documents contain ID/ IDREF. Otherwise, those graph-based approaches do not recognize instances of the same object. Therefore, they cannot find common descendants either. To the best of our knowledge, only [9] can detect such instances and find common descendants. Nevertheless, this work transfers an XML document to a graph which is similar to relational database, and follows Steiner tree semantics. Thus, it suffers from the inefficiency and may return meaningless answers because matching nodes may not be (or weakly) related (will be shown in Section 5).

A final answer obtained by LCA-based approaches includes two parts: an LCA node and a presentation of the answer, e.g., a subtree or a path. Arguably, the presentation as a subtree may contain common descendants. However, LCA-based approaches do not explicitly identify them and it may be hard to identify them because this presentation contains a great deal of irrelevant information. In contrast, our NCON semantics can and does clearly identify both common ancestors and common descendants.

1.3   Our Approach and Contributions

Like existing LCA-based approaches such as [20,22,10,13], we work with *data-centric* XML documents *without ID/IDREF*, in which objects may be duplicated as different instances due to $m : n$ relationships. We follow the *NCON semantics* so that both common ancestors and common descendants can be answers. Finding common descendants is much more challenging than finding common ancestors. Given a set of matching nodes, unlike a common ancestor which appears as only one node, a common descendant may appear as many different nodes. Therefore, it requires more complex techniques of indexing and searching to find common descendants.

We propose a new search strategy which uses XML object tree *(O-tree)* rather than the whole XML document for the search. An O-tree is extracted from an XML data tree by keeping only object nodes and associating all non-object nodes (e.g., attributes and values) to the corresponding object nodes. This helps reduce the number of meaningless answers because only object nodes are returned. Moreover, this reduces the search space greatly since the number of nodes in O-tree is much smaller than those in the whole XML document (due to not counting non-object nodes). To find common descendants, we use a *reversed O-tree*, whose paths from the root to leafs are reversed from those of the given O-tree. Then, HCODs of the given O-tree can be found as LCOAs of the reversed O-tree. Fig. 2(b) shows the reversed O-tree w.r.t. the O-tree in Fig. 2(a).

We do not use ID/IDREF to connect instances of the same object because if we do so, XML data will be modeled as an XML graph instead of an XML tree. Searching over graph-structured data has been known to be equivalent to the group Steiner tree problem, which is NP-Hard [3]. In contrast, with the duo of the original and the reversed O-trees, we can leverage the efficient computation of finding common ancestors based on common prefix of Dewey labels as the LCA-based approaches [20,22,10] do.

In brief, we make the following contributions.

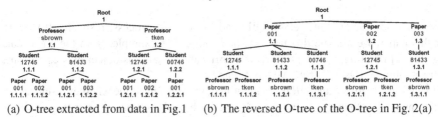

(a) O-tree extracted from data in Fig.1   (b) The reversed O-tree of the O-tree in Fig. 2(a)

Fig. 2. The original and reversed XML object trees (O-trees)

- Based on object identification, we introduce a novel semantics for XML keyword search, called NCON, which returns a more complete set of meaningful answers (Section 2).
- We propose an efficient search which uses the O-tree rather than the whole XML document, and reversed O-tree rather than ID/IDREF (Section 3 and Section 4).
- We have implemented all of our ideas in XRich system for evaluation. Although XRich has overhead from finding HCODs, experimental results show that it outperforms the state-of-the-art approaches in terms of both effectiveness and efficiency because it is still based on tree and works with O-trees rather than whole XML documents (Section 5).

2   Object Semantics

We propose to use the semantics of object for XML keyword search. In this section, we first recall the concept of object in XML and the identification of object in Section 2.1. Based on object, we introduce Nearest Common Object Node semantics for answering XML keyword queries in Section 2.2.

2.1   Object Identification

An *object* represents a real world entity. Recall that in XML, an object can be represented by multiple object *instances*, each of which corresponds to a group of nodes, starting at an *object nodes*, followed by *non-object nodes*. An object node may have other object nodes as its descendants. Among all nodes describing an object instance, the object node (i.e., the object class tagged node) is the most "important" because it is the root of the group containing these nodes. Hereafter, it is used as the representative for the entire object instance in unambiguous contexts.

Among non-object nodes, a special attribute (or a set of attributes) together with *object class* that can uniquely define an object is called *object identifier (OID)*. An OID can be a set of attributes because under several cases, a single attribute cannot uniquely identify an object (similar to a key in relational database may contain more than one attribute). OID is different from ID in XML schemas, e.g., DTD because an ID can always be an OID but not vice versa. With ID/IDREF, an OID is defined as an ID in XML schemas. However, in many cases, XML schemas may not have ID for objects such as lower objects in $m : n$ relationships in XML documents without ID/IDREF. Therefore, we need to identify OID in such cases.

Object identification has been studied by third party algorithms such as [14,15,12]. We apply the algorithm in [12] which has high accuracy (greater than 99%, 93% and

95% for discovering object class, OID and the overall process, respectively). Thus, for this paper, we assume this task has been done. For example, from the XML data Fig. 1, [12] identifies Professor, Student and Paper as object classes with the corresponding OIDs: StaffID, Stu_No and PID.

Once *object classes* and *OIDs* are identified, we can determine whether instances (or object nodes in other words) refer to the same object based on whether they have *the same object class and the same OID*. For example, object nodes Student(1.1.1) and Student(1.2.1) are of the same object <Student:12745> since they have the same object class Student and the same OID 12745. Multiple object nodes that refer to the same object are usually identical. However, XML does not enforce this. If we find two nodes that are not identical, they are still considered as referring to the same object as long as they have the same object class and OID value.

2.2   The Nearest Common Object Node (NCON) Semantics

Based on object identification, we introduce the NCON semantics, which includes both common ancestors and common descendants. The purpose of this inclusion is to provide a more complete set of answers for XML keyword search. Our proposed NCON semantics can be built on the LCA semantics [5] or any of its variants such as SLCA [20], VLCA [10] and ELCA [22]. For simplicity of presentation, we provide the following definitions which are based on the LCA semantics. It is straightforward to make the necessary minor modifications required if any of the variant semantics are preferred.

Let $u \prec_a$ ($\succ_a$, $\preceq_a$, or $\succeq_a$) v denote that object node u is an ancestor (a descendant, an ancestor-or-self, or a descendant-or-self respectively) of object node v. A keyword k matches an object node u if k is contained by u or by any of non-object nodes associated with u. The NCON semantics and its two components, i.e., LCOA (Lowest Common Object Ancestor) and HCOD (Highest Common Object Descendant) are defined as follows.

Definition 1 (LCOA of a set of object nodes). *Object node u is the LCOA of a set of object nodes $\{u_1, \ldots, u_n\}$ if (1) $u \preceq_a u_i \ \forall i = 1..n$ and (2) there exists no object node $v \succ_a u$ s.t. $v \preceq_a u_i \ \forall i = 1..n$.*

An LCOA is similar to an LCA. However, an LCOA must be an object node while an LCA can be an arbitrary node. This difference enables the NCON semantics to reduce the number of meaningless answers.

Definition 2 (HCOD of a set of object nodes). *Given a set of object nodes $\mathbb{S} = \{u_1, \ldots, u_n\}$, the set of object nodes $\mathbb{H} = \{h_1, \ldots, h_n\}$ is an HCOD of $\mathbb{S}$ if*
- *all h_i's refer to the same object and*
- *$u_i \preceq_a h_i \ \forall i = 1..n$ and*
- *there exists no set of object nodes $\mathbb{H}' = \{h'_1, \ldots, h'_n\}$ where $h'_i \prec_a h_i \ \forall i = 1..n$ which satisfies the above two conditions.*

HCOD is the distinguishing feature of the NCON semantics. An HCOD contains a set of object nodes which refer to the same object. Each of them is a descendant of the corresponding matching object node. Note that a set of object nodes has only one LCOA but may have several HCODs because a node has only one parent but several children.

Definition 3 (An NCON of a set of object nodes). *An NCON of a set of object nodes* $\mathbb{S}$ *is either an LCOA of* $\mathbb{S}$ *or an HCOD of* $\mathbb{S}$.

Definition 4 (An NCON of a query). *An NCON of a keyword query* $Q = \{k_1, \ldots, k_n\}$ *is an NCON of a set of object nodes* $\mathbb{S} = \{u_1, \ldots, u_n\}$ *where* u_i *matches* k_i.

3 Overview of Our Approach

The problem tackled in this paper is to find the set of NCONs for a keyword query issued against a data-centric XML document without IDREF. This section provides an overview of our approach, including the ideas about object orientation and reversal mechanism, and the overview of the process. Detailed techniques will be given in Section 4.

3.1 Object Orientation

Based on object nodes, we introduce the concept of XML object tree (O-tree) as follows.

Concept 1 (O-tree). *An O-tree OT is a tree extracted from an XML data tree DT by keeping all object nodes, and associating non-object nodes to the corresponding object nodes. For any object nodes u and v in DT having no other object nodes in between[7], there is a parent-child edge between u and v in OT.*

For example, the O-tree extracted from the XML data in Fig. 1 is shown in Fig. 2(a), in which in each node, Dewey label is used to identify object node while object class and OID are used to identify object.

O-tree brings two important benefits to XML keyword search. First, an answer is more likely to be meaningful since a returned node is an object node in O-tree and it represents a whole object rather than just an attribute or a value. Second, the search space is dramatically reduced because the number of nodes of the extracted O-tree is much smaller than that of the corresponding XML data tree. Suppose that the average number of attributes for an object class is N, then the number of nodes in the XML document is at least $2 \times N$ times larger than that of O-tree (due to not counting attributes and values for the O-tree). This extensively reduces the complexity of the search.

3.2 Reversal Mechanism

The set of NCONs includes LCOAs and HCODs. LCOAs can be found by any of existing LCA-based approaches such as [20,21]. To find HCODs, the idea is a *reversal mechanism* by which HCODs of the given O-tree are turned into LCOAs in its reversed O-tree, which is defined as follows.

Concept 2 (Reversed O-tree). *Given an O-tree OT, the reversed O-tree w.r.t. OT is an O-tree* OT_R *such that*
- *for each path of object nodes* $/u_1/u_2/\ldots/u_{n-1}/u_n$ *from the root to a leaf in OT, there is a corresponding reversed path* $/u'_n/u'_{n-1}/\ldots/u'_2/u'_1$ *in* OT_R *where each pair of object nodes* u_i *and* u'_i *refer to the same object, and*

---

[7] There may exist non-object nodes between them such as an aggregational node or a grouping node [12].

- *there does not exist any pairs of nodes in OT_R such that they refer to the same object and they have the same list of objects as their ancestors, and*
- *there is no other object node in OT_R.*

For example, Fig. 2(b) shows the reversed O-tree derived from the O-tree in Fig. 2(a).

To derive a reversed O-tree, we need to determine whether two object nodes refer to the same object based on their object class and OID. The reversed O-tree is used with the sole goal of finding HCODs. Although there may be duplication in O-trees, such duplication does not affect the efficiency thanks to our index and search techniques. We assume updating does not frequently happen as LCA-based approaches assume. Otherwise, adding or deleting one node can lead to change Dewey labels of all nodes in an XML document in those approaches.

For XML data containing only binary relationships, LCOAs can be found from original O-tree while HCODs can be found from the reversed O-tree. Thus, although other O-trees, apart from the reversed O-tree, may capture the same information with the original O-tree, only the duo of the original and reversed O-tree is *self-sufficient* to return the complete set of NCONs. For n-ary relationship ($n \geq 3$), using the reversed O-tree can return more answers than LCA-based approaches, but the results still may not be complete. We leave the improvement of this kind of relationships for future work. Fortunately, such relationships are rare in XML in practice.

3.3 Overview of the Process

The process of our approach, as shown in Fig. 3, comprises two components for *pre-processing* and *query processing*. Detailed techniques of these components will be discussed in Section. 4. For pre-processing, the three main tasks are extracting the O-tree from the input XML document, generating the reversed O-tree from the original O-tree and indexing.

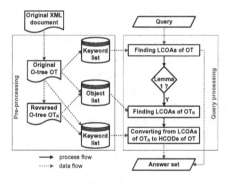

Fig. 3. Overview of the process

For query processing, we follow the reversal mechanism in which HCODs of the original O-tree are turned into LCOAs of the reversed O-tree. Therefore, our process has three steps: finding LCOAs in the original O-tree, finding LCOAs in the reversed O-tree, and converting LCOAs in the reversed O-tree to HCODs in the original O-tree. Our process is flexible in the sense that, it is independent of any LCA semantics adopted, and can be easily deployed to existing LCA-based approaches.

We observe that for several cases, using the reversed O-tree is not necessary because there is no HCOD. So we can optimize the processing with the following lemma.

Lemma 1. *Given an XML keyword query $Q = \{k_1, k_2, \ldots, k_n\}$, the reversed O-tree does not provide any new answer if any of the following holds:*
- *Q has only one keyword.*
- *All keywords of Q match the same object.*

– *Keywords of Q may have multiple matches. For a set of matching object nodes $S = \{u_1, u_2, \ldots, u_n\}$ where u_i matches keyword k_i, the reversed O-tree does not provide any new answer for S if there exist two different object nodes $u_i, u_j \in S$ which do not represent the same object such that they are the leaf nodes in the original O-tree.*

The first two conditions are intuitive. The rationale behind the third condition is that when u_i and u_j are the leaf nodes in the original O-tree, they become the highest nodes in the reversed O-tree with no ancestor beside the root. They do not have ancestor-descendant relationship for one of them to become a common ancestor either. Therefore, there is no common ancestor of these two nodes. Thus, there is no common ancestor of S. Hence, the reversed O-tree does not provide new answer.

4   Detailed Techniques of Our Approach

Following Section 3, this section presents detailed techniques of our approach.

4.1   Generating the Reversed O-tree

The process of generating the reversed O-tree from the original O-tree OT has two steps corresponding to two first conditions of Concept 2.

Step1: Reversing Object Node Paths. To reverse object node paths in OT, we traverse OT backward from each leaf node to the root to form a reversed path. Then, all reversed paths are connected to form the intermediate O-tree. Algorithm 1 presents this process. We use an *array-like-stack* S to store all object nodes in OT. An array-like-stack is an array in which *push* and *pop* operators are used in similar way to a stack while we still can access any element in S like an array. We traverse OT by depth first order and push visited object nodes into S. To handle the branches in the tree, we maintain the parent of each object node. Thus, we use the triple $\langle i, (objCls(i) : OID(i)), pre(i) \rangle$ to represent each object node i, where i is the *index* by depth first order (i starts from 1), $objCls(i)$ and $OID(i)$ are the *object class* and *OID* of i and $pre(i)$ is the *index* of the *parent* of i. Note that during the reversal, we associate relationship attributes (if any) to the lowest object node of the relationship it belongs to. Fig. 4 shows the intermediate O-tree w.r.t. the original O-tree in Fig. 2(a).

Step 2: Merging Object Nodes. To generate the reversed O-tree from reversed object node paths, we merge object nodes having the same set of ancestors. Particularly, at the first level of the intermediate O-tree, we merge branches where the starting object nodes refer to the same object. Then we recursively merge in the lower levels. Fig. 5 demonstrates merging processes w.r.t. the intermediate O-tree in Fig. 4.

Size of the Reversed O-tree. In the worst case where there exist $1 : m$ relationships, the size of the reversed O-tree is $\frac{N \times h}{2 \times l}$ where N, h, l are the number of nodes, the height, and the least number of attributes of an objects in the original XML document. The number of object nodes in the original O-tree is N/l. All leaf nodes ($(N/l)/2$ nodes) become the nodes in the first level (after the root node). In the worst case where there is no duplication among them, there will be maximum $\frac{N \times h}{2 \times l}$ nodes.

Algorithm 1. Reversing object node paths

Input: The original O-tree OT
Output: Intermediate O-tree OT_I

1 **Variables:** Array-like-Stack S: store object nodes in OT by DF order
2 **for** *visited object node* $i \in OT$ *by DF order* **do**
3 $\quad\llcorner$ S.Push $(\langle i, (objCls(i) : OID(i), pre(i))\rangle$
4 OT_I. Add (Root)
5 OT_I. NewBranch
6 **while** $S \neq \emptyset$ **do**
7 $\quad\mid$ $\langle i, (objCls(i) : OID(i), pre(i))\rangle \leftarrow S$.Pop
8 $\quad\mid$ OT_I. Add $(objCls(i) : OID(i))$
9 $\quad\mid$ $//pre(i) = 0$:parent is current top element
10 $\quad\mid$ **if** $pre(i) = 0$ **then**
11 $\quad\mid$ $\quad\llcorner$ OT_I. NewBranch
12 $\quad\mid$ $//pre(i) \neq 0$: parent has branches
13 $\quad\mid$ **if** $pre(i) \neq i - 1$ *and* $pre(i) \neq 0$ **then**
14 $\quad\mid$ $\quad\mid$ $k \leftarrow pre(i)$
15 $\quad\mid$ $\quad\mid$ **while** $k \neq 0$ **do**
16 $\quad\mid$ $\quad\mid$ $\quad\mid$ Access element k $\langle k, (objCls(k) : OID(k), pre(k))\rangle$
17 $\quad\mid$ $\quad\mid$ $\quad\mid$ OT_I. Add $(objCls(k) : OID(k))$
18 $\quad\mid$ $\quad\mid$ $\quad\mid$ **if** $pre(k) = 0$ **then**
19 $\quad\mid$ $\quad\mid$ $\quad\mid$ $\quad\llcorner$ OT_I. NewBranch
20 $\quad\mid$ $\quad\mid$ $\quad\mid$ **if** $pre(k) = k - 1$ **then**
21 $\quad\mid$ $\quad\mid$ $\quad\mid$ $\quad\llcorner$ k. Next
22 $\quad\mid$ $\quad\mid$ $\quad\llcorner$ $k \leftarrow pre(k)$

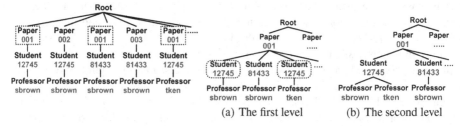

Fig. 4. The intermediate O-tree derived from the O-tree in Fig.2(a)

(a) The first level

(b) The second level

Fig. 5. Merging branches having the same set of ancestors

4.2 Indexes

Since a common descendant may appear as many different nodes, we need more complex kinds of index to accelerate finding common descendants.

Keyword List. Keyword list is to efficiently retrieve the set of matching objects[8] in the original XML document. Each keyword matches a list of objects ordered decreasingly by hierarchical level of objects. The space cost of the keyword list is $K \times M$ where K is the number of keywords in XML document and M is the maximum number of objects matching a keyword. Table 1 shows a part of keyword list of the XML data in Fig. 1.

---

[8] An object matches keyword k when any of its object node matches k.

Table 1. A part of keyword list of the XML data in Fig. 1

Keyword	Matching objects
Kennedy	`<Professor:tken>,<Student:12745>,<Paper:001>`
Clinton	`<Student:002>,<Paper:001>`
...	...

Object List. Object list is created for two purposes. It is used to determine whether two object nodes refer to the same object or not, and more importantly, to identify the set of object nodes in the reversed O-tree w.r.t. a given object. The latter will be used to find HCODs. Each object corresponds to a list of Dewey labels of its object nodes sorted by preorder numbering. The space cost of the object list is $M \times N$ where M is the total number of objects in the original O-tree and N is the maximum number of object nodes of an object. Part of the object list of the O-trees in Fig. 2 is given in Table 2.

Table 2. A part of object list of the O-trees in Fig. 2

Objects	Objects nodes in the original O-tree	Object nodes in the reversed O-tree
`<Professor:sbrown>`	1.1	1.1.1.1, 1.1.2.1, 1.2.1.1, 1.3.1.1
`<Student:12745>`	1.1.1, 1.2.1	1.1.1, 1.2.1
`<Student:81433>`	1.1.2	1.1.2, 1.3.1
`<Paper:001>`	1.1.1.1, 1.1.2.1, 1.2.1.1, 1.2.2.1	1.1
...	...	...

Reversed List. Given an object node in the reversed O-tree, reversed list is to trace back to corresponding object nodes in the original O-tree for final output presentation. It costs $N \times L$ where N is the number of object nodes in the reversed O-tree and L is the maximum number of object nodes in the original O-tree w.r.t. a given object node in the reversed O-tree. Table 3 shows a part of the reversed lists w.r.t. the O-trees in Fig. 2.

Table 3. A part of reversed list of the O-trees in Fig. 2

Object nodes in reversed O-tree	Corresponding object nodes in original O-tree
1.1	1.1.1.1, 1.1.2.1, 1.2.1.1, 1.2.2.1
1.1.1	1.1.1, 1.2.1
...	...

4.3 Query Processing

As shown in Fig. 3, to process a keyword query $Q = \{k_1, \ldots, k_n\}$, we have three steps.

Step1: finding LCOAs from the original O-tree OT. We can use any of existing LCA-based algorithms for this task. The list of object nodes matching keyword k_i can be retrieved from the *keyword list* and *object list*. Consider a set of matching object nodes $S = \{u_1, \ldots, u_n\}$ where u_i matches k_i. We denote $LCOA^O(S)$ and $LCOA^R(S)$ be the set of LCOAs for S w.r.t. the original O-tree OT and the reversed O-tree OT_R, respectively.

Based on Lemma 1, we determine whether we need to find $LCOA^R(S)$ or not. Since the reversed O-tree is used without users' awareness, $LCOA^R(S)$ will be converted to HCODs w.r.t. the original O-tree.

Step 2: finding LCOAs of the reversed O-tree OT$_R$. To find $LCOA^R(S)$, from S we identify the corresponding sets of object nodes on OT_R. To do this, we look up the *object list*. Note that, there may be more than one corresponding set in OT_R. After that, we can apply the same algorithm with the algorithm of finding $LCOA^O(S)$.

Step 3: converting LCOA^R(S) into HCODs of OT. An LCOAs v of OT_R corresponds a set $Superset$ of nodes in OT, which can be found by looking up *reversed list*. HCODs is the subset $H = \{h_1, \ldots, h_n\}$ of $Superset$ where h_i is a descendant of u_i.

All ideas discussed above about finding HCODs for a given set of matching object nodes $S = \{u_1, \ldots, u_n\}$ are presented in Algorithm 2.

Algorithm 2. Finding $HCODs(S)$

Input: A set of matching object nodes $S = \{u_1, \ldots, u_n\}$
    Object list
    Reversed list
Output: $HCODs(S)$
1 **for** *each* u_i **do**
2     $\quad S_i \leftarrow$ looking up object nodes of OT_R in object list
3 $LCOA^R(S) \leftarrow LCOA(S_1, \ldots, S_n)$
4 **for** *each* $v \in LCOA^R(S)$ **do**
5     $\quad Superset \leftarrow$ looking up object nodes of OT w.r.t. v
        in reversed list
6     $\quad$ **for** *each matching object node* u_i *in* S **do**
7         $\qquad h_i \leftarrow e \in Superset$ and $e \succeq_a u_i$
8     $\quad HCOD(S).\text{Add}(\{h_1, \ldots, h_n\})$

Complexity. The cost of finding HCODs(S) is dominated by the cost of looking up a node in object list and reversed list. In the worst case, it is $\log(m) \times \log(n)$ for the former, where m and n are the number of objects matched query keywords and the maximum number of object nodes w.r.t. an object. For the later, it is $\log(N) \times \log(L)$ where N and L has similar meanings in reversed list.

Presentation of an Answer. To avoid irrelevant information, we present an answer as a path from a returned NCON, i.e., a (set of) object node(s) to matching object nodes.

Fig. 6 shows the process and outputs for a set of matching nodes of query {Clinton, Kennedy} issued again the XML data in Fig. 1, in which one final output corresponds to an LCOA and the other corresponds to an HCOD.

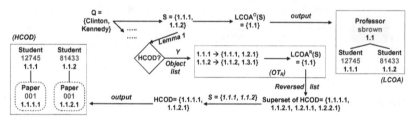

Fig. 6. Process and output of query {Clinton, Kennedy}

5   Experiment

We have developed XRich, a system for XML keyword search, based on our proposed approach. XRich was implemented using Java and was used for experimental evaluation. This section evaluates XRich on three aspects including efficiency, effectiveness and quality of the generated reversed O-tree.

5.1   Experimental Setup

Environment. Experiments were performed on a dual-core Intel Xeon CPU 3.0GHz running Windows XP operating system with 4GB of RAM and a 320GB hard disk.

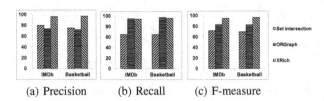

(a) Precision   (b) Recall   (c) F-measure

Fig. 7. Effectiveness Evaluation

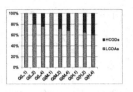

Fig. 8. Percentage of HCODs in NCONs

Datasets. We pre-processed two real datasets including **IMDb**[9], and **Basketball**[10]. We used the subsets with the sizes 150MB and 86MB for IMDb and Basketball respectively. IMDb dataset contains information about movies, actors, actresses, companies, and etc. An actor or actress can play for many movies, and a company can produce for several movies. Basketball dataset contains information about coaches, teams, players where a player and a coach can work for different teams in different years. Table 4 gives more statistics of the datasets.

Table 4. Statistics of datasets

Dataset	No. of nodes	No. of object nodes	No. of object classes	No. of keywords	Data size
IMDb	2,501,780	387,422	6	291,004	150M
Basketball	1,035,940	100,140	3	123,100	86M

Query Set. We randomly generated 120 queries from document keywords. To avoid meaningless queries, we filtered out generated queries which do not contain any value keyword, such as queries contains only tags, or prepositions, or articles, e.g., query {actor, the, to}. 87 remaining queries include 34 and 53 queries for Basketball and IMDb datasets respectively.

Compared Algorithms. We compared XRich with an LCA-based approach to show the advantages of our approach over LCA-based approaches. We chose Set-intersection [21] because it is recent and it outperforms other LCA-based approaches in term of efficiency. We also compare XRich with ORGraph[9] because it can also find common descendants. ORGraph converts XML document to a graph similar to relational database and is based on the Steiner tree semantics.

Metrics. To measure the efficiency, we compared the running time of approaches. We selected five (among 87) queries for each kind of queries, e.g., 2-keyword query. For each query, we ran it ten times to get the average response time. We finally reported the average response time of five queries for one kind of query.

To evaluate the effectiveness, we used standard *Precision* ($\mathcal{P}$), *Recall* ($\mathcal{R}$), and *F-measure* ($\mathcal{F}$) metrics. *F-measure* is the harmonic mean of precision and recall, and is calculated as $\mathcal{F}_\alpha = \frac{(1+\alpha^2) \times \mathcal{P} \times \mathcal{R}}{\alpha^2 \times \mathcal{P} + \mathcal{R}}$. Here we choose $\alpha = 1$ to evenly weight to precision and recall. Other values of α provides similar results. We randomly selected a subset (32 queries) of 87 generated queries for effectiveness evaluation. To compute precision

---

[9] http://www.imdb.com/interfaces
[10] http://www.databasebasketball.com/stats_download.htm

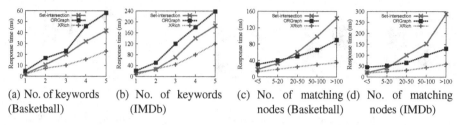

(a) No. of keywords (Basketball)   (b) No. of keywords (IMDb)   (c) No. of matching nodes (Basketball)   (d) No. of matching nodes (IMDb)

Fig. 9. Efficiency evaluation

and recall, we conducted surveys on the above 32 queries and the tested datasets. We asked 15 students in major of computer science to interpret 32 queries. Due to ambiguity of queries, a student may interpret a query in different ways. Common interpretations from at least 12 out of 15 (80%) students are considered as common intuitions. We then manually reformulate these interpretations into schema-aware XQuery queries and use their results as the ground truth.

5.2 Effectiveness Evaluation

Effectiveness. Fig. 7 shows the effectiveness of all compared approaches. As seen, XRich achieves high precision and recall (both are higher than 96%). Compared to Set-intersection, XRich outperforms Set-intersection both in term of recall and precision because XRich returns common descendants while Set-intersection does not; and a returned node of XRich corresponds an object node rather than an arbitrary node of Set-intersection. The difference in terms of recall (more than 25%) is higher than in term of precision. XRich improves both precision and recall, but the more important contribution is improving recall.

Compared to ORGraph, based on undirected Steiner tree, ORGraph has a lightly higher recall than XRich, however, XRich significantly outperforms ORGraph in term of precision because beside common descendants, ORGraph may also return many meaningless answers in sense that it is hard (or even impossible) to interpret such answers because the matching nodes have weak or no relationships. Therefore, if precision and recall is evenly weighed, the F-measure of XRich is higher than that of ORGraph as shown in Fig. 7(c).

Percentage of HCODs in NCONs. Fig. 8 shows the percentage of HCODs and LCOAs in NCONs for 9 queries containing $1-4$ keywords. Low (L), medium (M) and high (H) frequencies of keywords correspond to the number of matching objects between 1-100, 100-1000, and above 1000, respectively. $Q(f, k)$ denotes a query containing k keywords with frequency f. For 1-keyword queries, there is 0% HCOD because the reversed tree provides no new answers for such cases. For other queries, the high percentage of HCODs (20% - 40%) shows the importance of finding HCODs. The higher k and f are, the higher that percentage is.

5.3 Efficiency Evaluation

Efficiency. The response time of approaches is shown in Fig. 9, in which we varied the number of query keywords and the number of matching nodes. Although XRich

	Basketball	IMDb
Quality of extracted O-tree (%)	100	99.5
Quality of reversed O-tree (%)	100	99.5
Time to extract O-tree (min)	4	10.5
Time to generate reversed O-tree (min)	3.5	5.8

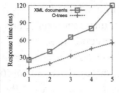

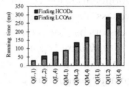

Fig. 10. Extracting original O-tree and generating reversed O-tree

Fig. 11. O-tree vs. XML data tree

Fig. 12. Overhead of finding HCODs

has overhead from finding HCODs, it still outperforms the other algorithms because it searches over the O-tree which is much smaller than the XML document and only uses the reversed O-tree when necessary. Set-intersection runs slower because it works with the whole large XML document. ORGraph runs also slower because ORgraph follows undirected Steiner tree semantics, which would lead exponential computation [3].

Overhead of Finding HCODs. Fig. 12 shows the overhead of finding HCODs for 9 queries discussed in Fig. 8. As shown, it is around 24.7% of the total time, which is not double thanks to Lemma 1.

Impact of Object on Efficiency. Fig 11 shows the response time of XRich when it searches over the O-trees versus the corresponding XML documents. It shows that it runs much faster with the O-trees, especially when the number of keywords increases because the size of the O-trees is much less than that of the XML documents. We randomly chose IMDb because Basketball dataset provides similar results.

5.4 Quality of the Extracted and Reversed O-Trees

To test the quality of the O-tree extracted from XML document, we check the accuracy of the object class and OID discovery. To test the reversed O-tree, we computed the ratio of the number of satisfied object nodes over the total number of object nodes in the reversed O-tree. The satisfied nodes are those in the reversed O-tree that satisfy the reversed schema (object class) which is manually generated. The results are given in Table 10. As can be seen, the quality of the reversed O-tree depends on the quality of the O-tree extracted from XML document, which is very high since our technique can discover object class and OID with high accuracy. Once the O-tree is extracted, the reversed O-tree can be derived accurately. The cost of these processes is not expensive since this computation is performed offline and only once.

6 Related Work

LCA-based approaches. XRANK [5] proposes a stack based algorithm to efficiently compute LCAs. XKSearch [20] defines Smallest LCAs (SLCAs) to be the LCAs that do not contain other LCAs. Meaningful LCA (MLCA) [13] incorporates SLCA into XQuery. VLCA [10] and ELCA [22] introduces the concept of valuable/exclusive LCA to improve the effectiveness of SLCA. MESSIAH [18] handles cases of missing values in optional attributes. Although extensive works have been done on improving the effectiveness, these works may return incomplete answer sets since they find only common ancestors but not common descendants.

Graph-based approaches. Graph-based approaches can be classified based on the semantics such as the Steiner tree [2], distinct root [6] and subgraph [11,7]. The Stener tree semantics can return common descendants if the XML document contains ID/IDREF, but it also returns meaningless answers as well. Distinct root semantics is similar to LCA semantics and cannot find common descendant. Sub-graph semantics provides more information for answers but still miss common descendants if ID/IDREF is not used in XML documents.

Object-oriented approaches. Object have been introduced in XSeek [14], XReal [1], and [19]. However, none of the above works considers OID. Thus, they may not distinguish an object and a composite attribute and/or a multi-valued attribute.

7   Conclusion

This paper shows advantages of object identification in XML keyword search. Based on object identification, we introduced the NCON semantics for XML keyword search, by which an answer corresponds to an object and the answer set includes not only common ancestors but also common descendants. We also proposed an approach based on the NCON semantics and use both the original and the reversed O-trees to find answers. Experimental results showed that XRich outperforms LCA-based and graph-based approaches in terms of both effectiveness and efficiency. Therefore, the approach could be a promising direction for XML keyword search to return a more complete set of meaningful answers. More broadly, this paper demonstrates the benefit of object orientation in XML. In future work, we will explore how other XML processing can similarly benefit and how to handle n-ary ($n \geq 3$) relationships.

References

1. Bao, Z., Ling, T.W., Chen, B., Lu, J.: Efficient XML keyword search with relevance oriented ranking. In: ICDE (2009)
2. Ding, B., Yu, J.X., Wang, S., Qin, L., Zhang, X., Lin, X.: Finding top-k min-cost connected trees in database. In: ICDE (2007)
3. Dreyfus, S.E., Wagner, R.A.: The steiner problem in graphs. Networks (1971)
4. Fong, J., Wong, H.K., Cheng, Z.: Converting relational database into XML documents with DOM. Information & Software Technology (2003)
5. Guo, L., Shao, F., Botev, C., Shanmugasundaram, J.: XRANK: Ranked keyword search over XML documents. In: SIGMOD (2003)
6. He, H., Wang, H., Yang, J., Yu, P.S.: BLINKS: Ranked keyword searches on graphs. In: SIGMOD (2007)
7. Kargar, M., An, A.: Keyword search in graphs: Finding r-cliques. PVLDB (2011)
8. Kim, J., Jeong, D., Baik, D.-K.: A translation algorithm for effective RDB-to-XML schema conversion considering referential integrity information. Journal Inf. Sci. Eng. (2009)
9. Le, T.N., Wu, H., Ling, T.W., Li, L., Lu, J.: From structure-based to semantics-based: Effective XML keyword search. In: Ng, W., Storey, V.C., Trujillo, J.C. (eds.) ER 2013. LNCS, vol. 8217, pp. 356–371. Springer, Heidelberg (2013)

10. Li, G., Feng, J., Wang, J., Zhou, L.: Effective keyword search for valuable LCAs over XML documents. In: CIKM (2007)
11. Li, G., Ooi, B.C., Feng, J., Wang, J., Zhou, L.: EASE: Efficient and adaptive keyword search on unstructured, semi-structured and structured data. In: SIGMOD (2008)
12. Li, L., Le, T.N., Wu, H., Ling, T.W., Bressan, S.: Discovering semantics from data-centric XML. In: Decker, H., Lhotská, L., Link, S., Basl, J., Tjoa, A.M. (eds.) DEXA 2013, Part I. LNCS, vol. 8055, pp. 88–102. Springer, Heidelberg (2013)
13. Li, Y., Yu, C., Jagadish, H.V.: Schema-free XQuery. In: VLDB (2004)
14. Liu, Z., Chen, Y.: Identifying meaningful return information for XML keyword search. In: SIGMOD (2007)
15. Ribeiro, L., Härder, T.: Entity identification in XML documents. In: Grundlagen von Datenbanken (2006)
16. Tao, Y., Papadopoulos, S., Sheng, C., Stefanidis, K.: Nearest keyword search in XML documents. In: SIGMOD (2011)
17. Termehchy, A., Winslett, M.: EXTRUCT: Using deep structural information in XML keyword search. PVLDB (2010)
18. Truong, B.Q., Bhowmick, S.S., Dyreson, C.E., Sun, A.: MESSIAH: Missing element-conscious slca nodes search in XML data. In: SIGMOD (2013)
19. Wu, H., Bao, Z.: Object-oriented XML keyword search. In: Jeusfeld, M., Delcambre, L., Ling, T.-W. (eds.) ER 2011. LNCS, vol. 6998, pp. 402–410. Springer, Heidelberg (2011)
20. Xu, Y., Papakonstantinou, Y.: Efficient keyword search for smallest LCAs in XML databases. In: SIGMOD (2005)
21. Zhou, J., Bao, Z., Wang, W., Ling, T.W., Chen, Z., Lin, X., Guo, J.: Fast slca and elca computation for XML keyword queries based on set intersection. In: ICDE (2012)
22. Zhou, R., Liu, C., Li, J.: Fast ELCA computation for keyword queries on XML data. In: EDBT (2010)

Large-Scale Similarity Join
with Edit-Distance Constraints

Chen Lin[1,2,*], Haiyang Yu[1], Wei Weng[3], and Xianmang He[4,*]

[1] School of Information Science and Technology, Xiamen University, Xiamen 361005, China
[2] Shenzhen Research Institute of Xiamen University, Shenzhen 518057, China
[3] School of Computer and Information Engineering, Xiamen University of Technology,
Xiamen 361024, China
[4] School of Information and Technology, Ningbo University, Ningbo,315122, China
chenlin@xmu.edu.cn, hexianmang@nbu.edu.cn

Abstract. In the age of big data, the data quality problem is more severe than ever. As an essential step in data cleaning, similarity join has attracted lots of attentions from the database community. In this work, to address the similarity join problem with edit-distance constraints, we first improve the partition-based join algorithm for small scale data. Then we extend the algorithm based on MapReduce framework for large-scale data. Extensive experiments on both real and simulated datasets demonstrate the efficiency of our algorithms.

Keywords: Similarity join, big data, Map Reduce, data cleaning.

1   Introduction

The essence of similarity join in cleaning large-scale data has been widely recognized in database community. The goal of similarity join is to find all "similar" instances. The degree of similarity can be computed by various measurements, including Jaccard distance[1], cosine similarity[4], edit distance[5][6][7], hamming distance[4][16] etc.

In this paper, we focus on similarity join problems with edit distance thresholds. Given two strings A and B, edit distance between A and B is the minimum number of edit operations (i.e. insertion, deletion and substitution) to transform A to B. Edit distance has two distinctive advantages over alternative distance or similarity measurements. On one hand, it reflects the ordering of tokens in the string. For example, "abc" and "bca" are not the same in term of edit distance, while they are equivalent using Jaccard distance or cosine distance. On the other hand, it is tolerant to trivial displacement, thus it is more robust to noisy data. For example, a spelling mistake "boby" can be easily recognized as a dirty record and transferred to the correct record "baby". These two properties make edit distance a good measure in many application domains such as getting typographical errors for text documents[17], capturing similarities for Homologous proteins or genes[18], and so on.

---

[*] Corresponding Authors.

S.S. Bhowmick et al. (Eds.): DASFAA 2014, Part II, LNCS 8422, pp. 328–342, 2014.
© Springer International Publishing Switzerland 2014

Most state–of-the-art techniques for similarity join problems with edit distance thresholds employ a filter-and-verify approach. To name a few, Part-Enum[3], All-Pairs-Ed[4], PP-Join[5] and ED-Join[6] generate substrings using Q-gam function, then verify candidate pairs. Efficiency of similarity join algorithms relies on filtering scheme where less similar instance pairs are pruned before verification. Therefore, a number of research efforts are conducted, including All-Pairs-Ed[4], PP-Join[5], ED-Join[6], and PassJoin[7]. Among them, a new filter way is presented in PassJoin[7] with preeminent experimental performance for both short and long strings.

The PassJoin algorithm follows the divide-and-conquer paradigm, in which partitions of strings are compared. Inspired by PassJoin, we propose a faster algorithm for small scale data: PassJoinK, in which the strings to be compared are divided into a flexible number of segments. As the string is divided into more segments, less candidate pairs are generated, the algorithm's performance is enhanced.

A major disadvantage of PassJoin is its incapability of handling large-scale data. As the volume of available datasets is growing increasingly higher nowadays, it is unlikely to store and/or process data in a single machine. PassJoin is designed for Self-Join, i.e. the strings to be compared are in the same file. In case there are multiple files to be computed (Multi-Join problem), efficient intra-file similarity join algorithms are needed. MapReduce framework[9] has been proven to be useful for analyzing large datasets in cluster environments. We extend the PassJoinK algorithm using Map-Reduce framework.

To summarize, we make the following contributions. (a)We present an algorithm PsssJoinK which extends the partition scheme of PassJoin to prune more candidate pairs. (b)We devise an algorithm called PassJoinKMRS using MapReduce to deal with similarity join problems for big data where memory consumption is not affordable in a single machine. Extensive experiments justify the efficiency of the proposed algorithm.

The rest of this paper is organized as follows. We introduce PassJoinK in Section 2. Section 3 introduces PassJoinKMRS based on PassJoinK using Hadoop for both Self-Join and Multi-Join. Experimental results are provided in Section 4. We review related work in Section 5 and make a conclusion in Section 6.

2   PassJoinK

2.1   Preliminaries

Definition 1 (String Similarity Joins). Given two sets of strings A, B and a threshold τ , the similarity join problem is to find all similar string pairs $< p, q >, p \in A, q \in B$ from the two collections if and only if $D(p,q) \leq \tau$ where $D(p,q)$ means the distance between p and q .

Definition 2 (Edit Distance). Given two strings p and q , the edit distance denoted by $ED(p, q)$ means the minimum number of edit operations (insertion, deletion, and substitution) to transform one string to another. For example, $ED(\text{baby}, \text{boy}) = 2$ cause "ab" can be transformed to "o" by first deleting "b" then substituting "a" to "o".

Definition 3 (Similarity Join with Edit Distance Threshold). Given two sets R, S and edit distance threshold τ, the similarity join problem with edit distance threshold is to find all pairs $<p,q>$ such that $p \in R$ and $q \in S$ satisfy $ED(p, q) \leq \tau$. For example, we have two strings $<s,r> = \,$<abcde,ace>. If the given threshold $\tau = 2$, s is similar with r because their edit distance is not bigger than 2.

The key idea of PassJoin algorithm is illustrated in Figure 1. Given two strings r and s, edit distance threshold τ. Suppose r and s are both divided into $\tau+1$ segments, if s is similar to r, then at least 1 segment of r and s are equivalent. As shown in Figure 1. $<s,r> = \,$<abcde,ace> are partitioned into $\tau+1$ segments, there is always at least one equivalent segment (substring match).

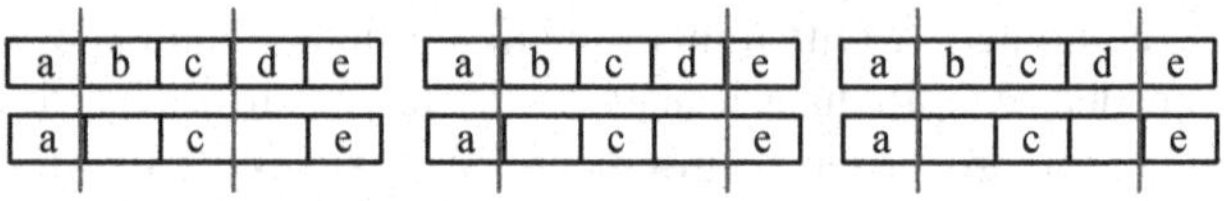

| a | b | c | d | e | | a | b | c | d | e | | a | b | c | d | e |

| a | | c | | e | | a | | c | | e | | a | | c | | e |

Fig. 1. Example of partitioning two strings to several segments

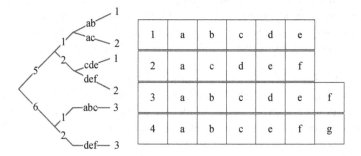

Fig. 2. An illustration of inverted index tree in PassJoin, given four strings (3 are indexed) and edit distance threshold $\tau = 1$

PassJoin consists of three stages: indexing, generation and verification. An inverted index tree is constructed to facilitate indexing and generation. In the stage of indexing, each string is evenly partitioned into substrings. Each substring is stored in a leaf node of the inverted index tree. For a new string, algorithm searches its matched substrings in the inverted index tree, and generates candidate string pairs. As illustrated in Figure 2, the top 3 strings shown in the right part of Figure 2 are partitioned and indexed. For example, "abcde" is partitioned into 2 segments{"ab","cde"}, and the length of the string (5) is stored in the branch of inverted index tree. For a new string "abcefg", the algorithm first gets the substring "ab" and finds a way "5->1->ab->1" from the inverted index tree, leading to a candidate string pair <4,1>.

In the stage of verification, two verification methods, length-aware verification method and extension-based verification method are adopted. In the dynamic programming algorithm to compute edit distance, the length-aware verification method utilizes the length difference to estimate the minimum number of edit operations. The extension-based verification method partitions r and s into three parts: the matching parts, the left parts and the right parts. A candidate pair could be pruned, if the left parts or the right parts of r and s are dissimilar within a deduced threshold[7].

2.2   Algorithm PassJoinK

Intuitively, if strings are divided into more segments, more candidate pairs can be pruned. Thus we introduce the following theorem.

[Theorem 1] Given edit distance threshold τ, two strings A, B. A and B are arbitrarily divided into $\tau + k$ segments. Then if two strings are similar which means $ED(p, q) \leq \tau$, at least k segments of A and B are equivalent.

[Proof] Suppose string s is similar to r, and s contains at most $m(m < k)$ unique substring that matches segments of string r. Thus there exists at least one edit operation between each remaining segment of r and s. Since there are $\tau + k - m$ mismatches, $ED(r, s)$ is larger than $\tau + k - m$. This is contradictory to our former statement that s is similar to r. Thus s must contain at least k substrings which matches the segments of r.

By enlarging the value of k, more unqualified pairs are filtered. We illustrate this property in Figure 3. Given $\tau = 2$ and given two strings $\{r, s\} = \{abcde, cabde\}$. We partition them into $\tau + 1$ and $\tau + 2$ segments. From the left part of Figure 3, $< r, s >$ is considered as a candidate pair, but from the right part of Figure 3, $< r, s >$ is filtered.

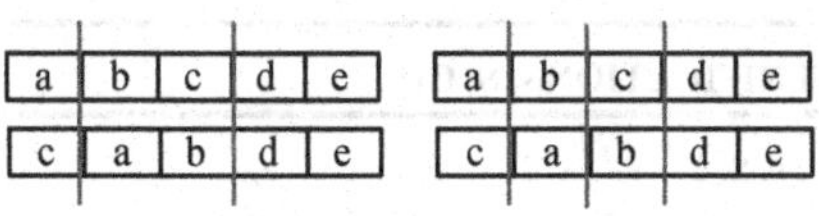

Fig. 3. More unqualified candidate pairs can be pruned by enlarging k

In partition scheme[7], each segment has a length of $\left\lfloor \dfrac{|s|}{\tau + k} \right\rfloor$ or $\left\lfloor \dfrac{|s|}{\tau + k} \right\rfloor + 1$. Let l
$= |s| - \left\lfloor \dfrac{|s|}{\tau + k} \right\rfloor * (\tau + k)$, and the first $\tau + k - l$ segments have length $\left\lfloor \dfrac{|s|}{\tau + k} \right\rfloor$.
The multi-match-aware substring selection method selects substring with start position in the range of :

$$[S_{start}, S_{end}] = [p_i - (i-1), p_i + (i-1)] \cap [p_i + \Delta - (\tau + k - i), p_i + \Delta + (\tau + k - i)].$$

Algorithm 1: Pass-Join-K(S, τ, k)

---

1   **Input : S, τ, K**

2   **Output :** $A = \{(s \in S, r \in s) \mid ED(s,r) \leq \tau\}$

3   **Begin**

4   Sort S by string length and second in alphabetical order;

5   **f or** $s \in S$

6   **for** $L_l^i(\mid s \mid - \tau \leq l \leq \mid s \mid)$

7   **for** $L_l^i(1 \leq i \leq \tau + k)$

8                   $W(s, L_l^i) = SUBSTRINGSELECTION(s, L_l^i)$;

9   **for** $w \in W(s, L_l^i)$

10   **if** w is in L_l^i **, then**

11   Add $L_l^i(w)$ into C(s);

12   **VERIFICATION(s , S , C);**

13 **End**

Function $SUBSTRINGSELECTION(s, L_l^i)$

---

1   **Input :** s : a string, L_l^i : Inverted index

2   **Output :** $W(s, L_l^i)$ **which contains selected substrings**

3   **Begin**

4       $W(s, L_l^i)$ = {w| w is a substring of s};

5   **End**

Function VERIFICATION(s, S, C)

---

1   **Input :** s, S, C

2   **Output :** $A = \{(s \in S, r \in S) \mid ED(s,r) \leq \tau\}$

3   **Begin**

4   **for** $c \in C$ **do**

5   **if** the number of c is bigger than k **then**

6   **If** $ED(S(s), s) \leq \tau$ **then**

7                   $A \leftarrow < S(c), s >$;

8   **End**

Algorithm 1. PassJoinK algorithm

The pseudo-code is shown in Algorithm 1. PassJoinK sorts strings first by length and then in alphabetical order. For each string, sequentially scan its substrings and then match it if they share the same path in the inverted index tree. If a candidate string is returned, add one to the counter. If the cumulative count is bigger than k, verify the candidate pair. The space complexity is $O(\max\limits_{l_{min} \leq j \leq l_{max}} \sum\limits_{l=j-\tau}^{j} (\tau + k) \times |S_l|)$ where $|S_l|$ means the number of strings with length l. The sort complexity is $O(\sum\limits_{l=l_{min}}^{l_{max}} |S_l| \times \log_2(|S_l|))$, selection complexity is $O(\tau^2 \times (\tau+k) \times |S|)$ and the verification complexity is $O(\sum_{s \in S} \sum_{|s|-\tau}^{|s|} \sum_{i=1}^{\tau+k} \sum_{w \in W(s,L_l^i)} \sum_{r \in L_l^i(w)\&\&C(r)>k} V(s,r))$.

3   PassJoinK Based on MapReduce

3.1   Intuition

MapReduce[9] is a popular paradigm for data-intensive parallel computation in shared-nothing clusters. In MapReduce, data is initially partitioned across the nodes of a cluster and stored in a distributed file system (DFS). Data is represented as (key, value) pairs. The computation is expressed using two functions: firstly a map function hash-partitions on the key. For each partition the pairs are sorted by their keys and then sent across the cluster in a shuffle phase. At each receiving node, all the received partitions are merged. All the pair values that share a certain key are passed to a single reduce call. Then the reduce task performs desired functions on the key-value pairs.

It is not a trivial problem to combine PassJoinK with MapReduce. PassJoinK keeps an inverted index tree which is not applicable in MapReduce. The main idea of our algorithms is to change the inverted index tree to a new data structure IIT-Record, and the substring to a new data structure SS-Record. Both IIT-Record and SS-Record are in the form: <k2,v2> = <[segmentString, segmentNumber, stringLength, Flag], ID> where segmentString is a string that is either indexed or to be compared, segment-Number suggests the ID of segment it belongs to, the stringLength indicates the length information and the Flag suggests the type of record. If the Flag equals to IIFLAG it means this record is an IIT-Record and otherwise if the Flag equals to SSFLAG it means this record is a SS-Record.

3.2   PassJoinKMR and PassJoinKMRS

Based on the above intuitions, we present PassJoinKMR and PassJoinKMRS, both of which consists of three stages: indexing, generation and verification. Indexing and generation are implemented by a Map phase, while verification is completed in a reduce phase.

- Stage 1: Indexing

In this stage, we first break every string into evenly $\tau + k$ parts, and for each substring an IIT record is constructed. For example, string "frequency" with ID r is broken into 4 segments expressed as {<[fre, 1, 9,IIFLAG], r >, <[qu, 2, 9, IIFLAG], r >, <[en, 3, 9, IIFLAG], r >,   <[cy, 4, 9, IIFLAG], r >}.

- Stage 2: Substrings Generation

In this stage the Map function generates SS-Records for substrings. For a string S ,a given threshold τ and k , the substring starts at position varies in :

$$[S_{start}, S_{end}] = [p_i - (i-1), p_i + (i-1)] \cap [p_i + \Delta - (\tau + k - i), p_i + \Delta + (\tau + k - i)]$$

Based on the length-filter, the possible matched string length is in the range of $[l - \tau, l + \tau]$. So for each string, we generate $(\tau + k) * 2\tau$ records <[substring, segmentNumber, stringLength, SSFlag], ID> where segmentNumber varies from $[1, \tau + k]$ and stringLength varies from $[l - \tau, l + \tau]$.

Now we have two different kinds of records IIT-Records and SS-Records for each string. All the substrings, together with length, segment information and Flag type are combined as key, their mapping ids are values in the (key, value) pairs. The Map phase outputs <[segmentString, segmentNumber, stringLength, Flag], ID> to Reduce phase and the Reduce task groups all the Records with their prefixes and generates the candidate pairs. For example, an IIT-Record <abc, 1, 9, 10, 0> and a SS-Record <abc, 1, 9, 12, 1> are grouped together and the string pair <10, 12> has one segment matched. If the pair has in total more than k segments matched, it is regarded as a candidate pair.

- Stage 3: Verification

We propose two different strategies to efficiently verify the candidate pairs. The first one is PassJoinKMR which keeps all the IIR-Record and SS-Record information and extension-based verification method to increase performance. The second one is PassJoinKMRS, in which the keys of IIR-Record and SS-Record are hashed to an integer (i.e. <[Hash(segmentString, segmentNumber, stringLength), Flag], ID>) and uses length-aware verification method. PassJoinKMR needs more bandwidth for the segment information, while PassJoinKMRS generates some additional candidate pairs for the hash table "conflict", but uses less bandwidth. It is difficult to say which algorithm is more efficient. However, since bandwidth is always a bottleneck in MapReduce, we argue that it will slow down computation if the verification time is decreased by using more transmission information. We will further study this problem in Section 4.

We show the whole process in Figue 4. The input record is <k1, v1>= {<R1, abcde>, <R2, bcfde>}. Mapper phase generates the IIT-Records and SS-Records and then the records are grouped together based on the key<segmentString, segmentNumber, stringLength>. Reducer phase verifies the candidate pairs. Compared with PassJoinK, sorting the strings is no longer needed in PassJoinKMR and PassJoinKMRS. We give the pseudo-code of our algorithm PassJoinKMR for Self-Join in Algorithm 2.

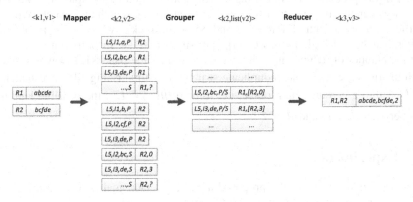

Fig. 4. The process of PassJoinKMR

Algorithm 2 PassJoinKMR with Self-Join(inDir,outDir,Tau,K)	
Input:	inDir(input directory with the records of dataset R), outDir (output directory), Tau(threshold), K(the number of addition segments)
Output	outDir contains all the results of the Similarity Join
1	Map phase(k1=line number ,v1=id:string)
2	Id ← v1.id
3	Str ← v1.string
4	L ← length(Str)
5	S_L^i ← partition record into segmentStrings
6	for each segment in S_L^i
7	Output([L, i, segmentString, IIFlag],id)
8	for $L-\tau \le l \le L, 1 \le i \le \tau+1$
9	Ψ ← SelectSubStrings(Str, l, i, τ)
10	for each substring in Ψ
11	Output([l, i, substring, SSFlag],id)
12	Reduce phase(k2=[l, i, substring, IIFlag/SSFlag], v2 =list(id))
13	IIlist ← list(id) with IIFlag
14	SSlist ← list(id) with SSFlag
15	for each rid in IIlist do
16	for each sid in SSlist
17	ed ← VerifyPair(Strings(rid), Strings(sid), τ)
18	if ed <=τ then
19	Output(k3 = [rid, sid], v3 = ed)

Algorithm 2. PassJoinKMR algorithm for Self-Join

For multi-join problems where the input strings are located in multiple files, we make necessary changes in generation stage to accelerate string search and match. We add the file information into IIT-Record and SS-Record <k2, v2> = <[segmentString, segmentNumber, stringLength, FLAG], [DirNumber, ID]>. In Self-Join case, to reduce the number of reduplicate pairs, we just consider the pair <s1, s2> where s1>s2. In Multi-Join case, to avoid reduplicate strings in distinguishing files, we match s with r within a certain length range, i.e. if $length(s) = L$, the possible length of r is between $L - \tau$ and $L + \tau$.

4   Experiments

In this section we evaluate the proposed algorithms. Our goal is to evaluate (1) the efficiency of PassJoinK with different k value, (2) the efficiency of PassJoinKMRS and PassJoinKMR with different k value, (3)the scalability of our algorithms.

The PassJoinK algorithm is implemented in C++. We run our programs on a Ubuntu machine with 4 i5-2300 2.8GHz processors and 32 GB memory. The PassJoinKMR algorithm uses Hadoop version 1.0 with 1 NameNode and 6 DataNodes. Up to 1 map and 4 reduce tasks are performed concurrently per node. Each node has 4 i5-2300 2.8GHz processors and 32 GB memory.

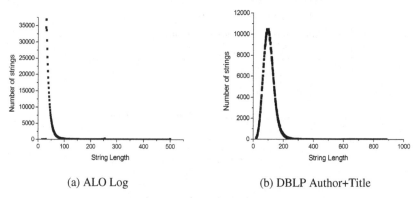

(a) ALO Log   (b) DBLP Author+Title

Fig. 5. String Length distribution

We used two real datasets: DBLP Author+Title[1] and AOL Query Log[2]. DBLP is an academic bibliography data set commonly used in database community. Query Log is a set of real query logs. The dataset's information is shown in Figure 5 .

4.1   Effect of different K

In this section we implement PassJoinK with different k values. Note that when $k = 1$, PassJoinK algorithm is essentially PassJoin. We report the join time of four

[1] http://www.informatik.uni-trier.de/~ley/db
[2] http://www.gregsadetsky.com/aol-data/

different k values with different edit distance thresholds in Figure 6. We have the following observations: (1)Compared to PassJoin, the join time reduces with larger k. For example, on Log dataset with edit distance threshold $\tau = 10$, the elapsed time for PassJoin(k=1) is 1502 seconds; when k= 2, the algorithm only used 490 seconds (3 times faster); when k=3, it only used 374 seconds (4 times faster). This suggests that PassJoinK with an appropriate value of k is superior than PassJoin.

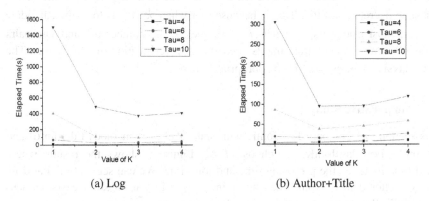

(a) Log   (b) Author+Title

Fig. 6. Time cost of PassJoinK with different k and τ

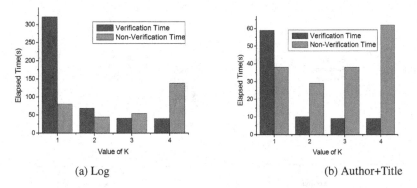

(a) Log   (b) Author+Title

Fig. 7. Verification and Non-Verification Time with different k

(2) The elapse time does not always decrease with increasing K. Consider the "U"-curves shown in Figure 6(a) and (b), the time cost slightly increases after a certain point (k =2 or k =3). This phenomena is more significant for larger edit distance threshold (e.g. τ =10). This is because with bigger k , the reduction of the verification time is less than the increased time building a more complicated inverted index tree.

We conduct another experiment to further analyze the verification time and non-verification time. We choose $\tau = 8$. We can see from Figure 7 that (1) on both datasets, the verification time is much more than non-verification time when $k=1$. However, for bigger K, the verification time is less than (or nearly equivalent to) non-verification time. (2) When $k=2$, the verification time decreases more than 5 time, the non-verification time decreases more than 30 percent on both data set. (3) With bigger K ($k = 3$ or K=4), the non-verification time increases apparently while verification time decreases a little. This is because that when k equals to 3, the candidate pairs are almost true pairs, so when k becomes 4, the number of candidate pairs remain still, more preparation time is needed so the verification time increases. The above analysis suggests $k=2$ is an appropriate value.

4.2   Comparative Study

In this section, we compare PassJoinK with state-of-the-art methods Ed-Join[6] and PassJoin[7]. For PassJoinK, we choose $k=2$. Figure 8 shows the results where elapsed time included the preparing time and join time. We can see that (1) PassJoin is always better than Ed-Join. The underlying reason is that PassJoin generates less substrings (2) PassJoinK outperforms PassJoin except for $\tau =4$ in Author+Title dataset. The reason why PassJoinK is better than PassJoin is that PassJoinK prunes more candidate pairs so the verification time is less.

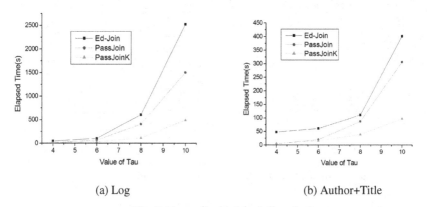

(a) Log   (b) Author+Title

Fig. 8. Comparison with other methods

4.3   Effect of Parameters

We first report the Join time of PassJoinKMR with four different k values and different edit distance threshold in Figure 9.

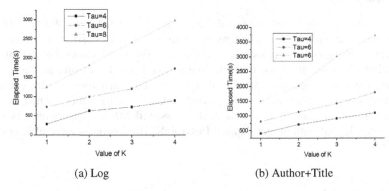

(a) Log   (b) Author+Title

Fig. 9. The efficiency of PassJoinKMR with different k

As shown in Figure 9: as k increases, the elapsed time increases too. This is different from PassJoinK algorithm because in RAM, building inverted index tree costs little time and the verification costs a big amount of time. But when we use Hadoop, building massive IIT-Records and SS-Records costs much time and also the transmission time is apparently more than that in RAM, so the portion of verification time is smaller. When k increases, the elapsed time increases. So we suggest that $k=1$ seems the best choice.

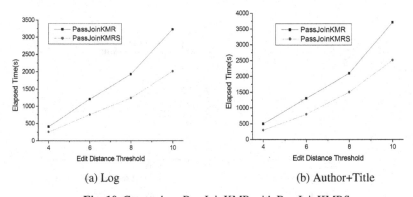

(a) Log   (b) Author+Title

Fig. 10. Comparison PassJoinKMR with PassJoinKMRS

Now we know that the transmission time is a big part of the elapsed time and even it is larger than verification time, we next show the experimental results of Pass-JoinKMRS (with less information transmission) in Figure 10. We can see that Pass-JoinKMRS almost reduces half of the time by PassJoinKMR. For example, on Log dataset, when tau = 6, PassJoinKMR costs 721 seconds while PassJoinKMRS costs only 392 seconds. Hence it verifies our assumption in section 3(b).

4.4   Scalability

Figure 11 shows the running time of different number of reducers. We find that the effectiveness becomes better when reduce number increases, so if we have more computers, the algorithm becomes more effective.   Figure 12 shows the running time of our algorithms on simulated datasets. We increase the size of original datasets by following the same rules of frequencies of symbols and distributions of string lengths. We can see our algorithms achieve very good scalability. Furthermore, when the dataset is small, for the limitation of bandwidth, PassJoinKMRS's performance is worse than PassJoinK, but with bigger dataset, PassJoinKMRS becomes more efficient than PassJoinK.

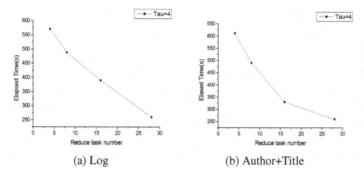

(a) Log   (b) Author+Title

Fig. 11. Comparison of Reduce task's number

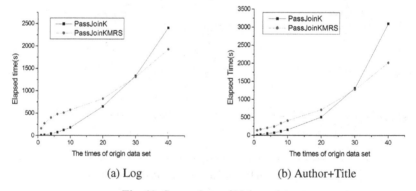

(a) Log   (b) Author+Title

Fig. 12. Comparison of N times datasets

5   Related Work

There are many related works on string similarity join[1,2,3,7,8] and similarity join in parallel[8,10,11] including exact similarity joins[1-16], approximate similarity joins[19] and top-k similarity joins[20]. The similarity measurements include cosine similarity[4], edit distance[5][6][7], Jaccard similarity[15], hamming distance[4][22] and synonym based similarity[23].

Most existing works with small data used filter-and-refine framework to solve string similarity join problem based on Q-gram[2] representation of strings such as Part-Enum[3], All-Pairs-Ed[4], PP-Join[5], ED-Join[6]. They first generate a set of signature for each string and ensure that every similar string pair must share at least one common signature, and then generate some pairs which share at least one signature and then verified the candidate pairs. Some recent works based on MapReduce framework[12,13,14] have been proposed. However most of them did not provide any experimental results on the performance of these strategies.

6   Conclusion

In this paper, we study the problem of string similarity joins with edit-distance constraints. We improve the PassJoin algorithm and propose a MapReduce algorithm PassJoinKMRS. Experiments show that our method outperforms the original PassJoin which has been proved better than other state-of-the-art join algorithms.

Acknowledgement. Chen Lin is partially supported by Shanghai Key Laboratory of Intelligent Information Processing under Grant No. IIPL-2011-004, China Natural Science Foundation under Grant Nos. NSFC61102136, NSFC61370010, NSFC81101115, the Natural Science Foundation of Fujian Province of China under Grant Nos. 2011J05158, 2011J01371, CCF-Tencent Open Research Fund under Grant No. CCF-Tencent 20130101, Base Research Project of Shenzhen Bureau of Science,Technology, and Information under Grand No. JCYJ20120618155655087. Xianmang He is supported by China Natural Science Foundation under Grant No. NSFC612020007.

References

1. Chaudhuri, S., Ganti, V., Kaushik, R.: A primitive operator for similarity joins in data cleaning. In: Proc of the 22nd International Conference on Data Engineering, ICDE, Washington (2006)
2. Gravano, L., Ipeirotis, P.G., Jagadish, H.V., Koudas, N., et al.: Approximate string joins in a database (almost) for free. In: Proc of the 27th International Conference on Very Large Data Bases, VLDB, pp. 491–500. Rome (2001)
3. Arasu, A., Ganti, V., Kaushik, R.: Efficient exact set-similarity joins. In: Proc of the 32nd International Conference on Very Large Data Bases, VLDB, pp. 918–929. Seoul (2006)
4. Bayardo, R.J., Ma, Y., Srikant, R.: Scaling up all pairs similarity search. In: Proc of the 16th International Conference on World Wide Web, pp. 131–140. ACM, Alberta (2007)
5. Xiao, C., Wang, W., Lin, X., et al.: Efficient Similarity Joins for Near Duplicate Detection. In: Proc of the 17th International Conference on World Wide Web, pp. 131–140. ACM, New York (2011)
6. Xiao, C., Wang, W., Lin, X.: Ed-join: An efficient algorithm for similarity joins with edit distance constraints. Proc of the VLDB Endowment 1(1), 933–944 (2008)
7. Li, G., Deng, D., Wang, J., et al.: Pass-join: A partition-based method for similarity joins. Proceedings of the VLDB Endowment 5(3), 253–264 (2011)

8. Jiang, Y., Deng, D., Wang, J., et al.: Efficient parallel partition-based algorithms for similarity search and join with edit distance constraints. In: Proceedings of the Joint EDBT/ICDT 2013 Workshops, pp. 341–348. ACM (2013)
9. Dean, J., Ghemawat, S.: MapReduce: Simplified data processing on large clusters. Communications of the ACM 51(1), 107–113 (2008)
10. Chaiken, R., Jenkins, B., Larson, P.Å., et al.: SCOPE: Easy and efficient parallel processing of massive data sets. Proceedings of the VLDB Endowment 1(2), 1265–1276 (2008)
11. Schneider, D.A., De Witt, D.J.: A performance evaluation of four parallel join algorithms in a shared-nothing multiprocessor environment. ACM (1989)
12. Blanas, S., Patel, J.M., Ercegovac, V., et al.: A comparison of join algorithms for log processing in mapreduce. In: Proceedings of the 2010 ACM SIGMOD International Conference on Management of Data, pp. 975–986. ACM (2010)
13. Olston, C., Reed, B., Silberstein, A., et al.: Automatic Optimization of Parallel Dataflow Programs. In: USENIX Annual Technical Conference, pp. 267–273 (2008)
14. Yang, H., Dasdan, A., Hsiao, R.L., et al.: Map-reduce-merge: Simplified relational data processing on large clusters. In: Proceedings of the 2007 ACM SIGMOD International Conference on Management of Data, pp. 1029–1040. ACM (2007)
15. Vernica, R., Carey, M.J., Li, C.: Efficient parallel set-similarity joins using MapReduce. In: Proceedings of the ACM SIGMOD International Conference on Management of Data, pp. 495–506 (2010)
16. Gionis, A., Indyk, P., Motwan, R.: Similarity Search in High Dimensions via Hashing. VLDB 1999, 518–529 (1999)
17. Graupmann, J., Schenkel, R., Weikum, G.: The spheresearch engine for unified ranked retrieval of heterogeneous XML and web documents. In: Proceedings of the 31st International Conference on Very Large Data Bases, VLDB Endowment, pp. 529–540 (2005)
18. Baxter, L., Tripathy, S., Ishaque, N., et al.: Signatures of adaptation to obligate biotrophy in the Hyaloperonospora arabidopsidis genome. Science 330(6010), 1549–1551 (2010)
19. Chakrabarti, K., et al.: An efficient filter for approximate membership checking. In: Proceedings of ACM SIGMOD International Conference on Management of Data 2008, pp. 805–818 (2008)
20. Xiao, C., et al.: Top-k set similarity joins. In: Proceedings of the 25th International Conference on Data Engineering, pp. 916–927 (2009)
21. Arasu, A., Chaudhuri, S., Kaushik, R.: Transformation-based framework for record matching. In: Proceedings of the 24th International Conference on Data Engineering, pp. 40–49 (2008)

Topical Presentation of Search Results on Database[*]

Hao Hu[1,2], Mingxi Zhang[1,2], Zhenying He[1,2], Peng Wang[1,2],
Wei Wang[1,2], and Chengfei Liu[3]

[1] School of Computer Science, Fudan University, Shanghai, China
[2] Shanghai Key Laboratory of Data Science, Fudan University
[3] Faculty of ICT, Swinburne University of Technology, Melbourne, Australia
{huhao,10110240025,zhenying,pengwang5,weiwang1}@fudan.edu.cn,
cliu@swin.edu.au

Abstract. Clustering and faceting are two ways of presenting search results in database. Clustering shows the summary of the answer space by grouping similar results. However, clusters are not self-explanatory, thus users cannot clearly identify what can be found inside each cluster. On the other hand, faceting groups results by labelling, but there might be too many facets that overwhelm users.

In this paper, we propose a novel approach, topical presentation, to better present the search results. We reckon that an effective presentation technique should be able to cluster results into reasonable number of groups with intelligible meaning, and provide as much information as possible on the first screen. We define and study the presentation properties first, and then propose efficient algorithms to provide real time presentation. Extensive experiments on real datasets show the effectiveness and efficiency of the proposed method.

1 Introduction

Database query results can be presented in a ranked list, in clusters or in facets. Ranked list is popular; however, it does not help navigate the answer space. In contrast, clustering and faceting are designed for users to quickly get a general picture of the whole answer space first, and then to locate relevant results. However, several issues remain to be studied. Let us first consider a simple example.

Example 1. *Consider a laptop database that maintains information like Brand, Screen Size, CPU type, etc. First, assume the results are presented in clusters. Table 1 shows a typical interface, similar laptops are grouped into the same cluster, and can be accessed by hyperlink in Zoom-in field. There are two shortcomings. (1) Clusters may be ambiguous (i.e., not self-explanatory). Users could be puzzled of what can be*

---

[*] This work was supported in part by NSFC grants (61170007, 60673133, 61033010, 61103009), the Key Project of Shanghai Municipal Science and Technology Commission (Scientific Innovation Act Plan, Grant No.13511504804), and ARC discovery project DP110102407.

S.S. Bhowmick et al. (Eds.): DASFAA 2014, Part II, LNCS 8422, pp. 343–360, 2014.

Table 1. Clustering example   **Table 2.** Results after applying *Zoom-in*

ID	Brand	Price	Color	CPU	ScreenSize	*Zoom-in*
05	Dell	539	Black	Intel i3	14	311 more laptops like this
06	Apple	1499	Silver	Intel i5	13	217 more laptops like this
07	HP	559	Black	Intel i7	13.3	87 more laptops like this
08	Sony	729	Pink	Intel i3	14	65 more laptops like this

Brand	Price	Color	CPU	...
Dell	539	Black	Intel i5	...
HP	549	Black	Intel i7	...
Dell	559	Silver	Intel i5	...
HP	559	Pink	Intel i3	...

*found through the hyperlink. We refer to this as **S1** (shortcoming 1, unknown label). (2) For thousands of results (**S2**, result overwhelming), if the number of clusters k is set small, each cluster may also contain many results, this might lead to other issues.*

(i) *Presenting each cluster may overwhelm users.*

(ii) *Results in one cluster may not be similar to each other. Table 2 shows the results after applying Zoom-in to the first cluster in Table 1. These laptops are considered* similar *according to the distance measure of the system. However, in fact, a novice user might think* Dell *and* HP *are similar for some reasons such as the prices are nearly the same, while a professional user may consider them quite different. This is a negative effect of **S1**, as the semantic similarity is unknown to users.*

Second, assume the results are presented by faceting, such as `amazon.com`*. Facets (i.e., labels) are listed on the left side of each returned web page. Actually, when submitting "laptop" as keyword, there are more than 100 facets, which cannot be fulfilled on the first screen. Each facet represents a group of results, and too many facets may lead to dissatisfaction as well (**S3**, facets overwhelming).* ■

Clustering solves the result overwhelming issue (**S2**) by setting a fixed number of clusters. However, each cluster is not self-explanatory. On the other hand, faceting groups results with labels; nevertheless, it might overwhelm users (**S3**).

In this paper, we try to combine the advantages of both sides. However, straightforward solutions may not be applicable.

On one hand, directly adopting label extraction methods [17,20] for clustering is improper. First, most label extraction methods are designed for document sets, while these methods are effective for texts, we cannot make the most of them for structured data. Second, label extraction methods often employ supervised learning algorithms, they emphasize the quality. However, result presentation requires real time interaction in practice.

On the other hand, selecting a few facets to avoid overwhelming may deliver insufficient information. Most of the facet selection methods [12,6] only consider result size, i.e., the goal is to select k facet groups that maximize the total result size. In the `amazon` example, top-3 facets are "windows 7", "windows 8", and "windows vista". Though too large facets could cover more tuples, they deliver not sufficient information on the first screen.

In this paper, we study how to present search results on database. Specifically, the goals that a good presentation approach should meet are as follows.

G1. (Goal 1, on **S1**) Each grouped results (packages) should be intelligible (i.e., packages should have labels), hence users can understand it easily.

G2. (On **S2**) Reasonable size of each package should be ensured, thus users are not overwhelmed. We do not prefer packages with too many tuples (overwhelming users) or too few tuples (providing insufficient information).

G3. (On **S3**) Due to the first screen size, only k packages can be presented, we should make these summarized k packages bring as much information as possible, so as to give users a more informative picture of the result set.

We propose a topical presentation (TP) approach for these goals. TP takes query result as input, and outputs k intelligible packages, where each package is neither too large nor too small, and collectively, the k packages aim to provide maximal information. To this end, we first need to generate all packages, and then summarize them.

We have identified three challenges. **First**, there are many packages in practice, we need to efficiently generate them and then summarize them. **Second**, the information of each package is hard to measure. Entropy-based measures are well studied in information theory. However, it is inappropriate to adopt these measures for packages with labels, because defining row wise entropy or defining the distribution of tuples is hard. Another difference is that entropy-based methods measure the information of the package, whereas we need to measure the information of both the package and its labels. **Third**, the k packages might overlap. For G3, TP needs to maximize the information without the overlapped tuples. Intuitively, the goal of summarizing packages is to expose users (1) as many tuples as possible and (2) as many labels as possible. However, the two aspects are inversely proportional (e.g., a facet with label "windows 8" consists more laptops than a facet with labels "windows 8" and "DELL").

To tackle the challenges, TP first analyzes how to get labelled groups (G1), then determines the acceptance for groups to achieve a reasonable group size (G2). Next, we introduce a metric to measure the information of each group. Finally, a fast algorithm is proposed to select k groups with maximal information (G3). We achieve this goal based on the *maximizing k-set coverage* principle.

For Example 1, four intelligible groups are shown in Table 3. We choose a representative tuple for each group. Each

Table 3. Topical presentation example

ID	Brand	Color	CPU	ScreenSize		Subtopic		
01	Dell	Black	Intel i5	14	186	Dell	Black	laptops
02	Sony	Pink	Intel i5	15	17	Sony	15 inches	laptops
03	Lenovo	Black	Intel i5	14	168	Intel i5	14 inches	laptops
04	MacBook	Silver	Intel i7	14	66	Silver	MacOS	laptops

group is assigned with several labels (highlighted in boxes) to describe the commonality of tuples in it. These groups help users to learn the answer space from different aspects. We aim to provide maximal information by choosing most representative labels in these groups.

Contributions. (1) We propose and define the TP problem. (2) We propose a metric to measure the information of each package. (3) We suggest a novel mechanism for summarizing many packages into k, and propose a fast algorithm. (4) Extensive experiments are performed to evaluate the proposed approach.

Roadmap. Section 2 defines the problem (G1). Section 3 generates acceptable packages (G2). In Section 4, we summarize the packages (G3). Section 5 shows the experiments, Section 6 and Section 7 discuss the related work and conclusion.

2   Problem Definition

We adopt the labels used in faceting to deliver clear meaning. Labels are from attribute values of tuples. This section begins with preliminaries, and then defines the problem. The terminology in OLAP is used. Table 4 lists some notations.

Table 4. Notations used in this paper

Symbol	Description	Symbol	Description	Symbol	Description
T	relation	l	label	$\mathcal{D}$	attribute space
mt	meta-topic	$OS(P_{st})$	overall score	$o(P_{st})$	overview ability
st	subtopic	$OS_m(\mathcal{P})$	informative score	$m(P_{st})$	meaningfulness
$\pi(mt)$	partition of mt	$\mathcal{M}$	set of packages that	$sup(P_{st})$	support of P_{st}
P_{st}	package of st		are from one mt	$\mathcal{I}_C$	summary set
$\mathcal{P}$	package set	R	query result set	$cha(t,l)$	character for label l in t

2.1   Preliminaries

Definition 1 (Meta-topic). *Let $T=(A_1,...,A_n)$ be a relation with attributes A_i. A **meta-topic** on T is a combination of attributes: $mt=(x_1,...,x_n)$ where $x_i=A_i$ or $x_i=*$ $(1\leq i\leq n)$, and $*$ is a meta symbol meaning that the attribute is generalized. The **partition** of mt (denoted as $\pi(mt)$) is a set of tuple sets, where each tuple set consists of tuples with same values on non-$*$ attributes of mt, i.e.,*

$$\pi(mt)=\{s|\forall t_i,t_j\in s:t_i[v_k]=t_j[v_k] \text{ if } x_k\neq*, 1\leq k\leq n\} \tag{1}$$

where $t_i[v_k]$ denotes the value of attribute A_k in tuple t_i.

For $mt=(x_1,...,x_n)$ and $mt'=(y_1,...,y_n)$, mt is an ancestor of mt' (i.e., $mt\succ mt'$) if $x_i=y_i$ for each $x_i\neq*$, and there exists $j\in[1,n]$ such that $x_j=*$ but $y_j\neq*$. ∎

Take Table 3 as an example, assume there are only 4 attributes, $T = (Brand, Color, CPU, ScreenSize)$, the partition of meta-topic $(*,Color,CPU,*)$ contains three tuple sets (we use IDs in Table 3 to represent each tuple), $\{01,03\},\{02\},\{04\}$. Moreover, $(*,*, CPU,*) \succ (*,Color, CPU, *)$.

Corollary 1 (Refinement). *Given mt, mt', if $mt'\succeq mt$, then $\pi(mt)$ refines $\pi(mt')$, i.e., $\forall s\in\pi(mt),\exists s'\in\pi(mt'):s\subseteq s'$.* ∎

In Table 3, tuple sets $\{\{01,03\},\{02\},\{04\}\}$ refines $\{\{01,02,03\},\{04\}\}$, where the latter is the partition of $(*,*,CPU,*)$. For tuple set $\{01,02,03\}$, tuples share common label `Intel i5`. All common labels of a tuple set consist of its subtopic.

Definition 2 (Subtopic). *Given a relation T, and a meta-topic $mt=(x_1,...,x_n)$, a **subtopic** for mt, denoted as $st\in mt$, is a combination of attribute values: $st=(z_1,...,z_n)$ where $z_i\in A_i$ (if $x_i=A_i$) or $z_i=*$ (if $x_i=*$). We call each non-$*$ z_i a **label** of st, and refer to all tuples of a subtopic st as a **package** (denoted as P_{st}), i.e.,*

$$P_{st}=\{t|t\in T, t[v_k]=z_k \text{ if } z_k\neq*, 1\leq k\leq n\} \tag{2}$$

*The **support** of P_{st} is defined as $sup(P_{st})=|P_{st}|/|T|$. Packages of all subtopics consist of $\pi(mt)$, if all the subtopics are from mt. i.e., $\cup_{\forall st\in mt}P_{st}=\pi(mt)$.* ∎

For example, $(*,*,\texttt{Intel i5},*)$ is a subtopic, it is an instance of meta-topic $(*,*,CPU,*)$. The package of this subtopic is $P_{(*,*,\texttt{Intel i5},*)} = \{01,02,03\}$. If there are only 4 tuples in T, its *support* is 0.75.

In this paper, we assume all tuples are in one table and attribute values are categorical. For numeric data, we assume it has been suitably discretized.

A subtopic can also be viewed as a label set. We use notations in set theory, such as $|st|$, $st_1 \cup st_2$, $st_1 \cap st_2$, to denote the number of labels in st, all distinct labels, and labels belonging to both st_1 and st_2, respectively.

2.2   Overview Ability and Meaningfulness

It is obvious that the more labels describing a package (more meaningful), the less tuples it contains (less overview ability). This is a trade-off between (1) the overview ability of the answer space; and (2) the meaningfulness of each package. **Overview Ability.** It is easy to see that, a package with more tuples has better overview ability. Therefore, the overview ability of P_{st} is defined as the proportion of the whole answer space in this paper.

$$o(P_{st}) = \Sigma_{t \in P_{st}} score(t) / \Sigma_{t \in R} score(t) \qquad (3)$$

where R is the query result set, and $score(t)$ is an adaptation of existing scoring techniques. For example, $score(t)$ can be the relevance between keyword query Q and t, or the feedback of t by users.

Meaningfulness. As discussed in Section 1, it is hard to quantify the meaningfulness. To this end, we assume that, a package P_{st} is more meaningful if there are more labels in st. This is a natural assumption because users are exposed with subtopics directly; they can learn more information if and only if there are more labels. The meaningfulness is defined as follows.

$$m(P_{st}) = \Sigma_{l \in st} weight(l) \qquad (4)$$

where $l \in st$ is a label, $weight(l)$ measures the importance of l. Many approaches (e.g., query log mining, frequency based scoring) have been proposed for label scoring. We assume the labels are independent for simplicity, hence the meaningfulness can be denoted as the sum of all label weights.

Overall Score. The overall score for overview ability and meaningfulness of a package is defined as follows.

$$OS(P_{st}) = o(P_{st}) \times m(P_{st}) \qquad (5)$$

We use the product of $o(P_{st})$ and $m(P_{st})$ because the two aspects are in inverse proportion. For a package, we prefer high score of both $o(P_{st})$ and $m(P_{st})$.

Remark 1. Note that $OS(P_{st})$ can be applied to many package ranking methods. For example, for frequency-based ranking [6] (where a facet/package is ranked by the number of tuples in it), we could set $score(t)=1$ and $m(P_{st})=1$. As a result, the ranking by $OS(P_{st})$ is equivalent to the frequency-based ranking.

Remark 2. There are many implementations for $score(t)$ and $weight(l)$, e.g., TFIDF, user feedback. Different scenarios require different scoring functions, thus we leave $score(t)$ and $weight(l)$ untouched, to make $OS(P_{st})$ adaptive.

Remark 3. Substituted by Eq. 3 and 4, $OS(P_{st}) = \Sigma_{t \in P_{st}} \Sigma_{l \in st} score(t) weight(l) / \Sigma_{t \in R} score(t) \propto \Sigma_{t \in P_{st}, l \in st} score(t) weight(l)$. Denote $cha(t,l) = score(t) weight(l)$ as the *character* for label l in tuple t, the overall score can be rewritten as $OS(P_{st}) \propto \Sigma_{t \in P_{st}, l \in st} cha(t,l)$.

2.3   Problem Definition

Problem 1 (TP). Given search results R, attribute space $\mathcal{D}=\cup_{i=1}^{|\mathcal{D}|} A_i$ and an integer k. Let $OS_m(\cdot)$ be the informative score of multiple packages, and $\mathcal{P}$ be all packages on $\mathcal{D}$, the topical presentation (TP) problem is to select a summary set I_C, that

$$\max_{I_C \subseteq \mathcal{P}, |I_C|=k} OS_m(I_C), s.t. \forall P_{st} \in I_C \begin{cases} P_{st} \text{ is acceptable} \\ st \text{ is on } \mathcal{D} \end{cases}$$

where $OS_m(I_C)=f(\cup_{P_{st} \in I_C} OS(P_{st}))$ measures the information of all packages in I_C.

Following the previous example, in frequency-based ranking, $OS_m(I_C) = \sum_{P_{st} \in I_C} OS(P_{st})$, and in set cover ranking, $OS_m(I_C) = |\cup_{P_{st} \in I_C} P_{st}|$.

In Problem 1, each $P_{st} \in I_C$ is required to be acceptable due to **G2**, and $OS_m(I_C)$ is required to be maximal due to **G3**. To compute I_C, we need all packages $\mathcal{P}$, informative score $OS_m(\cdot)$, and a fast algorithm. These issues are discussed next.

3   Package Generation

Given search results R and attribute space $\mathcal{D}$, by the definition of subtopic, we can use the *group-by* in SQL to generate packages. The DBMS could return all packages of a meta-topic mt, if we *group-by* all non-* values in mt. However, it needs to execute the *group-by* $2^{|\mathcal{D}|}$ times, since there are $2^{|\mathcal{D}|}$ meta-topics on $\mathcal{D}$. This is time consuming. Besides, not all packages are interesting, consider a package with $st = $ (DELL,Pink, Intel i5,13inches), actually, there is only one laptop in this package, thus presenting it on the first screen might not be desired.

We now describe acceptable packages and the package generation algorithm.

3.1   Acceptable Package

An acceptable package is valid, and has neither too many nor too few tuples.
Valid Packages Due to the *finer-than* relation between meta-topics (Corollary 1), if $\pi(mt_1)$ refines $\pi(mt_2)$, then the partition of their *meet* (a partition refines both mt_1 and mt_2), $\pi(mt_1 \wedge mt_2)$ is equivalent to $\pi(mt_1)$. In order to eliminate the redundancy, we propose the notion of valid package.

Definition 3 (Valid Package). *A package P_{st}, $st \in mt$, is valid if there is no descendant mt' of mt(i.e., more specific than mt) such that $P_{st' \in mt'}=P_{st}$.* ∎

The intuition of valid package is that if two packages are same in tuples, but different in subtopics, e.g., (*,Pink,Intel i5,*) and (*,Pink,Intel i5,13inches). We prefer the latter package, for it has more meaning.

Example 2 *The* finer-than *relation of meta-topics is a partial order and it is a complete lattice. Fig. 1 shows a table T and the lattice. $\pi((A,*,*))$ refines $\pi((*,B,*))$, thus the partition of their meet $\pi((A,B,*))$, is the same with $\pi((A,*,*))$. Therefore, packages from $\pi((A,*,*))$ are invalid. Besides, packages from $\pi((A,*,C))$ and $\pi((*,*,C))$ are also invalid.*

Table T

ID	A	B	C	D,....
1	a_1	b_1	c_1	w_1, w_2
2	a_1	b_1	c_2	w_1, w_2
3	a_2	b_1	c_3	w_1, w_2
4	a_2	b_2	c_3	w_1, w_2
5	a_1	b_2	c_3	w_1, w_2

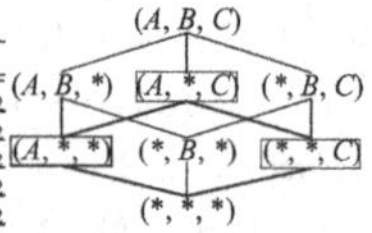

Fig. 1. Illustration on Lattice, the attribute space $\mathcal{D} = ABC$

Package Support Threshold. For packages with too many or too few tuples, we need to determine the maximal and minimal support (denoted by *min_sup* and *max_sup*) as thresholds. The thresholds can be set by system administrators according to experience, or by users. One can easily determine *min_sup* by setting a minimal package size (e.g., 5, 10, or 15). However, determining *max_sup* is hard, it depends on the result size $|R|$ and the distribution of tuples.

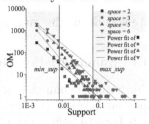

Fig. 2. *Zipf* distribution

In this work, we report that for real dataset, the support is described by a *Zipf* distribution over the meaningfulness of packages. Fig. 2 shows the meaningfulness is inversely proportional to the support. The packages are generated from 511 laptops with 11 attributes, and for all packages with the same support, we sum their meaningfulness as overall meaning, i.e., $OM(s)=\sum_{sup(P_{st})=s} m(P_{st})$. We can see that *Zipf* distribution is obeyed for different attribute space. Therefore, following [11], we estimate *max_sup* according to the transition point (denoted as *tp*) calculation of *Zipf*'s second law.

$$OM(tp) = \frac{1}{2}\left\{-1 + \sqrt{1 + 8count(OM(^1/_{|R|}))}\right\} \qquad (6)$$

where $count(OM(^1/_{|R|}))$ counts the number of packages with support $^1/_{|R|}$. This equation is adapted from [8]. Several support values s may satisfy $OM(s)=OM(tp)$, and we choose a minimal one as *max_sup*.

Remark 4. Determining the threshold may have other choices. For example, we can set an overall score *min_os* and mark packages with $OS(P_{st}) \leq min\_os$ unacceptable. However, it is hard to determine *min_os*. Extensive turning work needs to be done for a better *min_os*.

3.2 An Apriori Style Approach

A naive package generation method generates all packages first and then removes unacceptable ones. However, this `BaselineGeneration` algorithm is inefficient.

Algorithm 1. `FastPackageGeneration`

Input   : Query result set R, attribute space $\mathcal{D}$
Output: all acceptable packages
1 let $Mt_{(1)}=\{mt_1,mt_2,...,mt_{|D|}\}$ be 1-size meta-topics constructed from $\mathcal{D}$;
2 **for** $k=1$ to $|D|-1$ **do**
3 | generate packages from each meta-topic in $Mt_{(k)}$;
4 | remove invalid and too small packages, mark too big packages as "removed";
5 | generate $k+1$-size meta-topics $Mt_{(k+1)}$ with $Mt_{(k)}$;

Algorithm 1 (FG for short) is a fast generation algorithm using the valid and threshold conditions for filtering. To find invalid packages as soon as possible, we iterate meta-topics level-wise, where k-size meta-topics are used to explore $k+1$-size meta-topics. When iterating, we check the thresholds and the validity to avoid generating the unacceptable packages (Line 4), thus the searching space is reduced. When storing packages in each meta-topic, we use a heap to keep them ordered by overall score.

4 Summarization

We now define the summarization goal $OS_m(\cdot)$ and describe the algorithms.

4.1 Summarization Goal

Clustering is commonly adopted for summarization. We can define a distance measure between packages, and then cluster all packages into k clusters. However, it is hard to define a distance, e.g., Jaccard distance cannot tell the difference between two packages from one meta-topic, since there are no overlapped tuples.

This paper explores a different approach by leveraging the principle of *maximizing k-set coverage*. Specifically, we consider the goal of summarization as the following: maximizing overview ability and meaningfulness. Intuitively, this provides users the best balance between overview and understanding of summaries.

This principle is better illustrated in Fig. 3. Assume we want to pick two packages out of the four total packages (e.g., $k=2$). Selecting P_2 and P_3 allows users to view 9 of 12 tuples, and learn 19 units of tuple characters directly: 12 can be learned from 6-item package P_2 (2 characters per tuple) and 8 from 4-item package P_3, minus 1 character that is double counted because of the 1-item overlapping. In contrast, selecting the two non-overlapping packages P_1 and P_4 only gets 12 characters. We now define the informative score $OS_m(\cdot)$.

Package	Subtopic
P_1	$(a_1,*,c_1)$
P_2	$(*,b_1,c_1)$
P_3	$(a_2,*,c_1)$
P_4	$(*,b_1,c_2)$

Fig. 3. Summary with 4 packages. Each node is a tuple, the lines gather tuples into packages. The character $cha(t,l)$ is set to 1 for simplicity.

Definition 4 (Character Coverage). *Given a package set $\mathcal{P}=\cup_{i=1}^{n}P_{st_i}$, the informative score $OS_m(\mathcal{P})$ is defined as the Character Coverage, where the Character Coverage is the sum of all distinct characters.*

$$OS_m(\mathcal{P})=\sum_{i=1}^{n}\sum_{\substack{l\in st_i \\ t\in P_{st_i}}}cha(t,l)-\sum_{i\neq j}\sum_{\substack{l\in st_i\cap st_j \\ t\in P_{st_i}\cap P_{st_j}}}cha(t,l)+\sum_{i\neq j\neq k}\sum_{\substack{l\in st_i\cap st_j\cap st_k \\ t\in P_{st_i}\cap P_{st_j}\cap P_{st_k}}}cha(t,l)$$

$$-\cdots\pm\sum_{\substack{l\in st_1\cap\cdots\cap st_n \\ t\in P_{st_1}\cap\cdots\cap P_{st_n}}}cha(t,l) \tag{7}$$

This is an adaptation of the Inclusion-Exclusion Principle (a technique to compute the cardinality of the union of sets). $OS_m(\mathcal{P})$ has following properties.

Corollary 2. *Given a package set $\mathcal{P}$ and package P_{st}, if $\mathcal{P}\cap P_{st}=\varnothing$, then $OS_m(\mathcal{P}\cup P_{st})=OS_m(\mathcal{P})+OS(P_{st})$.*

Corollary 3. *Given a package set $\mathcal{P}$ and package P_{st}, $OS_m(\mathcal{P}\cup P_{st})\leq OS_m(\mathcal{P})+OS(P_{st})$.*

Corollary 2 and 3 are extended from the Inclusion-Exclusion Principle.

The TP problem now aims to find k packages with maximal character coverage $OS_m(\mathcal{P})$. This differs from previous ranking methods, since $OS_m(\mathcal{P})$ considers labels. We refer to ComputeCCov as the function to compute $OS_m(\mathcal{P})$.

4.2 Greedy Summarization Algorithm

Unfortunately, the objective function $OS_m(\cdot)$ is NP-hard to optimize.

Theorem 1. *The TP problem is NP-hard.*

Proof. *The basic idea is by reduction from the Weighted Maximum Coverage problem [9], which can be stated as follows. Given an integer k, and m sets $S=\cup_{i=1}^m S_i$ over a set of elements E, each element e_i is assigned with a weight $w(e_i)$. The goal is to find a k-set cover C ($C\subseteq S, |C|=k$) with maximum weight $\sum_{e_i\in C}w(e_i)$. The TP problems is, given a set of packages $\mathcal{P}=\cup_{i=1}^n P_{st_i}$ over a set of tuples R, each package is assigned with an overall scoring $OS(P_{st})\propto\sum_{t,l}cha(t,l)$. The goal is to find the set of k packages I_C ($I_C\subseteq\mathcal{P}, |I_C|=k$) with maximized $OS_m(I_C)$. Now, we transform an instance of the Weighted Maximum Coverage problem to an instance of the TP problem.*

*Assume the attribute space size is d, and we construct d elements for each tuple t_i, denoted as $\{e_{i1},...,e_{id}\}$. There are $d|R|$ elements. Each e_{ij} is assigned with weight $w(e_{ij})=cha(i,j)$, where $cha(i,j)$ is the character of label j for tuple t_i. For a package P_{st_r}, we construct a set $S_r=\{e_{ij}|t_i\in P_{st_r},j$ is a label in $st_r\}$. There are $|\mathcal{P}|$ sets in total. This transformation takes polynomial time. By Definition 4, $OS_m(I_C)$ calculates the sum of distinct characters, which is exactly $\sum_{e_{ij}\in C}w(e_{ij})$. It is now obvious that C maximizing $\sum_{e_{ij}\in C}w(e_{ij})$ **iff.** the set of k packages I_C maximizes $OS_m(I_C)$.* ∎

Algorithm 2 (BS for short) is a greedy summary algorithm. It starts by putting the package with the largest overall score into I (Line 1). At each iteration, it selects the package P_{st_i} that, together with the previously chosen packages I, produces the highest character coverage (Line 4). The algorithm stops after k packages have been chosen, and outputs I. Consider again the example in Fig. 3, when $k=2$, BS produces $\{P_2,P_3\}$; and when $k=3$, it produces $\{P_1,P_2,P_3\}$.

Algorithm 2. `BaselineSummarization`

Input: $\mathcal{P}=\cup_{i=1}^n P_{st_i}$ and k, the desired number of packages
Output: I
1 Initialize $I=\{\}$, and let package t be the largest overall score package in $\mathcal{P}$;
2 $I=I\cup\{t\}$, and remove t from $\mathcal{P}$;
3 **while** $|I|<k$ **do**
4 $\quad$ $t=\max_{t\in\mathcal{P}}(\texttt{ComputeCCov}(I\cup\{t\}))$;
5 $\quad$ $I=I\cup\{t\}$, remove t from $\mathcal{P}$;

BS is directly adapted from the greedy algorithm designed for *Maximum k-Set Cover problem*. It is known to have a $(1-1/e)$ approximation ratio [9].

BS computes the coverage in each iteration (Line 4), thus it can be expensive in practice. Function `ComputeCCov` has an exponential complexity, since each sub-part in Eq. 7 may require the summation of an exponential number of packages. As a result, summarization by maximizing $OS_m(\mathcal{P})$ turns to be hard.

4.3 Improved Summarization Algorithm

We now present pruning conditions to reduce the invocations of `ComputeCCov`. The key observation is that, there is no intersecting tuple between packages from one meta-topic. For example, in Fig. 3, if P_1 and P_3 are from meta-topic $(Brand,*,CPU)$, then $P_1\cap P_3=\varnothing$. This is obvious because each laptop has one brand and one CPU type, hence it belongs to only one package.

Formally, given a package set $\mathcal{P}$ with m meta-topics, $\mathcal{P}=\cup_{i=1}^{m}M^i$, let M^i be packages from the i-th meta-topic (i.e., $M^i=\cup_{x=1}^{|M^i|}P_{st_x^i}$, where each subtopic st_x^i belongs to the i-th meta-topic), we have $P_{st_x^i}\cap P_{st_y^i}=\varnothing$ for $x\neq y$. As described in Section 3.2, M^i is sorted in descending order of $OS(P_{st_x^i})$. Note that this package disjoint feature offers interesting information in each iteration of BS. We do not need to check all remaining packages to find a maximum character coverage. To give a better illustration, we first state the 3 filters and then show an example.

Specifically, for $\mathcal{P}=\cup_{i=1}^{m}M^i$, if $I=\cup_{i=1}^{m}I^i$ is the summary set in each iteration in Algorithm BS, $I^i=\{I_1^i,I_2^i,...\}$ consists of packages selected from M^i, then:

1. **InitialFilter** *Assume the initial largest package t in $\mathcal{P}$ comes from M^i, if $OS(P_{st_2^i})>OS(P_{st_1^r})$ holds for every $P_{st_1^r}$ ($r\in[1,m]$, $r\neq i$), then $P_{st_2^i}$ should be selected. This can be continued until there exists a package $P_{st_1^i}$ such that $OS(P_{st_1^i})<OS(P_{st_1^r})$ holds for some $r\neq i$. Moreover, at the iteration that initial filter fails, only packages in M^r with $OS(P_{st_j^i})<OS(P_{st_1^r})$ need to be checked, others can be skipped.*

2. **InclusiveFilter** *In the r-th iteration, for packages in M^i, starting from $P_{st_1^i}$, if $|I-\cup_{I_j^i\in I^i}I_j^i|=x$, then packages after the $(x+1)$-th package can be skipped.*

3. **ExclusiveFilter** *In the r-th iteration, for packages in M^i, starting from $P_{st_1^i}$, if $OS_m(P_{st_1^i}\cup I)-OS_m(I)=d$, then $P_{st_j^i}$ with $OS(P_{st_j^i})\leq d$ can be skipped.*

We omit the pseudo codes of the filters and refer to them as **InitialFilter**, **InclusiveFilter** and **ExclusiveFilter**. Consider the example in Fig. 4, assume $M^1=\{a_1,a_2,...\}$, $M^2=\{b_1,b_2,...\}$, $mt_1=(A,B,*)$ and $mt_2=(*,B,C)$. The intersection of every two packages in M^1(or M^2) is empty, i.e. $a_i\cap a_j=\varnothing(i\neq j)$. For simplicity, we use the term *"hit"* to denote the selection of a package in each iteration, and refer to $\delta(a_i)$ as the contribution of package a_i, $\delta(a_i)=OS_m(I\cup\{a_i\})-OS_m(I)$, thus to find a maximal character coverage can be restated as to find a package $P_{st_j^i}$ with maximal contribution $\delta(P_{st_j^i})$.

Initial Filter. The initial filter works at the beginning of BS. Consider the first selected package t with the largest overall score. Assume $t\in M^i$, then t is the first package in M^i (packages are sorted, i.e., $t=P_{st_1^i}$). In this case, if $OS(P_{st_1^i})$ is larger than all other $OS(P_{st_1^k})(i\neq k)$, we should select $P_{st_2^i}$ directly, for that the contribution $\delta(P_{st_2^i})$ is the largest.

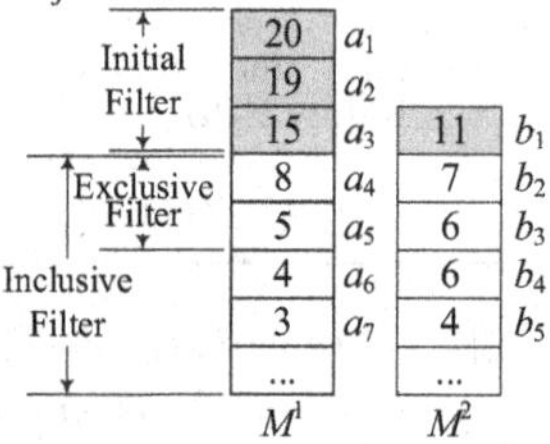

	M^1	M^2	
Initial Filter ↕	20 a_1		
	19 a_2		
	15 a_3	11 b_1	
Exclusive Filter ↕	8 a_4	7 b_2	
	5 a_5	6 b_3	
Inclusive Filter ↓	4 a_6	6 b_4	
	3 a_7	4 b_5	

Fig. 4. Illustration of filters. Packages a_i and b_i are from meta-topics M^1 and M^2, the number in each package is its overall score.

In Fig. 4, a_1 is the largest package, thus it is firstly picked. In the 2nd iteration, a_2 hits summary set I since $OS(a_2)=19>OS(b_1)=11$. Similarly, $|a_3|$ hits I in the 3rd iteration. With initial filter, we only need to check one package in each iteration.

In the 4th iteration, **InitialFilter** fails since $OS(a_4)<OS(b_1)$, thus we need to check all packages in M^1 and M^2 to find the next *hit*. If there is another meta-topic mt_3 with $M^3=\{c_1,c_2,...\}$, and $OS(c_1)=5$, then we can skip M^3 since $OS(a_4)>OS(c_1)$, the contribution of all packages in M^3 is less than $\delta(a_4)=8$.

Inclusive Filter. The initial filter answers the question, *"which meta-topic should be checked to find the hitting package?"* However, when it fails, we need to check all packages. Here, we ask another question, *"How many packages do we need to check in each meta-topic?"* Answering this leads us to inclusive filter.

Lemma 1. *Given* $M^i = \cup_{j=1}^{|M^i|} P_{st_j^i}$ *in descending order of* $OS(\cdot)$, *if* $P_{st_j^i} \cap I = \varnothing$, *then for* $k \in (j, |M^i|]$, *we have* $\delta(P_{st_k^i}) \leq \delta(P_{st_j^i})$.

Proof. By Corollary 2, 3, $\delta(P_{st_k^i}) = OS_m(I \cup P_{st_k^i}) - OS_m(I) \leq OS(P_{st_k^i}) \leq OS(P_{st_j^i}) = \delta(P_{st_j^i}).$ ∎

By Lemma 1, we can skip $P_{st_k^i}$ ($k \in (j, |M^i|]$) if $P_{st_j^i} \cap I = \varnothing$. In Fig. 4, assume the gray packages have been selected (i.e., $I = I^1 \cup I^2 = \{a_1 \cup a_2 \cup a_3\} \cup \{b_1\}$), and it is the 5th iteration. Assume $|I^2 - I^1| = 8$, thus for the remaining packages $\{a_4, a_5, ...\}$ in M^1, if there exists a package that has intersections with I, it can intersect 8 tuples at most, because it is disjoint with I^1. Therefore, we claim that in the next $8+1=9$ ordinal packages in M^1, there must exist one package a_k ($4 \leq k < 4+9$) such that $a_k \cap I = \varnothing$ (Pigeonhole principle), and by Lemma 1, packages $\{a_{k+1}, a_{k+2}, ...\}$ can be skipped. The scale for package checking is bounded into $|I - \cup_j I_j^i| + 1$.

Exclusive Filter. Exclusive filter calculates the overall score bound instead of scale bound. It works within each meta-topic. Consider the 5th iteration in Fig. 4, for a_4, if the contribution $\delta(a_4) = 5$, then we only need to check packages with its overall score larger than 5, because the rest packages cannot hit I with a less-than-5 overall score.

Algorithm 3. ImprovedSummarization (IS for short)

    Input : $\mathcal{P} = \cup_{i=1}^m M^i$, k is size of summary set
    Output: I

1 $upperBound = \{b_1, b_2, ..., b_m\}$;
2 $I = \cup_{i=1}^m I^i$, and initialize each I^i as $\varnothing$;
3 let package $t \in M^i$ be the largest package in $\mathcal{P}$;
4 $I^i = I^i \cup \{t\}$, remove t from $\mathcal{P}$;
5 $V = \text{InitialFilter}(t, M^i, \mathcal{P})$;   ▷ V records the necessity of checking M^i
6 **while** $|I| < k$ **do**
7       $upperBound = \text{UpdateBound}(I, V)$;
8   **forall the** $M^i \in \mathcal{P}$ **do** $t = \max_{t \in \{P_{st_1^i}, ..., P_{st_{b_i}^i}\}} \{\text{ComputeCCov}(I \cup \{t\})\}$ $I^i = I^i \cup \{t\}$,
       remove t from $\mathcal{P}$;

  Function UpdateBound(I, V)
    Input : $I = \{I^1, I^2, ..., I^m\}$, and $V = [V_1, V_2, ..., V_m]$ is the indicator for filtering
    Output: $upperBound$
1 $uBd_1 = \text{InclusiveFilter}(I, \mathcal{P}, V)$, $uBd_2 = \text{ExclusiveFilter}(I, \mathcal{P}, V)$;
2 $upperBound = \{b_1, b_2, ..., b_m\}$;
3 **forall the** b_i in $upperBound$ **do** $b_i = \min\{uBd_1[i], uBd_2[i]\}$; ▷ Choose a tighter bound

---

Improved Summarization Algorithm. InclusiveFilter and ExclusiveFilter both provide an upper-bound for package checking. When integrating them, the system can always choose a tighter bound to speed up the selection (see Function UpdateBound in Algorithm 3). Algorithm 3 summarizes packages with filters. In the first few selections, initial filter performs reduction by comparing the first packages in each meta-topic (Line 5). When InitialFilter fails, a boolean vector V is utilized to indicate whether each meta-topic requires checking (according to the last part of initial filter). In the following iterations, InclusiveFilter and ExclusiveFilter updates the upper-bound by `UpdateBound` (Line 7). Algorithm 3 skips many packages, hence is faster than BS.

Theorem 2. *Algorithm 3 and* BS *produce the same* I_C.   ∎
The proof is omitted due to the space constraint.

5 Experimental Study

This section reports evaluations on (1) the efficiency of package generation; (2) the efficiency of summarization; and (3) the quality of summarization.
Setup. We conducted all experiments on a Windows 2008 server, with a 2.83 GHz CPU, 8 GB memory, and 1TB hard disk. The program was coded in C++.
Datasets. We used two real datasets. The **first** is a laptop dataset. It contains 511 laptops with 11 attributes, such as `Brand`, `CPU`, `Memory`, etc. We assume the whole laptop dataset as a sample query result, and denote the query as QL. The **second** is the IMDB[1] dataset. We downloaded the raw IMDB data, and preprocessed it by removing duplicate movies and missing values. A subset of the raw data was converted into a large relational table. It has 14 attributes, e.g., `Year`, `Country`, `Producer`, `Genres`. Some attributes (e.g., "actor" and "actress") may have more than one values, following [22], we picked the most frequent value if multiple values exist. After preprocessing, we have 649,506 tuples.

Our test set for IMDB data consists of 8 queries (denoted by QI_1 to QI_8). Table 5 lists the queries and the query result

Table 5. Queries on IMDB dataset

QID	Query	$|R|$	QID	Query	$|R|$
QI_1	family, Christmas	366	QI_2	Revenge	507
QI_3	Legend, USA	577	QI_4	USA, Hero	1,009
QI_5	Magic	1,012	QI_6	Hong Kong Comedy	1,197
QI_7	Christmas	1,252	QI_8	short family Comedy	3,973

size. The result size is 1,236.625 on average. Note that the answers vary from different size, thus we can examine the effect when the number of tuples increases.

5.1 Package Generation

We test the efficiency of `BaselineGeneration` (BG) and FG (see Section 3.2) on three factors: (1) query result size $|R|$; (2) *threshold* (i.e., min_sup or max_sup); and (3) *attribute space size* $|\mathcal{D}|$. The support min_sup (max_sup) is proportional to package size, thus we use package size to denote max_sup and min_sup.

Fig. 5 shows the time cost for each query. Not surprisingly, FG outperforms BG, especially when $|R|$ gets larger (e.g., QI_7, QI_8). This is because FG could avoid producing the too small packages. The number of them gets larger when $|R|$ increases, hence the difference of time cost between FG and BG enlarges.

Fig. 6a gives the time cost on varying thresholds. We set $|\mathcal{D}|=5$ and use QI to denote the average time cost of QI_1 to QI_8. The red (blue) axis shows the time cost on varying max_sup_{size} (min_sup_{size}).

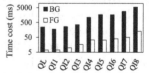

Fig. 5. Time cost on varying $|R|$. We set $|\mathcal{D}|=5$, $min\_sup_{size}=2$, and $max\_sup_{size}=100$.

---

[1] ftp://ftp.fu-berlin.de/pub/misc/movies/database/

The time cost of BG is almost unchanged, because BG generates all packages first and then removes the unacceptable ones, thus the generation time remains the same. However, the time cost of FG decreases as min_sup_{size} gets larger (see the blue lines). This is because FG does not generate the too-small-packages, and a larger min_sup_{size} often implies more too-small-packages. For max_sup_{size} (see the red lines), FG behaves the same as BG, the time cost of FG is stable. This is because when removing invalid packages, the too big ones are not removed physically (Line 4 of FG), thus the search space is not reduced.

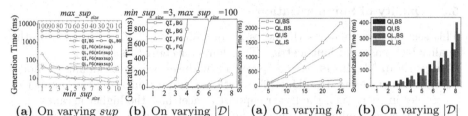

(a) On varying sup (b) On varying $|\mathcal{D}|$ (a) On varying k (b) On varying $|\mathcal{D}|$

Fig. 6. Package Generation Performance   **Fig. 8.** Summarization Performance

Fig. 6b shows the performance on varying $|\mathcal{D}|$. FG outperforms BG significantly especially for a larger $|\mathcal{D}|$. As we can see, BG fails to produce acceptable packages within a reasonable amount of time (1 second) as soon as $|\mathcal{D}|$ reaches 5 or 6. This is because more attributes often implies more packages, thus FG could avoid generating more unacceptable ones.

We can conclude that FG is sufficient for real time response in most cases. This is critical in our goal of presentation. In following experiments, we set $|\mathcal{D}|=$ 6, $min\_sup_{size}=3$ and $max\_sup_{size}=100$ by default.

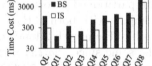

Fig. 7. Time cost on varying $|R|$

5.2   Efficiency of Summarization

We test the efficiency of Algorithm BS and IS on three factors: (1) result size $|R|$; (2) attribute space $|\mathcal{D}|$; and (3) the desired number of packages k.

Fig. 7 shows the summarization performance on varying $|R|$. The time cost is in log scale. We can see that IS outperforms BS, the time cost is reduced by 72.2% at most and 53.8% on average. Fig. 8a shows the time cost on varying the number of desired packages k. As we can see, the summarization time has been reduced 48.1% or more especially when k is large. The advantage of IS lies in the fact that it reduces the number of packages to be checked, thus the invocations of ComputeCCov are reduced. A larger $|R|$ (or k) often implies less packages for checking (compared to BS); hence the time cost is reduced greatly.

Fig. 8b shows the time cost on varying $|\mathcal{D}|$. We set $k = 5$ to reduce the advantage caused by larger k. As expected, IS still outperforms BS, especially for a larger $|\mathcal{D}|$. This is because more attributes often implies more acceptable packages, thus the number of skipping packages gets larger.

Section 5.1 and 5.2 show the efficiency of algorithms FG and IS. In this work, we analyze the properties of acceptable packages and meta-topics to perform TP

Table 6. # of acceptable packages

max_sup	10	20	**24**	50	100
QL	795	910	**928**	961	974
max_sup	10	20	**35**	50	100
QI_8	1,770	2,073	**2,208**	2,280	2,363

Table 7. Comparison of each method for QL

	# tuples	# labels	$OS_m(\cdot)$		# tuples	# labels	$OS_m(\cdot)$
BS	91	15	255	IS	91	15	255
MFR	106	10	226	SFR	98	5	105
MSC	115	11	252	SSC	99	5	105
MHS	113	10	227	SHS	64	5	105
MDS	101	15	237	SDS	98	5	105

task. Moreover, when computing, we remove (or skip) the unacceptable packages as soon as possible. Therefore, the time cost of FG and IS are reduced greatly.

5.3 Quality of Summarization

This section first validates the necessity of summarization, and then tests the quality by a case and four metrics. The character $cha(t,l)$ is set to 1.

Comparison Methods. We compared TP with several facet-ranking methods.

(1) **FR** Frequency-based ranking [6], where facets are ranked by the number of tuples in them (i.e., the larger support a facet has, the higher it ranks).

(2) **SC** Set-cover ranking [6], where k facets are selected to maximize the union of tuples in these facets.

(3) **HS** Hill-climbing Selection [12], where k facets are selected by hill climbing technique. We define the cost as the number of tuples exposed to users.

(4) **DS** Deterministic Selection, where a set of top-k largest overall scoring packages are chosen (an adaptation of frequency-based ranking with $OS(\cdot)$).

For faceting, there are single facet and multi facets. In single facet, the tuples are partitioned according to 1 attribute, and in multi-facets, m attributes. We compare with both of them. In total, we got 7 competitors: 4 ranking methods, each of them with 2 kinds of faceting. Note that for single facet, DS and FR are the same since $cha(t,l) = 1$. We denote the 7 competitors as SFR (equivalent to SDS), SSC, SHS, MFR, MSC, MHS, MDS. For a code 'XYZ', X indicates the size of facets, i.e., S for single facet, M for multi-facets. YZ indicates the ranking method. In following experiments, we also use YZ to denote both SYZ and MYZ.

Necessity. Table 6 describes the number of acceptable packages grows with max_sup_{size}. The support value 24 and 35 are estimated by Eq. 6. Note that even when $max\_sup_{size}=10$, the number of acceptable packages reaches into hundreds or thousands, it is too large for a user. This result clearly shows that obtaining a summary of packages is necessary for presentation.

Case study. Table 7 gives a case for QL, and Fig. 9 shows the subtopics returned by 5 methods. In this case, we chose 5 attributes: Brand, CPU, Memory, HardDrive, GraphicsCard, and set $min\_sup_{size}=3$, $max\_sup_{size}=24$, and $k=5$.

As we can see, IS returns more labels with a larger character coverage. Fig. 9a shows that labels returned by IS are more diverse than others, it contains labels from all 5 attributes. Moreover, if we measure the diversity of labels by averaging the number of unique labels in each attribute (i.e., $LabelDiv(I_C) = \Sigma_{A_i \in \mathcal{D}}$Number of unique labels in $A_i/|\mathcal{D}|$), we can find that IS has the largest $LabelDiv(I_C)$. The **label diversity** for IS, MDS, MSC, MFR and MHS are 2, 1.8, 1.6, 1.8 and 1.8, respectively. Therefore IS tends to produce informative packages. However, IS returns less tuples in Table 7. We compare these methods in detail next.

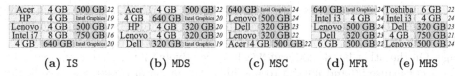

(a) IS	(b) MDS	(c) MSC	(d) MFR	(e) MHS

Fig. 9. A Case on QL, each facet is followed by the number of tuples in it

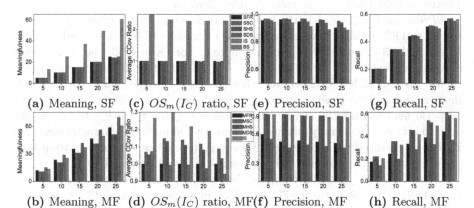

Fig. 10. Summarization quality on varying k. SF stands for 1-facet, and MF multi-facets

Metrics. For TP and its 7 competitors, we compare (1) meaningfulness, (2) character coverage, (3) precision and (4) recall, where,

$$precision(I_C)=\frac{\#\ \text{distinct tuples in } I_C}{\#\ \text{tuples presented to users}}, \quad recall(I_C)=\frac{\#\ \text{distinct tuples in } I_C}{|R|} \quad (8)$$

1.Meaningfulness Fig. 10a and 10b compare the meaningfulness. As we can see, IS and MDS have more meanings than FR, SC and HS. Recall that FR returns the top-k support facets, but such facets may often be less meaningful (e.g., in Fig. 9, MFR only returns 10 labels). On the other hand, the subtopics returned by IS are as different as possible from each other so as to get a higher character coverage, thus are likely to be meaningful (e.g., IS gets 15 labels for QL).

The SC selects facets by maximizing the union of tuples, thus it tends to return facets with more tuples and less overlapping (e.g., the facets of MSC have more tuples than IS and MDS in Fig. 9). Due to the inverse proportion of package size and meaningfulness, the meaning of SC is limited (e.g., MSC gets 11 labels). The HS finds a local optimal coverage for tuples, thus the meaning is also limited.

Note that MDS has more meaning than IS; this is because MDS selects packages with the highest overall scores, and the meaningfulness is a factor to be considered. However, there are significant overlapping between facets, hence the meanings are also overlapped (e.g., MDS has less label diversity than IS for QL).

2.Character coverage Fig. 10c and 10d show that BS and IS produce the highest character coverage. Note that IS achieves exactly the same character coverage as BS, which confirms the correctness of our filters in Section 4.3. Therefore, in following evaluations we compare only IS with other methods.

FR and DS return facets with largest support or overall scoring, which may neglect the overlapping between facets (e.g., the label diversity of FR and DS is less than IS for QL). Hence, they fail to return facets with high character coverage. On the other hand, MSC and MHS have relatively higher character coverage. This is because SC and HS aim to maximize the union of tuples, thus are likely to return disjoint facts (e.g., Lenovo 500GB and Lenovo 320GB in Fig. 9c). Therefore, they tend to have more distinct characters.

3.Precision Fig. 10e and 10f show the precision of each method. As we can see, SC and HS get higher score, because they are likely to have less overlaps than IS, FR and DS due to their optimization goal. For FR and DS, the overlapping is significant especially for MFR and MDS (e.g., in Fig. 9, MFR has 11 overlapped tuples, whereas IS has 3). Therefore, their precisions are low. On the other hand, the precision of IS is close to others especially when k is small (e.g., $k < 15$), and outperforms MFR and MDS significantly. This is because when k gets larger, the overlap between facets may increase (IS aims to find maximal character coverage instead of maximal tuple coverage). However, in practice, the presentation task prefers good quality with a relatively small k to avoid overwhelming.

4.Recall Fig. 10g and 10h compare the recall. We can see that IS is close to SC and HS, also it outperforms FR and DS on most cases. SC and HS are designed to get the largest tuple coverage, thus they have strong overview ability (e.g., the distinct tuples of MSC and MHS ranks the top 2 largest in Table 7). We can conclude that IS has slightly more overlapped tuples than SC and HS, hence its overview ability is close to them.

The four metrics test the quality differently. *Meaningfulness* evaluates the meaning of packages; *character coverage* tests both tuples and labels collectively; *precision* evaluates the overlaps and *recall* the overview ability. As we can see, most existing methods aim to optimize precision and recall, rather than the meaning. In this work, IS combines meaningfulness and overview ability by character coverage. The experiments show that IS improves the meaning by 49.1% on average, and losses overview ability slightly by 6.2% at most. Given the superior performance, we can conclude that IS produces promising summaries.

6   Related Work

Facets. Faceted search has been extensively studied in DB community [4,16,6,12]. Recently, facet-ranking methods are proposed for the facet overwhelming problem [6,12]. However, most of them focus the number of tuples in each facet; they neglect to rank facets with labels. Besides, [3] proposes a probabilistic ranking model, whereas it is designed for documents. On the other hand, [4,6,16] rank facets by a cost model to minimize the user efforts. They are different from TP, since TP does not involve the interaction of users. Bin [22] answers aggregate queries, which is similar to subtopics in this paper. However, no further summarizing was performed for the overwhelming problem.

Clustering. Clustering helps users search the answer space (see [2] for a survey). Various methods [20,13,17] present the results by finding the naturally close groups in answer space. However, the clusters are not intelligible. [17] proposes a labelling clustering method for web pages, whereas it is hard to adopt on relational data for real time response. [13] is another effective clustering method, it reduces the tuple overwhelming by further clustering results in each cluster.

Ranking. Ranking and top-k query answering [1,5,14,15] rank results by structural or statistical information. It is effective, but for navigational queries, users need to navigate the answer space, hence faceting and clustering are proposed.

Diversification. Diversification could provide more different results. In DB community, some pioneering approaches [13,7,21,10] have emerged. Bin [13] diversifies results by choosing representatives from each cluster. DivQ [7] aims to discover diversified schemas. BROAD [21] captures both structure and semantic information by a kernel distance metric. These approaches provide different results, whereas TP provides packages with maximal information.

User Interface. User interfaces (e.g., Skimmer[18], MusiqLens[13], and DataScope[19]) are designed to give a fast overview of answer space, and meanwhile feed users a small fraction of answers. Clustering or sampling is performed.

7   Conclusion

In this paper, we had proposed TP for presenting search results on database. To our best knowledge, TP is the first work that addresses the presentation from both package size and labels. First, we identified acceptable packages and designed a fast algorithm to generate them. Then we quantified the character coverage and proposed algorithms for summarization. The experimental results show that TP yields promising summaries efficiently. In future, we would like to integrate TP with existing ranking methods.

References

1. Agrawal, S., Chaudhuri, S., Das, G., Gionis, A.: Automated ranking of database query results. In: CIDR (2003)
2. Carpineto, C., Osinski, S., Romano, G., Weiss, D.: A survey of web clustering engines. ACM Comput. Surv. 41(3) (2009)
3. Carterette, B., Chandar, P.: Probabilistic models of ranking novel documents for faceted topic retrieval. In: CIKM, pp. 1287–1296 (2009)
4. Chakrabarti, K., Chaudhuri, S., won Hwang, S.: Automatic categorization of query results. In: SIGMOD, pp. 755–766 (2004)
5. Chaudhuri, S., Das, G., Hristidis, V., Weikum, G.: Probabilistic information retrieval approach for ranking of database query results. ACM TODS 31(3), 1134–1168 (2006)
6. Dakka, W., Ipeirotis, P.G., Wood, K.R.: Automatic construction of multifaceted browsing interfaces. In: CIKM, pp. 768–775 (2005)
7. Demidova, E., Fankhauser, P., Zhou, X., Nejdl, W.: DivQ: Diversification for keyword search over structured databases. In: SIGIR, pp. 331–338 (2010)
8. Donohue, J.C.: Understanding scientific literatures: A Bibliometric Approach. The MIT Press, Cambridge (1973)

9. Hochbaum, D.S. (ed.): Approximation algorithms for NP-hard problems. PWS Publishing Co., Boston (1997)

10. Hu, H., Zhang, M., He, Z., Wang, P., Wang, W.: Diversifying query suggestions by using topics from wikipedia. In: Web Intelligence, pp. 139–146 (2013)

11. Koller, D., Sahami, M.: Hierarchically classifying documents using very few words. In: ICML, pp. 170–178 (1997)

12. Li, C., Yan, N., Roy, S.B., Lisham, L., Das, G.: Facetedpedia: Dynamic generation of query-dependent faceted interfaces for wikipedia. In: WWW, pp. 651–660 (2010)

13. Liu, B., Jagadish, H.V.: Using trees to depict a forest. PVLDB 2(1), 133–144 (2009)

14. Luo, Y., Lin, X., Wang, W., Zhou, X.: Spark: Top-k keyword query in relational databases. In: SIGMOD, pp. 115–126 (2007)

15. Luo, Y., Wang, W., Lin, X., Zhou, X., Wang, J., Li, K.: Spark2: Top-k keyword query in relational databases. IEEE Trans. Knowl. Data Eng. 23(12), 1763–1780 (2011)

16. Roy, S.B., Wang, H., Das, G., Nambiar, U., Mohania, M.K.: Minimum-effort driven dynamic faceted search in structured databases. In: CIKM, pp. 13–22 (2008)

17. Scaiella, U., Ferragina, P., Marino, A., Ciaramita, M.: Topical clustering of search results. In: WSDM, pp. 223–232 (2012)

18. Singh, M., Nandi, A., Jagadish, H.V.: Skimmer: Rapid scrolling of relational query results. In: SIGMOD Conference, pp. 181–192 (2012)

19. Wu, T., Li, X., Xin, D., Han, J., Lee, J., Redder, R.: Datascope: Viewing database contents in google maps' way. In: VLDB, pp. 1314–1317 (2007)

20. Zeng, H.-J., He, Q.-C., Chen, Z., Ma, W.-Y., Ma, J.: Learning to cluster web search results. In: SIGIR, pp. 210–217 (2004)

21. Zhao, F., Zhang, X., Tung, A.K.H., Chen, G.: Broad: Diversified keyword search in databases. PVLDB 4(12), 1355–1358 (2011)

22. Zhou, B., Pei, J.: Answering aggregate keyword queries on relational databases using minimal group-bys. In: EDBT, pp. 108–119 (2009)

An Efficient Approach for Mining Top-k High Utility Specialized Query Expansions on Social Tagging Systems

Jia-Ling Koh and I-Chih Chiu

Department of Information Science and Computer Engineering,
National Taiwan Normal University, Taipei, Taiwan, R.O.C.
jlkoh@csie.ntnu.edu.tw

Abstract. A specialized query expansion consists of a set of keywords, which is used to reduce the size of search results in order to help users find the required data conveniently. The utility of a specialized query expansion represents the qualities of the top-N high quality objects matching the expansion. Given the search results of a keyword query on social tagging systems, we want to find k specialized query expansions with the highest utilities without redundancy. Besides, the discovered expansions are guaranteed to match at least N objects. We construct a tree structure, called an UT-tree, to maintain the tag sets appearing in the search results for generating the specialized query expansions. We first propose a depth-first approach to find the top-k high utility specialized query expansions from the UT-tree. For further speeding up this basic approach, we exploit the lower bound and upper bound estimations of utilities for a specialized query expansion to reduce the size of the constructed UT-tree. Only the tag sets of objects which are possibly decide the top-k high utility specialized query expansions need to be accessed and maintained. By applying this strategy, we propose another faster algorithm. The experiment results demonstrate that the proposed algorithms work well on both the effectiveness and the efficiency.

Keywords: specialized query expansion, social tagging system.

1   Introduction

Social tagging has become widely popular in many social sharing systems. Social tagging systems enable users to freely label web resources of interest with tags for describing the content or semantics of the resources, which provides a simple solution for annotations. In recent studies, it has been shown that tags can be used to improve search, navigation and recommendations effectively [1].

Keyword based search is a popular way for discovering required data of interest from a huge collection of sharing resources. The effectiveness of data retrieval mainly depends on whether the given queries properly describe the information needs of users. However, it is not easy to give a precise query because most queries are short (less than two words on average) and many query words are ambiguous. Using a general keyword with broad semantics as a query usually causes a huge amount of data returned. Most of the search services return search results as a ranked list, but it is

S.S. Bhowmick et al. (Eds.): DASFAA 2014, Part II, LNCS 8422, pp. 361–376, 2014.
© Springer International Publishing Switzerland 2014

difficult for users to explore and find objects satisfying their search needs from a long list of results. Consider a query "apple" on Flickr as an example. Because various semantics are represented by the word "apple", the search results may contain the images of apples, the products of apple company, the NewYork City etc., which are mixed together in the returned ranked list. It is costly for users to sequentially browse the objects in the ranked list to find a favorite image of red apples. Accordingly, how to automatically group search results into semantically "topics" has become a signifi- cant issue to improve the usability of the results.

Faceted search is a common feature of e-commerce sites, where the objects are structured data with attributes. Users can select an attribute and a corresponding attribute value, such as the price or category of products, to filter the search results. Recently, several search engines also provide a set of facets to the search interface, which are usually the structural attribute of data like location, time, size of document etc. Faceted interfaces provide convenient and efficient way to navigate the search results. However, most of the systems have to define the facets of their search inter- faces in advance or assume a prior taxonomy exists.

In a social tagging system, each object in the search results has the annotated tags. A tag can be viewed a binary facet attribute of objects. In spirit to recent works on automatic facets generation [6], we would like to dynamically find keyword sets from the tags of the search results to form more specific queries in order to help users find the required data easily. For example, {fruit}, {mac}, and {NewYork, City} are good keyword sets to further distinguish different information needs of the query "apple".

The problem of finding the top-k high quality expansions was first introduced in [7]. This work considered a search scenario in which each object is annotated with a set of keywords. Besides, it was assumed that each object has an associated score computed from its attributes to show its importance or quality, e.g. the popularity of a social sharing object or the score of an answer in a QA community site. An expansion of a query consists of a set of keywords. An object whose keyword set contains both the query and the expansion is called that it matches the expansion. According to a given constant value N, the utility of an expansion is computed by summing up the top-N highest utilities of the objects which match the expansion. Then the work proposed various algorithms to find the top-k expansions with the highest utilities. However, in a social tagging system, it is not computationally feasible to illustrate the combinations of keywords to generate expansions because the amount of various tags annotated for objects is huge. Besides, according to the utility of an expansion defined in [7], it is possible to find an expansion which appears rarely but is selected into the top-k expan- sions because it is matched by some little objects with high utilities. In this case, the expansion could not provide the desired number of search results (N objects).

In this paper, we name an expansion defined in [7] a *specialized query expansion* in order to denote its characteristic for query specialization. Our work is motivated from the problems mentioned above. Given the search results of a keyword query on social tagging systems, we want to find the k specialized query expansions with the highest utilities without redundancy. Besides, the discovered expansions are guaranteed to match at least N objects. We construct a tree data structure, called an UT-tree, to maintain the tag sets appearing in the search results, which are possible to generate the specialized query expansions. We first propose a depth-first approach to find the top-k high utility specialized query expansions from the UT-tree. For further speeding up this basic approach, we exploit the lower bound and upper bound estimations of utilities for a specialized query expansion to reduce the size of the constructed *UT-tree*. Only the tag sets of objects which are possibly decide the top-*k* high utility specialized query expansions need to be accessed and maintained. By applying this strategy, we propose another faster algorithm. A systematic performance study is performed to verify the effectiveness and the efficiency of the proposed algorithms.

The rest of the paper is organized as follows. In the next section, a brief overview of related works is introduced. The formal problem definition of the top-*k* high utility specialized query expansions discovering is given in Section 3. Section 4 introduces the proposed data structure and algorithms for solving the problem efficiently. The performance evaluation on the proposed algorithms is reported in Section 5. Finally, we conclude this paper and provide the directions for future studies in Section 6.

2   Related Works

In [10], a tag clustering approach was proposed to solve the syntactic and semantic tag variations during searching and browsing activities of users. The similarity between each pair of tags is computed for discovering the syntactic clusters of tags. Based on co-occurrences of tags, the cosine similarity is measured to find clusters of semantics related tags, which are used for searching and browsing the tag spaces. However, the computation cost of finding both syntactic and semantic tag clusters is very high. The tag ranking and selection strategies were proposed in [9, 11] to find tag clouds for describing groups of tagged resources. Although tag clouds provide a convenient way of navigating collections of tagged resources, most tag clouds only compose of single tags without considering tag sets. However, some tag sets can represent more clear semantics than single tags. Moreover, the requirement of finding high quality objects/resources was not considered.

Recently, many works studied the problem of query reformulation suggestions [2,4,8]. After submitting a query, query reformulations can help users change or modify the search direction in order to disambiguate the results and get better answers. It is also possible to help users modify their information needs by providing other related queries. Most methods are designed based on "The wisdom of crowds" by analyzing the search logs of past queries. Accordingly, these methods require a large set of search logs for learning the semantic relationships among the queries.

Furthermore, several works studied how to dynamically discover a small set of facets and values from search results in order to find the results most interesting to a user. The Facetedpedia system proposed in [6] was a faceted retrieval system designed for information discovery and exploration in Wikipedia. The system automatically and dynamically constructs facet hierarchy for navigating the set of Wikipedia articles resulting from a keyword query. A supervised method based on a graphical model was proposed in [5] for automatically extract facets of a query, which are groups of semantically related terms extracted from search results. In spirit to the success of facet interface, we would like to suggest keyword sets as facets for effectively navigating the search results with high qualities in a tagging system. However, because the facets are structured and have the form of attribute-value pairs, those approaches are not suitable or directly applicable for our problem.

3   Problem Definition

Let *WDB* denote a database of web resources, which contains a set of objects: $WDB = \{o_1, ..., o_n\}$. We assume that each object o_i in *WDB* has a set of annotated tags and a utility value representing its quality, denoted by $o_i.tagset$ and $o_i.utility$, respectively. Without loss of generality, we assume that $o_i.utility$ is in [0,1], and a larger value represents a higher quality. A query q consists of a nonempty set of keywords. The returned search results of q from *WDB*, denoted by O_q, is the set of objects whose tag sets contain all the keywords in q, i.e., $O_q = \{o \mid o \in WDB \wedge q \subseteq o.tagset\}$.

Let CT_q denote the union of the tag sets of the objects in O_q except the tags in q, i.e., $CT_q = \{t \mid t \in o.tagset \wedge o \in O_q\} - q$. We call a nonempty subset of CT_q a *specialized query expansion* of q. In the following, a specialized query expansion of a query is also expressed by an expansion for short. A specialized query expansion containing l items is named an *l-query expansion* (*l-QE*). For a given specialized query expansion *QE*, we say that an object o_i *matches QE* if and only if $QE \subseteq o_i.tagset$. The set of objects in O_q, which match *QE*, is denoted by $O_q(QE)$. The *popularity* of a specialized query expansion *QE* in O_q, denoted by $P_q(QE)$, is the number of the objects in O_q matching *QE*, i.e., $P_q(QE) = |O_q(QE)|$.

To prevent from discovering unpopular expansions, a positive integer called the *minimum popularity* threshold, denoted by N, is given to find the expansions *QE* with $P_q(QE) \geq N$. Besides, since users usually browse only a few top objects in the search results, the importance of a specialized query expansion *QE* is determined by the top-N utilities from $O_q(QE)$. Let $sort\_u(O_q(QE))$ denote a function on $O_q(QE)$ to get a sorted list of the objects in $O_q(QE)$ in descending order of their utilities and then in ascending order of the object identifiers. Let $top\text{-}N(O_q(QE))$ denote the first N objects in $sort\_u(O_q(QE))$. Accordingly, for the objects in $top\text{-}N(O_q(QE))$, their utilities are no less than the ones of the objects in $O_q(QE)$ - $top\_N(O_q(QE))$. The utility of a specialized query expansion *QE* of q, denoted by $u_q(QE)$, is defined as follows:

$$u_q(QE) = \sum_{o \in top\_N(O_q(QE))} o.utility \,. \tag{1}$$

Let QE_i and QE_j denote two expansions generated from CT_q. If QE_i is a subset of QE_j, $O_q(QE_i)$ is a superset of $O_q(QE_j)$. Therefore, $top\_N(O_q(QE_i)) \supseteq top\_N(O_q(QE_j))$ and $u_q(QE_i) \geq u_q(QE_j)$. When $top\_N(O_q(QE_i)) = top\_N(O_q(QE_j))$, providing both QE_i and QE_j is redundant because the top-N results of QE_i and QE_j are the same. Because QE_j represents more specific semantics than QE_i, QE_j is a better specialized query expansion than QE_i. Accordingly, when QE_j is selected to be one of the top-k high utility specialized query expansions, we should prevent from selecting QE_i into the top-k expansions.

Definition 1 (top-k high utility specialized query expansions discovering problem). Given a database of web resources *WDB*, a query q, and a minimum popularity threshold N, the top-k high utility specialized query expansions (QE_top_k) discovering problem is to find a set of k specialized query expansions of q such that
1)   For each specialized query expansion *QE* in QE_top_k, $P_q(QE) \geq N$.
2)   For each specialized query expansion *QE* in QE_top_k, it is a *closed expansion on the top_N result*. It means that there does not exist a specialized query expansion *QE'* such that $QE \subset QE'$ and $top\_N(O_q(QE)) = top\_N(O_q(QE'))$.

3) For each specialized query expansion QE in QE_top_k, $u_q(QE) \geq u_q(QE'')$ for each QE'' not in QE_top_k.

Table 1. A sample database WDB

Object id	Tagset	Utility
[952]	art band blue music	0.6
[995]	concern live music rock	0.8
[056]	concern microphone music	0.4
[402]	ticket	0.3
[609]	elephant tree animal nature band concern guitar live music	0.5

[**Example 3.1**] Assume a sample tag database is given as shown in Tab. 1. When $q=\{music\}$, the corresponding $O_q=\{952, 995, 056, 609\}$ and $CT_q =\{art, band, blue, concern, live, rock, microphone, ticket, guitar\}$. Suppose the minimum popularity threshold N is set to 2 and we would like to discover a solution of QE_top_2. Among the all specialized query expansions generated from the tags in CT_q, there are only four expansions whose popularities are no less than 2: $QE_1 =\{band\}$, $QE_2=\{concern\}$, $QE_3=\{live\}$, $QE_4=\{concern,live\}$. The $u_q(QE_2)=[995].utility+[609].utility=0.8+0.5 =1.3$. Similarly, we can get $u_q(QE_1)=1,1$, $u_q(QE_3)=1.3$, and $u_q(QE_4)=1.3$. Because both QE_2 and QE_3 are subsets of QE_4 and $top\_2(O_q(QE_2))=top\_2(O_q(QE_3))= top\_2(O_q(QE_4))$, QE_4 is selected into the result of QE_top_2 instead of QE_2 and QE_3. Finally, the discovered QE_top_2 includes QE_1 and QE_4.

4   Top-*k* High Utility Specialized Query Expansions Discovery

4.1   Utility Tree Structure

The frequent-pattern tree (FP-tree) proposed in [3] is an extended prefix-tree structure for storing compressed and crucial information in transactions about frequent itemsets. In our approach, we modify the FP-tree structure to propose a data structure called a *utility tree* structure (abbreviated as *UT*-tree) for maintaining the candidate tag sets of specialized query expansions. The popularities and utilities of an expansion can be computed from the information maintained in the *UT*-tree.

In a *UT*-tree, except the root node, each node represents a specialized query expansion consisting of the tags stored in the path from the root node reaching the node. In addition to the *children* links, each non-root node consists of four fields: *tag, count, obj-list*, and *node-link*, where *tag* field records the tag stored in the node, *count* field records the popularity of the expansion represented by this node, *obj-list* field records the object identifiers of the objects matching the expansion of the node, and *node-link* links to the next node in the tree structure which stores the same tag. Furthermore, a header table of the *UT*-tree structure is constructed, in which each entry consists of four fields: (1) *tag,* which stores the keyword of a tag *t*; (2) *count,* which is used to accumulate the number of objects whose tag sets contains the tag *t*; (3) *utility,* which is used to accumulate the top-N utilities of objects whose tag sets contains tag t and called the utility of t; and (4) *link*, which points to the first node of the linked list for chaining all the nodes storing tag t.

[Property 1] Given two query expansions QE_1 and QE_2, $P_q(QE_1) \geq P_q(QE_2)$ if $QE_1 \subset QE_2$.

Because of the anti-monotone property of popularity, if the popularity of a specialized query expansion composing a single tag t is less than N, any expansion composing t has its popularity less than N and can never be a top-k specialized query expansion. Consequently, the basic method constructs the UT-tree structure in two scans on the tag sets of O_q. In the first scan, the objects in O_q are sorted according to their utilities in descending order. Besides, the count for each tag in CT_q is accumulated. In the second scan, when constructing the UT-tree structure, the tags with counts less than N are ignored from the tag set of an object. Besides, the tags in a tag set are sorted in lexicographic order. Then the tag sets of the objects in O_q are inserted into the UT-tree one-by-one as constructing a FP-tree. However, the difference of constructing a UT-tree is that each node maintains a list of objects identifiers of the objects matching the specialized query expansion of the node. In addition, the header table of the tree structure has an additional field to compute the utility of each tag. The constructed tree is called the UT-tree of O_q.

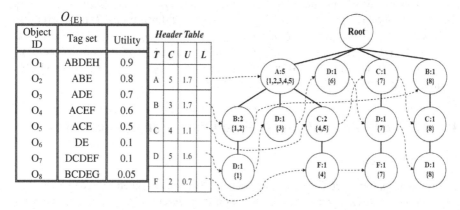

Fig. 1. An example of search results and the constructed UT- tree

[Example 4.1] Suppose the query is {E} and the search results of {E} in *WDB* is shown as the table in Fig. 1. Besides, N is set to be 2. After the first scan, the tags with their counts are obtained as the following: A:5, B:3, C:4, D:5, F:2, G: 1. Among the tags, the count of tag G is less than 2. Therefore, tag G in the tag sets are ignored when constructing the UT-tree. Accordingly, the constructed UT-tree for the search results is shown as Fig. 1.

In the constructed UT-tree of O_q, all the tag sets of the objects in O_q are maintained in the paths starting from the root node. From the paths rooted at the nodes of tag t_a, we can get the subsets contained in the tag sets of $O_q(\{t_a\})$, whose tags are with lexicographic order larger than t_a. This set of tag sets provides essential information for generating the specialized query expansions beginning with tag t_a, which is denoted

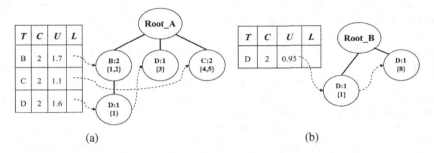

Fig. 2. The *CUT*-trees and the *CH*-table of {A} and {B} for the example

by $CT(\{t_a\})$. The *UT*-tree constructed for $CT(\{t_a\})$, which is called the *conditional UT-tree (CUT-tree)* of $\{t_a\}$, is obtained by assigning a root node pointing to all the sub-trees rooted at the nodes of tag t_a. Besides, the corresponding conditional header table *(CH-table)* is constructed. For each tag t_b stored in the *CH*-table of $\{t_a\}$, the *count* and *utility* field respectively represent the popularity and the utility of the expansion $\{t_a\}\cup\{t_b\}$. Accordingly, only the tags, whose count values in the *CH*-table of $\{t_a\}$ are no less than N, possibly generate the expansions in QE_top_k and are remained in the table. For an expansion *l-QE* with $l \geq 2$, the *CUT*-tree and the corresponding *CH*-table can be obtained by applying the method recursively.

[Example 4.2] According to the *UT-tree* constructed in Example 4.1, Fig. 2(a) shows the *CUT-tree* of {A} and the corresponding header table. The tag F is removed from the table because its count appearing together with A is less than N, i.e. 2. The *CUT*-tree and *CH*-table of {B} are shown in Fig. 2(b).

4.2   UT-Growth Algorithm

According to the constructed *UT*-tree, we propose an algorithm named the *UT-growth* to discover a solution of QE_top_k based on the following properties.

[Property 2] For a given $(l-1)$-query expansion QE_1 with $l \geq 2$, let QE_2 denote a *l*-query expansion and $QE_1 \subset QE_2$. If $P_q(QE_2) \geq N$, then $u_q(QE_1) \geq u_q(QE_2)$.

For any object o in O_q, if o matches QE_2, it is necessarily that it matches QE_1. Accordingly, $O_q(QE_2) \subseteq O_q(QE_1)$. The utilities of the objects in $top\_N(O_q(QE_2))$ are less than or equal to the ones of objects in $top\_N(O_q(QE_3))$. Therefore, the utilities of query expansions are anti-monotonic. Accordingly, we can get the following property.

[Property 3] For any query expansion QE, if $P_q(QE) \geq N$, then $u_q(QE) \leq min\{u_q(\{t\}) \mid t \in QE\}$.

[Property 4] If there are a set of tags t_1, t_2, ..., and t_s, $top\_N(O_q(\{t_1\})) = top\_N(O_q(\{t_2\})) = ... = top\_N(O_q(\{t_s\}))$, then any *j*-combination of the s tags with $0 < j < s$ are not *closed expansions* on the *top_N* results.

Notice that the tags in a tag set are sorted in lexicographic order when constructing a *UT*-tree. The proposed *UT*-growth algorithm adopts the prefix-based and depth-first approach to generate expansions. That is, the algorithm generates expansions starting from a specific tag. If the specialized query expansion QE has popularity no less than N, the tags with lexicographically order larger than the tags in QE are appended individually into QE for generating larger specialized query expansions. In order to perform the mining process efficiently, the algorithm maintains a list of discovered expansions which have the top-k high utilities temporarily. According to Property 2, the algorithm can terminate the depth-first growing of an expansion if its utility is less than the minimum utility in the temporary top-k result.

The pseudo code of the UT-growth algorithm is shown in Algorithm 1. According to Property 3, at the beginning of the algorithm, the tags in the header table of O_q are sorted in decreasing order of the utilities (line 3) and then selected one by one. Each selected tag forms a specialized query expansion and inserted into the temporary result, denoted by *TempResult*. Besides, according to Property 4, if there are s selected tags with the same *top_N* lists, any j-combination of the s tags with $0 < j < s$ cannot be selected into QE_top_k. Accordingly, these s tags are combined to generate an expansion and inserted into *TempResult*. The tag selection continues until *TempResult* contains k expansions (lines 4-11). The set C_{QE} maintains the tags which are components of the k expansions in *TempResult*. The tags in C_{QE} are selected one by one in decreasing order of the utilities to individually form a specialized query expansion. Then the depth-first approach to find a solution of QE_top_k is performed by calling the procedure *top_k_QE()* recursively (lines 14-19).

In the procedure *top_k_QE()*, for the given specialized query expansion *CurrentQE*, its *CUT*-tree and the *CH*-table are constructed (line24) to find the tag t which appears together with *CurrentQE* in at least N objects in O_q, i.e. $P_q(CurrentQE \cup \{t\}) \geq N$. Besides, only the tag t with utility no less than the minimum utilities of the tags in C_{QE}, denoted by *min_u*, is considered to generate the expansion *CurrentQE* $\cup \{t\}$ (line 25). The reason is that, for a tag t', whose utility is less than *min_u*, $u_q(CurrentQE \cup \{t'\})$ must be less than the utilities of the temporary top-k expansions according to Property 3. Therefore, *CurrentQE* $\cup \{t'\}$ is not possible in a solution of QE_top_k. In addition, two further criterion are checked to decide whether the specialized query expansion *CurrentQE* $\cup \{t\}$ needs to be generated. The first one is *top_N*(O_q(*CurrentQE* $\cup \{t\}$))= *top_N*(O_q(*CurrentQE*)), it indicates that *CurrentQE* is not a closed expansion on the *top_N* results. Another one is that the utility of *CurrentQE* $\cup \{t\}$ is no less than *min_u*, it means that *CurrentQE* $\cup \{t\}$ possibly becomes a member in the solution of QE_top_k (line 26).

After generating this new specialized query expansion as *CurrentQE*, it is necessarily to check whether there is an expansion in *TempResult* whose *top_N* results are the same with the ones of *CurrentQE*. If there exists such an expansion in *TempResult*, it means that *CurrentQE* is not a closed expansion on the *top_N* results so it is not inserted into *TempResult*. Otherwise, *CurrentQE* is inserted into *TempResult*. Besides, the expansions in *TempResult* which are not closed expansions on the *top_N* results are removed from *TempResult* (line 29-32). Then *top_k_QE(CurrentQE)* is performed recursively (line 33). After finishing the tasks of performing *top_k_QE(CurrentQE)* recursively, k specialized query expansions in *TempResult* with the highest utilities are outputted (line 20).

[**Example 4.3**] According to the UT-tree constructed in Example 4.1, given N is 2 and k is 3, the UT-growth algorithm is performed as follows. The tags and their utilities in

---

Algorithm 1. The *UT-growth* Algorithm

Input: the query result O_q, the minimum popularity N, the value of k
Output: the discovered QE_top_k
1. Sort the objects in O_q in decreasing order of the utilities;
2. Construct the *UT*-tree for O_q
3. Sort the tags in the header table in decreasing order of their utilities,
     Let $<t_1, t_2, ..., t_n>$ denote the sorted list of tags;
4. Set $i=0$ and $j = 0$;
5. While $((j < k) \land (i < n))$
6. { $i= i+1$;
7.   Insert tag t_i into C_{QE}.
8.   If there is a QE in *TempResult* such that $(top\_N(O_q(\{t_i\}))= top\_N(O_q(QE))$
9. $QE = QE \cup \{t_i\}$;
10. else { $j=j+1$; $QE_j = \{t_i\}$;
11.   Insert QE_j into *TempResult*;}}
12. $temp\_tags = C_{QE}$
13. $min\_u = min\{u_q(\{t\})| t \in C_{QE} \}$
14. Repeat
15. { Let t_{max} denote the tag in *temp_tags* with the highest utility value;
16.   Remove t_{max} from *temp_tag*,
17. $CurrentQE =\{t_{max}\}$;
18.   Call *top_k_QE(CurrentQE)*;
19. } until *(temp_tags = ϕ)*
20. Output k specialized query expansions in *TempResult* with the highest utilities
21. }
22.
23. *top_k_QE(CurrentQE)*
24. { Construct the *CUT*-tree of *CurrentQE*;
25.   For each tag t in the *CH*-table of *CurrentQE* and $u_q(t) \geq min\_u$;
26.   { If $(top\_N(O_q(CurrentQE \cup \{t\})) = top\_N(O_q(CurrentQE)))$ $\lor$
                       $(u_q(CurrentQE \cup \{t\}) \geq min\_u)$
27.   { $CurrentQE = CurrentQE \cup \{t\}$;
28. $contain = \{QE' \mid QE' \in TempResult \land CurrentQE \subset QE'$
                       $\land top\_N(O_q(QE')) = (top\_N(O_q(CurrentQE)))\}$
29.   If $(contain = \phi)$
30.   {Insert $CurrentQE$ into *TempResult*;
31. $non\_close =\{QE'' \mid QE'' \in TempResult \land QE'' \subset CurrentQE$
                       $\land(top\_N(O_q(QE''))=top\_N(O_q(CurrentQE))$
32. $TempResult = TempResult - non\_close;$ }
33.   Call *top_k_QE(CurrentQE)*;
34.   }
35.   return; }
36. }

---

the header table are A:1.7, B:1.7, C:1.1, D:1.6, and F:0.9. Because $top\_2(O_q(\{A\}))=top\_2(O_q(\{B\}))$, *TempResult* contains the temporarily discovered expansions with the top-3 utilities as follows: {A,B}:1.7, {D}:1.6, and {C}:1.1. Besides, the content in C_{QE} is <A, B, D, C>, where the tags are sorted in decreasing order of the utilities. Besides, $min\_u = 1.1$. Then each tag in C_{QE} is accessed to form a 1-*QE* and *top_k_QE(CurrentQE)* is performed to find the specialized query expansions composing *CurrentQ* as follows.

<1> *CurrentQE*={A}. The *CUT*-tree of {A} is shown as Fig. 2(a). Accordingly, we get $u_q(\{A, B\})=1.7$, $u_q(\{A, C\})=1.1$, and $u_q(\{A, D\})=1.6$. By performing *top_k_QE* ({A,B}) recursively, no expansion is generated by growing from {A,B} because the *CUT*-tree of {A, B} is empty. Besides, because $top\_2(O_q(\{A, C\})) = top\_2(O_q(\{C\}))$, {C} is removed from *TempResult* and {A,C} is inserted into *TempResult*. It is similar to replace {D} by {A, D} because $top\_2(O_q(\{A, D\})) = top\_2(O_q(\{D\}))$. The *CUT*-trees of {A, C} and {A, D} are both empty, thus, no expansion is generated by growing from these two query expansions.

<2> *CurrentQE*={B}: According to the *CUT*-tree of {B}, we get $u_q(\{B,D\})=0.95$, which is less than *min_u*. Therefore, {B,D} is not inserted into *TempResult*.

<3> *CurrentQE*={D}: The *CUT*-tree of {D} is empty.

<4> *CurrentQE*={C}: The *CH*-table of {D} shows that $u_q(\{C, D\})=0.15$ and $u_q(\{C, F\})=0.7$, which are less than *min_u*. Neither {C, D} and {C, F} are inserted into *TempResult*.

Finally, the *TempResult* contains the discovered top-3 high utility specialized query expansions: {A,B}, {A,D}, and {A,C}.

4.3   Dynamic UT-growth Algorithm

The *UT*-growth algorithm has to construct the complete *UT*-tree of O_q by inserting all the tag sets of the objects in O_q. However, most of the objects with low utilities do not influence the discovered solution of QE_top_k. For further speeding up the process of discovering a solution of QE_top_k, we propose another algorithm named the *dynamic UT-growth*, which constructs the *UT*-tree of O_q dynamically. The basic idea is that, when the tag sets of a part of the objects in O_q are known, we can estimate an upper bound and a lower bond of the utility of an expansion. The estimated bonds can be used to early terminate the construction of the *UT*-tree, so the mining process of finding QE_top_k can be performed more efficiently.

Let $UUB^i(QE)$ and $ULB^i(QE)$ denote an upper bound and a lower bound of the utility of a query expansion QE after the ith object in O_q is accessed. Notice that the objects in O_q are sorted in decreasing order of the utilities. Let $c^i(QE) = |\{o_j \mid QE \in o_j.tagset \wedge 1 \le j \le i\}|$. If $c^i(QE) \le N$, we use the following two equations to estimate the two bounds of utilities for QE:

$$UUB^i(QE) = \sum_{1 \le j \le i, o_j \in O_q(QE)}^{i} o_j.utility + (N - c^i(QE)) \times o_i.utility \qquad (1)$$

$$ULB^i(QE) = \sum_{1 \le j \le i, o_j \in O_q(QE)}^{i} o_j.utility + (N - c^i(QE)) \times o_{|O_q|}.utility \qquad (2)$$

Otherwise, $UUB^i(QE) = ULB^i(QE) = \sum_{1 \le j \le i, o_j \in top\_N(O_q(QE))}^{i} o_j.utility$.

In the dynamic UT-growth algorithm, when constructing the UT-tree, the header table has two additional fields: ULB and UUB, to maintain the estimated lower bound and upper bound of utility for a tag, respectively. Let top-tags$^{ULB}(i)$ and top-tags$^{UUB}(i)$ denote the top-i list of tags according to their $ULBs$ and $UUBs$, respectively. Whenever a tag set of an object in O_q is inserted into the UT-tree, the top-tags$^{ULB}(k)$ and top-tags$^{UUB}(k+1)$ are updated accordingly. When 1) the ULB value of rank k in top-tags$^{ULB}(k)$ is larger than the UUB value of rank $(k+1)$ in top-tags$^{UUB}(k+1)$ and 2) the tags in top-tags$^{ULB}(k)$ all have counts no less than N, it indicates that the tags in top-tags$^{ULB}(k)$ have higher utilities than the other tags. Therefore, the process of constructing the UT-tree is stopped and the following steps for finding a solution of QE_top_k are performed.

Step 1: Insert the tags in top-tags$^{ULB}(k)$ into C_{QE}.

Step 2: The tags in C_{QE} are selected one by one in decreasing order of the utilities to individually form a query expansion. Then the depth-first approach to find QE_top_k is performed by calling the procedure $top\_k\_QE()$ similar to the one used in Algorithm 1. The difference is that when constructing the CUT-tree of a specialized query expansion QE, the tags with counts less than N are remained because they may get more counts after inserting the tag sets of the following objects in O_q.

Step 3: If the number of the discovered expansions in $TempResult$ is less than k, the algorithm resumes the task of UT-tree construction until at least one more tag in the header table with its count achieving N. These tags are then inserted into C_{QE} and Step 2 continues. Otherwise, Step 4 is performed.

Step 4: Output k specialized query expansions in $TempResult$ with the highest utilities.

[**Example 4.4**] By performing the dynamitic UT-growth algorithm, after inserting the tag set of O_4 into the UT-tree, the constructed UT-tree is shown as Fig. 3(a). The tags in top-tags$^{ULB}(3)$ are A, B, and D with $ULB^4(\{A\})=1.7$, $ULB^4(\{B\})=1.7$, and $ULB^4(\{D\})=1.6$. The popularities of all these tree tags are no less than N, i.e. 2. Besides, the tag of rank 4 in top-tags$^{UUB}(4)$ is C with $UUB^4(\{C\})=1.2$. Therefore, the process of constructing the UT-tree is stopped. After performing the mining process on the partially constructed UT-tree, the discovered specialized query expansions in $TempResult$ are $\{A, B\}$ and $\{A, D\}$. Because the number of expansions in $TempResult$ is less than k, i.e. 3, the algorithm resumes the UT-tree construction by inserting the tag sets of the following objects into the UT-tree until the count of tag C achieves N. Fig. 3(b) shows the constructed UT-tree after the tag set of O_6 is inserted. Then the process continues to find other specialized query expansions by growing from tag A. Finally, the discovered QE_top_k includes $\{A, B\}$, $\{A, D\}$, and $\{A, C\}$, which is the same with the result of performing the UT-growth Algorithm. For the other possible specialized query expansions such as $\{A, B, D\}$ and $\{A, C, F\}$, after reading O_6, it is not necessarily that their popularities in O_q are less than N. However, we can guarantee that their utilities must be less than the minimum utility of the discovered QE_top_k, i.e. 1.1, because $UUB^6(\{A, B, D\})= 0.9 + 0.1 = 1$ and $UUB^6(\{A, C, F\})= 0.6+0.1=0.7$.

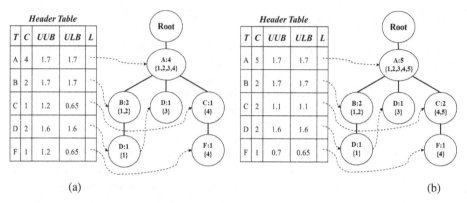

(a)   (b)

Fig. 3. The dynamically constructed *UT*-tree for the example

5   Performance Evaluation

5.1   Experiment Setup

A systematic study is performed to evaluate the effectiveness and efficiency of the proposed *UT-growth* and *dynamic UT-growth* algorithms. In the experiments, we used the web image dataset created by NUS's Lab. The dataset includes 269,641 images and the associated tags from Flickr. There are a total of 407,982 unique tags annotated for these images. The utility of each image is simulated by a random number between 0 and 1. Besides, the Delicious T140 data set is obtained from the site of NLP Group of UNED, which contains 144,574 page links and 67,218 various tags. The number of accessed times of each link is used to compute the utility of the web page. In data preprocessing, we used the Porter's stemming algorithm to transfer the tags which are plurals to their singular forms.

The tags in the datasets are separated into five groups: the frequency in 100~3000, 3000~6000, 6000~9000, 9000~12000, and >12000. Two tags are selected randomly from each group of tags and totally generate 10 test queries.

Furthermore, a set of simulation data sets is generated by the IBM Data Generator, where the number of various tags(*tnum*), the average size of a tag set(*tsize*), and the number of objects(*onum*) are controlled, respectively. The simulation data sets are used to systematically observe the effects of the various parameters for generating the data sets on the performance efficiency of the proposed algorithms.

5.2   Performance Evaluation

5.2.1   Effectiveness Evaluation

In order to evaluate the effectiveness of the discovered QE_top_k on search and navigation, three evaluation metrics provided in [9] are used: 1) coverage, 2) overlap, and 3) selectivity. The coverage value denotes the percentage of the objects in O_q which match any expansion in the discovered QE_top_k; the overlap value denotes the average degree of redundancy on the returned objects between each pair of expansions in the QE_top_k; the selectivity value denotes the average percentage of the filtered out objects for a query by the expansions in the discovered QE_top_k.

In order to ensure that the tags in the QE_top_k are semantically related to the query, we applied the Novelty method proposed in [9] to select 20 representative tags from O_q for each test query q. Only the representative tags are used to generate the specialized query expansions. The default values of k and N are 10 and 5, respectively.

[**Exp. 1-1**] In this experiment, the goal is to compare the search and navigation effectiveness of the discovered QE_top_k by performing representative tags selection before finding the QE_top_k with the one without tags selection.

The results of the three evaluation metrics on the two real data sets are shown in Tab. 2. The results indicate that, by performing representative tags selection, the discovered QE_top_k has higher coverage, lower overlap, and higher selectivity on both the real data sets. Accordingly, representative tags selection is performed before finding the QE_top_k in the following experiments. The coverage obtained on the Delicious data set is higher than that on the Flickr data set because the distribution of tags in the Delicious data set is denser than the one in the Flickr data set. That is also the reason that the overlap metric gets higher value on the Delicious data set.

Table 2. The experiment results of Exp. 1-1

	Flickr Data Set		Delicious Data Set	
	without tag selection	with tag selection	without tag selection	with tag selection
Coverage	0.11	**0.42**	**0.66**	**0.72**
Overlap	0.28	**0.18**	0.29	**0.26**
Selectivity	0.10	**0.38**	0.56	**0.62**

[**Exp. 1-2**] In this experiment, the goal is to observe the search and navigation effectiveness of the discovered QE_top_k on the top-*num* objects in O_q.

The results of the three evaluation metrics are computed for the discovered QE_top_k by varying the values of *num*, which are shown in Fig 4(a) and 4(b) for the Flickr and Delicious data sets, respectively. The top_*num* "all" means all the objects in O_q. The results show that, at least 80% of the top 50 high utility objects can be found through browsing the matched objects of the 10 expansions in QE_top_{10}.

5.2.2 Efficiency Evaluation

In addition to implement the proposed *UT-growth* and *dynamic UT-growth* algorithms, we also implemented the Naïve algorithm proposed in [7] as the baseline method. The execution times of the three algorithms are compared. The execution time of the algorithms on the Flickr data set, which include the tag selection time and the mining time for discovering the QE_top_k are shown separately in Tab. 3. Besides, the number of tags(*tnum*) and the average size of a tag set(*tsize*) are varied when generating the synthesis datasets to observe the execution times of the algorithms as shown in Fig. 4(c) and 4(d), respectively. Each synthesis data set consists of 5000 objects with utilities in [0, 1] randomly. Fig. 4(e) and 4(f) show the influence of the running time parameter k and N on the execution time of the algorithms.

Table 3. The execution time of the algorithms on the Flickr dataset (ms)

		Baseline	UT-growth	dynamic UT-growth
Flickr dataset	tag selection time	298.4	298.4	298.4
	mining time	623.8	69.5	48.2
	Total execution time	922.2	367.9	346.6

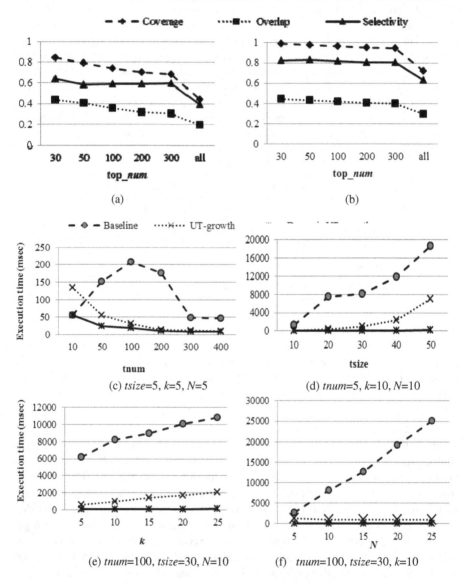

Fig. 4. The execution time of the proposed algorithms and the baseline method

Overall, the performance efficiencies of the *UT*-growth and dynamic *UT*-growth algorithms are better than the baseline method. The data sets become denser when *tnum* is smaller. In a dense dataset, the discovered expansions in QE_top_k tend to be larger than the ones got from a sparse data set. That is why the *UT*-growth and dynamic *UT*-growth algorithms requires more execution time when *tnum* is smaller. However, the number of tags is usually large in a social tagging system. The proposed algorithm can significantly improve the performance efficiency when *tnum* is no less than 50. Without surprising, the execution times of the three algorithms increase as *tsize*, *k*, and *N* increase respectively. It is worth noting that the baseline method has to read in more objects and enumerate more query expansions as *N* increases, which leads to the execution time of the baseline method increases. In contrast, the influence of changing *N* on the execution time is less obvious for both the *UT*-growth and dynamic *UT*-growth than the baseline method.

6   Conclusion and Future Work

In this paper, we formulate the top-*k* high utility specialized query expansions discovering problem. The *UT*-tree structure is designed to maintain the tag sets appearing in the search results, which are used to generate the specialized query expansions. The *UT*-growth algorithm is proposed to find the top-*k* high utility specialized query expansions from the *UT*-tree. Furthermore, we exploit the lower bound and upper bound estimations of utilities for a specialized query expansion to reduce the size of the constructed *UT*-tree. By applying this strategy, we propose another dynamitic *UT*-tree algorithm. A systematic performance study is performed to verify the effectiveness and the efficiency of the proposed algorithms.

Although the top-*N* results of the discovered specialized query expansions are not duplicate each other, there are some overlaps among these results. How to further reduce overlapping among the matching objects of the discovered expansions is under our current investigation.

References

1. Carmel, D., Roitman, H., Yom-Tov, E.: Social bookmark weighting for search and recommendation. The VLDB Journal 19, 761–775 (2010)
2. Dang, V., Kumaran, G., Troy, A.: Domain dependent query reformulation for web search. In: The ACM International Conference on Information and Knowledge Management, CIKM (2012)
3. Han, J., Pei, J., Yin, Y.: Mining frequent pattern without candidate generation. In: The ACM International Conference on Management of Data, SIGMOD (2010)
4. Kato, M.P., Sakai, T., Tanaka, K.: Structured query suggestion for specialization and parallel movement: Effect on search behaviors. In: The 21st International Conference on World Wide Web, WWW (2012)
5. Kong, W., Allan, J.: Extracting query facets from search results. In: The International Conference on Research and Development in Information Retrieval, SIGIR (2013)

6. Li, C., Yan, N., Roy, S.B., Lisham, L., Das, G.: Facetedpedia: Dynamic generation of query-dependent faceted interfaces for wikipedia. In: The 19th International Conference on World Wide Web (WWW), pp. 651–660 (2010)
7. Liang, X., Xie, M., Lakshmanan, L.V.S.: Adding structure to top-k: From items to expansions. In: The 20th International Conference on Information and Knowledge Management, CIKM (2011)
8. Ozertem, U., Chapelle, O., Donmez, P., Velipasaoglu, E.: Learning to suggest: A machine learning framework for ranking query suggestions. In: The ACM International Conference on Research and Development in Information Retrieval, SIGIR (2012)
9. Skoutas, D., Alrifai, M.: Tag clouds revisited. In: The 20th ACM International Conference on Information and Knowledge Management, ACM CIKM (2011)
10. Vandic, D., van Dam, J.-W., Hogenboom, F.: A semantic clustering-based approach for searching and browsing tag spaces. In: The 26th ACM Symposium on Applied Computing, SAC (2011)
11. Venetis, P., Koutrika, G., Garcia-Molina, H.: On the selection of tags for tag clouds. In: The ACM International Conference on Web Search and Data Mining, WSDM (2011)

Novel Techniques to Reduce Search Space in Periodic-Frequent Pattern Mining

R. Uday Kiran and Masaru Kitsuregawa

Institute of Industrial Science,
The University of Tokyo, Tokyo, Japan
{uday_rage,kitsure}@tkl.iis.u-tokyo.ac.jp

Abstract. Periodic-frequent patterns are an important class of regularities that exist in a transactional database. Informally, a frequent pattern is said to be periodic-frequent if it appears at a regular interval specified by the user (i.e., periodically) in a database. A pattern-growth algorithm, called PFP-growth, has been proposed in the literature to discover the patterns. This algorithm constructs a *tid*-list for a pattern and performs a complete search on the *tid*-list to determine whether the corresponding pattern is a periodic-frequent or a non-periodic-frequent pattern. In very large databases, the *tid*-list of a pattern can be very long. As a result, the task of performing a complete search over a pattern's *tid*-list can make the pattern mining a computationally expensive process. In this paper, we have made an effort to reduce the computational cost of mining the patterns. In particular, we apply greedy search on a pattern's *tid*-list to determine the periodic interestingness of a pattern. The usage of greedy search facilitate us to prune the non-periodic-frequent patterns with a sub-optimal solution, while finds the periodic-frequent patterns with the global optimal solution. Thus, reducing the computational cost of mining the patterns without missing any knowledge pertaining to the periodic-frequent patterns. We introduce two novel pruning techniques, and extend them to improve the performance of PFP-growth. We call the algorithm as PFP-growth++. Experimental results show that PFP-growth++ is runtime efficient and highly scalable as well.

Keywords: Data mining, pattern mining and periodic behaviour.

1 Introduction

Since the introduction of periodic-frequent patterns in [1], the problem of finding these patterns has received a great deal of attention in data mining [2,3,4,5]. The classic application is market basket analysis, where these patterns can provide useful information pertaining to the sets of items that were not only sold frequently, but also purchased regularly by the customers. The basic model of periodic-frequent patterns is as follows [1].

Let $I = \{i_1, i_2, \cdots, i_n\}$ be the set of items. A set $X = \{i_j, \cdots, i_k\} \subseteq I$, where $j \leq k$ and $j, k \in [1, n]$, is called a pattern (or an itemset). A transaction $t = (tid, Y)$ is a tuple, where tid represents a transaction-id (or timestamp) and

S.S. Bhowmick et al. (Eds.): DASFAA 2014, Part II, LNCS 8422, pp. 377–391, 2014.

Y is a pattern. A transactional database TDB over I is a set of transactions, i.e., $TDB = \{t_1, t_2, \cdots, t_m\}$, $m = |TDB|$, where $|TDB|$ represents the size of TDB in total number of transactions. If $X \subseteq Y$, it is said that t contains X or X occurs in t and such transaction-id is denoted as tid_j^X, $j \in [1, m]$. Let $TID^X = \{tid_j^X, \cdots, tid_k^X\}$, $j, k \in [1, m]$ and $j \leq k$, be the set of all transaction-ids where X occurs in TDB. The **support** of pattern X, denoted as $S(X)$, represents the number of transactions containing X in TDB, i.e., $S(X) = |TID^X|$. Let tid_p^X and tid_q^X, $p, q \in [1, m]$ and $p < q$, be the two consecutive transaction-ids where X has appeared in TDB. The number of transactions (or the time difference) between tid_p^X and tid_q^X can be defined as a **period** of X, say p_i^X. That is, $p_i^X = tid_q^X - tid_p^X$. Let $P^X = \{p_1^X, p_2^X, \cdots, p_r^X\}$, $r = |TID^X| + 1$, be the complete set of all periods of X in TDB. The **periodicity** of X, denoted as $Per(X) = max(p_1^X, p_2^X, \cdots, p_r^X)$. (It was argued in the literature that the largest *period* of a pattern can provide the upper limit of its periodic occurrence characteristic.) The pattern X is said to be **frequent** if $S(X) \geq minSup$, where $minSup$ represents the user-defined *minimum support*. The frequent pattern X is said to be **periodic-frequent** if $Per(X) \leq maxPer$, where $maxPer$ represents the user-defined *maximum periodicity*. **Please note that the *support* and *periodicity* of a pattern can also be expressed in percentage of $|TDB|$.** We now explain the model using the transactional database shown in Table 1.

Table 1. Transactional database

TID	Items	TID	Items	TID	Items	TID	Items	TID	Items
1	a, b	3	c, e, f, j	5	b, c, d	7	a, b, i	9	a, e, f, g
2	a, c, d, i	4	a, b, f, g, h	6	d, e, f	8	c, d, e	10	a, b, c

Example 1. The database shown in Table 1 contains 10 transactions. Therefore, $|TDB| = 10$. Each transaction in this database is uniquely identifiable with a transaction-id (*tid*), which also represents the timestamp of corresponding transaction. The set of items, $I = \{a, b, c, d, e, f, g, h, i, j\}$. The set of items '$a$' and '$b$', i.e., $\{a, b\}$ is a pattern. It is a 2-pattern. For the purpose of simplicity, we represent this pattern as 'ab'. The pattern 'ab' occurs in *tids* of $1, 4, 7$ and 10. Therefore, the list of *tids* containing 'ab' (or *tid*-list of 'ab'), i.e., $TID^{ab} = \{1, 4, 7, 10\}$. The *support* of '$ab$', i.e., $S(ab) = |TID^{ab}| = 4$. The complete set of all periods for this pattern are: $p_1 = 1 \ (= 1 - tid_i)$, $p_2 = 3 \ (= 4 - 1)$, $p_3 = 3 \ (= 7 - 4)$, $p_4 = 3 \ (= 10 - 7)$ and $p_5 = 0 \ (= 10 - tid_l)$, where $tid_i = 0$ represents the initial transaction and $tid_l = |TDB| = 10$ represents the *tid* of last transaction in the transactional database. The *periodicity* of 'ab', denoted as $Per(ab) = max(1, 3, 3, 3, 0) = 3$. If the user-specified $minSup = 2$, then 'ab' is a frequent pattern because $S(ab) \geq minSup$. If the user-specified $maxPer = 3$, then the frequent pattern 'ab' is a periodic-frequent pattern because $Per(ab) \leq maxPer$.

The periodic-frequent patterns satisfy the *anti-monotonic property*. That is, "all non-empty subsets of a periodic-frequent pattern are also periodic-frequent."

Tanbeer et al. [1] have proposed a pattern-growth algorithm, called Periodic-Frequent Pattern-growth (PFP-growth), to mine the patterns. Briefly, this algorithm compresses the database into a Periodic-Frequent tree (PF-tree), and mines it recursively to discover the patterns. The nodes in PF-tree do not maintain the support count as in FP-tree. Instead, they maintain a list of *tids* (or a *tid*-list) in which the corresponding item has appeared in a database. These *tid*-lists are later aggregated to derive the final *tid*-list of a pattern (i.e., TID^X for pattern X). A complete search on this *tid*-list provides the *support* and *periodicity*, which are later used to determine whether the corresponding pattern is a periodic-frequent or a non-periodic-frequent pattern. In other words, the PFP-growth performs a complete search on a pattern's *tid*-list to determine whether it is a periodic-frequent or a non-periodic-frequent pattern.

In very large databases, the *tid*-list of a pattern can be very long. As a result, the task of performing a complete search on a pattern's *tid*-list can make the pattern mining a computationally expensive process (or the PFP-growth a computationally expensive algorithm).

In this paper, we have made an effort to reduce the computational cost of mining the periodic-frequent patterns. The contributions of this paper are as follows:

1. In this paper, we apply greedy search on a pattern's *tid*-list to determine whether it is a periodic-frequent or a non-periodic-frequent pattern.
2. A novel concept known as *local-periodicity* has been proposed in this paper. For a pattern, the *local-periodicity* corresponds to a sub-optimal solution (i.e., maximum *period* found in a subset of *tid*-list), while the *periodicity* corresponds to the global optimal solution. If the *local-periodicity* of a pattern fails to satisfy the *maxPer*, then we immediately determine the corresponding pattern as a non-periodic-frequent pattern and avoid further search on the *tid*-list to measure its *periodicity*. This results in reducing the computational cost of mining the patterns.
3. Using the concept of *local-periodicity*, we introduce two novel pruning techniques and extend them to improve the performance of PFP-growth. We call the algorithm as PFP-growth++. The proposed techniques facilitate the PFP-growth++ to prune the non-periodic-frequent patterns with a sub-optimal solution, while finds novel periodic-frequent patterns with a global optimal solution. Thus, we do not miss any knowledge pertaining to periodic-frequent patterns.
4. Experimental results show that PFP-growth++ is runtime efficient and highly scalable as well.

Since the real-world is non-uniform, it was observed that mining periodic-frequent patterns with a single *minSup* and *maxPer* constraint leads to the "rare item problem." At high *minSup*, we miss the patterns involving rare items, and at low *minSup*, combinatorial explosion can occur producing too many patterns. To confront this problem, an effort has been made in [4] to mine the patterns using multiple *minSup* and *maxPer* thresholds. Amphawan et al. [5] have extended PFP-growth algorithm to mine top-k periodic-frequent patterns in a database. As the real-world is imperfect, it was observed that the periodic-frequent pattern mining algorithms fail to discover those interesting frequent patterns whose appearances were almost periodic in the database. Uday and Reddy [2] have introduced *periodic-ratio* to capture the almost periodic behavior of the frequent patterns. Alternatively, Rashid et al. [3] have employed *standard deviation* to assess the periodic behavior of the patterns. The discovered patterns are known as regular frequent patterns. The algorithms used in all of the above works are the extensions of PFP-growth, and therefore, perform a complete search on the *tid*-list of a pattern. Thus, all these algorithms are computationally expensive to use in very large databases. The pruning techniques that are going to be discussed in this paper can be extended to improve the performance of all these algorithms. In this paper, we confine our work to finding periodic-frequent patterns using *minSup* and *maxPer* thresholds.

The rest of the paper is organized as follows. Section 2 describes the PFP-growth algorithm and its performance issues. Section 3 describes the basic idea and introduces the proposed PFP-growth++ algorithm. Section 4 presents the experimental evaluation on both PFP-growth and PFP-growth++ algorithms. Finally, Section 5 concludes the paper.

2   PFP-Growth and Its Performance Issues

2.1   PFP-Growth

The PFP-growth involves two steps: (*i*) Construction of PF-tree and (*ii*) Recursive mining of PF-tree to discover the patterns. Before explaining these two steps, we describe the structure of PF-tree as we also employ similar *tree* structure to discover the patterns.

Structure of PF-tree. The structure of PF-tree contains PF-list and prefix-tree. The PF-list consists of three fields – item name (i), support (f) and periodicity (p). The structure of prefix-tree is same as that of the prefix-tree in FP-tree. However, please note that the nodes in the prefix-tree of PFP-tree do not maintain the support count as in FP-tree. Instead, they explicitly maintain the occurrence information for each transaction in the tree by keeping an occurrence transaction-id, called *tid*-list, only at the last node of every transaction. Two types of nodes are maintained in a PF-tree: ordinary node and *tail*-node. The former is the type of nodes similar to that used in FP-tree, whereas the latter is the node that represents the last item of any sorted transaction. The *tail*-node structure is of form $I[tid_1, tid_2, \cdots, tid_n]$, where I is the node's item name and tid_i, $i \in [1, n]$, (n be the total number of transactions from the *root* up to the node) is a *tid* where item I is the last item.

Construction of PF-tree. The PFP-growth scans the database and discover periodic-frequent items (or 1-patterns) using Algorithm 1. Figure 1(a), (b) and (c) respectively show the PF-list generated after scanning the first, second and every transaction in the database (lines 2 to 11 in Algorithm 1). To reflect

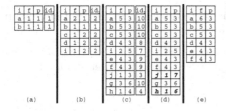

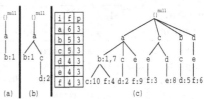

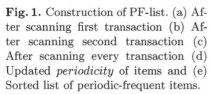

Fig. 1. Construction of PF-list. (a) After scanning first transaction (b) After scanning second transaction (c) After scanning every transaction (d) Updated *periodicity* of items and (e) Sorted list of periodic-frequent items.

Fig. 2. Construction of PF-tree. (a) After scanning first transaction (b) After scanning second transaction and (c) After scanning every transaction.

the correct *periodicity* for an item, the p_{cur} value of every item in PF-list is re-calculated by setting $t_{cur} = |TDB|$ (line 12 in Algorithm 1). Figure 1(d) shows the updated *periodicity* of items in PF-tree. It can be observed that the *periodicity* of 'j' and 'h' items have been updated from 3 and 4 to 7 and 6, respectively. The items having $f < minSup$ or $p > maxPer$ are considered as non-periodic-frequent items and pruned from the PF-list. The remaining items are considered as periodic-frequent items and sorted in descending order of their f (or support) value (line 13 in Algorithm 1). Figure 1(e) shows the sorted list of periodic-frequent items. Let PI denote the sorted set of periodic-frequent items.

Using the FP-tree construction technique, only the items in the PI will take part in the construction of PF-tree. The *tree* construction starts by inserting the first transition, '$1 : a, b$', according to PF-list order, as shown in Figure 2(a). The tail-node '$b : 1$' carries the *tid* of the transaction. After removing the non-periodic-frequent item 'i', the second transaction is inserted into the *tree* with node '$d : 2$' as the tail-node (see Figure 2(b)). After inserting all the transactions in the database, we get the final PF-tree as shown in Figure 2(c). For the simplicity of figures, we do not show the node traversal pointers in trees, however, they are maintained in a fashion like FP-tree does.

Mining PF-tree. To discover the patterns from PF-tree, PFP-growth employs the following steps:

i. Choosing the last item 'i' in the PF-tree as an initial suffix item, its prefix-tree (denoted as PT_i) constituting with the prefix sub-paths of nodes labeled 'i' is constructed. Figure 3(a) shows the prefix-tree for the item 'f', say PT_f.

ii. For each item 'j' in PT_i, we aggregate all of its node's *tid*-list to derive the *tid*-list of the pattern 'ij', i.e., TID^{ij}. Next, we perform a complete search on TID^{ij} to measure the *support* and *periodicity* of the pattern 'ij'. Next, we determine whether 'ij' is a periodic-frequent pattern or not by comparing its *support* and *periodicity* against $minSup$ and $maxPer$, respectively. If 'ij' is a periodic-frequent pattern, then we consider 'j' is periodic-frequent in PT_i.

Algorithm 1. PF-list (TDB: transactional database, $minSup$: minimum support and $maxPer$: maximum periodicity)

---

1. Let t_{cur} denote the tid of current transaction. Let id_l be a temporary array that explicitly records the $tids$ of last occurring transactions of all items in the PF-list.
2. **for** each transaction t_{cur} in TDB **do**
3.   **if** t_{cur} is i's first occurrence **then**
4.   Set $f = 1$, $id_l = t_{cur}$ and $p = t_{cur}$.
5.   **else**
6.   Set $f = f + 1$, $p_{cur} = t_{cur} - id_l$ and $id_l = t_{cur}$.
7.   **if** $p_{cur} > p$ **then**
8. $p = p_{cur}$.
9.   **end if**
10.   **end if**
11. **end for**
12. Calculate p_{cur} value of every item in the list as $|TDB| - id_l$. Next, update the p value of every item with p_{cur} if $p_{cur} > p$.
13. Prune the items in the PF-list that have $f < minSup$ or $p > maxPer$. Consider the remaining items as periodic-frequent items and sort them with respect to their f value.

Example 2. Let us consider the last item 'e' in the PT_f. The set of $tids$ containing 'e' in PT_f is $\{3, 6, 9\}$. Therefore, the tid-list of the pattern 'ef', i.e., $TID^{ef} = \{3, 6, 9\}$. A complete search on TID^{ef} gives $S(ef) = 3$ and $Per(ef) = 3$. Since $S(ef) \geq minSup$ and $Per(ef) \leq maxPer$, the pattern 'ef' is considered as a periodic-frequent pattern. In other words, 'e' is considered as a periodic-frequent item in PT_f. Similar process is repeated for the other items in PT_f. The PF-list in Figure 3(a) shows the *support* and *periodicity* of each item in PT_f.

iii. Choosing every periodic-frequent item 'j' in PT_i, we construct its conditional tree, CT_i, and mine it recursively to discover the patterns.

Example 3. Figure 3(b) shows the conditional-tree, CT_f, derived from PT_f. It can be observed that the items 'a', 'b', 'c' and 'd' in PT_f are not considered in the construction of CT_f. The reason is that they are non-periodic-frequent items in PT_f.

iv. After finding all periodic-frequent patterns for a suffix item 'i', we prune it from the original PF-tree and push the corresponding nodes' tid-lists to their parent nodes. Next, once again we repeat the steps from i to iv until the PF-list $= \emptyset$.

Example 4. Figure 3(c) shows the PF-tree generated after pruning the item 'f' in Figure 2(c). It can be observed that the tid-list of all the nodes containing 'f' have been pushed to their corresponding parent nodes.

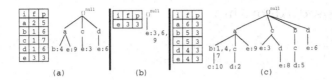

Fig. 3. Mining periodic-frequent patterns using 'f' as a suffix item. (a) Prefix-tree of f, i.e., PT_f (b) Conditional tree of 'f', i.e., CT_f and (c) PF-tree after removing item 'f'.

2.2   Performance Issues

We have observed that PFP-growth suffers from the following two performance issues:

1. The PFP-growth scans the database and constructs the PF-list with every item in the database. The non-periodic-frequent items are pruned from the list only after the scanning of database. We have observed that this approach can cause performance problems, which involves the increased updates and search costs for the items in the PF-list.

   Example 5. In Table 1, the items 'g' and 'h' have initially appeared in the transaction whose $tid = 4$. Thus, their first period is going to be 4, which is greater than the $maxPer$. In other words, these two items were non-periodic-frequent by the time they were first identified in the database. Thus, these two items need not have been included in the construction of PF-list. However, PFP-growth considers these items in the construction of PF-list. This results in the performance problems, which involves increased updates and search cost for the items in the PF-list.

2. Another performance issue of PFP-growth lies at the Step ii of mining PF-tree. That is, for every item 'j' in PT_i, PFP-growth performs a complete search on its tid-list to determine whether it is periodic-frequent or not. In very large databases, the tid-list of a pattern (or for an item 'j' in PT_i) can be generally long. In such cases, performing a complete search on a pattern's tid-list to determine whether it is periodic-frequent or not can be a computationally expensive process. Thus, PFP-growth is a computationally expensive algorithm.

   Example 6. Let us consider the item 'a' in PT_f. The tid-list of 'a' in PT_f, i.e., $TID^{af} = \{4, 9\}$. Its periods are: $p_1^{af} = 4 \ (= 4 - tid_i)$, $p_2^{af} = 5 \ (= 9 - 4)$ and $p_3^{af} = 1 \ (= tid_l - 9)$. The PFP-growth measures $periodicity = 5 \ (= max(4, 5, 1))$, and then determines '$a$' is not a periodic-frequent item in PT_f. In other words, PFP-growth performs a complete search on the tid-list of 'af' to determine it is not a periodic-frequent pattern. However, such a complete search was not necessary to determine 'af' as a non-periodic-frequent pattern. It is because its first period, i.e., $p_1^{af} > maxPer$.

In the next section, we discuss our approach to address the above two performance issues of PFP-growth.

3   Proposed Algorithm

In this section, we first describe our basic idea. Next, we explain our PFP-growth++ algorithm to discover the patterns.

3.1   Basic Idea: The Local-Periodicity of a Pattern

Our idea to reduce the computational cost of mining the patterns is as follows.

> " Apply greedy search on a pattern's *tid*-list to derive its *local-periodicity*. For a pattern, the *local-periodicity* represents a sub-optimal solution, while the *periodicity* corresponds to the global optimal solution. If the *local-periodicity* of a pattern fails to satisfy the *maxPer*, then we immediately determine it as a non-periodic-frequent pattern, and avoid further search on the *tid*-list to measure its actual *periodicity*. Thus reducing the computational cost of mining the patterns."

Definition 1 defines the *local-periodicity* of a pattern X. Example 7 illustrates the definition. The correctness of our idea is shown in Lemma 1, and illustrated in Example 8.

Definition 1. *(Local-periodicity of pattern X.) Let $P^X = \{p_1^X, p_2^X, \cdots, p_n^X\}$, $n = S(X) + 1$, denote the complete set of periods for X in TDB. Let $\widehat{P^X} = \{p_1^X, p_2^X, \cdots, p_k^X\}$, $1 \le k \le n$, be an ordered set of periods of X such that $\widehat{P^X} \subseteq P^X$. The local-periodicity of X, denoted as loc-per(X), refers to the maximum period in $\widehat{P^X}$. That is, $loc\text{-}per(X) = max(p_1^X, p_2^X, \cdots, p_k^X)$.*

Example 7. Continuing with Example 1, the set of all *periods* for 'ab', i.e., $P^{ab} = \{1, 3, 3, 0\}$. Let $\widehat{P^{ab}} = \{1, 3\} \subset P^{ab}$. The local-periodicity of 'ab', denoted as $loc\text{-}per(ab) = max(p_j^{ab} | \forall p_j^{ab} \in \widehat{P^{ab}}) = max(1, 3) = 3$.

Lemma 1. *For the pattern X, if loc-per(X) > maxPer, then X is a non-periodic-frequent pattern.*

Proof. For the pattern X, $Per(X) \ge loc\text{-}per(X)$ as $\widehat{P^X} \subseteq P^X$. Therefore, if $loc\text{-}per(X) > maxPer$, then $Per(X) > maxPer$. Hence proved.

Example 8. In Table 1, the pattern 'af' occurs in *tids* of 4 and 9. Therefore, $TID^{af} = \{4, 9\}$. The first *period* of 'af,' i.e., $p_1^{af} = 4 \ (= tid_i - 4)$. At this point, the $loc\text{-}per(af) = p_1^{af} = 4$. Since $loc\text{-}per(af) > maxPer$, it is clear that 'af' is a non-periodic-frequent pattern as $Per(af) \ge loc\text{-}per(af) > maxPer$.

If X is a periodic-frequent pattern, then its *local-periodicity* equals the *periodicity*. Thus, we do not miss any knowledge pertaining to periodic-frequent patterns. The correctness of this argument is based on Property 1, and shown in Lemma 2.

Property 1. For a pattern X, $loc\text{-}per(X) \le Per(X)$ as $\widehat{P^X} \subseteq P^X$.

Lemma 2. *If X is a periodic-frequent pattern, then $loc\text{-}per(X) = Per(X)$.*

Proof. If X is a periodic-frequent pattern, then we perform a complete search on its *tid*-list, i.e., TID^X. Thus, $\widehat{P^X} = P^X$. From Property 1, it turns out that $loc\text{-}per(X) = Per(X)$. Hence proved.

Overall, our idea of using the greedy search technique facilitates the user to find the periodic-frequent patterns with an optimal solution, while pruning the non-periodic-frequent patterns with a sub-optimal solution. Thus, our idea reduces the computational cost of mining the patterns without missing any knowledge pertaining to periodic-frequent patterns.

Two novel pruning techniques have been developed based on the concept of *local-periodicity*. The first pruning technique addresses the issue of pruning non-periodic-frequent items (or 1-patterns) effectively. The second pruning technique addresses the issue of pruning non-periodic-frequent k-patterns, $k \geq 2$. These two techniques have been discussed in subsequent subsection.

3.2 PFP-Growth++

The proposed algorithm also involves the following steps: (i) construction of PF-tree++ and (ii) Mining PF-tree++ recursively to discover the patterns. We now discuss each of these steps.

Construction of PF-tree++. The structure of PF-tree++ consists of two components: (i) PF-list++ and (ii) prefix-tree. The PF-list++ consists of three fields – item name (i), total support (f) and *local-periodicity* (p^l). **Please note that PF-list++ do not explicitly store the** *periodicity* **of an item** i **as in the PF-list.** The structure of prefix-tree in PF-tree++, however, remains the same as in PF-tree. The structure of prefix-tree has been discussed in Section 2.

Since periodic-frequent patterns satisfy the anti-monotonic property, periodic-frequent items (or 1-patterns) play a key role in discovering the patterns effectively. To discover these items, we employ a pruning technique which is based on the concepts of 2-Phase Locking [6] (i.e., 'expanding phase' and 'shrinking phase'). We now discusses the pruning technique:

- **Expanding phase:** In this phase, we insert every new item found in a transaction into the PF-list++. Thus, expanding the length of PF-list++. This phase starts from the first transaction (i.e., $tid = 1$) in the database and ends when the tid of the transaction equals to $maxPer$ (i.e., $tid = maxPer$). If the tid of a transaction is greater than the $maxPer$, then we do not insert any new items found in the corresponding transaction into the PF-list++. It is because these items are non-periodic-frequent items as their first *period* (or local-periodicity) fails to satisfy the user-defined $maxPer$ threshold.
- **Shrinking phase:** In this phase, we delete the non-periodic-frequent items from the PF-list++. Thus, shrinking the length of PF-list++. The non-periodic-frequent items are those items that have a *period* (or *local-periodicity*)

greater than the $maxPer$. This phase starts right after the completion of expansion phase (i.e., when the tid of a transaction equal to $maxPer + 1$) and continues until the end of database.

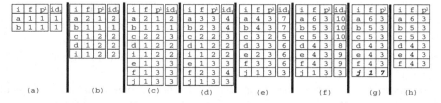

Fig. 4. The construction of PF-list++. (a) After scanning the first transaction (b) After scanning the second transaction (c) After scanning the third transaction (d) After scanning the fourth transaction (e) After scanning the seventh transaction (f) After scanning all transactions (g) Measuring actual *periodicity* for items and (h) The sorted list of periodic-frequent items.

Algorithm 2 shows the construction of PF-list++ using the above two phases. We illustrate the algorithm using the database shown in Table 1. The 'expanding phase' starts from the first transaction. The scan on the first transaction with $t_{cur} = 1$ results in inserting the items 'a' and 'b' in to the list with $f = 1$, $p^l = 1$ and $id_l = 1$ (lines 4 to 6 in Algorithm 2). The resultant PF-list++ is shown in Figure 4(a). The scan on the second transaction with $t_{cur} = 2$ results in adding the items 'c', 'd' and 'i' into the PF-list++ with $f = 1$, $id_l = 2$ and $p^l = 2$. Simultaneously, the f and id_l values of 'a' are updated to 2 and 2, respectively (lines 7 and 8 in Algorithm 2). The resultant PF-list++ was shown in Figure 4(b). Similarly, the scan on the third transaction with $t_{cur} = 3$ results in adding the items 'e', 'f' and 'j' into the PF-list++ with $f = 1$, $id_l = 3$ and $p^l = 3$. In addition, the f, p and id_l values of 'c' are updated to 2, 2 and 3, respectively. Figure 4(c) shows the PF-list++ generated after scanning the third transaction. Since $maxPer = 3$, the expanding phase ends at $t_{cur} = 3$. The 'shrinking phase' begins from $t_{cur} = 4$. The scan on the fourth transaction, '$4 : a, b, f, g, h$', updates the f, p^l and id_l of the items 'a', 'b' and 'f' accordingly as shown in Figure 4(d). Please observe that we have **not added** the new items 'g' and 'h' in the PF-list++ as their *period* (or *local-periodicity*) is greater than the $maxPer$ (lines 11 to 16 in Algorithm 2). Similar process is repeated for the other transactions in the database, and the PF-list++ is constructed accordingly. Figure 4(e) shows the PF-list++ constructed after scanning the seventh transaction. It can be observed that the non-periodic-frequent item 'i' was pruned from the PF-list++. It is because its *local-periodicity* (or $p^l = 5 \, (= t_{cur} - id_l)$) has failed to satisfy the $maxPer$ (line 15 in Algorithm 2). Figure 4(f) shows the **initial PF-list++** constructed after scanning the entire database. Figure 4(g) shows the updated *periodicity* of all items in the PF-list++ (line 19 in Algorithm 2). It can be observed that the p^l value of 'j' has been updated from 3 to 7. Figure 4(h) shows the set of periodic-frequent items sorted in descending order of their frequencies. Let PI denote this sorted list of periodic-frequent items.

Using FP-tree construction technique, we perform another scan on the database and construct prefix-tree in PI order. The construction of prefix-tree has been discussed in Section 2. Figure 5 shows the PF-tree++ generated after scanning the database shown in Table 1. Since the *local-periodicity* of a periodic-frequent item (or 1-pattern) is same as its *periodicity* (see Lemma 2), the constructed PF-tree++ resembles that of the PF-tree in PFP-growth. However, there exists a key difference between these two *trees*, which we will discuss while mining the patterns from PF-tree++.

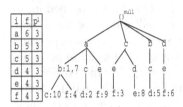

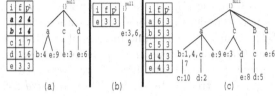

Fig. 5. The PF-tree++ generated after scanning every transaction in the database

Fig. 6. Mining periodic-frequent patterns using 'f' as suffix item. (a) Prefix-tree of suffix item 'f', i.e., PT_f (b) Conditional tree of suffix item 'f', i.e., CT_f and (c) PF-tree++ after pruning item 'f'.

Mining PF-tree++. Choosing the last item 'i' in the PF-list++, we construct its prefix-tree, say PT_i, with the prefix sub-paths of nodes labeled 'i' in the PF-tree++. Since 'i' is the bottom-most item in the PF-list++, each node labeled 'i' in the PF-tree++ must be a *tail*-node. While constructing the PT_i, we map the *tid*-list of every node of 'i' to all items in the respective path explicitly in a temporary array. It facilitates the construction of *tid*-list for each item 'j' in the PF-list++ of PT_i, i.e., TID^{ij}. The length of TID^{ij} gives the *support* of 'ij'. Algorithm 3 is used to measure the *local-periodicity* of 'ij.' This algorithm applies greedy search on the *tid*-list to identify whether j is periodic-frequent or not in PT_i. The pruning technique used in this algorithm is as follows:

"If $loc\text{-}per(X) > maxPer$, then X is a non-periodic-frequent pattern."

If j is periodic-frequent in PT_i, then its p^l denotes its actual *periodicity* in the database. However, if 'j' is non-periodic-frequent in PT_i, then its p^l denotes the *local-periodicity* (or the *period*) which has failed to satisfy the $maxPer$. The correctness of this technique has already been shown in Lemma 2.

Figure 6(a) shows the prefix-tree of item 'f', PT_f. The set of items in PT_f are 'a', 'b', 'c', 'd' and 'e.' Let us consider an item 'a' in PT_f. The *tid*-list of the nodes containing 'a' in PT_f gives $TID^{af} = \{4, 9\}$. The *support* of 'af', $S(af) = 2(= |TID^{af}|)$. As $S(af) \geq 2$ $(= minSup)$, we determine 'af' as a frequent pattern. Next, we pass TID^{af} as an array in Algorithm 3 to find whether 'af' is a periodic-frequent or a non-periodic-frequent pattern. The first period, $p_{cur} = 4 (= 4 - 0)$ (line 1 in Algorithm 3). As $p_{cur} > maxPer$, we determine 'af' as a non-periodic-frequent pattern and return $p^l = p_{cur} = 4$ (lines 2 to 4 in Algorithm 3). Thus, we prevent the complete search on the tid-list of a non-periodic-frequent pattern. Similar process is applied for the remaining items in the PT_f. The PF-list++ in Figure 6(a) shows the *support* and *local-periodicity* of items in PT_f. It can be observed that the p^l value of non-periodic-frequent items, 'a' and 'b', in PT_f are set to 4 and 4, respectively. Please note that these values are not their actual *periodicity* values. The actual *periodicity* of 'a' and 'b' in PT_f are 5 and 6, respectively (see Figure 3(a)). This is the key difference between the PFP-tree++ and PFP-tree.

The conditional tree, CT_i, is constructed by removing all non-periodic-frequent items from the PT_i. If the deleted node is a *tail*-node, its tid-list is pushed up to its parent node. Figure 6(b), for instance, shows the conditional tree for 'f', say CT_f, from PT_f. The same process of creating prefix-tree and its corresponding conditional tree is repeated for the further extensions of 'ij'. Once the periodic-frequent patterns containing 'f' are discovered, the item 'f' is removed from the original PF-tree++ by pushing its node's tid-list to its respective parent nodes. Figure 6(c) shows the resultant PF-tree++ after pruning the item 'f'. The whole process of mining for each item in original PF-tree++ is repeated until its PF-list++$\neq \emptyset$. The above bottom-up mining technique on support-descending PF-tree++ is efficient, because it shrinks the search space dramatically with the progress of mining process.

4   Experimental Results

The algorithms, PFP-growth and PFP-growth++, are written in Java and run with Ubuntu 10.04 operating system on a 2.66 GHz machine with 4GB memory. The runtime specifies the total execution time, i.e., CPU and I/Os. We pursued experiments on synthetic (T10I4D100K and T10I4D1000K) and real-world (Retail and Kosarak) datasets. The T10I4D100K dataset contains 100,000 transactions with 1000 items. The T10I4D1000K dataset contains 1,000,000 transactions with 1000 items. The Retail dataset [7] contains 88,162 transactions with 16,470 items. The Kosarak dataset is a very large dataset containing 990,002 transactions and 41,270 distinct items.

The $maxPer$ threshold varies from 0% to 100%. In this paper, we vary the $maxPer$ threshold from 1% to 10%. The reason is as follows. Very few (almost nil) periodic-frequent patterns are discovered in these databases for the $maxPer$ values less than the 1%. Almost all frequent patterns are discovered as periodic-frequent patterns when the $maxPer$ values are greater than the 10%.

4.1   Discovering Periodic-Frequent Patterns

Figure 7(a) and (b) respectively show the number of periodic-frequent patterns generated in T10I4D100K and Retail datasets at different $maxPer$ thresholds.

Algorithm 2. PF-list++ (*TDB*: transactional database, *minSup*: minimum support and *maxPer*: maximum periodicity)

1. **for** each transaction $t \in TDB$ **do**
2.   **if** $t_{cur} < maxPer$ **then**
3.   /*Expanding phase*/
4.   **if** t_{cur} is i's first occurrence **then**
5.   Insert i into the list and set $f = 1$, $id_l = t_{cur}$ and $p^l = t_{cur}$.
6.   **else**
7.   Calculate $p_{cur} = t_{cur} - id_l$. Set $f = f + 1$, $id_l = t_{cur}$ and $p^l = (p_{cur} > p^l)?p_{cur} : p^l$.
8.   **end if**
9.   **else**
10.   /*Shrinking phase*/
11.   Calculate $p_{cur} = t_{cur} - id_l$.
12.   **if** $p_{cur} < maxPer$ **then**
13.   Set $f = f + 1$, $id_l = t_{cur}$ and $p^l = (p_{cur} > p^l)?p_{cur} : p^l$.
14.   **else**
15.   Remove i from PF-list++.
16.   **end if**
17.   **end if**
18. **end for**
19. For each item in the PF-list++, we re-calculate p_{cur} value as $|TDB| - id_l$, and prune the non-periodic-frequent items.

The *minSup* values are set at 0.01% and 0.01% in T10I4D100K and Retail datasets, respectively. The usage of a low *minSup* value in these datasets facilitate us to discover periodic-frequent patterns involving both frequent and rare items. It can be observed that the increase in *maxPer* has increased the number of periodic-frequent patterns. It is because some of the periods of a patterns that are earlier considered aperiodic have been considered periodic with the increase in *maxPrd* threshold.

4.2   Runtime for Mining Periodic-Frequent Patterns

Figure 8(a) and (b) shows the runtime taken by PFP-growth and PFP-growth++ algorithms to discover periodic-frequent patterns at different *maxPer* thresholds in T10I4D100K and Retail datasets, respectively. The following three observations can be drawn from these figures: (*i*) Increase in *maxPer* threshold has increased the runtime for both the algorithms. It is because of the increase in number of periodic-frequent pattern with the increase in *maxPer* threshold. (*ii*) At any *maxPer* threshold , the runtime of PFP-growth++ is no more than the runtime of PFP-growth. It is because of the greedy search technique employed by the PFP-growth++ algorithm. (*iii*) At a low *maxPer* value, the PFP-growth++ algorithm has outperformed the PFP-growth by an order of magnitude. It is because the PFP-growth++ has performed only partial search on the *tid*-lists of non-periodic-frequent patterns.

Algorithm 3. CalculateLocalPeriodicity (TID: an array of tid's containing X.)

1. Set $p^l = -1$ and $p_{cur} = TID[0]$ $(= TID[1] - 0)$.
2. **if** $p_{cur} > maxPer$ **then**
3.   return p_{cur}; /*(as p^l value).*/
4. **end if**
5. **for** $i = 1;\ i < TID.length - 1; ++$i **do**
6.   Calculate $p_{cur} = TID[i + 1] - TID[i]$.
7. $p^l = (p_{cur} > p^l)?p_{cur} : p^l$
8.   **if** $p^l > maxPer$ **then**
9.   return p^l;
10.   **end if**
11. **end for**
12. Calculate $p_{cur} = |TDB| - TID[TID.length]$, and repeat the steps numbered from 7 to 10.

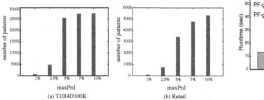

Fig. 7. Periodic-frequent patterns discovered at different $maxPer$ thresholds in various datasets.

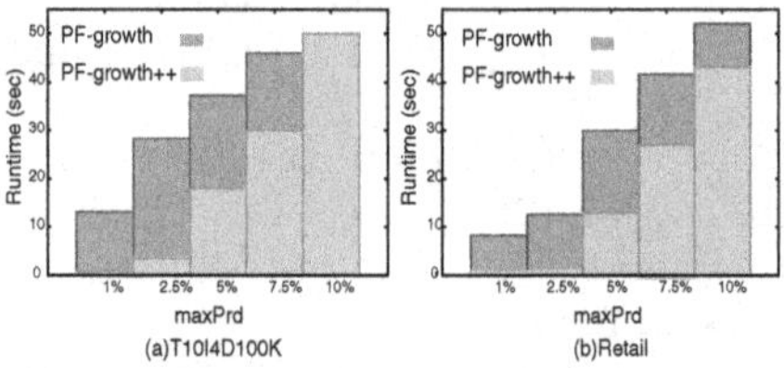

Fig. 8. Runtime comparison of PF-growth and PF-growth++ algorithms at different $maxPer$ thresholds in various datasets.

4.3 Scalability Test

We study the scalability of PFP-growth and PFP-growth++ algorithms on execution time by varying the number of transactions in $T10I4D1000K$ and $Kosarak$ datasets. In the literature, these two datasets were widely used to study the scalability of algorithms. The experimental setup was as follows. Each dataset was divided into five portions with 0.2 million transactions in each part. Then, we investigated the performance of both algorithms after accumulating each portion with previous parts. We fixed the $minSup = 1\%$ and $maxPer = 2.5\%$.

Figure 9(a) and (b) shows the runtime requirements of PFP-growth and PFP-growth++ algorithms in $T10I4D1000K$ and $Kosarak$ datasets, respectively. It can be observed that the increase in dataset size has increased the runtime of both the algorithms. However, the proposed PFP-growth++ has taken relatively less

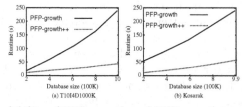

Fig. 9. Scalability of PFP-growth and PFP-growth++ algorithms

runtime than the PFP-growth. In particular, as the dataset size increases, the proposed PFP-growth++ algorithm has outperformed PFP-growth by an order of magnitude. The reason is as follows. The *tid*-list of a pattern gets increased with the increase in the database size. The complete search on the *tid*-list of both periodic-frequent and non-periodic-frequent patterns by PFP-growth has increased its runtime requirements. The partial search on the *tid*-list of a non-periodic-frequent pattern has facilitated the PFP-growth++ to reduce its runtime requirements.

5 Conclusions and Future Work

In this paper, we have employed greedy search technique to reduce the computational cost of mining the periodic-frequent patterns. The usage of greedy search technique facilitated the user to find the periodic-frequent patterns with the global optimal solution, while pruning the non-periodic-frequent patterns with a sub-optimal solution. Thus, reducing the computational cost of mining the patterns without missing any useful knowledge pertaining to periodic-frequent patterns. Two novel pruning techniques have been introduced to discover the patterns effectively. A pattern-growth algorithm, known as PFP-growth++, has been proposed to discover the patterns. Experimental results show that the proposed PFP-growth++ is runtime efficient and scalable as well.

As a part of future work, we would like to extend the proposed concepts to mine partial periodic-frequent patterns in a database. In addition, we would like to investigate alternative search techniques to further reduce the computational cost of mining the patterns.

References

1. Tanbeer, S.K., Ahmed, C.F., Jeong, B.-S., Lee, Y.-K.: Discovering periodic-frequent patterns in transactional databases. In: Theeramunkong, T., Kijsirikul, B., Cercone, N., Ho, T.-B. (eds.) PAKDD 2009. LNCS, vol. 5476, pp. 242–253. Springer, Heidelberg (2009)
2. Kiran, R.U., Reddy, P.K.: An alternative interestingness measure for mining periodic-frequent patterns. In: Yu, J.X., Kim, M.H., Unland, R. (eds.) DASFAA 2011, Part I. LNCS, vol. 6587, pp. 183–192. Springer, Heidelberg (2011)
3. Rashid, M. M., Karim, M. R., Jeong, B.-S., Choi, H.-J.: Efficient mining regularly frequent patterns in transactional databases. In: Lee, S.-g., Peng, Z., Zhou, X., Moon, Y.-S., Unland, R., Yoo, J. (eds.) DASFAA 2012, Part I. LNCS, vol. 7238, pp. 258–271. Springer, Heidelberg (2012)
4. Uday Kiran, R., Krishna Reddy, P.: Towards efficient mining of periodic-frequent patterns in transactional databases. In: Bringas, P.G., Hameurlain, A., Quirchmayr, G. (eds.) DEXA 2010, Part II. LNCS, vol. 6262, pp. 194–208. Springer, Heidelberg (2010)
5. Amphawan, K., Lenca, P., Surarerks, A.: Mining top-k periodic-frequent pattern from transactional databases without support threshold. In: Papasratorn, B., Chutimaskul, W., Porkaew, K., Vanijja, V. (eds.) IAIT 2009. CCIS, vol. 55, pp. 18–29. Springer, Heidelberg (2009)
6. Gray, J.: Notes on data base operating systems. In: Advanced Course: Operating Systems, pp. 393–481 (1978)
7. Brijs, T., Goethals, B., Swinnen, G., Vanhoof, K., Wets, G.: A data mining framework for optimal product selection in retail supermarket data: The generalized profset model. In: KDD, pp. 300–304 (2000)

Inferring Road Type in Crowdsourced Map Services

Ye Ding[1], Jiangchuan Zheng[1], Haoyu Tan[1], Wuman Luo[1], and Lionel M. Ni[1,2]

[1] Department of Computer Science and Engineering
[2] Guangzhou HKUST Fok Ying Tung Research Institute
[3] The Hong Kong University of Science and Technology
{valency,jczheng,hytan,luowuman,ni}@cse.ust.hk

Abstract. In crowdsourced map services, digital maps are created and updated manually by volunteered users. Existing service providers usually provide users with a feature-rich map editor to add, drop, and modify roads. To make the map data more useful for widely-used applications such as navigation systems and travel planning services, it is important to provide not only the topology of the road network and the shapes of the roads, but also the types of each road segment (e.g., highway, regular road, secondary way, etc.). To reduce the cost of manual map editing, it is desirable to generate proper recommendations for users to choose from or conduct further modifications. There are several recent works aimed at generating road shapes from large number of historical trajectories; while to the best of our knowledge, none of the existing works have addressed the problem of inferring road types from historical trajectories. In this paper, we propose a model-based approach to infer road types from taxis trajectories. We use a combined inference method based on stacked generalization, taking into account both the topology of the road network and the historical trajectories. The experiment results show that our approach can generate quality recommendations of road types for users to choose from.

1 Introduction

In recent years, crowdsourced map services has become a powerful competitor to public and commercial map service providers such as Google Maps. Different from commercial map services in which maps are produced from remote sensing images and survey data by a small group of professionals, crowdsourced maps are maintained by tens of thousands of registered users who continuously create and update maps using sophisticated map editors. Therefore, crowdsourced map services can be better in keeping up with recent map changes than existing commercial map services. For instance, it has been reported that OpenStreetMap (OSM) [1], the world's largest crowdsourced mapping project, can provide richer and more timely-updated map data than comparable proprietary datasets [2].

Similar to other crowdsourcing applications, crowdsourced map services rely on lots of volunteered works which are error-prone and can have severe consistency problems. In fact, providing quality maps is far more challenging than

S.S. Bhowmick et al. (Eds.): DASFAA 2014, Part II, LNCS 8422, pp. 392–406, 2014.

most crowdsourcing applications such as reCAPTCHA [3]. One major reason is that map objects (e.g., roads and regions) are usually complex, which makes it difficult to make map editors both feature-rich and user-friendly. To address this issue, a recent work proposed a map updating system called CrowdAtlas[4] to probe map changes via a large number of historical taxi trajectories. CrowdAtlas reduces the cost of drawing roads by generating the shapes of new/changed roads from trajectories automatically. The generated shapes of roads can be used as recommendations in a map editor. A contributor can then directly use the generated roads or slightly adjust them based on his/her own knowledge and experience.

Existing works only focus on generating the shapes of roads automatically. However, the metadata of roads is also important to many map-based applications such as navigation systems and travel planning services. Typical metadata of roads includes width, speed limit, direction restriction and access limit. These metadata can be effectively reflected by the type of road segments [1], which often includes: motorway, primary / secondary way, residential road, etc. For example, the speed limit is often higher of a motorway than a secondary way; a motorway or a primary way is often a two-way street, while a residential road may be a single-way street. Therefore, to contribute to a quality crowdsourced map service, users need to provide not only the shapes of the roads, but also the types of the roads. Consequently, to further reduce the cost of updating crowdsourced maps for users, it is necessary to automate the process of labeling road types.

There are many challenges of inferring the types of road segments. The types may be directly inferred from the topology of the road network, e.g., road segments with the same direction may have the same type. However, it is often not accurate, as in our experiments. Hence, in this paper, we combine the inference based on the topology of the road network, with the real trajectories of vehicles driving on the road segments. Trajectories can effectively show the types of road segments. For example, it is generally believed that vehicles drive faster on motorways than small roads, thus it is possible to infer the type of a road segment as motorway if the average driving speed of it is much faster than other road segments. However, the method of combining two inference methods, as well as the weights between them are very difficult to conduct. Moreover, the types between each other may be ambiguous, and there are no exact definitions to draw the borders of different types, which makes it difficult to find an accurate inference result.

Since it is private and difficult to obtain private vehicle data, we use the trajectories of taxis in this paper. There are many challenges using the trajectories of taxis. For example, the taxi data is very sparse due to the inaccuracy of the GPS device, and we have to filter the inaccurate data through preprocessing. Moreover, since taxis are only a part of all the vehicles in the city, the trajectories of taxis are biased, thus 1) not every road segment has been traversed; and 2) the density of taxis cannot directly show the traffic of the road segment [5].

---

[1] A road consists of a number of road segments which could be of different types. Therefore, the meaning of *road type* is essentially road segment type.

394   Y. Ding et al.

Table 1. Specifications of Trajectory Data

Data Type	Description
Taxi ID	Taxi registration plate number.
Timestamp	Timestamp of the sample point.
Latitude / Longitude	GPS location of the sample point.
Speed	Current speed of the taxi.
Angle	Current driving direction of the taxi.
Status	Indicator of whether the taxi is occupied or vacant.

In this case, we also use the topology of the road network to cover the shortage of using only trajectories. At last, a taxi has its own characteristics which may be different from other vehicles, and we have to consider them in the inference. In this paper, as many works do [6], we consider taxi drivers are experienced drivers, who often choose the fastest routes rather than shortest routes.

The main contributions of this paper are listed as follows:

- First, we propose a novel problem, as inferring the types of road segments;
- Second, we conduct a combined inference method considering both the topology of the road network, and the trajectories of taxis. The results show that our method is much better than baseline methods;
- Third, our method is flexible and scalable, where the models in our method can be replaced by other suitable models when handling different types of data;
- At last, we introduce large-scale real-life trajectories, and a real road network in a large city of China in our experiments.

The rest of this paper is organized as follows. Section 2 describes the dataset we use and formally defines the problem of road type inference. Section 3 presents the methodology in details. Section 4 presents the evaluations results. Section 5 outlines the related works and Section 6 concludes the paper.

2 Data Description and Problem Definition

2.1 Data Description

Our trajectory data is collected in Shenzhen, China in September, 2009 [7]. The data contains the trajectories of around 15,000 taxis for 26 days, and the sampling rate is around 20 seconds. A trajectory is represented as a series of sample points. The details are shown in Table 1.

Our road network data is provided by the government. There are 27 different types of road segments in our data, and they can be generally classified into 7 categories, based on the meanings of types defined by the government. The types

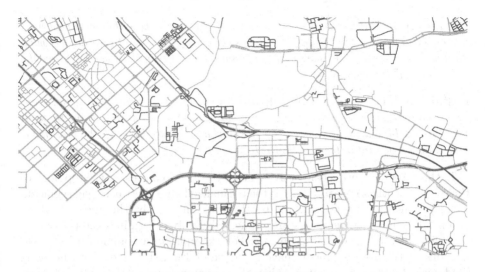

Fig. 1. The types of road segments from our data, the notations of colors are shown in Table 2

Table 2. Types of Road Segments

Type	Type ID	Color	Description
National Expressway	1-100	Red	National limited-access expressway.
City Expressway	100-200	Green	City expressway, often with relief roads.
Regular Highway	200-300	Blue	Regular highway.
Large Avenue	300-400	Orange	Direction-separated large avenue.
Primary Way	400-500	Yellow	The road that connects regions of the city.
Secondary Way	600-700	Aqua	The road that connects blocks of a region.
Regular Road	800-900	Purple	The road that constructs within a block.

are as shown in Table 2 and Figure 1. These types are used as ground truth to verify the accuracy of our model.

2.2 Problem Definition

Before introducing the problem definition, let us first introduce some terminologies used in this paper.

Definition 1 (Road segment). *A road segment τ is the carriageway between two intersections. An expressway or a large avenue may have two different road segments between two intersections, because they are different directions with limited-access.*

Definition 2 (Road network). *A road network $\{\tau_i\}_{i=1}^{n}$ consists of a set of road segments.*

Definition 3 (Taxi status). *The status of a taxi can be either occupied by a passenger or vacant for hiring. The status changes from vacant to occupied when the taxi picks up a passenger, and it changes from occupied to vacant when the taxi drops off the passenger.*

Definition 4 (Pick-up event / drop-off event). *A pick-up event is the event when the taxi picks up a passenger, and changes its status from vacant to occupied. Similarly, a drop-off event is the event when the taxi drops off the passenger, and changes its status from occupied to vacant.*

In this paper, we use a connectivity matrix $M^{n \times n}$ to represent the topology of the road network, where $m_{ij} \in M$ is the normalized angle between road segments i and j if they are connected, and 0 otherwise. Intuitively, the angle at which two neighboring road segments are connected highly determines the relation of the types of these two road segments. For example, in a common urban road network, if two road segments have an angle of 180°, they are often the same road with the same road name. When the angle becomes smaller, like 90°, they are often two different roads with different road names. Similarly, if we drive along a road, the road segments we traveled are often straight, i.e., with large angles up to 180°. When we change to another road, we will have to change the angle of our driving direction, and which will result in a smaller degree than the previous driving angle.

Nevertheless, such information may not always be accurate, since some road segments with the same type are also with 90° angle. Hence, the connectivity matrix is not a unique feature that can be used directly to identify the types, and it is necessary to combine the usage of the connectivity matrix with other features.

The problem, road type inference, is defined as inferring the types of road segments based on the road network and correspondent trajectories. The formal definition is shown in Definition 5.

Definition 5 (Road type inference). *Given a road network $R = \{\tau_i\}_{i=1}^n$, each road segment τ_i is associated with a feature vector $f_i = \langle f_i^1, f_i^2, \ldots, f_i^k \rangle$, and a connectivity variable $m_{i,j} = \Pr(\tau_i = y_{1...l}|\tau_j = y_{1...l})$ towards another road segment τ_j, infer the type $y_i \in Y$ for each τ_i, where $Y = \{y_i\}_{i=1}^l$ is the set of types.*

3   Methodology

An overview of our method is shown in Figure 2. In this paper, we firstly develop two weak predictors for the task of road type inference which exploit two different sources of information separately, and then introduce an ensemble approach that combines these two predictors to produce a strong predictor. In particular, we design a number of features to characterize each road segment, including topological features computed from the road network, and statistical features obtained from the historical trajectory data. A logistic regression model is then

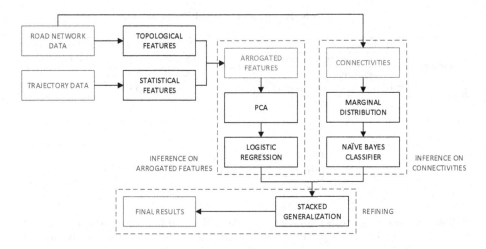

Fig. 2. An overview of our method

built upon these features to make a preliminary prediction of road segment types. However, due to the data sparsity problem, the statistical data for certain road segments may be too limited to build a reliable feature representation for those road segments. To overcome this problem, we note that there exists latent constraints on the relations of the types of two neighboring road segments depending on their connection angles. This motivates us to exploit the types of the neighboring road segments as auxiliary information for accurate road type inference. We realize this approach using a naive Bayes classifier based on the connectivity matrix built before. Finally, we combine these two predictors via stacked generalization so that their respective predictions can complement each other to achieve a more reliable prediction.

3.1   Inference on Arrogated Features

In this paper, we consider the trajectories as traveling along road segments, rather than roaming in an open area. However, the raw data of the trajectories collected from GPS devices are represented as a latitude-longitude pair with timestamp, without any road network information. Hence, we have to map the trajectories onto the road network via map matching. In this paper, we use the map matching method *ST-Matching* proposed in [4].

ST-Matching considers both the spatial geometric / topological structures of the road network, and the temporal features of the trajectories. ST-Matching is suitable to handle low-sampled trajectories, such like the taxi trajectories in this paper. It first constructs a candidate graph based on the spatial locations of the sample points of trajectories, and then generate the matched path based on the temporal features of trajectories. If any two matched road segments are not connected, it uses path-finding methods, such like shortest path method or most frequent path method [8], to generate an intermediate path that connects

Table 3. Features of a Road Segment

Topological	Road length	Statistical	Average speed of occupied taxis
	Cumulative flutter value		Density of occupied taxis
	♮ of neighbors		Density of vacant taxis
	♮ of adjacent road segments		♮ of pick-up events

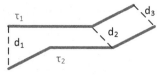

Fig. 3. The distance of two road segments. We will try to pair the road segment with fewer vertices (τ_1) towards the road segment with more vertices (τ_2). The distance of the two road segments is thus $avg(d_1, d_2, d_3)$.

them. The efficiency of ST-Matching is close to $O(nm \log m)$ using weak Fréchet distance.

In this paper, the features of road segments we use consist of two set of features: the topological features and the statistical features. The topological features are extracted from the road network, while the statistical features are extracted from the trajectory data. The details are shown in Table 3.

For the topological features in Table 3, road length and cumulative flutter value can effectively show the type of a road segment. For example, a large avenue is often limited-access, and there are few intersections within it during a long distance. Hence, according to the definition of a road segment in Definition 1, a road segment with long length is more likely to be a large avenue, or an expressway. Similarly, a road segment is more likely to be a large avenue when it is straighter, and less when it is twisted, based on our experience. Hence, we can use cumulative flutter value to show the types of road segments. For the neighbors in Table 3, we consider two road segments are neighbors when they are topologically connected. If a road segment has many neighbors, it is less likely to be a large avenue, because a large avenue, or an expressway, often has one or two neighbors as the entrance / exit of it. For the adjacent road segments in Table 3, we define adjacent as the distance between two road segments is less than a small distance (10 meters in this paper). The distance of two road segments is calculated via the average distance between each vertex of the polylines of the road segments, as shown in Figure 3. According to Definition 1, two adjacent road segments may have the same type especially when they have opposite directions.

The statistical features in Table 3 include the statistics of taxi trajectories that may reflect the types of road segments, based on our experience. These features are often time-dependent. For example, Figure 4 shows a clear difference

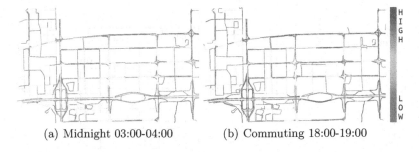

(a) Midnight 03:00-04:00   (b) Commuting 18:00-19:00

Fig. 4. Density of taxis on road segments in different time of a day [7]. It is obvious that the density when commuting is higher than midnight.

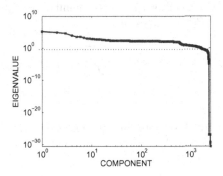

Fig. 5. Eigenvalues of the principal components in our experiments. The density of points on the plot represents the density of components with the correspondent eigenvalue.

in the density of taxis between different time slots. To account for such time-varying nature, instead of constructing a single value for each feature, we split the time domain into several time slots, and calculate the statistical features for each time slot.

To reduce the dimensionality of the feature vector, we apply principal component analysis (PCA) to find the low dimensional subspace that can well account for most variance in the data, and project each feature vector to this subspace to get its low-dimensional representation. The basis vectors of the subspace are those eigenvectors of the covariance matrix of the data set that correspond to large eigenvalues. The eigenvalues of the components of our data are shown in Figure 5. In Figure 5, it is clear that most of the components have a eigenvalue less than 1, thus we tend to select those principal components with eigenvalues larger than 1.

Based on the principal components, we can now apply a logistic regression model to infer the types of road segments. The details of the settings of our logistic regression model in the experiments are described in Section 4.

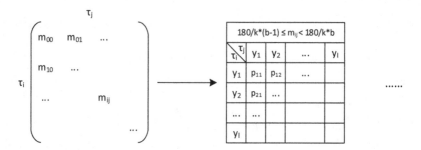

Fig. 6. The conversion from connectivity matrix to marginal distribution. Each multi-nomial distribution $180/k*(b-1) \leq m_{ij} < 180/k*b$ indicates bucket b with the range of $(180/k*(b-1))°$ to $(180/k*b)°$, where k is the number of buckets. p_{ij} in each bucket means the probability of $P(\tau_i = y_i | \tau_j = y_j)$.

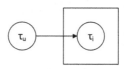

Fig. 7. The model of naive Bayes classifier in this paper. τ_u is the road segment we want to infer, and τ_i is a set of observed road segments, which are the neighbors of τ_u.

3.2   Inference on Connectivities

As mentioned before, to overcome the sparsity problem, we exploit the connectivity relationships between road segments as auxiliary information to help infer the types of road segments. Intuitively, it is observed that the type of a particular road segment has a strong indication of the possible types that its neighboring road segments can take, depending on the connection angles. Inspired by this observation, for a pair of road segments connected with each other, we model the type of one road segment as a multinomial distribution conditioned on the type of the other road segment as well as the connection angle. Equivalently, for each possible connection angle, we define for the target type a set of multinomial distributions, one for each source type. The relevant parameters can be framed as a matrix shown in Figure 6, where each row of the matrix specifies the target type distribution conditioned on a particular source type, and each matrix corresponds to a connection angle.

The graphical representation of our model is given in Figure 7, which turns out to be a naive Bayes classifier. The parent node specifies the type of the source road segment, and the child nodes are the types of neighboring road segments, one for each neighbor.

Our task then is to learn these matrix automatically from the data using maximum likelihood approach. After learning, the inference of the type of a particular road segment given the types of its neighboring road segments can be done using Bayes rule, as shown in Formula 1.

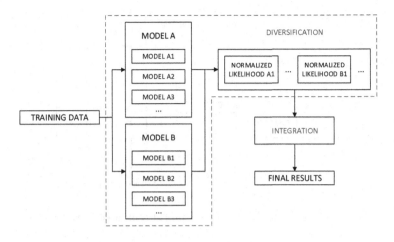

Fig. 8. The stacked generalization model used in this paper

$$
\begin{aligned}
&P(\tau_u = y_u | \tau_1 = y_1, \tau_2 = y_2, \dots) \\
&= \frac{P(\tau_1 = y_1, \tau_2 = y_2, \dots | \tau_u = y_u) * P(\tau_u = y_u)}{\sum_{i=1}^{l} P(\tau_1 = y_1, \tau_2 = y_2, \dots | \tau_u = y_i) * P(\tau_u = y_i)} \\
&= \frac{P(\tau_1 = y_1 | \tau_u = y_u) * P(\tau_2 = y_2 | \tau_u = y_u) * \cdots * P(\tau_u = y_u)}{\sum_{i=1}^{l} P(\tau_1 = y_1 | \tau_u = y_i) * P(\tau_2 = y_2 | \tau_u = y_i) * \cdots * P(\tau_u = y_i)}
\end{aligned} \tag{1}
$$

3.3 Refining

We use stacked generalization [9,10] based on logistic regression in this paper to perform refining, as shown in Figure 8. There are two phases of stacked generalization:

– Level 0: diversification through the use of different models. In this paper, we use logistic regression and naive Bayes classifier;
– Level 1: integration through meta-learning. In this paper, we use logistic regression.

Stacked generalization is one example of hybrid model combination, and it can effectively improve the accuracy of cross-validated models. Nevertheless, it is also possible to use other model combination techniques, such like random forests [11].

4 Evaluation

4.1 Metrics

In this paper, we evaluate our model via both the *accuracy* and the *expected reciprocal rank* [12]. For accuracy, it takes the maximum value of the likelihood

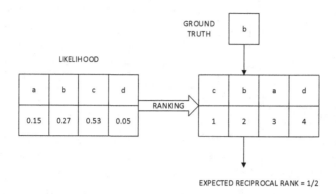

Fig. 9. An example of expected reciprocal rank used in this paper

of each road segment as the type of the road segment. For expected reciprocal rank, it evaluates each prediction as the derivative of the rank of its maximum likelihood, as an example shown in Figure 9.

In Figure 9, there are four likelihoods for the four types (a, b, c, and d) of a road segment. We rank the four likelihoods and then we get the ranks of c, b, a, and d are 1, 2, 3, and 4, respectively. Since the ground truth of the road segment is type b, we can find that the rank of type b is 2. Thus the expected reciprocal rank of the road segment is the derivative of 2, which is 1/2.

Expected reciprocal rank can be considered as a fairer metric comparing with accuracy. For example, if the likelihoods of the types of a road segment are $\langle a : 0.49, b : 0.48, c : 0.03 \rangle$, it is actually difficult to determine whether the type of the road segment is a or b, but it is clear that the type is not c. However, using accuracy cannot give the bonus of such observation, since no matter the ground truth is a or b, the accuracy is similar (0.49 and 0.48). If we evaluate the likelihoods using expected reciprocal rank, it will give us a comprehensive distribution of the accuracy (either 1 or 1/2). Hence, the evaluation of expected reciprocal rank is fairer, which widens the gap between different likelihoods. Nevertheless, in our experiments, we will conduct both metrics.

Besides the two metrics introduced before, in this paper, we use random guess as the baseline method, and we assume the probability of guessing any type of a road segment is $1/l$. For the accuracy, the expectation of random guess is $1/l$. For the expected reciprocal rank metric, the expectation is $(1*1/l+1/2*1/l+1/3*1/l+\cdots+1/l*1/l)/(l*1/l) = \sum_{i=1}^{l} 1/i/l$. In this paper, we have 7 different types of road segments as introduced in Table 2, thus the expectations are around 0.1429 and 0.3704, respectively.

4.2   Experiments

In this paper, we use four topological features and four statistical features as shown in Table 3. Since the statistical features are time-dependent, we split the statistical features by hours, as 24 features per day for a total of 26 days. Hence there are total 2,500 features.

There are total 44,793 road segments in this paper. In order to clearly show the accuracy of our model, we use 10-fold cross-validation to conduct the experiments, and each fold contains around 4,479 randomly selected road segments. In our experiments, we take one fold as test data, and the rest as training data.

In order to clearly show the differences between different settings of different models, in this paper, we have adopted the following eight settings of models in level 0 diversification of stacked generalization:

- L0-LR-T10M: the multinomial logistic regression using the features through principal component analysis. This model uses the principal components with the top ten eigenvalues, but not the top one.
- L0-LR-T10/20/30: similar as L0-LR-T10M, but uses the principal components with the top 10/20/30 eigenvalues, including the top one.
- L0-BAYES-A: the naive Bayes classifier using the connectivity as introduced in Section 3.2 where $0 \leq m_{ij} < 180$. This model sets the equal initial probabilities of $P(\tau_u = y_u) = 1/l$ for each type.
- L0-BAYES-A-D: similar as L0-BAYES-A, but sets the initial probabilities of $P(\tau_u = y_u)$ being the statistical distribution of the types of training data. That is, if there are k road segments with type u among a total of n road segments, it sets $P(\tau_u = y_u) = k/n$.
- L0-BAYES-B: similar as L0-BAYES-A, but uses $0 \leq m_{ij} < 90$. If $m_{ij} > 90$, it sets $m_{ij} \leftarrow 180 - m_{ij}$. This metric is based on the assumption that two road segments tend to have the same type if they have a smaller acute angle.
- L0-BAYES-B-D: similar as L0-BAYES-B, but sets the initial probabilities of $P(\tau_u = y_u)$ being the statistical distribution of the types of training data as L0-BAYES-A-D.

In this paper, we use multinomial logistic regression model in level 1 integration of stacked generalization. The comparison of the evaluations of both level 0 and level 1 models is shown in Figure 10.

In Figure 10, it is clear that our method is always better than single level 0 model. Moreover, the average expected reciprocal rank of the final prediction reaches 0.81859, which is a very accurate result, and it is far better than baseline method. Some models in our results are not performing better than the baseline method, such like L0-BAYES-A and L0-BAYES-B. It shows that using the statistical distribution of the types of training data is a good choice, and it can indeed increase the accuracy.

In order to show the scalability of our model, we conduct 5 experiments based on different sizes of the training data, as shown in Figure 11. In Figure 11, it is clear that the performance of our model is not dropping dramatically when shrinking the size of the training data, comparing with the scalability of logistic regression and naive Bayes classifier. Thus, our model can be considered scalable upon the size of the training data.

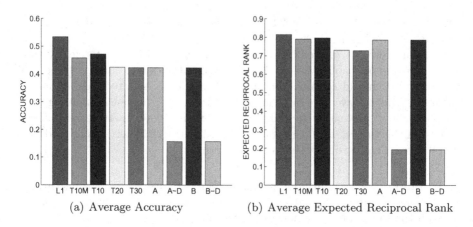

(a) Average Accuracy   (b) Average Expected Reciprocal Rank

Fig. 10. The comparison of level 1 prediction and all level 0 predictions, evaluated via different metrics. From left to right: L1, L0-LR-T10M, L0-LR-T10/20/30, L0-BAYES-A, L0-BAYES-A-D, L0-BAYES-B, L0-BAYES-B-D.

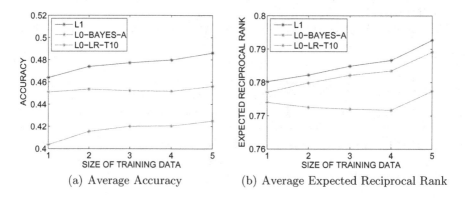

(a) Average Accuracy   (b) Average Expected Reciprocal Rank

Fig. 11. The comparison of the experiments using different sizes of training data, based on L0-REG-T10 and L0-BAYES-A as level 0 models

5   Related Works

The information of a road network is an essential requirement to enable further analysis of urban computing [13]. The inference of a road network generally consists of two categories: inference from aerial imagery, and inference from trajectories. Both inference method aims to discovery the missing road segments in the data. The inference from aerial imagery is often based on pattern recognition methods [14], and it is often very difficult to find those road segments in the shadow of skyscrapers or forests. Hence, such methods often require a very high resolution color orthoimagery [15].

The inference from trajectories is an effective, efficient, and inexpensive method comparing with the inference from aerial imagery. Many works in this category are often based on the clustering of trajectories, such like k-means [16,17,18], kernel density estimation [19,20,21], trace merging [22], and some other methods, like TC1 [23]. These methods often first identify the trajectories in different clusters, and then apply fitting to the trajectories in the clusters. Some works also split the entire map into small grids to increase performance [20]. As mentioned in Section 1, most of these works only focus on identify the missing road segments that are not existed in the current road network, but few of them focus on the inference of the properties of road segments, such like the types introduced in this paper.

6 Conclusion

In this paper, we propose a novel problem, as identify the type of a road segment. To solve the problem, we introduce a combined model based on stacked generalization, using both the topology of the road network, and the knowledge learned from taxi trajectories. For level 0 diversification, we use 1) a multinomial logistic regression model on a set of arrogated features consists of both the topological features from the road network, and the statistical features from the taxi trajectories; and 2) a naive Bayes classifier based on the connectives of road segments. The experimental results show that our method is much better than the baseline method.

The model proposed in this paper is highly flexible and scalable. For level 0 diversification, it is possible to use different models despite of the models proposed in this paper, such like decision tree, expectation-maximization algorithm, kernel density estimation, etc. For level 1 integration, it is also possible to use different models, like support vector machine. Moreover, since the taxi trajectories we use in this paper are sparse and bias, it is also eligible to use other measurement methods, such like PageRank values [24], rather than the connectivities in level 0 models. A comparison of different models upon different trajectory datasets will be our future work.

Since the road network is often partially available in a crowd-sourcing platform, we only adopt supervised models in this paper. However, in some cases, we may not have any information of the road network besides the topology. Hence, inferring the types of road segments based on unsupervised / semi-supervised models is also a challenging problem.

Acknowledgment. This paper was supported in part by National Basic Research Program of China (973 Program) under grant no. 2014CB340304; Hong Kong, Macao and Taiwan Science & Technology Cooperation Program of China under grant no. 2012DFH10010; Nansha S&T Project under grant no. 2013P015; and NSFC under grant no. 61300031.

References

1. Haklay, M.M., Weber, P.: Openstreetmap: User-generated street maps. IEEE Pervasive Computing 7(4), 12–18 (2008)
2. Neis, P., Zielstra, D., Zipf, A.: The street network evolution of crowdsourced maps: Openstreetmap in germany 2007-2011. Future Internet 4(1), 1–21 (2012)
3. von Ahn, L., Maurer, B., Mcmillen, C., Abraham, D., Blum, M.: Recaptcha: Human-based character recognition via web security measures, 1465–1468 (2008)
4. Lou, Y., Zhang, C., Zheng, Y., Xie, X., Wang, W., Huang, Y.: Map-matching for low-sampling-rate gps trajectories. In: GIS, pp. 352–361 (2009)
5. Liu, S., Liu, Y., Ni, L.M., Fan, J., Li, M.: Towards mobility-based clustering. In: KDD, pp. 919–928 (2010)
6. Yuan, J., Zheng, Y., Xie, X., Sun, G.: Driving with knowledge from the physical world. In: KDD, pp. 316–324 (2011)
7. Ding, Y., Liu, S., Pu, J., Ni, L.M.: Hunts: A trajectory recommendation system for effective and efficient hunting of taxi passengers. In: MDM, pp. 107–116 (2013)
8. Luo, W., Tan, H., Chen, L., Ni, L.M.: Finding time period-based most frequent path in big trajectory data. In: SIGMOD Conference, pp. 713–724 (2013)
9. Wolpert, D.H.: Stacked generalization. Neural Networks 5(2), 241–259 (1992)
10. Ting, K.M., Witten, I.H.: Stacked generalizations: When does it work? In: IJCAI (2), pp. 866–873 (1997)
11. Breiman, L.: Random forests. Machine Learning 45(1), 5–32 (2001)
12. Chapelle, O., Metlzer, D., Zhang, Y., Grinspan, P.: Expected reciprocal rank for graded relevance. In: CIKM, pp. 621–630 (2009)
13. Zheng, Y., Liu, Y., Yuan, J., Xie, X.: Urban computing with taxicabs. In: Ubicomp, pp. 89–98 (2011)
14. Hu, J., Razdan, A., Femiani, J., Cui, M., Wonka, P.: Road network extraction and intersection detection from aerial images by tracking road footprints. IEEE T. Geoscience and Remote Sensing 45(12-2), 4144–4157 (2007)
15. Chen, C.C., Shahabi, C., Knoblock, C.A.: Utilizing road network data for automatic identification of road intersections from high resolution color orthoimagery. In: STDBM, pp. 17–24 (2004)
16. Edelkamp, S., Schrödl, S.: Route planning and map inference with global positioning traces. In: Klein, R., Six, H.-W., Wegner, L. (eds.) Computer Science in Perspective. LNCS, vol. 2598, pp. 128–151. Springer, Heidelberg (2003)
17. Agamennoni, G., Nieto, J.I., Nebot, E.M.: Robust inference of principal road paths for intelligent transportation systems. IEEE Transactions on Intelligent Transportation Systems 12(1), 298–308 (2011)
18. Schrödl, S., Wagstaff, K., Rogers, S., Langley, P., Wilson, C.: Mining gps traces for map refinement. Data Min. Knowl. Discov. 9(1), 59–87 (2004)
19. Wang, Y., Liu, X., Wei, H., Forman, G., Chen, C., Zhu, Y.: Crowdatlas: Self-updating maps for cloud and personal use. In: MobiSys, pp. 27–40 (2013)
20. Davies, J.J., Beresford, A.R., Hopper, A.: Scalable, distributed, real-time map generation. IEEE Pervasive Computing 5(4), 47–54 (2006)
21. Biagioni, J., Eriksson, J.: Map inference in the face of noise and disparity. In: SIGSPATIAL/GIS, pp. 79–88 (2012)
22. Cao, L., Krumm, J.: From gps traces to a routable road map. In: GIS, pp. 3–12 (2009)
23. Liu, X., Biagioni, J., Eriksson, J., Wang, Y., Forman, G., Zhu, Y.: Mining large-scale, sparse gps traces for map inference: Comparison of approaches. In: KDD, pp. 669–677 (2012)
24. Yang, B., Kaul, M., Jensen, C.S.: Using incomplete information for complete weight annotation of road networks - extended version. CoRR abs/1308.0484 (2013)

Rights Protection for Trajectory Streams

Mingliang Yue[1], Zhiyong Peng[1], Kai Zheng[2], and Yuwei Peng[1,*]

[1] School of Computer, Wuhan University,
Wuhan, Hubei, 430072, China
ywpeng@whu.edu.cn
[2] School of Information Technology and Electrical Engineering,
The University of Queensland, Brisbane, Australia
kevinz@itee.uq.edu.au

Abstract. More and more trajectory data are available as streams due to the unprecedented prevalence of mobile positioning devices. Meanwhile, an increasing number of applications are designed to be dependent on real-time trajectory streams. Therefore, the protection of ownership rights over such data becomes a necessity. In this paper, we propose an online watermarking scheme that can be used for the rights protection of trajectory streams. The scheme works in a finite window, single-pass streaming model. It embeds watermark by modifying feature distances extracted from the streams. The fact that these feature distances can be recovered ensures a consistent overlap between the recovered watermark and the embedded one. Experimental results verify the robustness of the scheme against domain-specific attacks, including geometric transformations, noise addition, trajectory segmentation and compression.

Keywords: Rights protection, trajectory streams, watermarking, robustness.

1 Introduction

Recent advances in mobile positioning devices such as smart phones and in-car navigation units made it possible for users to collect large amounts of GPS trajectories. After expensive and laborious data acquiring process, companies and institutions frequently outsource their trajectory data for profit or research collaborations. Therefore, the rights protection for the owner over the precious data becomes an important issue.

Watermarking is one of the most important techniques that can be used for the rights protection of digital data. Basically, watermarking means slightly modifying the host data and forcing it to imply certain secret information. The information (i.e., watermark) is identifiable and can be detected from the (possibly modified) data for ownership assertion [1]. In fact, watermarking is predominantly used for the rights protection of digital contents, e.g., images [2], audios [3], videos [4] and vector maps [5][6][7]. For trajectory databases, two watermarking schemes have been proposed [8][9]. Given a trajectory set, both schemes

---

* Corresponding Author.

S.S. Bhowmick et al. (Eds.): DASFAA 2014, Part II, LNCS 8422, pp. 407–421, 2014.

embed watermark by slightly modifying coordinates of locations in trajectories. And both schemes are robust against certain data modifications (attacks) such as geometric transformations and noise addition.

However, as other kinds of data collected from sensors, trajectory data often come to databases in a streaming fashion. As Sion et. al noted [10][11], batched watermarking schemes, such as schemes in [8] and [9], are not applicable to streaming data. Such schemes can embed and detect watermark only when the entire dataset is available, while attacks may come before the entire dataset is collected. For example, suppose taxi company A obtains trajectories from its affiliated taxies and sells the data to service provider B for real-time passenger-hunting recommendation (for taxi drivers) [12]. If the data is not watermarked, and there is another service provider C who needs the data to support the passenger ridesharing service [13], then B can simply re-direct the trajectory stream to C for profit. In this scenario, online watermarking schemes [10][11] should be employed to embed watermark immediately after the taxi company receiving the data.

Moreover, any (batched or online) watermarking scheme should be robust against certain domain-specific transformations (attacks). For trajectory data, attacks like geometric transformations and noise addition have been considered and handled in [8] and [9]. However, in those two schemes, two important data operations are not considered: trajectory segmentation and trajectory compression. Simply speaking, trajectory segmentation is used to filter out subsets of trajectory locations for sub-trajectory mining [14][15], while trajectory compression aims to reduce the size of trajectory data to resolve the inefficiency in data transmission, querying, mining and rendering processes [16][17][18]. Both operations (attacks), may not influence the usability of trajectory data in certain scenarios, however, can modify the data in a significant magnitude, and in turn, harm the watermark embedded in the data. A watermarking scheme designed for trajectory data should have the ability to handle those two attacks.

Considering the problems mentioned, in this paper, we propose for the first time an online watermarking scheme for the rights protection of trajectory streams. The scheme operates in a finite window, single-pass streaming model. The main idea is to embed watermark into the feature distances, i.e., distances between pairs of feature locations in trajectories, and the feature locations are identified using the proposed Time Interpolated Feature Location Selection algorithm. In addition to traditional attacks (i.e., geometric transformations and noise addition), our scheme is also robust against trajectory segmentation and compression. Our contributions include (1) the identification of the problem of watermarking trajectory streams and major types of attacks on the data; (2) the design and analysis of new watermarking scheme for the data; (3) a proof of the scheme's robustness against the considered attacks based on experimental evaluation.

The remainder of this paper is organized as follows. Section 2.4 outlines the major challenges. It introduces the scenario and the underlying processing model, discusses associated attacks and overviews related work. Then, the proposed online watermarking scheme is given in Section 3. Finally, performance study and conclusions are given in Section 4 and 5 respectively.

2   Challenges

2.1   Scenario

Fig. 1 shows the general scenario of watermarking streaming data, which was firstly demonstrated in [10]. In the scenario, streaming data is modeled as an (almost) infinite timed sequence of values of a particular type (e.g., temperature, stock market data). The watermarking technique is used to deter a malicious customer (licensed data user B in Fig. 1), with direct stream access, to re-sell possibly modified trajectory streams to others (unlicensed data user C for example) for profit. The challenges are that the underlying watermarking scheme should operate in a finite window, single pass model (i.e., online model), and needs to be robust against domain-specific attacks.

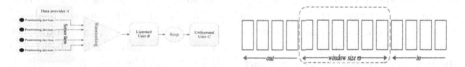

Fig. 1. Stream Watermarking Scenario   **Fig. 2.** Space Bounded Processing Window

2.2   Processing Model

The stream processing should be both time and space bounded [10]. The time bound derives from the fact that it has to keep up with incoming data. For space bound, as demonstrated in Fig. 2, we model the space by the concept of a *processing window* of size ϖ: no more than ϖ of locations can be stored locally at the processing time. As more incoming data become available, the scheme should push older locations out to free up space for new locations.

2.3   Attack Model

The attacks that can be applied to the watermarked data are domain-specific. In this paper, since our purpose is to watermark trajectory streams, we identify several meaningful attacks on trajectory data as: ($A1$) geometric transformations (i.e., translation, rotation and scaling), ($A2$) noise addition, ($A3$) trajectory segmentation, and ($A4$) trajectory compression. $A1$ and $A2$ are intuitive [8][9] hence no further explanation will be outlined here. While $A3$ and $A4$ are more complicated, before explaining these two attacks, we give three definitions as follows [23].

A **Trajectory Stream** T is a possibly unbounded sequence of time-stamped locations, denoted as $T = \{\langle x_1, y_1, t_1 \rangle, \langle x_2, y_2, t_2 \rangle, \ldots, \langle x_n, y_n, t_n \rangle, \ldots\}$, where t_i is an element of a numerable, disjoint *Time Schedule* defined as $TS = \{0, 1, 2, \ldots\}$, x_i, y_i represent geographic coordinates of the moving object at sampling time t_i.

The (a, b)-**Segmented Trajectory** $T_{a:b}$ of a trajectory stream T is a subset of consecutive locations of T, denoted as $T_{a:b} = \{\langle x_i, y_i, t_i \rangle | a \leq t_i \leq b, \langle x_i, y_i, t_i \rangle \in T\}$.

Given a distance threshold ε, trajectory T^ε is a ε-**Simplified Trajectory** of a trajectory stream T, if $T^\varepsilon \subsetneqq T$, $\forall L_{t_i} = \langle x_i, y_i, t_i \rangle \in T - T^\varepsilon$, $dist(L_{t_i}, T^\varepsilon) \leq \varepsilon$, where $dist()$ is a certain distance function. $\forall L_{t_i} \in T^\varepsilon$, we call it a **feature location**.

Trajectory segmentation (compression) attack means extracting segmented (simplified) trajectories and re-streaming them for profit. The implementation of trajectory segmentation is straightforward. While for trajectory compression, many online trajectory simplification algorithms can be employed for attack [16][18] [19][20][21]. Those algorithms also take compression ratio (i.e., the size of the original trajectory divided by the size of the compressed trajectory) as threshold, and return simplified trajectories satisfying compression ratio constraints as compressed trajectories. Both attacks, may not influence the usability of trajectory data in certain scenarios [14][15][19][20], however, can modify the data in a significant magnitude, and in turn, harm the watermark embedded in the data. In this paper, all the attacks ($A1$ to $A4$) will be properly considered and handled. To the authors' knowledge, this is the *first piece of rights protection work* that takes trajectory segmentation and compression into account as attacks.

2.4   Related Work

An Online Watermarking Scheme (OLWS) has been proposed in [11] (which is extended based on the work in [10]) to watermark streaming data. The scheme takes general data streams (e.g., temperature, stock market data, etc.) as input. The attacks considered are summarization, extreme sampling, segmentation, geometric transformations and noise addition. During the embedding, the scheme firstly initializes a data normalization process, and then identifies major extremes (which are composed of values at and close to local minimums or maximums) based on the normalized data. Finally, the scheme embeds one watermark bit into all the values in a major extreme if the extreme satisfies a certain selection criterion.

The scheme works well for general data streams, however, it is not suitable for trajectory streams. On the one hand, the attacks faced by an OLWS for trajectory streams are different (see Section 2.3). For example, the scheme may be ineffective under trajectory compression: locations with local minimum or maximum coordinates are not necessarily the feature locations in a trajectory, and may be deleted after compression. On the other hand, the robustness against geometric transformations of the scheme mainly depends on the data normalization process, which can only be done based on a known data distribution. If the distribution is unknown, an additional initial *discovery* run should be employed to learn one. For trajectories that are composed of locations representing the motion of real objects, such distribution may not exist or is difficult to find out. Therefore, the technique of this work cannot be employed in this paper to handle geometric transformation attacks for trajectory streams.

The resistance to geometric transformations and noise addition have been achieved in two batched trajectory watermarking schemes [8][9]. The scheme in [8] and [9] embeds watermark into distance ratios and Fourier coefficients of locations respectively. Since both the cover data are geometrical invariants, the schemes can properly withstand geometric transformations. And the resistance of the schemes to noise addition is achieved by embedding the watermark bits into the properly selected bits of the cover data. These two schemes also have two problems for trajectory streams. First, none of the schemes operates in online setting, which makes the schemes not applicable for the scenario. Second, both schemes embed watermark based on all the locations in the trajectories. Hence, they are fragile to trajectory compression attack (work in [9] is also fragile to trajectory segmentation).

In summary, due to the unique properties of streaming trajectory data and the corresponding attacks, an online scheme for watermarking trajectory streams that can handle the attacks outlined in Section 2.3 is in need.

3   Watermarking Trajectory Streams

The basic idea of the proposed watermarking scheme is based on two observations. First, feature locations of trajectories will survive trajectory compression attack, if the malicious attacker wants to preserve the usability of the trajectories. Second, distances between pairs of locations will not change if the trajectory is translated or rotated as a whole.

Therefore, the proposed watermarking scheme embeds watermark into feature distances (i.e., distances between pairs of feature locations) with the following process: (1) identify the feature locations in the trajectory stream as processing window advancing; (2) if two consecutive feature locations appear in a same processing window, calculate the (feature) distance between the locations; (3) determine a watermark bit of the global watermark based on the distance according to a selection criterion, and embed the bit into the distance. The fact that these feature distances can be recovered ensures a consistent overlap (or even complete identity) between the recovered watermark and the embedded one (in the un-attacked data). In the watermark detection process, (4) the feature locations are identified and the feature distances are calculated; for every feature distance, (5) the selection criterion in (3) is used once again to match the distance and the watermark bit, the corresponding 1-bit watermark is extracted, and ultimately the global watermark is gradually reconstructed, by using majority voting.

In the following we firstly introduce the proposed *Time Interpolated Feature Location Selection* (TIFLS) algorithm in Section 3.1. The algorithm can identify stable feature locations from an attacked trajectory stream. Then, we give the watermark embedding method in Section 3.2, by employing TIFLS for feature location identification. The watermark detection method is introduced in Section 3.3. At last, parameters used in the scheme are discussed in Section 3.4.

3.1   Time Interpolated Feature Location Selection

To guarantee that the embedded watermark is still detectable after various attacks, the feature locations used for watermark embedding should be recovered after attack. Many online simplification algorithms have been proposed to identify feature locations in a trajectory stream [16][18][19][20][21]. The basic intuition behind those algorithms is that the locally characteristic locations may also be characteristic in a global view. Hence, the process of feature location selection can be described as, (1) fill up the processing window with incoming locations; (2) select feature locations in the current window based on a certain batched simplification method (e.g., Douglas-Peucher algorithm in [22]), free the window and repeat the process. The intuition is reasonable and the results are satisfactory. However, these solutions for feature location selection are ineffective in our scenario: for a possibly compressed and/or segmented trajectory, the selected feature locations may change (i.e., cannot be recovered), since the locations filled in every processing window may be very different from those fetched from the original trajectory.

Therefore, in this paper, we identify feature locations based on *Time Window* defined as follows. A **Time Window** TW_m is a ω-sized consecutive subset of TS, denoted as $TW_m = \{t_s, \ldots, t_e\}$, where $m \in \{0, 1, 2, \ldots\}$, $t_s, t_e \in TS$, $t_s = \omega * m$, $t_e = t_s + \omega - 1$. Then, given a trajectory stream T, and a time window TW_m, we can deduce a projection of T on TW_m as $T_m = \{< x_i, y_i, t_i > | t_i \in TW_m, < x_i, y_i, t_i > \in T\}$

Based on the notations, given a threshold ϵ, for every TW_m, our *Time Interpolated Feature Location Selection* (TIFLS) algorithm identifies a location as feature location from its corresponding T_m if (1) among other locations in T_m, the location has the largest Synchronous Euclidean Distance (SED) according to the reference line; (2) its SED is larger than ϵ. The reference line is the line connecting the locations with boundary times of TW_m as sampling times (i.e., L_{t_s} and L_{t_e}, we call these locations boundary locations hereafter).

TIFLS is not a simplification algorithm. Rather, it identifies the location which meets the characteristic conditions in each T_m as feature location for watermark embedding. Hence, these feature locations are very likely to be recovered during watermark detection, even after attack. One can deduce that if a segmented trajectory covers more than two time windows, then in the worst case, trajectory segmentation only influences the first and the last T_m with respect to the segmented trajectory stream. The feature location in each *inner* T_m can be properly recovered. Meanwhile, after trajectory compression (denote T^c the compressed trajectory stream), every T_m^c is a subset of the original T_m. Compared with other locations, the feature locations will more likely get through trajectory compression, and be identified by TIFLS.

The only problem left is that the boundary locations may not exist in the input trajectory. We solve the problem by interpolating a virtual location for every missing boundary location, based on the locations that exist in the trajectory and adjacent to the boundary location. Given two locations L_a and L_c, $a < c - 1$, interpolating a location L_b ($a < b < c$) means finding the L_b with $\overrightarrow{v}_{ab} = \frac{b-a}{c-a} \overrightarrow{v}_{ac}$, where $\overrightarrow{v}_{ij}$ denote the vector from L_i to L_j in the Euclidean plane. The virtual boundary location can excellently simulate the real location, since the SED between any boundary location deleted due to compression and the compressed trajectory should be less than a certain threshold (see the notation of compression attack in Section 2.3).

3.2 Watermark Embedding

To synchronize processing window and time window, we define the time coverage of a processing window as follows. The **Time Coverage** TC of a processing window is a consecutive subset of TS, denoted as $TC = \{t_f, \ldots, t_l\}$, where t_f and t_l is the sampling time of the first and the last location in the processing window respectively.

Let TC_c denote the time coverage of the *current* processing window, our watermark embedding algorithm can be described as follows. In the **first step**, (1) forward the processing window, apply TIFLS on every T_m with $TW_m \subsetneqq TC_c$, until a feature location L_f is identified; (2) push out the locations previous to L_f, i.e., the first location in the current processing window is L_f, a feature location.

Then, in the **second step**, as locations streaming in, as long as a time window TW_m is completely covered by TC_c (i.e., $TW_m \subsetneqq TC_c$), apply TIFLS on its corresponding T_m for feature location identification. If a feature location is identified, go to the third step. Otherwise, if no feature location is identified until the processing window is full, push out the locations in and previous to the last T_m with $TW_m \subsetneqq TC_c$, and return to the first step.

Let W be the watermark to be embedded, each $w_i \in \{0,1\}$ be the i-th bit of W, and L_s be the feature location identified in the second step. In the **third step**, we embed one watermark bit into L_s as follows. (1) calculate the distance d_{fs} between L_f and L_s as $d_{fs} = ||\overrightarrow{v}_{fs}||$, where $|| \circ ||$ signifies the L_2 norm of a vector. Let β control the bitwise position in d_{fs} in which the watermark bit will be embedded, and $msb(d, \beta)$ return the bits of d that are higher than β. (2) calculate i as $i = H(msb(d_{fs}, \beta), k) \ mod \ l$, where $H()$ is a cryptographic hash function such as MD5, l is the bit length of W, and k is a secret key given by data owner. (3) replace the β-th bit of d_{fs} as the ith bit of W, w_i, to get the watermarked distance d_{fs}^w. (4) find the L_s^w satisfying $\overrightarrow{v}_{fs}^w = \frac{d_{fs}^w}{d_{fs}} \overrightarrow{v}_{fs}$, where L_s^w is exactly the watermarked location of L_s, and $\overrightarrow{v}_{fs}^w$ is the vector from L_f to L_s^w in in the Euclidean plane. **Finally**, push out the locations previous to L_s^w from the current processing window, denote L_s^w as the current L_f, return to the second step.

To guarantee at least two complete time windows can be synchronized in a same processing window, ω should be set to be less than or equal to $\varpi/3$. The

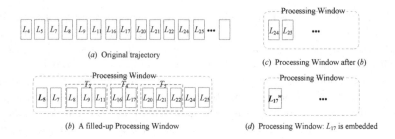

(a) Original trajectory

(b) A filled-up Processing Window

(c) Processing Window after (b)

(d) Processing Window: L_{17} is embedded

Fig. 3. Watermark Embedding Process

cryptographic hash function is employed in (2) of the third step to force the malicious attacker into a guessing with respect to the watermark bit embedded. Its power derives strength from both the one-wayness and randomness properties [24]. While the reason behind the use of the most significant bits of d_{fs} is resilience to minor alterations and errors due to watermark embedding and noise addition attack. The embedding strategy employed in (3) of the third step can be easily extended to use quantization index modulation [25] (as we did in the experiments). The method can provide the watermarking scheme with resistance to noise distortion. One can refer to [25] for details about the method.

Fig. 3 illustrates the watermark embedding process, in which ϖ is set to 12, ω to 4. Fig. 3 (a) shows the original trajectory. Fig. 3 (b) demonstrates the filled-up processing window, by assuming the feature location identified in the first step is L_5, and no feature location is identified in the second step. The time coverage of this processing window is $TC_c = \{5, 6, \ldots, 25\}$. Hence TW_2, TW_3, TW_4 and TW_5 are completely covered by TC_c, and T_2, T_4 and T_5 has been considered by TILFS for feature location identification. Fig. 3 (c) demonstrates the next processing window corresponding to Fig. 3 (b): locations previous to L_{24} are pushed out. Fig. 3 (d) demonstrates the processing window after a feature location (L_{17} in this particular example) is identified and a watermark bit is embedded.

3.3   Watermark Detection

In the detection process, the watermark is gradually reconstructed as more and more locations are processed. The reconstruction relies on an array of majority voting buckets as follows. For each bit w_i in the original watermark W, let B_i^0 and B_i^1 be the buckets (unsigned integers) which are incremented each time a corresponding 0/1 bit w_i^d is recovered from the streams. The actual w_i will be estimated by the difference between B_i^0 and B_i^1, i.e., if $B_i^0 - B_i^1 > \vartheta$, then the estimated value for this particular bit becomes $w_i^e = 0$, otherwise, if $B_i^1 - B_i^0 > \vartheta$, $w_i^e = 1$, where $\vartheta \geq 0$. The rationale behind this mechanism is that for unwatermarked streams, the probability of detecting $w_i^d = 0$ and $w_i^d = 1$ would be equal, thus yielding virtually identical values for B_i^0 and B_i^1. In this case, w_i will be undefined and the object will be regarded as un-watermarked [1].

Detection starts by identifying feature locations and calculating feature distances. As long as a feature distance d_i is calculated, the corresponding $j = H(msb(d_i, \beta), k) \bmod l$ is determined, then d_i was likely used for the embedding of the j-th bit of W. The bit is then extracted from the β-th bit of d_i depending on its value. Then, the corresponding bucket B_j^0 or B_j^1 is incremented by 1.

3.4   Discussion on Parameters

Except watermark W and secret key k, four parameters should be determined and used in both watermark embedding and detection, i.e., ϖ, ω, ϵ and β. W and k are common to all watermarking schemes. They have been discussed in-depth in many previous works [3, 6, 16]. Therefore, we focus on the latter four parameters. The size of the processing window, ϖ, should be large enough so that it is possible to identify at least two feature locations in most processing windows. ω and ϵ control how characteristic the identified feature locations can be. More characteristic locations have greater chance to be preserved after compression. We will discuss the selection of ω and ϵ in Section 4.1. Consequently, ϖ can be determined based on the ω and ϵ used. As to β, it controls the trade-off between the error introduced and the robustness against noise addition. Obviously, modification on higher bits of distances will result in larger errors on the trajectory data. On the other hand, watermark embedded in the lower bits tends to be more fragile to noise addition. Same as [8], we set β around the bit position of the data's tolerance precision to ensure data fidelity.

4   Experimental Results

In this section, we give an empirical study of the proposed watermarking scheme. The watermarking algorithm was implemented in C++ and run on a computer with Intel Core CPU (1.8GHz) and 512M RAM. The dataset used in the evaluation is obtained from the T-Drive trajectory data sample provided by Microsoft T-Drive project [26][27]. The dataset contains a one-week trajectories of 10,357 taxis. The total number of locations in this dataset is about 15 million and the total distance of the trajectories reaches 9 million kilometers. Each file of this dataset, which is named by the taxi ID, contains the trajectories of one taxi.

Due to the nature of the data, in the experiments, we simulated for each taxi, a trajectory stream with one location per 5-second as sampling rate (incoming data per time unit). During the embedding and detection, β was set as -5 since the precision tolerance of the location's coordinates is $\tau = 10^{-5}$. In the following we verify in turn (1) the performance of TIFLS under various ω and ϵ; (2) the robustness of our method against various attacks; (3) the time overhead and the impact on data quality of the scheme.

---

[1] Considering also the associated random walk probability [11], we set ϑ as $\sqrt{\frac{B_i^0 + B_i^1}{\pi}}$.

4.1 Evaluation on ω and ϵ

As we have stated in Section 3.4, the feature locations identified for watermark embedding should be characteristic enough to withstand trajectory compression. However, how characteristic the feature locations should be is highly related to the compression ratio that the scheme aims to resist. As verified in [20], existing trajectory compression algorithms can compress trajectory data in a compression ratio up to 10 without resulting in significant SED error on the data. The compression ratio is defined as the size of the original trajectory divided by the size of the compressed trajectory [20]. That is, to ensure the usability of the proposed scheme in real applications, (1) 10% (or less) locations should be identified as feature locations for embedding, (2) the feature locations identified should be actually (or positively) characteristic to withstand compression.

In the experiment, we model the two requirements by the concepts of Selection Ratio (SR) and Characteristic Ratio (CR). SR represents the ratio of the number of feature locations identified by TIFLS (denoted as f) with respect to the total number of locations in the trajectories (denoted as n). CR represents the ratio of the number of positive feature locations (denoted as f_p) with respect to the total number of feature locations identified by TIFLS. While by setting the compression ratio of Douglas-Peucker algorithm to 10, the positive feature locations are the locations reported by both TIFLS and Douglas-Peucker algorithm. Namely, we have $SR = f/n$ and $CR = f_p/f$. We use the feature locations selected by Douglas-Peucker algorithm as reference to evaluate the precision of TIFLS since as an optimization algorithm, the performance of Douglas-Peucker algorithm is widely acknowledged. And the algorithm has been extensively employed in almost all the trajectory compression work to evaluate the performance of their methods [16][18][19][20][21].

We applied TIFLS on 1000 trajectories (each containing 10000 locations) for feature location identification, and recorded the SR and CR with respect to different pairs of ω and ϵ. The results are presented in Table 1. From the results, the following conclusions can be drawn: (1) the size of time window ω can be employed to control the SR of the identified locations. Since given a certain ω, the upper bound of SR is $1/\omega$. (2) in general, the lower the SR is, the higher CR will be. However, CR is primarily determined by ϵ. A larger ϵ can lead to a higher CR. If the ϵ is big enough, even a CR of 1 can be achieved. In real world applications, ω can be set as the compression ratio the scheme needs to resist. For ϵ, an initial determination process can be carried out on a small data sample (a one-day trajectory streams for example) to ensure a high CR (e.g., CR$\geq$ 0.9).

4.2 Evaluation on Robustness

In the following experiments, watermarks were first embedded into the trajectory streams. Then, taking the attacked trajectory streams as input, the watermarks were reconstructed using the detection method. For a watermarking scheme, robustness means the scheme can detect a reasonable amount of watermark bits

Table 1. Evaluation on Different ω and ϵ

ω	ϵ	SR	CR	ω	ϵ	SR	CR
	0	0.1	0.481		0	0.05	0.568
	0.00005	0.069	0.866		0.00005	0.045	0.863
10	0.0001	0.061	0.894	20	0.0001	0.039	0.906
	0.0005	0.042	0.957		0.0005	0.023	0.951
	0.01	0.001	1		0.01	0.001	1

from watermarked data after attacks. Hence, after watermark extraction, the match rate (MR) is calculated to assess robustness of the scheme, where MR is defined as the ratio of the number of identical bits between the extracted and the embedded watermark to the watermark length. Generally, a dataset can be regarded as watermarked if MR is larger than a given threshold ζ [28]. One can refer to [28] for the selection of ζ. In the following demonstration, the ω and ϵ used in the experiments was 10 and 0.0001 respectively. The size of processing window, ϖ, was set as 30 since averagely a feature location can be reported from every 15 locations (since the CR is 0.061).

In the experiments, one important fact that should be noted is that due to the infinite nature of stream data, watermark detection is a continuous process. It takes time for the watermark detection to be convergent [10]. Our experimental results show that the detection converges when averagely each bit receives 60 votes [2]. In the following experiments, we always record the converged MR under various attacks. For a certain attack, the final MR is the average of the converged MRs reported in different processes using randomly generated watermarks with 64, 128 and 256 as watermark length.

4.2.1 Geometrical Attacks

In this experiment, we applied translation, rotation and scaling to the water-marked trajectory streams. The magnitudes of the attacks are measured with relative coordinate offset, rotation angle and scaling factor, and were set to [-100%, 100%], [-180°, 180°] and [0.1, 2]. The experimental results show that if the trajectories are only translated and rotated (i.e., Scaling = 1), the MRs are the same and equal to 1. For scaled trajectories, the scheme needs to transform the coordinates back into their values on the original scale before feature locations identification and feature distance calculation. This re-scaling and detection has been employed in various batched watermarking schemes [6][7]. According to our experimental results, the method is feasible, and the MRs after re-scaling are equal to 1. The experiment verifies the good preference of the proposed scheme under geometrical attacks.

---

[2] This means averagely 2000 locations can support a convergent detection of 1-bit, hence a total of 7500 bits can be embedded into the whole dataset.

4.2.2 Noise Addition

To verify the resilience of the proposed scheme to noise addition, we insert random noise to the locations' coordinates in the watermarked trajectory streams. The assumption is that the introduction of noise should not degrade data usability. Hence, the noise added is assumed to be uniformly distributed in the $[0, \chi]$ interval, where $\chi \leq \tau$. The experimental results with respect to different value of χ are presented in Fig. 4. The figure demonstrates that (1) if the attack magnitude χ is less than 0.2τ, the MR remains as 1. This is guaranteed by the power of quantization index modulation. (2) MR decreases with an increasing χ when χ exceeds 0.2τ. However, even when the χ reaches τ, the watermark can still be recovered with a considerably high MR.

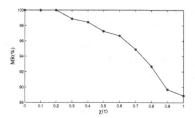

Fig. 4. Resistance to Noise Addition   **Fig. 5.** Resistance to Compression

4.2.3 Segmentation and Compression

Due to the processing model of our scheme, an *active segment* that contains more than ϖ locations or covers 2ω or larger time interval is detectable. Hence, our scheme can naturally resist trajectory segmentation. The conclusion was also verified by our experiments: with sufficient active segments, all the watermark bits can be properly recovered. Namely, as long as the active segments are sufficient enough to support an convergent detection, a MR of 1 can always be achieved.

As to trajectory compression, we compressed the watermarked trajectory streams using Douglas-Peucker algorithm, by setting compression ratio as 0, 5, 10, 15, 20, 25 and 30 respectively. The detection results corresponding to various compressed sets are demonstrated in Fig. 5. As we can see, (1) the entire watermark can be properly recovered, when the compression ratio is less than 10. The reason is that the majority of feature locations survived from the compression attack and can be recovered during the detection. The majority voting ensures a consistent overlap between the detected watermark and the embedded one. (2) MR decreases to 0.88 (a considerably high level) when compression ratio reaches 15, since after compression, only 6.67% locations are left in the compressed trajectories. These locations are not very consistent with the feature locations reported by TIFLS, which makes a small portion of the feature locations used during the embedding unrecoverable in the detection. (3) MR decreases dramatically when compression ratio reaches 20. The reason is intuitive: larger compression ratio causes fewer recovered feature locations. When compression ratio is 30, only 3.33% locations are left for detection. Even all these locations were used for embedding and recovered for detection, the feature distances may change since the consecutive relations among feature locations change. In this case, the parameters used for the feature location identification (i.e., ω and ϵ) should be enlarged to select even more characteristic locations for watermark embedding.

Another more implicit fact to be noted is that boundary location interpolation in TIFLS may also cause the selection of wrong feature locations. However, in the perspective of watermark detection, this is only an extra 'bad' consequence caused by trajectory compression. The good performance of TIFLS for selecting stable feature locations has been implied by the perfect MR facing reasonable compression attacks.

4.3   Overhead and Impact on Data Quality

We performed experiments aimed at evaluating the computation overhead of the proposed watermarking scheme. By far the most computationally intensive operation is the feature location selection. Fortunately, the selection is a single pass process, which means the time complexity of the scheme is $O(n)$. In this experiment, to avoid the time consumed by waiting for the sampling of new locations, the sampling rate was set to infinity: during the processing, all the locations are assumed available as soon as they need to be pushed into the processing window. We tested the embedding and detection on 100 trajectories, each has 10000 locations. The process cost $3340ms$, which means averagely the processing of each processing window needs approximately $50\mu s$ (67000 processing windows have been processed in the embedding and detection). The time consumed is negligible with respective to the time waiting for a processing window to be filled up ($5s * 30 = 150s$). Note that our scheme will introduce a maximum transmission delay of $5 * \varpi$ seconds due to the watermark embedding. However, according to our experimental results, the average transmission delay of the scheme is only $2.5 * \varpi$. According to the User Guide of T-Drive Data [27], this delay only introduces an average error of 282 meters for the stream receiver (when $\varpi = 30$). The error can be handled by many trajectory prediction methods introduced in [29].

We also performed experiments evaluating the impact of our watermark embedding on data quality. We adopt the relative error ξ defined in [9]: $\xi = \sum_{T} \frac{||T - T^w||}{||T||}$, where T and T^w represents the original and the corresponding watermarked trajectory respectively. Our experimental results show that the average value of ξ over a large number (1000+) of runs is 0.23%, much less than 1%, the bound that can lead to a visible error [9].

5 Conclusions

In this paper, we investigated the problem of rights protection for trajectory streams. We proposed an online watermarking scheme that is resilient to various common trajectory transformations. We implemented the proposed watermarking algorithm and evaluated it experimentally on real data. The method proves to be resilient to all the considered transformations, including geometric transformations, noise addition, trajectory segmentation and compression. For future work we plan to consider the online content authentication for trajectory streams, which is another important issue related to streaming data security.

Acknowledgments. This work is supported by the National Natural Science Foundations of China under grant (No.61100019 and No. 61262021). Our appreciation also goes to Liao Zhang for his contribution in the experiment of this paper.

References

1. Petitcolas, F.A.P., Anderson, R.J., Kuhn, M.G.: Information Hiding: A survey. Proceedings of the IEEE 87(7), 1062–1078 (1999)
2. Wang, Y., Doherty, J.F., Van Dyck, R.E.: A Wavelet-based Watermarking Algorithm for Ownership Verification of Digital Images. IEEE Transaction on Image Processing 11(2), 77–88 (2002)
3. Kirovski, D., Malvar, H.S.: Spread Spectrum Watermarking of Audio Signals. IEEE Transactions on Signal Processing 51(4), 1020–1033 (2003)
4. Langelaar, G.C., Lagendijk, R.L.: Optimal Differential Energy Watermarking of DCT Encoded Images and Video. IEEE Transactions on Image Processing 10(1), 148–158 (2001)
5. Doncel, V.R., Nikolaidis, N., Pitas, I.: An Optimal Detector Structure for the Fourier Descriptors Domain Watermarking of 2D Vector Graphics. IEEE Transactions on Visualization and Computer Graphics 13(5), 851–863 (2007)
6. Yan, H., Li, J., Wen, H.: A Key Points-based Blind Watermarking Approach for Vector Geospatial Data. Computers, Environment and Urban Systems 35(6), 485–492 (2011)
7. Wang, C.J., Peng, Z.Y., Peng, Y.W., Yu, L., Wang, J.Z., Zhao, Q.Z.: Watermarking Geographical Data on Spatial Topological Relations. Multimedia Tools Applications 57(1), 67–89 (2012)
8. Jin, X., Zhang, Z., Wang, J., Li, D.: Watermarking Spatial Trajectory Database. In: Zhou, L., Ooi, B., Meng, X. (eds.) DASFAA 2005. LNCS, vol. 3453, pp. 56–67. Springer, Heidelberg (2005)
9. Lucchese, C., Vlachos, M., Rajan, D., Yu, P.S.: Rights Protection of Trajectory Datasets with Nearest-neighbor Preservation. The VLDB Journal 19, 531–556 (2010)
10. Sion, R., Atallah, M., Prabhakar, S.: Resilient Rights Protection for Sensor Streams. In: IEEE International Conference on Very Large Databases, vol. 30, pp. 732–743 (2004)
11. Sion, R., Prabhakar, S.: Rights Protection for Discrete Numeric Streams. IEEE Transactions on Knowledge and Data Engineering 18(5), 699–714 (2006)

12. Yuan, N.J., Zheng, Y., Zhang, L., Xie, X.: T-Finder: A Recommender System for Finding Passengers and Vacant Taxis. IEEE Transactions on Knowlege and Data Enginerring 25(10), 2390–2403 (2013)
13. Ma, S., Zheng, Y., Wolfson, O.: T-Share: A Large-Scale Dynamic Taxi Ridesharing. In: IEEE International Conference on Data Engineering, pp. 410–421 (2013)
14. Aung, H.H., Guo, L., Tan, K.-L.: Mining sub-trajectory cliques to find frequent routes. In: Nascimento, M.A., Sellis, T., Cheng, R., Sander, J., Zheng, Y., Kriegel, H.-P., Renz, M., Sengstock, C. (eds.) SSTD 2013. LNCS, vol. 8098, pp. 92–109. Springer, Heidelberg (2013)
15. Lee, J.G., Han, J., Whang, K.Y.: Trajectory Clustering: A Partition-and-Group Framework. In: Proceedings of the ACM SIGMOD International Conference on Management of Data, pp. 593–604 (2007)
16. Muckell, J., Hwang, J.H., Lawson, C.T., Ravi, S.S.: Algorithms for Compressing GPS Trajectory Data: An Empirical Evaluation. In: Proceedings of the ACM SIGSPATIAL International Conference on Advances in Geographic Information Systems, pp. 402–405 (2010)
17. Chen, M., Xu, M., Franti, P.: A Fast O(N) Multiresolution Polygonal Approximation Algorithm for GPS Trajectory Simplification. IEEE Transactions on Image Processing 21, 2770–2785 (2012)
18. Potamias, M., Patroumpas, K., Sellis, T.: Sampling Trajectory Streams with Spatiotemporal Criteria. In: International Conference on Scientific and Statistical Database Management, pp. 275–284 (2006)
19. Muckell, J., Hwang, J.H., Patil, V., Lawson, C.T., Ravi, S.S.: SQUISH: An Online Approach for GPS Trajectory Compression. In: International Conference on Computing for Geospatial Research & Applications, p. 13 (2011)
20. Muckell, J., Olsen, P.W., Hwang, J.H., Lawson, C.T., Ravi, S.S.: Compression of Trajectory Data: A Comprehensive Evaluation and New Approach. GeoInformatica (2013), doi:10.1007/s10707-013-0184-0
21. Keogh, E., Chu, S., Hart, D., Pazzani, M.: An Online Algorithm for Segmenting Time Series. In: Proceedings of IEEE International Conference on Data Mining, pp. 289–296 (2001)
22. Douglas, D.H., Peucker, T.K.: Algorithms for the Reduction of the Number of Points Required to Represent a Digitized Line or Its Caricature. Canadian Cartographer 10, 112–122 (1973)
23. Lee, W.C., Krumm, J.: Trajectory preprocessing. In: Computing with Spatial Trajectories, pp. 3–33. Springer, New York (2011)
24. Schneier, B.: Applied Cryptography. John Wiley and Sons (1996)
25. Chen, B.: Quantization Index Modulation: A Class of Provably Good Methods for Digital Watermarking and Information Embedding. IEEE Transactions on Information Theory 47(4), 1423–1443 (2001)
26. Yuan, J., Zheng, Y., Zhang, C., Xie, W., Xie, X., Sun, G., Huang, Y.: T-drive: Driving directions based on taxi trajectories. In: Proc. of the 18th ACM SIGSPATIAL International Conference on Advances in Geographic Information Systems, pp. 99–108 (2010)
27. Yuan, J., Zheng, Y., Xie, X., Sun, G.: Driving with knowledge from the physical world. In: Proc. of the 17th ACM SIGKDD International Conference on Knowledge Discovery and Data Mining, pp. 316–324 (2011)
28. Cox, I., Miller, M., Bloom, J.: Digital watermarking. Morgan Kaufmann (2001)
29. Srinivasan, S., Dhakar, N.S.: Route-choice Modeling using GPS-based Travel Surveys (2011-008) (2013)

Efficient Detection of Emergency Event from Moving Object Data Streams

Limin Guo[1], Guangyan Huang[2], and Zhiming Ding[1]

[1] Institute of Software, Chinese Academy of Sciences, China
{limin,zhiming}@nfs.iscas.ac.cn
[2] Centre for Applied Informatics, School of Engineering & Science,
Victoria University, Australia
guangyan.huang@vu.edu.au

Abstract. The advance of positioning technology enables us to online collect moving object data streams for many applications. One of the most significant applications is to detect emergency event through observed abnormal behavior of objects for disaster prediction. However, the continuously generated moving object data streams are often accumulated to a massive dataset in a few seconds and thus challenge existing data analysis techniques. In this paper, we model a process of emergency event forming as a process of rolling a snowball, that is, we compare a size-rapidly-changed (e.g., increased or decreased) group of moving objects to a snowball. Thus, the problem of emergency event detection can be resolved by snowball discovery. Then, we provide two algorithms to find snowballs: a clustering-and-scanning algorithm with the time complexity of $O(n^2)$ and an efficient adjacency-list-based algorithm with the time complexity of $O(n\log n)$. The second method adopts adjacency lists to optimize efficiency. Experiments on both real-world dataset and large synthetic datasets demonstrate the effectiveness, precision and efficiency of our algorithms.

1 Introduction

With the advance of mobile computing technology and the widespread use of GPS-enabled mobile devices, the location-based services have been improved greatly. Positioning technologies enable us to online collect abundant moving object data streams for many applications in the fields of traffic analysis, animal studies, etc.

One of the most significant applications is to detect emergency event through observed abnormal behavior of objects for disaster prediction. The forming process of an emergency event can be compared to the process of rolling a snowball, that is a group of abnormally-behaved objects whose size is increased or decreased rapidly and continuously for a period of time. Thus, we model the problem of emergency event detection as the snowball discovery problem. Snowball discovery can be widely used for emergency event detection to predict disasters. The detection of rapidly gathering crowd can stop an illegal riot. The early aware of the rapidly decreasing of animal population can save extinct animal, for example, the drastic reduction of herring may be caused by ecological damage. Detecting a continually and dramatically growing of a group of vehicles in real time can navigate to avoid traffic jams and locate the traffic

S.S. Bhowmick et al. (Eds.): DASFAA 2014, Part II, LNCS 8422, pp. 422–437, 2014.
© Springer International Publishing Switzerland 2014

accidents. Discovering animals' abnormal and rapid gathering can assist predict the earthquake. Therefore, snowball discovery has wide applications.

We must efficiently discover snowball patterns, and leave enough time for people to take immediate actions; most applications need online processing of moving object data streams. However, there are few techniques for emergency event detection; since the continuously generated moving object data streams are often accumulated to a massive dataset in a few seconds and thus challenge existing data analysis techniques.

The state-of-the-art studies of moving object pattern mining are generally classified into two categories: clustering moving objects and clustering trajectories. The former studies group patterns, while the later aims to group similar trajectories. However, these methods cannot discover the abnormal trends of moving object clusters.

We summarize the most related work into two classes: moving together groups and gathering as shown in Fig. 1.

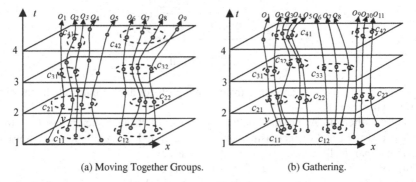

(a) Moving Together Groups.   (b) Gathering.

Fig. 1. Two Most Related Works

The first class focuses on finding object groups that move together, including moving cluster[1], flock[2], convoy[3], companion[4] and swarm[5]. Moving cluster finds size-dynamically-changed groups of moving objects which share a percentage of common objects, while flock, convoy, companion and swarm study on object groups that move together for a period of time. In Fig. 1(a), for example, $<c_{11}, c_{21}, c_{31}, c_{41}>$ is a moving cluster, $<o_7, o_8>$ is a convoy (or flock, companion), and $<o_6, o_7, o_8, o_9>$ is a swarm. They all require clusters to contain the same set of objects or share a percentage of common objects. However, none of these techniques can detect snowballs— the continuously and rapidly increase (or decrease) of clusters.

The second class aims to discover a sequence of clusters without strong coherent membership, such as gathering[6] where the locations of two consecutive clusters are close. Also, objects participate in gathering lasting for a time period. In Fig. 1(b), $<c_{11}, c_{21}, c_{32}, c_{41}>$ is a gathering example. However, the limitations of gathering are (1)the clusters of gathering maybe not related to each other, and (2)the clusters of gathering cannot reflect abnormal changes. For instance, $<c_{12}, c_{23}, c_{33}, c_{42}>$ is a gathering, but there is no relationship between any two consecutive clusters (e.g., c_{12} and c_{23}), so, the gathering cannot detect the abnormal trend of clusters.

In this paper, we aim at continuously detecting emergency event from moving object data streams, which refer to a series of clusters that increase or decrease rapidly for a period of time. As we have modeled an emergency event forming as a snowball rolling, emergency event detection and snowball discovery are exchangeable. We propose a novel method, called E^3D (Efficient Emergency Event Detection), to discover *snowball* patterns. We define two forms of snowball: *snowball+* and *snowball–*. Snowball+ (or snowball–) denotes a group of objects with a continuously increased (or decreased) size. Fig. 2 shows an example of snowball+. More and more objects join in the group and move together with time.

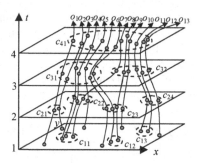

Fig. 2. Snowball

However, discovery of snowball from moving object data streams challenges us in the following aspects.

Strong Coherent. A snowball pattern is a series of strong-coherent clusters, which steadily lead the trend of increasing (or decreasing) and thus exclude those noise objects that dynamically change (in and out the cluster).

Durability. The snowball patterns must be tracked for a period of time.

Portion. The clusters in a snowball pattern should share a part of common objects.

Efficiency. High efficiency is vital to process large scale data, since moving object data streams are generated continuously and accumulated to a massive dataset quickly.

Effectiveness. To avoid mining redundant snowballs in large amount of numbers, we should discover long-lasting snowballs instead of short-time ones.

To approach the above challenges, we provide two algorithms for snowball discovery: a clustering-and-scanning algorithm and an adjacency-list-based algorithm. There are two steps in clustering-and-scanning algorithm: clustering step aims to cluster objects with strong coherent, and scanning step aims to discover qualified snowballs. The adjacency-list-based algorithm adopts adjacency lists to optimize efficiency. In summary, the main contributions of this paper are as followings.

(1) A new concept of snowball is proposed.

(2) A general method, clustering-and-scanning algorithm, is proposed with time complexity of $O(n^2)$.

(3) An adjacency-list-based algorithm is proposed to improve the efficiency of the general method. The time complexity is $O(n\log n)$.

(4) The effectiveness, precision and efficiency of our methods are demonstrated on both real and synthetic moving object data streams.

The rest of the paper is organized as follows. Sect. 2 briefs the related work. Sect. 3 defines the problem. Sect. 4 proposes two algorithms to discover snowball. Sect. 5 evaluates the experiment's performances. Finally we conclude the paper in Sect. 6.

2   Related Work

In this section, we present a survey of methods for discovering common patterns from moving object data streams, which are classified into two classes: clustering moving objects and clustering trajectories.

2.1   Clustering Moving Objects

Clustering moving objects aims at catching interesting common patterns or behaviors of individual moving objects, typically, to find a group of objects moving together.

Existing work on finding patterns of travelling together requires clusters to contain the same set or a part of objects in the whole trajectory. Flock[2], convoy[3] and companion[4] discover a group of objects moving together for a fixed consecutive time. Swarm[5] is an extension of above, which relaxes the consecutive time constraint. Micro-clustering[7] finds a group of objects which are close to each other and are treated as a whole. Moving cluster[1] is a group of objects which are partly overlapped at two consecutive timestamps. Gathering[6] aims to find a group of clusters moving closer to each other and desires some objects to occur in gathering for a time period.

In addition to the most related work mentioned above, some other work detects the relationships based on the movement directions and locations, such as leadership[8], convergence[9], etc. Leadership finds a group of objects that move together in the same direction, and at least one object is a group leader for a fixed consecutive time. Convergence finds a group of objects that pass through or in the same circular region.

However, the above mentioned methods cannot be used to discover snowball patterns, since snowball is a very different new type of patterns, which is used to capture the size-rapidly-changed groups of objects moving forwards instead of groups of moving objects with common behavior in the most related work such as gathering, moving cluster and convoy. Snowball is actually used to find changes or trends of clusters that comprise common behaved moving objects.

2.2   Clustering Trajectories

The methods for clustering trajectories focus on the spatial-temporal characteristics of trajectories, which refer to the common paths for a group of moving objects.

The key problem of the research is to define similarity between two trajectories. Most earlier methods of trajectory clustering take a trajectory as a whole, such as probabilistic method[10], distance based method[11], etc. But in recent years, more and more works detect similar portions of trajectories (e.g., common sub-trajectory). For example, partition-and-group framework [12] partitions each trajectory into a set of sub-trajectories, and then groups clusters using distance function. In [13], both the spatial and temporal criteria are considered for trajectory dividing and clustering. In [14-16], regions-of-interest are detected from trajectories to find frequent patterns.

While above mentioned methods for clustering trajectories often group static similar trajectories into the same cluster, our snowball discovering dynamically and continuously detect emergency event to satisfy our goal.

3   Problem Definition

In this section, we give the definitions of all necessary concepts and state the snowball discovery problem. Table.1 lists the notations used throughout this paper.

Table 1. List of Notations

Notation	Explanation	Notation	Explanation	Notation	Explanation
T	the time series	t, t_i, t_j	the timestamps	ε	the distance threshold
O	the object set	o, o_i, p, q	the objects	μ	the density threshold
S	the trajectory stream	s_i, s_j	the snapshots in stream	δ_{speed}	the speed threshold
C	the cluster set	$c_i, c_{t,i}, c_{t,j}$	the clusters	δ_θ	the integrity threshold
SN	the snowball set	sn, sn'	the snowballs	δ_t	the duration threshold
SN_{cand}	the candidate set	sn_{cand}, sn'_{cand}	the candidates	h	the time window size
Ar	the affected area	AL	the adjacency list	G,V,E,W	the graph

3.1   Basic Concept

Before we define our snowball pattern, we introduce the preliminary knowledge about density-based clustering (e.g., DBSCAN[17]), which discovers clusters with arbitrary shapes and specified density. The further studies of [17] can refer to [18] and [19].

Let $T = \{t_1, t_2, ..., t_n\}$ be a time series. Let $O = \{o_1, o_2, ..., o_m\}$ be a collection of objects that move during $[t_1, t_n]$. Let $S = \{s_1, s_2, ..., s_n\}$ be a sequence of snapshots, where s_i is a set of objects and their locations at timestamp t_i.

[17] defines **ε-neighborhood** of an object p as $N_\varepsilon(p) = \{q \in O | dist(p, q) \le \varepsilon\}$, where $dist(p, q)$ is the distance between p and q. An object p is **density-reachable** to q if there is a chain of $\{p_1, p_2, ..., p_n\}$, where $p_1 = q$, $p_n = p$, p_{i+1} and p_i satisfying $|N_\varepsilon(p_i)| \ge \mu$ and $p_{i+1} \in N_\varepsilon(p_i)$ w.r.t ε and μ. An object p is **density-connected** to q if there is o such that p and q are density-reachable from o. c is a **cluster** if: (1) $\forall p, q \in c$: p is density-reachable to q w.r.t ε and μ; (2) $\forall p, q \in c$: p is density-connected to q w.r.t ε and μ.

DBSCAN may produce noisy objects that dynamically move in and out the cluster, but we require a more coherent cluster without noises. Snowball only focuses on the major objects that steadily lead the trend of increasing (or decreasing), since the historical behaviors of objects affect the cluster. Therefore, in this paper, we define a new dynamic neighborhood, which considers the historical behaviors of the objects, to replace ε-neighborhood in DBSCAN. We assume that the cluster is influenced by behaviors of objects over a fixed **time window**. Suppose the size of time window is h, for a coming timestamp t_i the time window is formed as $(t_{i-h}, t_i]$.

Definition 1 (Dynamic Neighborhood): Given a distance threshold ε, the dynamic neighborhood of an object p, denoted by $N_\varepsilon(p, t_i)$, is defined as following:
$$N_\varepsilon(p, t_i) = \{q \in O | dist(p, q, t_i, h) \le \varepsilon\}$$
where t_i is the current timestamp, h is the size of time window, and $dist(p, q, t_i, h) = \max\{dist(p, q, t_{i-k}) | 0 \le k < h\}$, $dist(p, q, t_{i-k})$ is the distance between p and q at t_{i-k}.

Dynamic neighborhood describes a set of strong coherent objects based on distance that can be observed within time windows. We compare dynamic neighborhood with ε-neighborhood in Fig. 3. Let $h = 3$, the dynamic neighborhood of o_3 at t_3 is $\{o_2, o_4\}$, while the ε-neighborhood of o_3 is $\{o_1, o_2, o_4\}$.

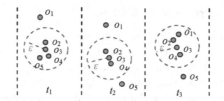

Fig. 3. Dynamic Neighborhood and ε-Neighborhood

Different from density-based clustering[17], we modify the neighborhood function and use dynamic neighborhood to replace ε-neighborhood to cluster moving objects. Thus, dynamic neighborhood can satisfy our goal of snowball discovery, since objects are clustered with stronger coherence without noise.

3.2   Problem Definition

In this subsection, we define snowball and model the problem of snowball discovery.

Definition 2 (Cluster-Connected): Let c_{t_i} and $c_{t_{i+1}}$ be two clusters at consecutive timestamps, δ_{speed} be a speed threshold, δ_θ be an integrity threshold . $c_{t_{i+1}}$ is cluster-connected to c_{t_i} in two forms: cluster-connected+ and cluster-connected−.

$c_{t_{i+1}}$ is cluster-connected+ to c_{t_i} if:

(1) $|c_{t_i} \cap c_{t_{i+1}}|/|c_{t_i}| \geq \delta_\theta$

(2) $(|c_{t_{i+1}}| - |c_{t_i}|)/|c_{t_i}| \geq \delta_{speed}$

or

$c_{t_{i+1}}$ is cluster-connected− to c_{t_i} if:

(1) $|c_{t_i} \cap c_{t_{i+1}}|/|c_{t_{i+1}}| \geq \delta_\theta$

(2) $(|c_{t_i}| - |c_{t_{i+1}}|)/|c_{t_i}| \geq \delta_{speed}$

The first condition indicates the cluster shares a part of common objects from the previous one w.r.t δ_θ, and the second condition describes the cluster is increase (or decrease) by the speed w.r.t δ_{speed}.

Definition 3 (Snowball): Let $sn = \{c_{t_1}, c_{t_2}, ..., c_{t_n}\}$ be a sequence of clusters at consecutive timestamps, δ_t be a duration threshold, sn is called snowball in two forms: snowball+ and snowball−. sn is snowball+/− if:

(1) $t_n - t_1 \geq \delta_t$

(2) $c_{t_{i+1}}$ is cluster-connected+/− to c_{t_i}, where $t_1 \leq t_i < t_n$.

The first condition implies the duration of sn is at least δ_t, and the second condition describes the cluster-connected relationship between each two consecutive clusters.

To avoid mining redundant snowballs, we define a ***maximal snowball***. A snowball sn is a maximal snowball if sn cannot be enlarged, which means $\nexists sn'$ s.t. sn' is a snowball and $sn \subset sn'$. As shown in Fig.2, let $\delta_t = 4$, $\delta_{speed} = 0.3$ and $\delta_\theta = 0.5$, suppose clusters are depicted with dotted lines, then $<c_{12}, c_{23}, c_{31}, c_{41}>$ is a maximal snowball.

Therefore, the problem that will be resolved in this paper can be defined as follows:

Snowball Discovery Problem. Given a sequence of snapshots $S = \{s_1, s_2, ..., s_n\}$, each snapshot $s_i = \{c_{t_i 1}, c_{t_i 2}, ..., c_{t_i m}\}$, where $c_{t_{ij}}$ is a cluster discovered at timestamp t_i. When the snapshot s_i arrives, the task is to detect emergency event, called E^3D (Efficient Emergency Event Detection) or snowball discovery problem, which discovers snowball set SN containing all the maximal snowballs.

3.3   Overview of the Proposed Algorithms

The general framework of E^3D can be divided into two steps: clustering moving objects and snowball discovery. In this subsection, we brief an overview of our two algorithms: clustering-and-scanning algorithm and adjacency-list-based algorithm.

The clustering-and-scanning algorithm clusters moving objects for each snapshot first, and then scans each cluster in the coming snapshot to check whether it can be extended. In this way, all snowballs are discovered gradually. In order to improve efficiency, we introduce pruning rules to reduce unnecessary process.

The adjacency-list-based algorithm adopts adjacency lists to further optimize efficiency, where we use adjacency lists to connect relationships between clustering and scanning steps.

4   Snowball Discovery

4.1   Clustering-and-Scanning Algorithm

In this subsection, we provide details of the clustering-and-scanning algorithm. As mentioned, there are two phases: clustering and scanning. In the first phase, we perform density-based clustering which is similar to DBSCAN at each timestamp to discover clusters. However, DBSCAN considers the locations of objects at current timestamp, not historical locations which also influence the dynamic neighborhood, so we maintain locations of objects for the last h timestamps with R-tree, the dynamic neighborhood over time window is obtained by searching the locations for the last h timestamps. The time complexity of clustering is $O(n\log n)$. The detail of this phase is omitted due to space limitation. The second phase aims to discover snowballs.

Let $sn_{cand} = \{c_{t_1}, c_{t_2}, ..., c_{t_n}\}$ be a set of consecutive clusters, sn_{cand} is a **snowball candidate** if $c_{t_{i+1}}$ is cluster-connected to c_{t_i} $(t_1 \leq t_i < t_n)$. Intuitively, the discovery of snowball can extend a shorter snowball candidate into a longer one until it cannot be extended. Once the duration of the candidate exceeds δ_t, it's a qualified snowball.

Algorithm 1 presents the pseudo code of snowball discovery. We first initialize sets (Line 1-2). At each coming snapshot, we cluster the objects (Line 3-5). Then we scan the last cluster of each snowball candidate to see whether there is a set of clusters that can be appended to the candidate (Line 6-9). If not, the candidate with enough duration is reported as maximal snowball (Line 10-12). Otherwise, the candidate is extended by appending each cluster-connected cluster (Line 13-17). After inserting new candidates, we remove old ones (Line 18). Since the remaining clusters that cannot be appended to any candidate may generate new candidates, they also should insert into candidate set (Line 19-22). Finally, the algorithm returns results (Line 23).

Function *DiscoverCluster* discovers clusters at current timestamp. Function *SearchValidCluster* searches clusters from cluster set C_{t_i} at current timestamp, which are cluster-connected to the last cluster $c_{t_{i-1}}$ of snowball candidate. A naïve method is to calculate all conditions in definition 2 for each $c_{t_i} \in C_{t_i}$. The most time-consuming part is to intersect each two consecutive clusters, suppose n and n_1 are the size of moving objects and $c_{t_{i-1}}$, then *SearchValidCluster* requires $O(n_1 \times n)$ time complexity, and it will be performed iteratively which is time consuming.

Algorithm 1: Clustering-and-Scanning Algorithm	
Input: $\varepsilon, \mu, h, \delta_{speed}, \delta_\theta, \delta_t, S$	
Output: SN	
1. $SN \leftarrow \emptyset$;	//set of maximal snowballs
2. $SN_{cand} \leftarrow \emptyset$;	//set of current snowball candidates
3. **for** $t_i = t_1$ to t_n **do**	
4. $R \leftarrow \emptyset$;	
5. $C_{t_i} \leftarrow DiscoverCluster(S, \varepsilon, \mu, h)$;	//cluster moving objects at t_i
6. **for each** snowball candidate $sn_{cand} \in SN_{cand}$ **do**	
7. $c_{t_{i-1}} \leftarrow$ the last cluster of sn_{cand};	
8. $C'_{t_i} \leftarrow SearchValidCluster(c_{t_{i-1}}, C_{t_i}, \delta_{speed}, \delta_\theta)$;	//find a set of clusters cluster-connected to $c_{t_{i-1}}$
9. $R \leftarrow R \cup C'_{t_i}$;	
10. **if** $C'_{t_i} = \emptyset$ **then**	//sn_{cand} cannot be extended
11. **if** $sn_{cand}.duration \geq \delta_t$ **then**	
12. $SN \leftarrow SN \cup sn_{cand}$;	//sn_{cand} is a maximal snowball
13. **else**	
14. **for each** $c_{t_i} \in C'_{t_i}$ **do**	
15. $sn'_{cand} \leftarrow$ append c_{t_i} to sn_{cand};	
16. $SN_{cand} \leftarrow SN_{cand} \cup sn'_{cand}$;	
17. $sn'_{cand}.duration \leftarrow sn'_{cand}.duration + |t_i - t_{i-1}|$;	
18. remove sn_{cand} from SN_{cand};	
19. **for each** $c_{t_i} \in C_{t_i} \backslash R$ **do**	//remaining clusters are new snowball candidates
20. $sn'_{cand} \leftarrow c_{t_i}$;	
21. $SN_{cand} \leftarrow SN_{cand} \cup sn'_{cand}$;	
22. $sn'_{cand}.duration \leftarrow 0$;	
23. **return** SN;	

Due to expensive computation, we introduce pruning rules to improve efficiency. Actually, it's no need to intersect each two clusters between consecutive snapshots.

Definition 4(Affected Area): Let $c_{t_{i-1}} = \{o_1, o_2, ..., o_n\}$ be a cluster at timestamp t_{i-1}, the affected area of $c_{t_{i-1}}$ is defined as a circle Ar at timestamp t_i satisfying:

(1) The center of Ar is $cent(Ar) = \{\sum_{o_k \in c_{t_{i-1}}} o_k(t_i).x_l / n\}_{l=1}^{dim}$.

(2) The radius of Ar is $r(Ar) = \max\{dist(cent(Ar), o_k(t_i)) | o_k \in c_{t_{i-1}}\}$.

where dim is the dimension of object, $o_k(t_i)$ and $o_k(t_i).x_l$ are the position and the lth dimensional position of o_k at t_i, $dist(a, b)$ is the distance between a and b.

As shown in Fig. 4(a), the affected area Ar of c_{12} is depicted with red dotted circle. Ar is determined by positions of $\{o_3, o_4, o_5\}$ at timestamp t_2, it is obvious that Ar only overlaps with cluster c_{22} and c_{23}, while c_{21} need not consider.

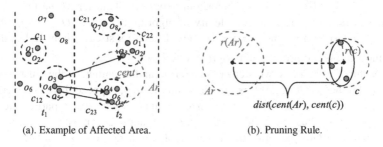

(a). Example of Affected Area.   (b). Pruning Rule.

Fig. 4. Affected Area

Lemma 1: Let c be a cluster, and Ar be an affected area, suppose $cent(c)$ is the geometry center of c and $r(c)$ is the distance from $cent(c)$ to the farthest object in c. If $dist(cent(Ar), cent(c)) > r(Ar) + r(c)$, then Ar and c are disjoint.

Proof : As Fig.4(b) shows, if $dist(cent(Ar), cent(c)) > r(Ar) + r(c)$, for $\forall\, o_i \in c$, $dist(o_i, cent(Ar)) > r(Ar)$, so o_i doesn't intersect with Ar, then Ar and c are disjoint.

Lemma 2: Let c_{t_i} and $c_{t_{i-1}}$ be two clusters at t_i and t_{i-1}, and δ_θ be an integrity threshold, if there are more than $(1 - \delta_\theta) \times |c_{t_{i-1}}|$ objects of $c_{t_{i-1}}$ appearing in other clusters at t_i, then c_{t_i} will not cluster-connected+ to $c_{t_{i-1}}$.

Proof: Since each object only belongs to one cluster in a snapshot, then even if all the remaining objects of $c_{t_{i-1}}$ appear in c_{t_i}, condition (1) of definition 2 will not satisfy, as $|c_{t_i} \cap c_{t_{i-1}}|/|c_{t_{i-1}}| < (|c_{t_{i-1}}| - (1 - \delta_\theta) \times |c_{t_{i-1}}|)/|c_{t_{i-1}}| < \delta_\theta$, c_{t_i} won't cluster-connected+ to $c_{t_{i-1}}$.

Similarly, if there are more than $|c_{t_{i-1}}| - \delta_\theta \times |c_{t_i}|$ objects of $c_{t_{i-1}}$ appearing in other clusters at t_{i-1}, then c_{t_i} will not cluster-connected− to $c_{t_{i-1}}$.

Lemma 1 and 2 can prune useless process that intersections will be stop earlier if the two lemmas are satisfied. Algorithm 2 shows an improved *SearchValidCluster*.

Algorithm 2: SearchValidCluster Algorithm

Input: $c_{t_{i-1}}$: the cluster at timestamp t_{i-1}, C_{t_i}: the set of clusters at timestamp t_i, δ_{speed}, δ_θ

Output: C

1. $C \leftarrow \emptyset$;   //set of valid clusters
2. $R \leftarrow \emptyset$;   //set of intersected objects from $c_{t_{i-1}}$
3. $Ar \leftarrow AffectArea(c_{t_{i-1}})$;   //generate affect area of $c_{t_{i-1}}$
4. **for each** $c_{t_i} \in C_{t_i}$ **do**
5. | **if** $dist(cent(Ar), cent(c_{t_i})) > r(Ar) + r(c_{t_i})$ **then**   continue;   //lemma 1
6. | **if** cluster-connected+ and $|R| > (1 - \delta_\theta) \times |c_{t_{i-1}}|$ **then**   break;   //lemma 2
7. | **if** cluster-connected− and $|R| > |c_{t_{i-1}}| - \delta_\theta \times |c_{t_i}|$ **then**   break;   //lemma 2
8. | **if** c_{t_i} is cluster-connected to $c_{t_{i-1}}$ **then**   //definition 2
9. | | $C \leftarrow C \cup c_{t_i}$;
10. | └ $R \leftarrow R \cup$ intersected objects between $c_{t_{i-1}}$ and c_{t_i};
11. **return** C;

In algorithm 2, we initialize and generate the affected area (Line 1-3). Then for each current cluster, before checking the cluster-connected relationship, we check lemma 1 and 2 first (Line 4-10). Finally, the algorithm returns valid clusters (Line 11).

In the worst case, the time complexity of algorithm 2 is still $O(n_1 \times n)$, where n_1 and n are the size of $c_{t_{i-1}}$ and moving objects. But in most case, lemma 1 and 2 can prune most of invalid process and improve efficiency.

Algorithm 1 comprises two steps of clustering and scanning. In the clustering step, the algorithm needs $O(n\log n)$ time. In the scanning step, suppose there are average m_1 clusters, and n_1 is the average cluster size. Actually each cluster participate in the iterative process, thus the time complexity of scanning step is $m_1 \times O(n_1 \times n) = O(m_1 \times n_1 \times n) = O(n^2)$ and the total time complexity is $O(n\log n + n^2) = O(n^2)$.

4.2   Adjacency-List-Based Algorithm

Although clustering-and-scanning algorithm uses pruning rules to prune any useless process, the time complexity is still $O(n^2)$, where n is the number of objects. The two steps, clustering and scanning, are separated from each other. But in fact there is some connection between them. In this subsection, we use the adjacency list to connect the relationship. Adjacency list is a list of clusters that are cluster-connected to a given cluster. For each cluster, a linked list of clusters cluster-connected to it can be set up in the clustering step, after that, it's no need to process the most costly part of intersection which can be obtained from the adjacency list.

Given a sequence of snapshots S, we use a weighted graph to represent the relationship between clusters. A weighted graph G is modeled as the form $G = (V, E, W)$, where V is a set of vertices, E is a set of edges and W is a set of values mapping to edges. A vertex $v \in V$, defined as $v = c_i$, describes a cluster c_i in S. An edge $e \in E$, defined as $e = <c_i, c_j>$, from c_i to c_j means c_i is cluster-connected to c_j. A weight $w \in W$, defined as $w(c_i,c_j) = num$, represents there are num intersected objects between c_i and c_j. Then, we can represent the graph by the adjacency lists of all vertices, an adjacency list of vertex v is a sequence of vertices cluster-connected to v.

We convert Fig. 2 to a weighted graph, the graph and the corresponding adjacency lists are as shown in Fig. 5. Each vertex in Fig. 5(a) is a cluster in Fig. 2, an edge connecting c_i to c_j means c_i is cluster-connected to c_j, the weight of each edge represents the number of intersected objects between two vertices. Fig. 5(b) shows the adjacency lists corresponding to the graph of Fig. 5(a).

In order to set up adjacency list for each cluster in the clustering step, we modify clustering algorithm by building adjacency lists, and snowballs can be directly detected using adjacency lists. The adjacency-list-based algorithm is presented in algorithm 3.

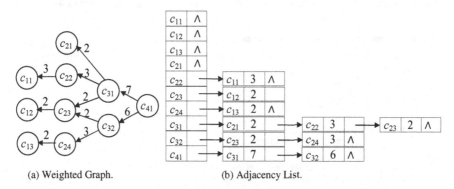

(a) Weighted Graph.   (b) Adjacency List.

Fig. 5. Weighted Graph and Adjacency List

In algorithm 3, we first initialize sets to store snowballs and adjacency lists (Line 1-2). At each coming snapshot, we cluster moving objects similar to DBSCAN, meanwhile, adjacency list is created or adjusted when every new cluster creates (Line 3-13). Then, we check the current link in adjacency list to see whether it satisfies cluster-connected relationship, if not, it will be removed (Line 14-16). Finally, we search links from adjacency lists and output qualified snowballs (Line 17-18).

Algorithm 3: Adjacency-List-Based Algorithm

Input: ε, μ, h, δ_{speed}, δ_θ, δ_t, S

Output: SN

1.	$SN \leftarrow \emptyset$;	//set of maximal snowballs
2.	$AL \leftarrow \emptyset$;	//set of adjacency lists
3.	**for** $t_i = t_1$ to t_n **do**	
4.	$C_{t_i} \leftarrow \emptyset$;	//set of clusters at t_i
5.	**for each** unvisited object $o \in s_i$ **do**	
6.	$o \leftarrow$ mark as visited;	
7.	$N \leftarrow GetNeighbours(o, \varepsilon, h, S)$;	//find the dynamic neighborhood of o w.r.t ε, h and S
8.	**if** $|N| < \mu$ **then**	
9.	$o \leftarrow$ mark as noise;	
10.	**else**	
11.	$c_{t_i} \leftarrow \emptyset$;	//create a new cluster
12.	$ExpandCluster(o, N, c_{t_i}, \varepsilon, \mu, AL)$;	//expand cluster and adjacency list
13.	$C_{t_i} \leftarrow c_{t_i} \cup C_{t_i}$;	
14.	**for each** link $c_{t_i} \rightarrow c_{t_{i-1}} \in AL$ **do**	//prune invalid links in AL
15.	**if** c_{t_i} is not cluster-connected to $c_{t_{i-1}}$ **then**	
16.	remove the link from AL;	
17.	$SN \leftarrow SearchList(AL, \delta_t, t_i)$;	//search links from AL and output maximal snowballs
18.	**return** SN;	

Function *GetNeighbours* finds dynamic neighborhood of the given object. Function *ExpandCluster* expands a new cluster and adjusts adjacency lists with the new cluster. We present the details of *ExpandCluster* in algorithm 4. When a new object is inserted into the cluster, the algorithm checks whether a link exists in adjacency lists, which points from the cluster to the previous cluster that the object belongs to at previous time. If so, the weight of the link is added, otherwise, a new link is created.

Algorithm 4: ExpandCluster Algorithm

Input: o: an object, N: the dynamic neighborhood of o, c_{t_i}: a new cluster, ε, μ, AL: adjacency lists

1.	$c_{t_i} \leftarrow o \cup c_{t_i}$;	
2.	$c_{ot_{i-1}} \leftarrow$ retrieve the cluster which o belongs to at t_{i-1};	
3.	$weight(c_{t_i}, c_{ot_{i-1}}) \leftarrow 1$;	//create a new link in AL
4.	**for each** $p \in N$ **do**	
5.	**if** p is unvisited **then**	
6.	$p \leftarrow$ mark as visited;	
7.	$N_p \leftarrow GetNeighbours(p, \varepsilon, h, S)$;	
8.	**if** $|N_p| \geq \mu$ **then**	
9.	$N \leftarrow N \cup N_p$;	
10.	**if** p is not member of any cluster **then**	
11.	$c_{t_i} \leftarrow p \cup c_{t_i}$;	
12.	$c_{pt_{i-1}} \leftarrow$ retrieve the cluster which p belongs to at t_{i-1};	
13.	**if** there is an link from c_{t_i} to $c_{pt_{i-1}}$ in AL **then**	
14.	$weight(c_{t_i}, c_{pt_{i-1}})$++;	
15.	**else**	
16.	$weight(c_{t_i}, c_{pt_{i-1}}) \leftarrow 1$;	
17.	**return**;	

Function *SearchList* searches links from adjacency lists, and outputs qualified snowballs. We provide midterm results in Fig. 6 to explain adjacency list search process for Fig. 5 Suppose the current timestamp is t_4 and a duration threshold δ_t is 4, we start search from c_{41}, and search forward according to the links in adjacency lists until there is no link. Then it is easy to see that $<c_{11}, c_{22}, c_{31}, c_{41}>$, $<c_{12}, c_{23}, c_{31}, c_{41}>$, $<c_{12}, c_{23}, c_{32}, c_{41}>$ and $<c_{13}, c_{24}, c_{32}, c_{41}>$ are qualified snowballs.

time	path				
t_4	c_{41}				
t_3	$<c_{41}, c_{31}>$			$<c_{41}, c_{32}>$	
t_2	$<c_{41}, c_{31}, c_{21}>$	$<c_{41}, c_{31}, c_{22}>$	$<c_{41}, c_{31}, c_{23}>$	$<c_{41}, c_{32}, c_{23}>$	$<c_{41}, c_{32}, c_{24}>$
t_1		$<c_{41}, c_{31}, c_{22}, c_{11}>$	$<c_{41}, c_{31}, c_{23}, c_{12}>$	$<c_{41}, c_{32}, c_{23}, c_{12}>$	$<c_{41}, c_{32}, c_{24}, c_{13}>$
duration	3	4	4	4	4

Fig. 6. An Adjacency List Search for the Weighted Graph in Fig. 5

Algorithm 3 includes two steps of clustering and searching adjacency list. Suppose there are average m_1 adjacency lists, and n_1 is the average size for each one. In the clustering step, the time complexity is $O(n \times (n_1 + \log n))$, since n_1 is significantly smaller than n, the time complexity is approximately equal to $O(n \log n)$. In the searching step, the time complexity is $O(m_1)$ and the total time complexity is $O(n \log n + m_1)$, due to m_1 is also significantly smaller than n, the time complexity is approximately equal to $O(n \log n)$.

5   Performance Evaluation

In this section, we conduct experiments to evaluate the effectiveness, precision and efficiency of our algorithms based on both real and synthetic moving object dataset.

The real data is a buffalo dataset (D1) retrieved from MoveBank.org[1]. Meanwhile, we simulate two datasets to show the precision and scalability of our algorithms; one dataset (D2) is simulated on the map of Beijing; another dataset (D3) is generated by a network-based generator of moving objects[20] on an open road map of Oldenburg.

All the experiments are implemented in C++ and run on a computer with Intel Pentium CPU(2.0GHZ) and 4GB memory. The main parameters are listed in Table 2.

Table 2. Experiment Settings

Name	Value	Meaning	Name	Value	Meaning
μ	20-110	the density threshold	ε	600-2000 (meter)	the distance threshold
δ_{speed}	0.05-0.6	the speed threshold	δ_θ	0.75-0.95	the integrity threshold
δ_t	6-30 (minute)	the duration threshold	h	2-3	the time windows size
Data Set	Objects#	Record#	Duration	Frequency	Data size
D1	740	26,609	2023 days	3-4 days	5.72 M
D2	6,000	36,000	1 hour	1 minutes	13.4M
D3	10,000	2,707,128	1 day	1 minutes	79.2 M

5.1   Effectiveness

In this subsection, we compare a snowball pattern with a corresponding convoy pattern that are extracted from D1 in Google Map.

We preprocess D1 by linear interpolation of the missing data with a time gap of one "day", each buffalo contains 901 days on average. Then we set $\mu = 5$, $\varepsilon = 0.001$

---

[1] http://www.movebank.org

(about 111 meters), $h = 2$, $\delta_{speed} = 0.6$, $\delta_\theta = 0.6$ and $\delta_t = 6$. Fig. 7 shows one snowball where each cluster has been lasting for two days. The discovery of snowballs enables us to capture the trends of moving behavior of buffalo and thus detect emergency events, such as migration. From the figure, we can see that a snowball+ starts with 5 objects from region A, then gathers 3 more in region B to increase its size to 8 objects, and finally arrives at region C with size of 15 objects, which lasts for 6 days, during the moving, more and more objects gather and keep moving.

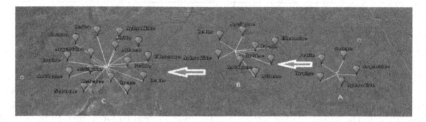

Fig. 7. One Snowball Discovered in D1

We also discover convoy patterns on the same data set D1 to compare with our snowballs. We set $\mu = 5$, $\varepsilon = 0.001$ (about 111 meters), $k = 6$ and $m = 3$, where m and k are minimum size of cluster and number of consecutive timestamps in convoy, and k is equal to δ_t in snowball. Fig. 8 shows one convoy pattern—three buffaloes which move together from region A to region C lasting 6 days where each cluster samples for two days. This convoy pattern is actually comprised in our snowball+ in Fig. 7; but the size of snowball+ is increased while the size of convoy keeps the same.

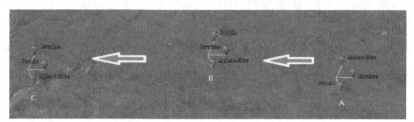

Fig. 8. One Convoy Discovered in D1

The comparison shows the difference between snowball and convoy, the snowball is more meaningful in detecting abnormal trend of clusters, while the convoy aims to discover moving together group that cannot capture tendency of abnormal group.

5.2   Precision

In this subsection, we simulate 20 emergency events and the corresponding behavior of 6000 moving objects to evaluate the precision of snowball patterns. To simulate the real-world emergency events, we rank emergency events into 5 grades based on the duration of an event; the higher the grade, the longer the event lasts for. Once the event occurs (e.g., traffic accident), the behaviors of moving objects will be impact, for example, traffic accident that brings traffic jams can be detected from analysis of moving object data streams through discovering of snowball patterns.

We use two metrics to evaluate the precision of our methods: $accuracy = \xi_{ESN}/\xi_{SN}$ and $recall = \xi_{ESN}/\xi_{EVENT}$, where ξ_{ESN} is the number of events discovered by snowballs, ξ_{SN} is the number of snowballs, and ξ_{EVENT} is the number of events. Fig. 9 plots the accuracy and recall changing with speed δ_{speed} and duration δ_t, the default parameters used in the experiments are: $\mu = 20$, $\varepsilon = 2000$, $h = 2$ and $\delta_\theta = 0.8$. Fig. 9(a) shows with the increase of δ_{speed}, the recall is decreased, the reason is that the condition of snowball becomes stricter with the increase of δ_{speed}, then only the most important events can be discovered and ξ_{ESN} is obviously reduced, while ξ_{EVENT} is remain the same, so the recall is decrease. In Fig. 9(b), there is no direct relationship between recall and δ_t, since with the increase of δ_t, both ξ_{ESN} and ξ_{EVENT} are decrease, thus the relationship between them is no longer clear. However, the recall is acceptable in most cases, it's almost 60% even in the worst case. On the other hand, the accuracy almost remains stable in both two figures, which is around 90%.

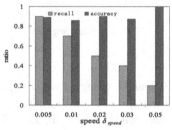
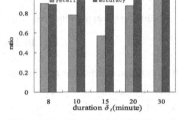

(a) Precision Changing with Speed δ_{speed}.   (b) Precision Changing with Duration δ_t.

Fig. 9. Precision Evaluation

5.3 Efficiency

In this subsection, we validate the efficiency of proposed algorithms and study their performance under different parameter settings using synthetic dataset (D3). The proposed algorithms are Clustering-and-Scanning algorithm without/with prune rule (CS/CSP) and Adjacency-List-based algorithm (AL). We show the runtime of CS, CSP and AL changing with different parameters in Fig. 10.

Fig. 10(a) and (b) shows the performance of all three algorithms changing with μ and ε, the default parameters are: $h = 3$, $\delta_{speed} = 0.1$, $\delta_\theta = 0.8$ and $\delta_t = 6$. From Fig. 10(a), we can see that with the increase of μ, runtime of CS and CSP are all decreased, while runtime of AL is nearly stable. The reason is that even the runtime of clustering step is stable, the number of clusters is decreased with the increase of μ, the number of intersection process is also decreased in scanning step of CS and CSP, therefore runtime of CS and CSP are reduced. While the runtime of AL is mainly affected by clustering step, so it keeps stable with the increase of μ. Fig. 10(b) shows that the runtime of all three algorithms are increased by the increase of ε. Because the costly part of clustering step is distance calculation, when ε is increased, the time cost of CS, CSP and AL are all increased. In addition, AL is more efficient than CS and CSP, while CSP is more efficient than CS in both two figures; these experimental results are consistent with our theoretical algorithm complexity analysis in section 4.

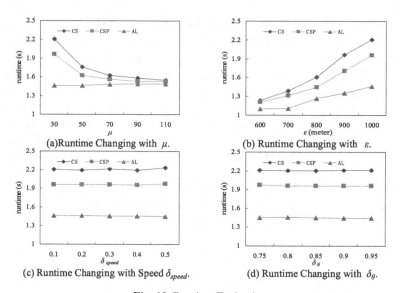

Fig. 10. Runtime Evaluation

Fig.10(c) and (d) show the runtime of CS, CSP and AL changing with δ_{speed} and δ_θ respectively, by setting $\mu = 30$, $\varepsilon = 1000$, h = 3 and $\delta_t = 6$. From Fig. 10(c) and Fig. 10(d) we can see that AL is the most efficient one, followed by CSP, and CS is the worst. In addition, the performances of all three algorithms do not change with the speed threshold δ_{speed} and δ_θ, since the variations only affect the results of snowballs not the efficiency of algorithms. The runtime of AL is around 35% less than CS and 27% less than CSP; this demonstrates that AL is far more efficient than CS and CSP, just as theoretical algorithm complexity analysis in section 4.

In summary, experiments on both real dataset and large synthetic datasets demonstrate the performance of our algorithms in terms of effectiveness, precision and efficiency. Compared with convoy pattern, E^3D can effectively discover rapidly increase (or decrease) clusters to detect emergency events, while convoy has no such a capability. Meanwhile, E^3D can discover snowballs in real time with high precision.

6   Conclusion

In this paper, we define a new snowball pattern and provide two algorithms for snowball discovery from moving object data streams. Different from most related existing patterns that aim to find a size-fixed group of objects moving together for a long time, snowball patterns assist to discover emergency event where the size of a cluster of objects with common behaviors is rapidly changed (e.g., increased or decreased) and the change is lasted for a certain period of time. We then proposed a clustering-and-scanning algorithm and an adjacency-list-based algorithm to discover snowball patterns. The effectiveness is demonstrated on a real dataset, while the precision and efficiency are validated using two large synthetic datasets.

Acknowledgement. This work was supported by three Chinese NSFC projects under grand number 91124001, 61202064 and 91324008.

References

[1] Kalnis, P., Mamoulis, N., Bakiras, S.: On discovering moving clustering moving clusters in spatio-temporal data. In: Medeiros, C.B., Egenhofer, M., Bertino, E. (eds.) SSTD 2005. LNCS, vol. 3633, pp. 364–381. Springer, Heidelberg (2005)

[2] Al-Naymat, G., Chawla, S., Gudmundsson, J.: Dimensionality reduction for long duration and complex spatio-temporal queries. In: SAC (2007)

[3] Jeung, H., Shen, H.T., Zhou, X.: Convoy queries in spatio-temporal databases. In: ICDE (2008)

[4] Tang, L.A., Zheng, Y.: On discovery of traveling companions from streaming trajectories. In: ICDE (2012)

[5] Li, Z., Ding, B., Han, J., et al.: Swarm: Mining relaxed temporal moving object clusters. In: VLDB (2010)

[6] Zheng, K., Zheng, Y., Yuan, N.J.: On Discovery of Gathering Patterns from Trajectories. In: ICDE (2013)

[7] Zhang, T., Ramakrishnan, R., Livny, M.: BIRCH: An efficient data clustering method for very large databases. In: SIGMOD (1996)

[8] Andersson, M., Gudmundsson, J., et al.: Reporting leadership patterns among trajectories. In: SAC (2007)

[9] Laube, P., van Kreveld, M., Imfeld, S.: Finding REMO-detecting relative motion patterns in geospatial lifelines. In: SDH 2005, pp. 201–215. Springer, Heidelberg (2005)

[10] Ni, J., Ravishankar, C.V.: Indexing spatio-temporal trajectories with efficient polynomial approximations. In: TKDE, vol. 19, pp. 663–678 (2007)

[11] Anagnostopoulos, A., Vlachos, M., Hadjieleftheriou, M., et al.: Global distance-based segmentation of trajectories. In: SIGKDD, pp. 34–43 (2006)

[12] Lee, J.-G., Han, J.: Trajectory clustering: A partition-and-group framework. In: SIGMOD (2007)

[13] Wu, H.-R., Yeh, M.-Y., Chen, M.-S.: Profiling Moving Objects by Dividing and Clustering Trajectories Spatiotemporally. In: TKDE (2012)

[14] Palma, A.T., Bogorny, V., Kuijpers, B., et al.: A clustering-based approach for discovering interesting places in trajectories. In: SAC, pp. 863–868 (2008)

[15] Giannotti, F., Nanni, M., Pinelli, F., et al.: Trajectory pattern mining. In: SIGKDD, pp. 330–339 (2007)

[16] Zheng, Y., Zhang, L., Xie, X., et al.: Mining interesting locations and travel sequences from gps trajectories. In: WWW, pp. 791–800 (2009)

[17] Ester, M., Kriegel, H.-P., Sander, J., et al.: A density-based algorithm for discovering clusters in large spatial databases with noise. In: SIGKDD (1996)

[18] Lelis, L., Sander, J.: Semi-supervised density-based clustering. In: ICDM, pp. 842–847 (2009)

[19] Yang, D., Rundensteiner, E.A., Ward, M.O.: Summarization and Matching of Density-Based Clusters in Streaming Environments. In: VLDB, pp. 121–132 (2012)

[20] Brinkhoff, T.: A framework for generating network-based moving objects. GeoInformatica 6(2) (2002)

Cost Reduction for Web-Based Data Imputation

Zhixu Li[1], Shuo Shang[2], Qing Xie[1], and Xiangliang Zhang[1]

[1] King Abdullah University of Science and Technology, Saudi Arabia
[2] Department of Software Engineering, China University of Petroleum-Beijing, P.R. China
{zhixu.li,qing.xie,xiangliang.zhang}@kaust.edu.sa,
jedi.shang@gmail.com

Abstract. Web-based Data Imputation enables the completion of incomplete data sets by retrieving absent field values from the Web. In particular, complete fields can be used as keywords in imputation queries for absent fields. However, due to the ambiguity of these keywords and the data complexity on the Web, different queries may retrieve different answers to the same absent field value. To decide the most probable right answer to each absent filed value, existing method issues quite a few available imputation queries for each absent value, and then vote on deciding the most probable right answer. As a result, we have to issue a large number of imputation queries for filling all absent values in an incomplete data set, which brings a large overhead. In this paper, we work on reducing the cost of Web-based Data Imputation in two aspects: First, we propose a query execution scheme which can secure the most probable right answer to an absent field value by issuing as few imputation queries as possible. Second, we recognize and prune queries that probably will fail to return any answers a priori. Our extensive experimental evaluation shows that our proposed techniques substantially reduce the cost of Web-based Imputation without hurting its high imputation accuracy.

Keywords: Web-based Data Imputation, Imputation Query, Cost Reduction.

1 Introduction

Data incompleteness is a pervasive problem in all kinds of databases [19]. The process of filling in absent field/attribute values is well-known as *Data imputation* [3]. Recently, a new generation of Web-based Data Imputation approaches were proposed to retrieve the absent field values from the web [16, 9, 23, 4, 13, 20]. In particular, external data sources on the Web are typically rich enough to answer absent fields in a wide range of incomplete data sets, while complete fields in a data set can be utilized as keywords in imputation queries for absent fields in the same data set.

However, the *high cost* remains a big problem with the Web-based Data Imputation approaches. In particular, due to the ambiguity of existing information about absent values in the local data set, various answers might be retrieved for each absent value. In order to find out the most probable correct answer for an absent value, existing method has to trigger all available imputation queries to this value, and then decides the most probable correct answer according to the voting results proposed by all issued queries. In particular, existing method relies on a confidence scheme to estimate a confidence score for every issued query [15, 16] indicating the probability that the query

S.S. Bhowmick et al. (Eds.): DASFAA 2014, Part II, LNCS 8422, pp. 438–452, 2014.

Table 1. An Example Incomplete Table with Absent Values

Title	Name	Uni.	Street	City	Zip
Prof.	Jack Davis	NYU	W. 4th Str.	NYC	10012
Dr.	Tom Smith	NYU		NYC	10012
Mr.	Bill Wilson	Columbia	Broadway	NYC	10027
Mr.	Wei Wang	Cornell	East Ave.	Ithaca	14850
Ms.	Ama Jones		W. 37th Str.		90089
	Lank Hanks	UCLA	W. Boulevard	LA	90024
	Wei Wang	UIUC	W. Illinois Str.	Urbana	61801

can retrieve a right answer. Then it employs a confidence combination scheme, such as Noisy-All [17, 1], to decide a joint confidence score to each distinct retrieved answer. For example, for the absent Uni Value in the 5th tuple in Table 1, assume we have multiple queries with different existing attribute values as keywords, which are listed with their probabilistic scores and retrieved answers as follows:

Query (Keywords → Target Absent Value)	Confidence Score	Answer
"Ms. + Ama Jones + W. 37th Str."→ Uni.	0.90	"USC"
"Ms. + Ama Jones + 90089" → Uni.	0.85	"UCLA"
"Ms. + Ama Jones + LA" → Uni.	0.80	"UCLA"
"Ms. + Ama Jones" → Uni.	0.65	"USC"

According to Noisy-All, the joint confidence score of "USC" is: $1 - (1 - 0.90) * (1 - 0.65) = 0.965$, while that of "UCLA" is: $1 - (1 - 0.85) * (1 - 0.80) = 0.97$. After all, the one with highest confidence score (i.e., "UCLA") will be taken as the most probable right answer, or what we call as the *winner answer*, to an absent value.

Given the above, to decide the winner answer to an absent value, existing method has to fire quite a few available imputation queries for each absent value in order for calculating the joint confidence of each distinct answer, which, on the other hand, brings a large overhead. In addition, among all fired queries, a significant part of them probably will not return any answer, which we call as *empty-result queries*. For example, for the absent Uni value in the fifth tuple in Table 1, the query "Ms. + Ama Jones + W. 37th Str + 90089" → Uni may not return any results, as there is no Web documents mentioning all these values together with the Uni value. Apparently, issuing these empty-result queries without getting any results is totally a waste of resources.

In this paper, we work on reducing the number of issued queries by Web-based Data Imputation based on two intuitions below: First, an answer can be secured as the winner answer to an absent value as long as we can ensure that the joint confidence of this answer must be higher than that of all the other answers. Particularly, we can calculate a *lower-bound joint confidence* and *upper-bound joint confidence* to each distinct answer, indicating the lowest and highest joint confidence that this answer can achieve respectively, based on the results of already issued queries and the number and confidence of the left un-issued queries. Once we have the lower-bound joint confidence of an answer higher than the upper-bound joint confidence of all the other answers, we can secure this answer as the winner answer to the absent value. As a result, we don't need to issue the left un-issued queries. In addition, in order for securing the winner answer as early as possible (i.e., with as few issued queries as possible), queries with higher confidence should be issued earlier than those with lower confidence, given that the cost of each query is more or less the same, but those with higher confidence have greater impact to the final imputation result to an absent value.

Secondly, a query w.r.t. an absent value in one tuple can be predicted as an empty-result query a priori, if it already failed to retrieve any answer to absent values in several other tuples, especially when these tuples share some common attribute values. The intuition here is, queries of the same format, that is, leveraging the same set of attributes (with complete values) as keywords and target at absent values under the same attribute, tend to have similar performance to absent values in similar instances. For example, after we learned that queries leveraging values under attributes `Title` + `Name` + `Zip` as keywords returned no answer to the absent `Street` value in several processed instances, it is very possible that query of the same format is also an empty-result query to an absent `Street` value in a new instance, when this new instance shares the same `Uni` values with the processed instances. Based on this intuition, we rely on a supervised learning method to construct an empty-result query prediction model for each query format. However, the drawback of supervised learning method lies on the requirement of a large training set. Fortunately, we can automatically generate training sets with complete instances of the data set.

In summary, the main contributions of this paper are as follows:

1) We propose to secure the winner answers to an absent value with as few issued queries as possible in Web-based Data Imputation through working out the condition for securing an answer as the winner answer to an absent value.
2) We devise empty-result queries predictors to predict empty-result queries a priori. Besides, an empty-result queries pre-pruning rule is also proposed for further identifying empty-result queries.
3) We perform experiments on several real-world data collections. The results demonstrate that the proposed two techniques could greatly decrease the number of executed queries by up to 85% comparing to existing method.

In the rest of this paper, we give preliminaries and then define our problems in Section 2. We present how we secure the winner answers early in Section 3, and then introduce the failure queries predictors in Section 4. The experiments are reported in Section 5, followed with related work in Section 6. We conclude in Section 7.

2 Preliminaries and Problems Definition

In this section, we give preliminaries on existing Web-based Data Imputation approaches, and then define our problems in this paper.

2.1 Preliminaries on Web-Based Data Imputation

Web-based Data Imputation formulates *Imputation Queries* to retrieve absent values from the Web [15, 16]. Specifically, for each absent value, imputation queries are issued to retrieve relevant web documents, from which we then extract the target absent value with Named Entity Extraction techniques [8, 5].

Definition 1. (Imputation Query). *Given an instance with an absent value under attribute Y and complete values under attributes $\mathbb{X}$, an imputation query to the absent*

Y value can leverage values under a set of attributes $\mathbf{X} \subseteq \mathbb{X}$ *as keywords, denoted as* $\mathbf{X} \rightarrow Y$, *in order for retrieving an answer for the absent* Y *value from the Web.*

Theoretically, any subset of existing attribute values in an instance can be used as keywords in an imputation query to any absent value in the same instance. In reality, however, only those queries with a relatively high *confidence* (which was set to ≥ 0.4 in [15, 16]) are qualified to be used for imputation.

In particular, the confidence of an imputation query format q: $(\mathbf{X} \rightarrow Y)$ indicates its probability in retrieving the right answer for an absent value under Y. The confidence of each query can be easily estimated based on a set of complete instances in the same data set [15]. For example, if queries of the same format $(\mathbf{X} \rightarrow Y)$ can retrieve correct answer for 8 complete instances out of 10, then its confidence can be calculated as 0.8. In reverse, the confidence of a retrieved answer equals to its corresponding imputation query, given the premise that all initially existing attribute values in the data set are correct.

However, due to the data complexity on the Web and the ambiguity of the used keywords in imputation queries, different queries may return different answers to the same absent value. In order to find out the most probable right answer to an absent value, the existing method has to trigger all imputation queries for this absent value, and then employs a confidence combination scheme to decide a probabilistic score to each distinct retrieved answer. Finally, only the answer with the highest *Joint Confidence* will be taken as the most probable right answer, or what we call the *Winner Answer*, to the absent value.

More specifically, assume an answer y was returned by a number of queries $Q(y)$ to an absent value under attribute Y, three different confidence combination schemes can be utilized as follows:

- *Max-Conf:* The highest confidence of a retrieved answer is taken as its joint confidence, that is,

$$C(y) = \max_{q \in Q(y)} C(q, y) \tag{1}$$

  where $Q(y)$ is the set of all imputation queries that retrieve y as the answer to the absent value, and $C(q, y)$ is the confidence of y retrieved by q.
- *Sum-Conf:* The sum of the confidence of a retrieved answer is taken as its joint confidence, that is,

$$C(y) = \mathcal{N} * \sum_{q \in Q(y)} C(q, y) \tag{2}$$

  Here $\mathcal{N}$ is a normaliser to prevent the confidence getting larger than 1. Usually, we set $\mathcal{N} = \dfrac{1}{\sum_{y' \in V(Y)} C(y')}$, where $V(Y)$ is the set of retrieved answers for this absent value.
- *Noisy-All:* We can also adopt the noisy-all scheme [17, 1] in estimating the joint confidence of a retrieved answer. That is,

$$C(y) = 1 - \prod_{q \in Q(y)} (1 - C(q, y)) \tag{3}$$

Example 1. Assume two distinct values y_1 and y_2 are retrieved for one absent value, where y_1 is retrieved by a query with confidence 0.8, and y_2 is retrieved by three other queries with confidence 0.5, 0.4 and 0.4 respectively. The joint confidences of y_1 and y_2 according to different calculation schemes are listed in Table 2. After all, with the Max-Conf scheme, the winner answer is y_1, while with the other two schemes, the winner answer is y_2.

Table 2. Computing Joint Confidence for an Absent Value

Distinct Values with Retrieved Confidences	Max-Conf	Sum-Conf	Noisy-All
y_1: 0.8	0.80	0.38	0.80
y_2: 0.5; 0.4; 0.4	0.50	0.62	0.82
The Winner Answer	y_1	y_2	y_2

2.2 Problems Definition

In order to find the winner answer for an absent value, the existing method triggers all available imputation queries to the value, which, as a result, brings a large overhead. In addition, among all fired queries, a part of them probably will not return any answer, which we call as *empty-result queries*. In order to reduce the cost of Web-based Data Imputation, our work presented in this paper addresses and proposes solutions to the following two challenging problems:

1) How we secure the winner answer for an absent value by issuing as few queries as possible, when any one of the three confidence combination schemes above is applied? (Section 3)
2) How we decrease the number of issued empty-result queries? (Section 4)

3   Securing Winner Answers Early

This sections focus on securing the winner answer to an absent value with as few issued queries as possible. We first introduce how we secure winner answer with Max-Conf, and then present how we secure winner answer with Sum-Conf or Noisy-All.

3.1   Securing Winner Answers with Max-Conf: Confidence-Prior Execution

When the Max-Conf scheme is employed, we can secure the winner answer for an absent value as long as we find out the non-empty-result query with the highest confidence to this absent value. According to Max-Conf scheme, the answer retrieved by this particular query must be the winner answer to this absent value.

In this light, we can employ a simple *Confidence-Prior Execution Strategy*, which always selects the query with the highest confidence for execution at a time. Following this execution strategy, the first retrieved answer must be the winner answer to an absent value according to the Mac-Conf scheme, and as a result, we do not need to issue/execute the left un-issued queries. Given the above, we have:

Lemma 1. *With the confidence-prior execution strategy, the first retrieved answer to an absent value must be the winner answer to this absent value when the Max-Conf scheme is employed.*

Proof. Let q_1 be the first issued query with a returned answer y_1 for an absent value under Y. According to Max-Conf scheme, the joint confidence of y_1 must be: $C(y_1) = C(q_1, y_1) = C(q_1)$. For all the other queries, since their confidence is lower than $C(q_1)$, so they are impossible to return an answer with a confidence higher than $C(y_1)$. Thus, y_1 must be the winner answer to the absent value.

3.2 Securing Winner Answers with Sum-Conf and Noisy-All

Different from Max-Conf, when either Sum-Conf or Noisy-All is employed, the joint confidence of every answer to an absent value can only be calculated when all corresponding queries are executed. Nonetheless, we can secure an answer, say y_1, as the winner answer if no other answer can beat y_1 with a higher joint confidence w.r.t. a target absent value. This condition requires us to compute the *upper-bound joint confidence* and *lower-bound joint confidence* of every distinct answer w.r.t. an absent value. Specifically, the upper-bound joint confidence of an answer w.r.t. an absent value is the highest possible confidence that this answer can achieve after all imputation queries to this absent value are executed, while the lower-bound joint confidence of an answer w.r.t. an absent value is the lowest possible confidence that this answer can achieve after all imputation queries to this absent value are executed.

Based on the definitions above, we can have the following condition for securing the winner answer to an absent value without executing all corresponding queries:

Lemma 2. *Once the lower-bound joint confidence of an answer y_1 is higher than the upper-bound joint confidence of all the other answers w.r.t. an absent value, we can secure y_1 as the winner answer to this absent value a priori.*

Note that the upper-bound and lower-bound joint confidence of every answer w.r.t. an absent value will be updated every time a new query to this absent value is executed. Once the winner answer to an absent value is secured, we don't need to execute the left yet executed queries for this absent value.

In the following, we introduce how we calculate the upper-bound and lower-bound join confidence to an answer with either Sum-Conf or Noisy-All scheme.

1. Calculating Upper-bound and Lower-bound with Sum-Conf: Let Q_t denote the set of executed queries to an absent value under Y, and $Q_t(y) \subset Q_t$ denote the set of executed queries that retrieve the answer y. Let Q_u denote the set of yet executed queries to the absent value. The upper-bound joint confidence of answer y w.r.t. this absent value can be achieved when all the unexecuted queries in Q_u will retrieve y. Hence, we have the upper-bound joint confidence to y as:

$$\overline{C(y)} = \frac{\sum_{q \in Q_t(y)} C(q) + \sum_{q \in Q_u} C(q)}{\sum_{q \in Q_t} C(q) + \sum_{q \in Q_u} C(q)} \tag{4}$$

where $C(q)$ denotes the confidence of a query q.

On the contrary, the lower-bound joint confidence of answer y w.r.t. the absent value can be gotten when all the unexecuted queries in Q_u will not retrieve y for the absent value. Hence, we have the lower-bound joint confidence to y as:

$$\underline{C(y)} = \frac{\sum_{q \in Q_t(y)} C(q)}{\sum_{q \in Q_t} C(q) + \sum_{q \in Q_u} C(q)} \tag{5}$$

Let $C_t(y) = \sum_{q \in Q_t(y)} C(q)$ denote the *Temporary Confidence* that an answer y holds w.r.t. an absent value according to the Sum-Conf scheme. Then we have the condition for securing the winner answer as:

Lemma 3. *Assume answer y_1 holds the highest temporary confidence $C_t(y_1)$ w.r.t. an absent value at a moment, and answer y_2 holds the second highest temporary confidence $C_t(y_2)$, we say y_1 can be secured as the winner answer if $C_t(y_1) - C_t(y_2) > \sum_{q \in Q_u} C(q)$ when the Sum-Conf scheme is employed.*

Proof. Let $\overline{C(y_2)}$ denote the upper-bound join confidence of y_2, and $\underline{C(y_1)}$ denote the lower-bound join confidence of y_1, both of which will be achieved when all the unexecuted queries retrieve y_2 for the absent value. According to Lemma 2, as long as we have $\overline{C(y_2)} < \underline{C(y_1)}$, we can secure y_1 as the winner answer. Finally, we have: $C_t(y_1) - C_t(y_2) > \sum_{q \in Q_u} C(q)$ as the condition to secure y_1 as the winner answer.

2. Calculating Upper-bound and Lower-bound with Noisy-All: With the Noisy-All scheme, the upper-bound joint confidence to an answer y w.r.t. an absent value can also be achieved when all the unexecuted queries in Q_u will retrieve y for the absent value. That is,

$$\overline{C(y)} = 1 - \prod_{q \in Q_t(y)} (1 - C(q)) * \prod_{q \in Q_u} (1 - C(q)) \tag{6}$$

, while the lower-bound joint confidence to y is also gotten when all the unexecuted queries in Q_u will not retrieve y for the absent value. More specifically,

$$\underline{C(y)} = 1 - \prod_{q \in Q_t(y)} (1 - C(q)) \tag{7}$$

Let $C_t(y) = 1 - \prod_{q \in Q_t(y)} (1 - C(q))$ denote the temporary confidence that an answer y holds w.r.t. an absent value, we have the condition for securing the winner answer with the Noisy-All scheme as:

Lemma 4. *With the Noisy-All scheme, an answer y_1 can be secured as the winner answer to an absent value if $\dfrac{1 - C_t(y_1)}{1 - C_t(y_2)} < \prod_{q \in Q_u} (1 - C(q))$, where y_1 is the answer with the highest temporary confidence $C_t(y_1)$, while y_2 is the answer with the second highest temporary confidence $C_t(y_2)$.*

Proof. Let $\overline{C(y_2)}$ denote the upper-bound join confidence of y_2, and $\underline{C(y_1)}$ denote the lower-bound join confidence of y_1, both of which will be achieved when all the unexecuted queries retrieve y_2 for the absent value. According to Lemma 2, as long as

we have $\overline{C(y_2)} < C(y_1)$, we can secure y_1 as the winner answer. Finally, we have:
$$\frac{1 - C_t(y_1)}{1 - C_t(y_2)} < \prod_{q \in Q_u} (1 - C(q))$$ as the condition to secure y_1 as the winner answer.

Note that Lemma 3 and Lemma 4 should work in conjunction with the confidence-prior execution strategy, such that the winner answer to an absent value can be identified as early as possible.

4 Pre-pruning Empty-Result Queries

The confidence-prior execution strategy proposed in Section 3.1 tends to give priority to queries with high confidence. However, queries with high confidence also have high risks to become *empty-result queries*, as these queries usually involve many attribute values which might not appear together with the absent value in one web page. As a result, the existing method may have to bear the cost of executing a large number of empty-result queries, while garnering non of the benefits.

In this section, we propose to reduce the number of issued empty-result queries by identifying them a priori with two orthogonal techniques below: First, we build an *Empty-Result Queries Predictor* for each query format in Section 4.1, which is actually a classification model that can classify queries into empty-result category and effective category. If the prediction result is positive (i.e., an effective query), we will execute this query, otherwise, we skip the query. Second, we introduce a heuristic *Empty-Result Queries Pre-pruning Rule* developed based on the relationship between queries in Section 4.2.

4.1 Empty-Result Queries Predictor

In this subsection, we introduce how we construct empty-result query predictors. Intuitively, queries targeting at the same target attribute Y by leveraging the same set of leveraged attribute $\mathbf{X}$ tend to have similar performance when applying them to similar instances.

Based on this intuition, we rely on a supervised model to predict empty-result queries a priori. More specifically, for all queries of the same format: q: $(\mathbf{X} \rightarrow Y)$, one supervised empty-result query predictor will be formulated, which will predict the imputation result of the query to an instance as either (1) *Negative*, which means the query is an empty-result query to the instance; or (2) *Positive*, which means the query is an effective query to the instance.

In the following, before presenting how we perform supervised learning to a query predictor, we first introduce the features that designed for each instance, and describe how we generate the training set with complete instances in the data set.

1. Features: Basically, we utilize two groups of features for each instance based on the intuitions (or observations) below: (1) instances having complete values under the same set of attributes are more likely to be in the same category. (2) instances sharing some complete attribute values are more likely to be in the same category.

Firstly, instances having complete values under the same set of attributes are covered by the same set of queries, which is the basic information required by the query predictor. Therefore, the first group of features are defined on whether an instance t has an existing value under each attribute A. If $t[A]$ is not absent, then the feature value corresponding to this attribute is "TRUE", otherwise "FALSE". This group of features provide the basic information of each instance.

Secondly, sharing some attribute values means instances having some common properties in some aspects, which may have more or less impact on whether a query can retrieve any answer for an instance. For example, if two instances in Table 1 share the same University value, which means the two people are from the same university, then there is perhaps some unified format for home pages of all staffs in this university, and as a result, their home pages may always mention the street address and Zip code of the university. However, some other shared attribute values may be less important such as Title values. On the other hand, taking every attribute value as a feature will let the feature space be very large, and the feature space be very sparse.

In order to find out "useful" shared attribute values (or their combinations) for constructing the second group of features, we propose to learn some special kind of "association rules" holding between attribute values under $\mathbf{X}$, and the retrieving result (Positive or Negative) of a query format to an instance. For example, if we learn that when City="NYC", queries of the format { Name, Uni, City} $\rightarrow$ Zip are not empty-result queries to 95% of training instances, then "NYC" will be taken as a "useful" shared attribute value to the predictor. Here we can set a confidence threshold (such as 75% in our experiments which is demonstrated as a good one) to divide "useful" attribute values (or combinations of them) from the others. That is, when an attribute value or a combination of attribute values appears in an "association rule" like the example one above, it will be taken as an "useful" attribute values, and the confidence of the association rule will be taken as the weight of the value (or the combination of values) in training predictors. Finally, each detected "useful" attribute value or combination of values will be taken as a feature. For each instance, if it contains an "useful" value or combination of values, it will have a "TRUE" value under corresponding feature, otherwise, "FALSE".

2. Training set: Now we introduce how we automatically generate the training set for a query format q: $\mathbf{X} \rightarrow Y$ based on a set of complete instances. For each complete instance t, we take its attribute values under $\mathbf{X}$ attributes, denoted as $t[\mathbf{X}]$, as keywords to formulate a web-based imputation query q like $(\mathbf{X} = t[\mathbf{X}]) \rightarrow (Y =?)$. We execute this query to get retrieved answers $Y_{t[\mathbf{X}]}$. If $Y_{t[\mathbf{X}]} = \emptyset$, this instance t will be labeled as "Negative". Otherwise, "Positive".

Note that when the table is large, we probably have a large number of complete instances to a query format. To save the cost, we only sample a small number of instances for training. However, simple sampling methods are not sufficient, given that the unbalanced distributions of $\mathbf{X}$ values in the data set. Let instances sharing the same $\mathbf{X}$ attribute values as one group. There might be several very large groups, and many small groups. As a result, either uniform row sampling or uniform group sampling will generate a biased training set. Here we adopt a *Two-Pass Sampling Scheme* [6] for estimating the confidence of each query pattern in a local table. In the first pass, we uniformly sample a number of rows and put into a sample set $\mathcal{S}$. In the second pass, we put instances sharing the same $\mathbf{X}$ attribute values in $\mathcal{S}$ into one group, and then uniformly sample some rows from each group. With this two-pass sampling, the contributions of large groups are not lost in the first pass, and enough information is collected about each of the sampled small groups to correctly compute their contributions to the overall training set [6].

3. Predictor Learning: We start with a small set of training data set, and then update query predictors with new completed instances. In particular, each time we execute a query to an instance, we will see if this query is a positive or negative one to this instance. This information can then be used to update the corresponding predictor according to an on-line learning algorithm.

Here we adopt a popular on-line learning algorithm, *winnow* [18], which is simple but effective when it is applied to many on-line learning problems. In our problem, all features are boolean-valued features, that is, $f_i(t) \in \{1, 0\}$ ($1 \le i \le n$), and the label of each instance is also boolean-valued, that is, $label_p(t) = \{\text{True, False}\}$. The algorithm maintains non-negative weights w_i for the i-th feature. Initially, the first group of features have uniformed weights, and the second group of features have their calculated weights respectively. For each instance t with a feature vector $[f_1(t), f_2(t), ..., f_n(t)]$, the algorithm uses a linear classifier for prediction as:

$$label_p(t) = \begin{cases} True & (\sum_{i=1}^{n} w_i f_i(t) > \delta) \\ \\ False & (otherwise) \end{cases} \qquad (8)$$

where δ is a threshold that defines a dividing hyperplane in the instance space. Good bounds are obtained if $\delta = n/2$.

Then, for each instance that the classifier has predicted, after executing the query corresponding the query pattern to this instance and get the true label $label(t)$ for it, we apply the following update rule to the classifier:

- If the instance is correctly classified, that is, $label(t) = label_p(t)$, we do nothing to the classifier.
- If the instance is incorrectly classified, and $label(t) = True$, but $label_p(t) = False$, all of the weights of the features with value 1 in the mistake instance are multiplied by a value α (promotion step).
- If the instance is incorrectly classified, and $label(t) = False$, but $label_p(t) = True$, all of the weights of the features with value 1 in the mistake instance are divided by a value α (demotion step).

We set a typical value 2 to α. Initially, we use the generated training data set to build the predictor for each query pattern. Later, in the imputation process, each time we executed an imputation query which returns empty result, we will update its corresponding predictor according to the rules above.

4.2 Query Pre-Pruning Rule

Besides the empty-result queries predictors, an empty-result query pre-pruning rule can also be developed based on the relationship between queries to the same absent value as follows:

Lemma 1. *For one absent value, if a query* q_1: $(\mathbf{X}_1 \rightarrow Y)$ *fails to retrieve any value for an instance, then another query* q_2: $(\mathbf{X}_2 \rightarrow Y)$, *where* $\mathbf{X}_1 \subseteq \mathbf{X}_2$, *will also be an empty-result query to this instance.*

Proof. We assume q_2 can also retrieve an answer, say y', for the absent value in an instance, there should be some web pages containing both y' and values under $\mathbf{X}_2$ in this instance. Since $\mathbf{X}_1 \subseteq \mathbf{X}_2$, in no doubt, these web pages should also contain values under $\mathbf{X}_1$ in this instance, thus q_1 should also return y' as an answer to the absent value in this instance. However, since we know that q_1 fails to retrieve any answer to this absent value, thus the assumption that q_2 can retrieve an answer for the absent value in this instance is incorrect.

According to Lemma 1, each time a query q fails to retrieve any answer for an instance, we could prune some other queries for the same instance, whose leveraged attribute set is a superset of q's leveraged attribute set.

5   Experiments

We have implemented a Web-based Data Imputation prototype in Java which uses Google API to answer data imputation queries [15, 16]. To demonstrate the effectiveness of our techniques, we have experimented with two large real-world data sets and one small but interesting real-world data set. We say the 3rd data set is interesting since it is a multilingual data set which could evaluate the performance of our techniques in multilingual scenario:

- *Personal Information Table (PersonInfo):* This is a 50k-tuples, 9-attributes table, which contains contact information for academics including name, email, title, university, street, city, state, country and zip code. This information is collected from more than 1000 different universities in the USA, UK, Canada and Australia.
- *DBLP Publication Table (DBLP):* This is a 100k-tuples, 5-attributes table. Each tuple contains information about a published paper, including its title, first author, conference name, year and venue. All papers are randomly selected from DBLP.
- *Multilingual Disney Cartoon Table (Disney):* This table contains names of 51 classical disney cartoons in 8 different languages collected from Wikipedia.

All the three data sets are complete relational tables. To generate incomplete tables for our experiments, we remove attribute values at random positions from the complete table, while making sure that at least one key attribute value will be kept in each tuple. For the PersonInfo dataset, the name and email are key attributes. For the DBLP dataset, the paper title is the only key attribute. For the Disney dataset, all attributes are key attributes.

Each reported result is the average of 5 evaluations, that is, for each absent value percentage (1%, 5%, 10%, 20%, 30%, 40%, 50%, 60%), 5 incomplete tables will be generated with 5 random seeds, and the experimental results we present are the average results based on the 5 generated incomplete tables. We then impute these generated incomplete tables using our methods and evaluate the performance of our solutions by using the original complete table as the ground truth.

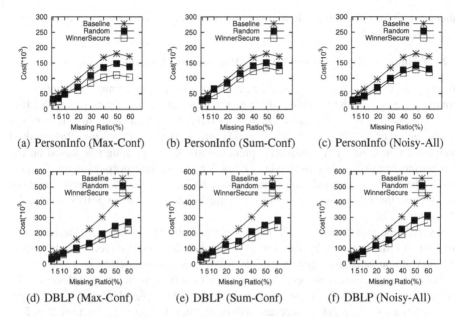

Fig. 1. The Effectiveness of Winner Answer Securing Techniques

5.1 Evaluating Winner Answers Securing Techniques

To evaluate the effectiveness of the winner answer securing techniques, we compare the imputation cost of the following three methods based on the PersonInfo and DBLP data sets: (1) a *Baseline* method which issues all available queries for each absent value; (2) the *Random* method which randomly decides the next query to execute, and also applies the condition for securing winner answers; and (3) the *WinnerSecure* method, which uses the conditions for securing winner answers together with the confidence-prior execution strategy.

As demonstrated in Figure 1, with the winner answer securing condition only, the Random method can decrease around 30%-40% issued queries. In addition, with the confidence-prior execution strategy, the WinnerSecure method can further decrease about 15% issued queries when the Max-Conf scheme is applied, and can decrease

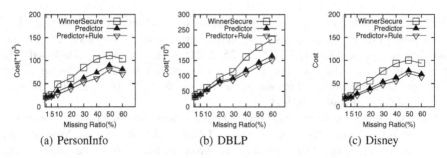

Fig. 2. The Effectiveness of Empty-Answer Queries Pre-Pruning Techniques

about 5-10% issued queries when either one of the other two confidence combination schemes is applied.

5.2   Evaluating Empty-Result Queries Pre-pruning Techniques

To evaluate the effectiveness of the empty-result queries pre-pruning techniques, we compare the cost of the following three methods on all the three data sets with Max-Conf scheme: (1) *WinnerSecure* method; (2) *Predictor* method which uses the empty-result queries predictor based on the *WinnerSecure* method; (3) *Predictor+Rule* method which uses both empty-result queries predictor and query pruning rule based on the *WinnerSecure* method. As described in Figure 2, by using the empty-result queries predictor, the *Predictor* method can decrease about 20% cost of the *WinnerSecure* method. In addition, the *Predictor+Rule* method can further decrease about 5-10% cost of the *Predictor* method.

Overall, the combination of *WinnerSecure* and *Predictor+Rule* can decrease the number of issued queries by up to 85% of the *Baseline* method. As a result, imputing each absent value averagely costs only 1.2-1.5 imputation queries.

Finally, we evaluate whether the empty-result queries predictors will decrease the imputation accuracy by killing some positive queries. As depicted in Figure 3, by using the predictors, the imputation accuracy will be slightly decreased by no more than 5%. Apparently, compared with the improvement of the imputation efficiency, it is cost-efficient to use the predictors in doing web-based imputation.

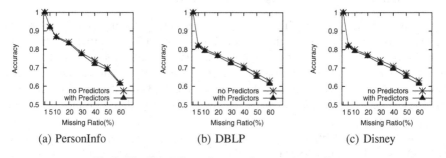

(a) PersonInfo   (b) DBLP   (c) Disney

Fig. 3. The Side-Effect of Imputation Results Predictors

6   Related Work

The process of filling in absent attribute values is well-known as Data Imputation [3]. While most of the previous efforts focus on recovering absent attribute values from the complete part of the data set [11, 24, 26, 25, 2, 21], which we call as table-based data imputation methods, there are also many recent work conducted on retrieving absent attribute values from external resources, such as the Web [15, 14, 7, 23, 13, 16], i.e., web-based data imputation methods.

Existing table-based data imputation methods can be divided into two categories: (1) substitute-based data imputation; and (2) model-based data imputation. More specifically, the substitute-based data imputation approaches find a substitute value for the absent one from the same data set [11, 24, 25], such as selecting the most common attribute value [11] or the k-Nearest Neighbor [10, 25] as the absent value. Besides, association rules between attribute values are also learned and used to find a close-fit value for an absent one from a similar context [22]. However, they are unlikely to reach a high imputation precision and recall, since the right absent value can only be inferred when the value is in the complete part of the data set. Differently, the model-based data imputation approaches [2, 21, 26] aim at building a prediction model based on the complete part of the data set for estimating an answer to each absent value. However, close estimations can not replace the absent original values, especially for non-quantitive string attribute values like email addresses.

Web-based data imputation methods resort to the World Wide Web for answering absent values. Many work has been conducted to harvest absent values from either web lists [13, 7], or web tables [14, 12, 23, 20]. More recently, a general Web-based approach, WebPut, was proposed to retrieve missing data from all kinds of web documents [15, 16]. In principle, WebPut utilizes the available information in an incomplete database in conjunction with the data consistency principle. It extends popular Information Extraction (IE) methods for the purpose of formulating data imputation queries that are capable of effectively retrieving absent values with high accuracy. However, the cost of web-based retrieving is much higher than that of the table-based inference, as all absent values need to be retrieved from the web. Although some schedule algorithms were proposed in WebPut, they are insufficient to solve the high cost problem. This paper is an extended work on WebPut, which aims at using much fewer queries to reach the same high imputation accuracy.

7 Conclusions and Future Work

In this paper, we mainly focus on minimizing the number of issued imputation queries in doing Web-based Data Imputation. We not only propose to secure the winner answers by executing as few imputation queries as possible in Web-based Data Imputation through working out the conditions for securing an answer as the winner answer to an absent value, but also devise empty-result queries predictors to predict empty-result queries a priori. The experiments based on several real-world data collections demonstrate that we can greatly decrease the number of issued queries by up to 85% comparing to the baseline.

An underlying assumption of the work is that all existing attribute values are faultless (i.e., clean data), meanwhile we leave the problem of data imputation in the presence of incorrect and dirty data as part of our future work. We will also work on a hybrid approach that combines and integrates our web-based approach with previous model-based data imputation methods.

References

[1] Agichtein, E., Gravano, L.: Snowball: Extracting relations from large plain-text collections. In: ACM DL, pp. 85–94 (2000)
[2] Barnard, J., Rubin, D.: Small-sample degrees of freedom with multiple imputation. Biometrika 86(4), 948–955 (1999)
[3] Batista, G., Monard, M.: An analysis of four missing data treatment methods for supervised learning. Applied Artificial Intelligence 17(5-6), 519–533 (2003)
[4] Chaudhuri, S.: What next?: A half-dozen data management research goals for big data and the cloud. In: PODS, pp. 1–4 (2012)

[5] Cohen, W., Sarawagi, S.: Exploiting dictionaries in named entity extraction: Combining semi-markov extraction processes and data integration methods. In: ACM SIGKDD, pp. 89–98 (2004)

[6] Cormode, G., Golab, L., Flip, K., McGregor, A., Srivastava, D., Zhang, X.: Estimating the confidence of conditional functional dependencies. In: SIGMOD, pp. 469–482 (2009)

[7] Elmeleegy, H., Madhavan, J., Halevy, A.: Harvesting relational tables from lists on the web. PVLDB 2(1), 1078–1089 (2009)

[8] Finkel, J., Grenager, T., Manning, C.: Incorporating non-local information into information extraction systems by gibbs sampling. In: ACL, pp. 363–370 (2005)

[9] Gomadam, K., Yeh, P.Z., Verma, K.: Data enrichment using data sources on the web. In: 2012 AAAI Spring Symposium Series (2012)

[10] Grzymala-Busse, J., Grzymala-Busse, W., Goodwin, L.: Coping with missing attribute values based on closest fit in preterm birth data: A rough set approach. Computational Intelligence 17(3), 425–434 (2001)

[11] Grzymała-Busse, J.W., Hu, M.: A comparison of several approaches to missing attribute values in data mining. In: Ziarko, W.P., Yao, Y. (eds.) RSCTC 2000. LNCS (LNAI), vol. 2005, pp. 378–385. Springer, Heidelberg (2001)

[12] Gummadi, R., Khulbe, A., Kalavagattu, A., Salvi, S., Kambhampati, S.: Smartint: Using mined attribute dependencies to integrate fragmented web databases. Journal of Intelligent Information Systems, 1–25 (2012)

[13] Gupta, R., Sarawagi, S.: Answering table augmentation queries from unstructured lists on the web. PVLDB 2(1), 289–300 (2009)

[14] Koutrika, G.: Entity Reconstruction: Putting the pieces of the puzzle back together. HP Labs, Palo Alto, USA (2012)

[15] Li, Z., Sharaf, M.A., Sitbon, L., Sadiq, S., Indulska, M., Zhou, X.: Webput: Efficient web-based data imputation. In: Wang, X.S., Cruz, I., Delis, A., Huang, G. (eds.) WISE 2012. LNCS, vol. 7651, pp. 243–256. Springer, Heidelberg (2012)

[16] Li, Z., Sharaf, M.A., Sitbon, L., Sadiq, S., Indulska, M., Zhou, X.: A web-based approach to data imputation. In: WWW (2013)

[17] Lin, W., Yangarber, R., Grishman, R.: Bootstrapped learning of semantic classes from positive and negative examples. In: ICML Workshop on The Continuum from Labeled to Unlabeled Data, vol. 1, p. 21 (2003)

[18] Littlestone, N.: Learning quickly when irrelevant attributes abound: A new linear-threshold algorithm. Machine Learning 2(4), 285–318 (1988)

[19] Loshin, D.: The Data Quality Business Case: Projecting Return on Investment. Informatica (2006)

[20] Pantel, P., Crestan, E., Borkovsky, A., Popescu, A., Vyas, V.: Web-scale distributional similarity and entity set expansion. In: EMNLP, pp. 938–947 (2009)

[21] Quinlan, J.: C4. 5: Programs for machine learning. Morgan kaufmann (1993)

[22] Wu, C., Wun, C., Chou, H.: Using association rules for completing missing data. In: HIS, pp. 236–241 (2004)

[23] Yakout, M., Ganjam, K., Chakrabarti, K., Chaudhuri, S.: Infogather: Entity augmentation and attribute discovery by holistic matching with web tables. In: SIGMOD, pp. 97–108. ACM (2012)

[24] Zhang, S.: Shell-neighbor method and its application in missing data imputation. Applied Intelligence 35(1), 123–133 (2011)

[25] Zhang, S.: Nearest neighbor selection for iteratively knn imputation. Journal of Systems and Software 85(11), 2541–2552 (2012)

[26] Zhu, X., Zhang, S., Jin, Z., Zhang, Z., Xu, Z.: Missing value estimation for mixed-attribute data sets. IEEE TKDE 23(1), 110–121 (2011)

Incremental Quality Inference in Crowdsourcing

Jianhong Feng, Guoliang Li, Henan Wang, and Jianhua Feng

Department of Computer Science, Tsinghua University, Beijing 100084, China
{fengjh11,whn13}@mails.thu.edu.cn, {liguoliang,fengjh}@tsinghua.edu.cn

Abstract. Crowdsourcing has attracted significant attention from the database community in recent years and several crowdsourced databases have been proposed to incorporate human power into traditional database systems. One big issue in crowdsourcing is to achieve high quality because workers may return incorrect answers. A typical solution to address this problem is to assign each question to multiple workers and combine workers' answers to generate the final result. One big challenge arising in this strategy is to infer worker's quality. Existing methods usually assume each worker has a fixed quality and compute the quality using qualification tests or historical performance. However these methods cannot accurately estimate a worker's quality. To address this problem, we propose a worker model and devise an incremental inference strategy to accurately compute the workers' quality. We also propose a question model and develop two efficient strategies to combine the worker's model to compute the question's result. We implement our method and compare with existing inference approaches on real crowdsourcing platforms using real-world datasets, and the experiments indicate that our method achieves high accuracy and outperforms existing approaches.

1 Introduction

Crowdsourcing has attracted widespread attention from many communities such as database and machine learning. The primary idea of Crowdsourcing is to take advantage of human intelligence to solve problems which are still difficult for computers, such as language translation, image recognition [6,17]. Several Crowdsourcing-based database systems have been proposed recently, e.g., CrowdDB [5,4], Qurk [10,11] and DECO [13]. These systems embedded in traditional relational database implement complicated crowdsourcing-based operations. Crowdsourcing platforms, such as Amazon Mechanical Turk (AMT) [2] and CrowdFlower [1] , provide APIs to facilitate these systems to accomplish crowdsourcing tasks. Task publishers (called requesters) can easily publish a large number of tasks on these crowdsourcing platforms, and obtain the answers completed by many human labors (called workers). Workers receive the pre-set financial rewards if their answers are accepted by the requester.

Workers on Crowdsourcing platforms typically have different backgrounds (e.g., age and education), coming from different countries or regions [12], and thus the answers may be affected by the various subjective experiences. Besides, spam workers provide answers randomly to get financial rewards. Therefore the

S.S. Bhowmick et al. (Eds.): DASFAA 2014, Part II, LNCS 8422, pp. 453–467, 2014.

answers collected from crowdsourcing platforms are usually not accurate. In order to achieve high quality of final results, a typical solution is to assign each task to multiple workers and infer the final results from the received answers[7,8].

To infer the final results, Majority Vote (MV) is the most popular inference method which has already been employed by CrowdDB [5,4] and DECO [13]. In MV, workers are assumed to have the same quality, and the answer provided by majority workers is taken as the result. Obviously MV ignores the fact that workers with different background and experience may have different quality, thus it leads to a low-quality inference result. In order to address the problem, the inference methods in [9,7,14,16] consider different qualities for each worker. The strategies used to reflect the quality of each worker can be categorized into two types. The first one is a fixed strategy adopted by CDAS [9], with the quality of each worker estimated by the worker's historical performance or qualification tests. This strategy is simple but not precise enough. For example, a worker's quality may increase as she learns more about questions through the answering procedure, or her quality may decrease when she is a little bit tired of answering questions. Therefore, modeling the quality for each worker is necessary for inferring the final results. The second inference method [7,14,16] is an iterative strategy. This strategy is based on the Expectation-Maximization [3] algorithm which can improve the results' accuracy by modeling each worker's quality dynamically. However, this inference method is rather expensive, because whenever it receives a new answer submitted by workers, it uses all received answers to re-estimate every worker's quality.

In summary, CDAS [9] can rapidly return the inference results, at expense of low quality. EM obtains results with higher quality while involving large inference time. To overcome these limitations, we propose an incremental quality inference framework, called INQUIRE, which aims to make a better tradeoff between the inference time and result quality. We devise a novel worker model and a question model to quantify the worker's quality and infer the question's result, respectively. When a worker submits her answer, INQUIRE can incrementally update the worker model and the question model, and return the inference results instantly. We compare INQUIRE with existing inference methods on real crowdsourcing platforms using real-world datasets, and the experiments indicate that our method achieves high accuracy and outperforms existing approaches.

This paper makes the following contributions:

- We formulate the incremental quality inference problem, and propose the INQUIRE framework to solve this problem.
- We devise a novel worker model to quantify the worker's quality, and a question model to infer the question's result instantly.
- We propose two incremental strategies to effectively update the question model and an incremental strategy to update the worker model.
- We compare INQUIRE with MV, CDAS and EM on real crowdsourcing platforms using real-world datasets. Our experimental results illustrate that INQUIRE can achieve a better tradeoff between the inference's time and accuracy.

This paper is organized as follows. We formulate the problem in Section 2 and introduce the INQUIRE's framework in Section 3. Question model and worker model are discussed in Section 4 and we discuss how to update the two models incrementally in Section 5. In Section 6, we show our experiment results and provide result analysis. Section 7 concludes the paper.

2   Problem Formulation

Since workers do not want to answer complicated questions, the tasks on crowd-sourcing platforms are usually very simple and most of them are binary choices questions. For example, in entity resolution, each question contains two entities and asks workers to decide whether the two entities refer to the same entity [15]. In this paper, we also focus on these binary questions with only two possible choices. For ease of presentation, we assume there is only one correct choice for each question. It is worth noting that our method can be easily extended to support the questions with multiple choices.

Formally, a requester has a set of n binary questions $Q = \{Q_1, Q_2, \ldots, Q_n\}$ where each question asks workers to select the answer from two given choices. To achieve high quality, each question will be assigned to m workers. The true result for each question Q_i is denoted as R_i. R_i is 1 (or 0) indicating the returned result is the first choice (the second choice) for question Q_i. For example, if each question has two pictures, and workers are required to decide whether the people in two pictures are the same person. The two possible choices for this question is "same" (first choice) or "different" (second choice). If they are the same person, then $R_i=1$, otherwise $R_i=0$. After the requester published questions on the crowdsourcing platform, workers' answers are returned in a streaming manner. We use $\langle Q_i, W_k, L_{ik} \rangle$ to denote the result received from worker W_k for question Q_i with answer L_{ik} where $L_{ik} \in \{0,1\}$. Every time an answer $\langle Q_i, W_k, L_{ik} \rangle$ receives, we infer the result of question Q_i based on the current answer L_{ik}, the accuracy of worker W_k, and previous results of Q_i.

3   INQUIRE Framework

The goal of INQUIRE is to accurately and efficiently infer the final results of each question. To achieve this goal, we design two models, question model and worker model. The framework of INQUIRE is illustrated in Figure 1.

INQUIRE publishes all the questions to a crowdsourcing platform. Interested workers answer the questions. Each time a worker W_k completes a question Q_i, INQUIRE gets the corresponding answer $\langle Q_i, W_k, L_{ik} \rangle$.

We build a question model for each question, denoted by QM_i, which is designed to decide the inference result. INQUIRE updates QM_i based on both the worker's accuracy and the newly received answers. Section 5.1 gives how to incrementally update question model.

We construct a worker model for each worker, denoted by WM_k, to capture the quality of each worker. The accuracy of worker W_k can be directly derived from WM_k and INQUIRE updates WM_k based on the worker model QM_i and the answer of the worker $\langle Q_i, W_k, L_{ik} \rangle$. We present the strategy of incrementally updating the worker model in section 5.2.

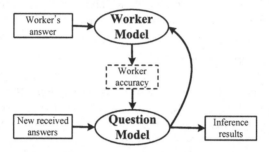

Fig. 1. INQUIRE Framework

The process of INQUIRE includes two steps:

In the first step, when W_k submits an answer of Q_i, QM_i is updated according to the worker model WM_k and the new answer L_{ik}. Then INQUIRE returns the inference result.

In the second step, if Q_i has already been answered m times, INQUIRE respectively updates the workers' model for those workers who have answered Q_i.

Following the process above, Algorithm 1 illustrates the pseudo code of our algorithm. The triple $\langle Q_i, W_k, L_{ik} \rangle^j$ is the j-th received answer of Q_i answered by W_k with answer L_{ik}. The inference result given by INQUIRE in the j-th round is denoted as $\langle Q_i\_Result \rangle^j$.

Example 1. Assume that a requester publishes a set of questions Q and asks three workers to answer each question. Take Q_1 as an example. Q_1 has received two answers $\langle Q_1, W_1, L_{11} \rangle^1$, $\langle Q_1, W_9, L_{19} \rangle^2$ and has a result $\langle Q_1\_Result \rangle^2$. For the arrival of $\langle Q_1, W_4, L_{14} \rangle^3$, the QM_1 is updated with the current accuracy of W_4 and QM_1. Then INQUIRE returns the new result $\langle Q_1\_Result \rangle^3$ depending on QM_1 (line 3 to line 5). After that, three answers of Q_1 have been completely received and then INQUIRE updates WM_1, WM_4, and WM_9 (line 7).

Compared INQUIRE to CDAS, the main difference between them is that the workers' accuracy never changes in CDAS, while the variation of workers' accuracy is expressed by updating worker model in INQUIRE. Differing from INQUIRE, every time EM receives a new answer, it re-estimates every worker's quality and infers new results relying on all received answers. For example, in example 1, when EM receives $\langle Q_1, W_4, L_{14} \rangle^3$, in addition to three answers of Q_1, EM applies other questions' answers $\langle Q_i, W_k, L_{ik} \rangle^j$ collected so far to obtain every question's result and every worker's accuracy.

4   Question Model and Worker Model in INQUIRE

In this section, we introduce the question model and worker model. The question model is utilized to infer the result of questions and the worker model is used to evaluate the quality of workers.

Algorithm 1. INQUIRE

    Input: $\langle Q_i, W_k, L_{ik} \rangle^j$
    Output: $\langle Q_i\_Result \rangle^j$

1 **begin**
2   **for** *arriving* $\langle Q_i, W_k, L_{ik} \rangle^j$ **do**
3          $W_k\_accuracy \leftarrow WM_k$;
4          $QM_i \leftarrow (QM_i, W_k\_accuracy)$;
5          $\langle Q_i\_Result \rangle^j \leftarrow QM_i$;
6   **for** W_k *has answered* Q_i **do**
7             $WM_k \leftarrow (QM_i, WM_k, L_{ik})$;

4.1 Question Model

INQUIRE builds question model QM_i: $(p_i, 1 - p_i)$ for question Q_i where p_i is the probability that question Q_i's true result is the first choice and $1 - p_i$ is the probability that question Q_i's true result is the second choice. For each question Q_i, INQUIRE compares the value of p_i with $1\text{-}p_i$ and chooses the choice with larger probability as the inference result. That is, if $p_i > 1 - p_i$, INQUIRE takes the first choice as the result, otherwise INQUIRE returns the second choice. The initial value of p_i is 0.5.

4.2 Worker Model

The key part of achieving high-quality inference result is to estimate workers' quality in time. The fixed-quality strategy in CDAS would make the inference results not very accurate since setting a fixed value as each worker's quality neglects the change of each worker's quality with time. To address this problem, some algorithms [7,14,16] propose to use confusion matrix to calculate workers' quality. Confusion matrix is built by comparing worker's answers to inference results that EM returns. However, because EM probably infers results with low accuracy when it receives a small number of answers, the values of confusion matrix may be inaccurate.

Different from previous work, this paper proposes a more accurate worker model to compute worker's quality. INQUIRE builds a worker model (WM_k) for each worker

$$\begin{bmatrix} c_{00} & c_{01} \\ c_{10} & c_{11} \end{bmatrix}$$

The row subscript (the first 0/1 number) means the answer that worker gives and the column subscript (the second 0/1 number) means the true result of the question. Let c_{ij} denote the total contribution of the worker's answers to questions. To a question, the worker's contribution is represented by the value of QM_i based on the worker's answer. For example, suppose that QM_1 is (0.6, 0.4) and L_{11} is 1. Then the contribution of W_1 to that Q_1's true result is first choice is 0.6. Our worker model is different from confusion matrix. In confusion matrix, the value is the number of times a question which inference result is j was answered as i. For example, if $p_2 = 0.54$ and $L_{23} = 0$, that is Q_2's inference result is 1. Then c_{01} in W_3's confusion matrix plus one. Because p_2 is just 0.54, the inference result is incorrect with high probability. That means confusion matrix can not precisely represent actual worker's performance sometimes.

We can easily use the worker model to calculate the accuracy of a worker. There are two methods we can compute the worker's accuracy. The first one is that the worker's accuracy is computed separately when the worker gives different choice. If the answer L_{ik} is 1, the worker W_k's accuracy is denoted by α_k. Let β_k denote the W_k's accuracy if L_{ik} is 0. α_k and β_k can be computed with Formulas 1 and 2, respectively.

$$\alpha_k = p(R_i = 1 | L_{ik} = 1) = \frac{c_{11}}{c_{11} + c_{10}} \tag{1}$$

$$\beta_k = p(R_i = 0 | L_{ik} = 0) = \frac{c_{00}}{c_{00} + c_{01}} \tag{2}$$

Some workers show biases for certain types of questions and their answers tend to one choice [7], so α_k and β_k are not accurate for bias workers. In this paper, if the difference between α_k and β_k is more than 50%, we consider this worker as a bias one.

The second method uses a general accuracy, which is that no matter what answers the worker returns, the W_k's accuracy is calculated as (called γ_k) :

$$\gamma_k = \frac{c_{11} + c_{00}}{c_{11} + c_{10} + c_{00} + c_{01}} \tag{3}$$

We can initialize each worker model by qualification test or the worker's historical records. If there is not any pre-information of workers, then c_{ij} is 0, and α_k, β_k, γ_k are all set to 50%.

Example 2. Assume that, in example 1, the worker models for three workers who answered Q_1 are the following.

$$WM_1 : \begin{bmatrix} 11 & 6 \\ 7 & 12 \end{bmatrix}, WM_4 : \begin{bmatrix} 3 & 15 \\ 2 & 9 \end{bmatrix}, WM_9 : \begin{bmatrix} 11 & 6 \\ 3 & 10 \end{bmatrix}$$

We calculate these workers' accuracy respectively by Formulas 1, 2 and 3. The results are shown in Table 1. From Table 1, we observed that W_4 tends to choose first choice as the answer. The difference between α_4 and β_4 is more than 50%, so W_4 is a biased worker.

Table 1. Workers' accuracy

WorkerID	α	β	γ
W_1	0.632	0.647	0.639
W_4	0.818	0.167	0.414
W_9	0.769	0.647	0.7

5   Updating Question Model and Worker Model

In this section, we discuss how to incrementally update the two models in IN-QUIRE.

5.1   Updating Question Model

Whenever a new answer returns, QM_i is updated. We propose two different updating strategies for QM_i: Weighted Strategy and Probability Strategy. From the voting perspective, we design the first strategy and from the probabilistic perspective, we design the second strategy.

Weighted Strategy. Weighted Strategy (called *WS*) is a weighted voting method. The current inference result is gotten by weighted voting, and QM_i is the weight of this vote. The new answer is another vote, and its weight is the accuracy of the worker. Since the accuracy of a worker can be expressed in two patterns $\alpha\beta$ and γ (as discussed in Section 4.2), updating QM_i can be calculated by Formulas 4 or 5. Formula 4 is called *WS-$\alpha\beta$* for short and Formula 5 is called *WS-γ*:

$$p_i = \frac{p_i + (\alpha_k \cdot L_{ik} + (1 - \beta_k) \cdot (1 - L_{ik}))}{2} \tag{4}$$

$$p_i = \frac{p_i + (\gamma_k \cdot L_{ik} + (1 - \gamma_k) \cdot (1 - L_{ik}))}{2} \tag{5}$$

The example 3 illustrates how the *WS* works.

Example 3. Assume that, in example 1, the received answers are set as $\langle Q_1, W_1, 1 \rangle^1$, $\langle Q_1, W_9, 0 \rangle^2$ and $\langle Q_1, W_4, 1 \rangle^3$. We take these three answers and the workers' accuracy in Table 1 as an example. Figure 2 and Figure 3 describe *WS* more specifically. Figure 2 is the process of updating QM_1 by *WS-$\alpha\beta$*. Since $\langle Q_1, W_1, 1 \rangle^1$ is the first received answer of Q_1, based on Formula 1 which is expressed "(1)" in the figure, the accuracy of W_1 is $\alpha_1 = 0.632$. Taking $\alpha_1 = 0.632$ and $p_1 = 0.5$ of current QM_1 into Formula 4, new p_1 is calculated as 0.656. Then QM_1 is (0.656, 0.434). At this time, $p_1 > 1 - p_1$, so the first inference result of Q_1 is $\langle Q_1\_1 \rangle^1$. The second answer of Q_1 is $\langle Q_1, W_9, 0 \rangle^2$ and the accuracy of W_9 is calculated by Formula 2, so $\beta_9 = 0.647$. Formula 4 calculates new p_1 using $\beta_9 = 0.647$ and $p_1 = 0.656$. Then QM_1 is (0.459, 0.541). Since $p_i < 1 - p_i$, the second inference result of Q_1 is $\langle Q_1\_0 \rangle^2$. Received the third answer $\langle Q_1, W_4, 1 \rangle^3$, based on $\alpha_4 = 0.818$ and $p_1 = 0.459$, Formula 4 calculates new $p_1 = 0.639$. QM_1 is (0.639, 0.361). $p_1 > 1 - p_1$, so the third inference result of Q_1 is $\langle Q_1\_1 \rangle^3$.

Figure 3 illustrates the process of *WS-γ*. The updating process of *WS-γ* is similar to that of *WS-$\alpha\beta$*. There are only two differences between them: the first is that the worker's accuracy is calculated by Formula 3 and the second is that Formula 5 updates p_i, in *WS-γ*.

As seen in Figure 2 and Figure 3, the third inference result of *WS-$\alpha\beta$* is different from the result of *WS-γ*. According to these workers' accuracy in Table 1, because W_4 is bias as mentioned in Section 4.2, it is reasonable that the third

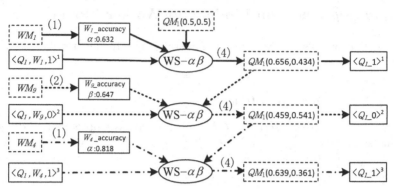

Fig. 2. Updating process of $WS\text{-}\alpha\beta$

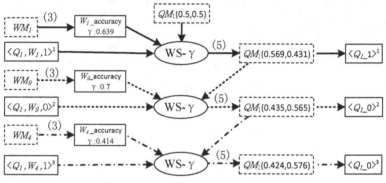

Fig. 3. Updating process of $WS\text{-}\gamma$

inference result is 0. That is, when some specific workers are biased, the inference results are more accurate using $WS\text{-}\gamma$ to update question model. We compare these two strategies through experiments in Section 6.

Probability Strategy. Probability strategy (called PS) applies Bayesian methods to update the question model. Let A_1 denote the new arriving answer which is L_{ik}. And A_2 represents the current inference result. Given $A = \{A_1, A_2\}$, we can compute the probability that the first choice is the true result (i.e. p_i) by Bayesian formula, following Formula 6.

$$p(R_i = 1|A) = \frac{p(A|1) \cdot p(1)}{p(A)} = \frac{p(A|1) \cdot p(1)}{p(A|1) \cdot p(1) + p(A|0) \cdot p(0)} \tag{6}$$

The prior probability of first choice is equal with the prior probability of second choice, so Formula 6 can be simplified as:

$$p(R_i = 1|A) = \frac{p(A|1)}{p(A|1) + p(A|0)} \tag{7}$$

We know that $p(A|R_i) = p(A_1|R_i) \cdot p(A_2|R_i)$ $(R_i=0,1)$. Formulas 8 and 9 calculate $p(A_1|R_i)$ as $\alpha\beta$ expresses the workers' accuracy. Formula 10 calculates $p(A_1|R_i)$ as γ expresses the worker's accuracy. We denote $I(cond)$ as the decision function. That is, the result is 1 if the $cond$ is true, otherwise the result is 0.

$$p(A_1|R_i = 1) = \alpha^{I(A_1=R_i)} \cdot (1 - \alpha)^{I(A_1 \neq R_i)} \tag{8}$$

$$p(A_1|R_i = 0) = \beta^{I(A_1=R_i)} \cdot (1 - \beta)^{I(A_1 \neq R_i)} \tag{9}$$

$$p(A_1|R_i) = \gamma^{I(A_1=R_i)} \cdot (1 - \gamma)^{I(A_1 \neq R_i)} \tag{10}$$

$P(A_2|R_i) = \frac{p(R_i|A_2) \cdot p(A_2)}{p(R_i)}$. The prior probabilities of A_2 and R_i are equal. Then $P(A_2|R_i)$ is calculated as:

$$p(A_2|R_i = 1) = \frac{P(R_i = 1|A_2) \cdot p(A_2)}{p(R_i = 1)} = p(R_i = 1|A_2) = p_i \tag{11}$$

$$p(A_2|R_i = 0) = \frac{P(R_i = 0|A_2) \cdot p(A_2)}{p(R_i = 0)} = p(R_i = 0|A_2) = 1 - p_i \tag{12}$$

As $\alpha\beta$ expresses the worker's accuracy, Formulas 8, 9, 11 and 12 are substituted into Formula 7. We have Formulas 13 which is called PS-$\alpha\beta$.

$$p(R_i = 1|A) =$$
$$\frac{\alpha^{I(A_1=1)} \cdot (1 - \alpha)^{I(A_1 \neq 1)} \cdot p_i}{\alpha^{I(A_1=1)} \cdot (1 - \alpha)^{I(A_1 \neq 1)} \cdot p_i + \beta^{I(A_1=0)} \cdot (1 - \beta)^{I(A_1 \neq 0)} \cdot (1 - p_i)} \tag{13}$$

As γ expresses the worker's accuracy, combining Formulas 10, 11 and 12, we have Formulas 14 (called PS-γ).

$$p(R_i = 1|A) =$$
$$\frac{\gamma^{I(A_1=1)} \cdot (1 - \gamma)^{I(A_1 \neq 1)} \cdot p_i}{\gamma^{I(A_1=1)} \cdot (1 - \gamma)^{I(A_1 \neq 1)} \cdot p_i + \gamma^{I(A_1=0)} \cdot (1 - \gamma)^{I(A_1 \neq 0)} \cdot (1 - p_i)} \tag{14}$$

Example 4 shows how PS works.

Example 4. We use the answers and workers in Example 3 as an example. Figure 4 illustrates the process of PS-$\alpha\beta$, and Figure 5 shows the process of PS-γ. The process of PS is similar to that of WS in Example 3. The only difference between PS and WS is that they apply different formulas to update p_i.

The results in Figures 4 and 5 are similar to the results in Figure 2 and 3. The reason that Figure 4 and 5 return different third inference result is similar to the analysis of Figure 2 and 3. In Section 6, we compare WS with PS through experiments.

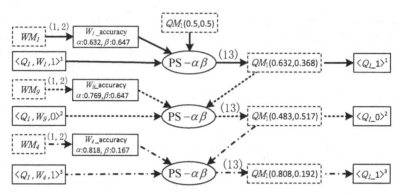

Fig. 4. Updating process of PS-$\alpha\beta$

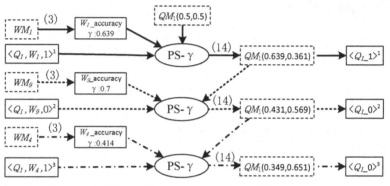

Fig. 5. Updating process of PS-γ

5.2 Updating Worker Model

As discussed in Section 3, INQUIRE has to compute the change of workers' accuracy in time, and update m workers' model when a question has received m answers. We use QM_i to calculate c_{ij} as mentioned in Section 4.2. According to L_{ik}, WM_k is updated as:

$$c_{00} = c_{00} + (1 - p_i), c_{01} = c_{01} + p_i \qquad \textbf{if } L_{ik} = 0 \tag{15}$$

$$c_{10} = c_{10} + (1 - p_i), c_{11} = c_{11} + p_i \qquad \textbf{if } L_{ik} = 1 \tag{16}$$

The example 5 shows the process of updating worker model.

Example 5. We take the results in Figure 5 as an example. Since L_{11}=1 and L_{14}=1, we can apply Formula 16 to update WM_1 and WM_4. WM_9 is updated by Formula 15 since the answer returned by W_9 is 0. The new workers' models are as follows.

$$WM_1 : \begin{bmatrix} 11 & 6 \\ 7 + 0.651 & 12 + 0.349 \end{bmatrix}, WM_4 : \begin{bmatrix} 3 & 15 \\ 2 + 0.651 & 9 + 0.349 \end{bmatrix}, WM_9 : \begin{bmatrix} 11 + 0.651 & 6 + 0.349 \\ 3 & 10 \end{bmatrix}$$

6   Experiments

In this section, we conduct experiments to evaluate our method. Our experimental goal is to test the efficiency and quality of different inference methods.

6.1   Experiment Setting

Platform. We conducted our experiments on real crowdsourcing platform Amazon Mechanical Turk. We implemented our method using python. All the experiments were run on a PC with Intel core i5 duo 2.6GHz CPU and 4GB RAM.

Datasets. We used two sets of binary choices tasks: Filmpair and Animal. (a) Filmpair is a dataset of movie poster. Each question contains two movie posters and workers decide which movie is released earlier. The ground truth of Filmpair is the actual released time of movies. We choose the most well-known movie posters in 1996-2006 from IMDB. There are 2000 questions and each question is answered three times. A total number of 146 workers answered these questions. (b) Animal is a dataset of animal pictures. Each question contains two animal pictures and workers decide which animal's size is larger. The ground truth of Animal is based on Animal-Size-Comparison-Chart[1]. Animal contains 300 questions and each question is completed by five workers. 36 workers participated in Animal. We use these datasets in all experiments because these are different types of questions and different number of questions.

Comparison Methods. Firstly, we choose 600 answers and 150 answers respectively from Filmpair and Animal. These answers are used as historical information to initialize workers' model. The remaining answers are the validation dataset. Then we evaluate INQUIRE on the validation dataset from the following three aspects: (1) Section 6.2 evaluates the effectiveness of worker model; (2) Section 6.3 compares the inference results' accuracy between INQUIRE and other inference methods (MV,CDAS and EM); (3) Section 6.4 compares the runtime between INQUIRE and other three existing methods mentioned above. To be general, all these methods run three times and the final experimental results are the average of the three results.

6.2   Worker Model Analysis

We verify the effectiveness of the worker model by comparing the similarity between worker model and real confusion matrix (called CM_k). This paper applies the cosine distance to measure the similarity. The value of cosine distance is between 0 to 1. The bigger value means that WM_k and CM_k are more similar. Let define CM_k:

$$\begin{bmatrix} b_{00} & b_{01} \\ b_{10} & b_{11} \end{bmatrix}$$

---

[1] http://myuui.deviantart.com/art/Animal-Size-Comparison-Chart-109707959

The similarity between WM_k and CM_k is calculated as follows:

$$Sim(WM_k, CM_k) = \frac{(c_{00} \cdot b_{00} + c_{01} \cdot b_{01} + c_{10} \cdot b_{10} + c_{11} \cdot b_{11})}{\sqrt{c_{00}^2 + c_{01}^2 + c_{10}^2 + c_{11}^2} \cdot \sqrt{b_{00}^2 + b_{01}^2 + b_{10}^2 + b_{11}^2}}$$

As mentioned in Section 4.2, we need worker's historical performance to initialize the value of worker model. We define the percentage of workers with historical-information in total workers as `worker ratio`. Because we cannot guarantee that every worker has historical performance, we study the effect of `worker ratio` on similarity between WM_k and CM_k. We vary `worker ratio` from 0%, 10%, 25%, 50%, 70%, 85% to 100% and plot the average similarity of four strategies ($WS\text{-}\alpha\beta$, $WS\text{-}\gamma$, $PS\text{-}\alpha\beta$, $PS\text{-}\gamma$) as addressed in Section 5.1.

From Figure 6, we can see that the similarity grows with `worker ratio` increase regardless of the PS or WS. The worker model can achieve relatively similar results to the real confusion matrix when `worker ratio` is higher than 50%. Especially, when the `worker ratio` is 100%, i.e., all the workers has initial information, the similarity is above 90%. This indicates that our worker model is more reasonable for the workers' quality with a higher `worker ratio`.

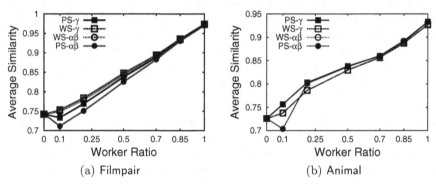

(a) Filmpair   (b) Animal

Fig. 6. Average similarity between worker model and real confusion matrix

6.3   Accuracy Analysis

We first compare the results' accuracy of four strategies ($WS\text{-}\alpha\beta$, $WS\text{-}\gamma$, $PS\text{-}\alpha\beta$, $PS\text{-}\gamma$). The results' accuracy is calculated by comparing the inference results with the ground truth. Figure 7 shows the effect of `worker ratio` on the average accuracy of four strategies. We can see that the results of $PS\text{-}\gamma$ and $WS\text{-}\gamma$ are better than that of $PS\text{-}\alpha\beta$ and $WS\text{-}\alpha\beta$ no matter how `worker ratio` varies. The main reason is that some workers subjectively want to choose certain choice as an answer as mentioned in Section 4.2. This results indicate that, to reflect workers' accuracy, γ method is better than $\alpha\beta$. Thus in the following experiments, we only evaluate γ method. In addition, the accuracy with 70% `worker ratio` is close to that with 100% `worker ratio`, so we use 70% `worker ratio` in the following experiments.

Next we evaluate the online results of MV, CDAS, EM and INQUIRE. The accuracy of the results are shown in Figure 8. We can observe that, on both

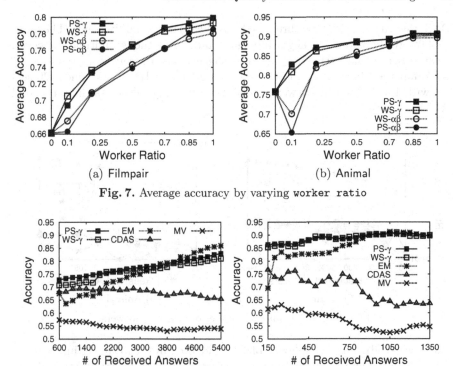

Fig. 7. Average accuracy by varying worker ratio

Fig. 8. Accuracy on the number of received answers

datasets, the accuracy of INQUIRE with $WS\text{-}\gamma$ or $PS\text{-}\gamma$ is always higher than that of MV and CDAS, because INQUIRE can reflect the change of workers' accuracy in time. Since the workers' quality are equal in MV, just like the analysis in section 1, the results of MV is worse than those of CDAS. We also observed that the results of EM changes a lot. For example, on Filmpair dataset, the accuracy is only about 0.65 when receiving 600 answers. Once received all answers, the accuracy can achieve 0.85 which is 0.03 higher than those of INQUIRE with $PS\text{-}\gamma$. However, the accuracy of INQUIRE increased gradually with increased answers. When receiving less than 70% of answers INQUIRE performs better than EM. On Animal dataset, EM and INQUIRE works similarly when receiving 100% answers, since the amount of answers is not large. As the results in Filmpair, EM also performs poorly in the beginning. We addressed in section 4.2 that the values of confusion matrix maybe not precise when the number of answers is small. And then the EM's results are possible not accurate. That is the main reason why EM has worse performance than INQUIRE with fewer answers.

6.4   Runtime Analysis

In this section, we compare the runtime of our method to other three methods mentioned in Section 6.3. Figure 9 shows the effect of the number of received answers on runtime. Compared the runtime of INQUIRE with that of MV and

CDAS, we can see that these three methods always run in milliseconds. The number of received answers has slight influence on these methods' runtime. However, we can observe that EM's runtime rises sharply when received answers grows. For example, on Filmpair dataset, the running time of EM costs almost 20 seconds with 100% received answers, while its processing takes less than a second in the beginning.

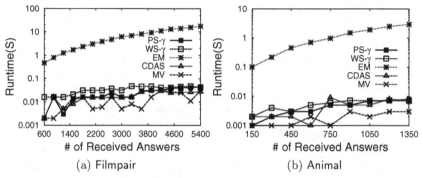

(a) Filmpair   (b) Animal

Fig. 9. Runtime on the number of received answers

Next we study time complexity of EM and INQUIRE. EM algorithm is iteratively calculated by two steps. In the E-step, it calculates the results' probability of every question by known workers' accuracy. In the M-step, it calculates every worker's accuracy according to the results obtained in the E-step. The time complexity of the E-step is $O(2n+2a)$, n is the number of questions and a is the number of received answers. The time complexity of the M-step is $O(max(2n, 2a))$. Thus the running time of EM increases linearly with the increase of n and a. So when the number of answers is large EM does not perform well. According to Algorithm 1 in Section 3, time complexity of INQUIRE is $O(1)$, that is, neither the number of questions nor received answers influence the runtime of INQUIRE.

In summary, when the accuracy and runtime are considered together, INQUIRE is obviously superior to MV, CDAS and EM. Comparing the results of *PS-γ* with *WS-γ* in Figure 7 and 8, we can see that the accuracy of *PS-γ* is better than that of *WS-γ* with more answers. So we propose to adopt *PS-γ* in INQUIRE.

7   Conclusion

In this paper, we studied the problem of incremental quality inference in Crowdsourcing. We presented an incremental inference method, INQUIRE, which contains the question model and worker model. For the question model, we proposed two different updating strategies to efficiently infer results. For the worker model, it can dynamically represent workers' quality, and we proposed an incremental strategy to update worker model. We evaluated our method on real-world datasets. Compared to MV, CDAS and EM, INQUIRE achieved a good trade-off between accuracy and time, and thus INQUIRE is more effective and efficient.

Acknowledgement. This work was partly supported by the National Natural Science Foundation of China under Grant No. 61373024, National Grand Fundamental Research 973 Program of China under Grant No. 2011CB302206, Beijing Higher Education Young Elite Teacher Project under grant No. YETP0105, a project of Tsinghua University under Grant No. 20111081073, Tsinghua-Tencent Joint Laboratory for Internet Innovation Technology, the "NExT Research Center" funded by MDA, Singapore, under Grant No. WBS:R-252-300-001-490, and the FDCT/106/2012/A3.

References

1. http://crowdflower.com/
2. http://www.mturk.com
3. Dempster, A.P., Laird, N.M., Rubin, D.B.: Maximum likelihood from incomplete data via the em algorithm. J.R.Statist.Soc.B 30(1), 1–38 (1977)
4. Feng, A., Franklin, M., Kossmann, D., Kraska, T., Madden, S., Ramesh, S., Wang, A., Xin, R.: Crowddb: Query processing with the vldb crowd. Proceedings of the VLDB Endowment 4(12) (2011)
5. Franklin, M.J., Kossmann, D., Kraska, T., Ramesh, S., Xin, R.: Crowddb: Answering queries with crowdsourcing. In: Proceedings of the 2011 ACM SIGMOD International Conference on Management of Data, pp. 61–72. ACM (2011)
6. Howe, J.: Crowdsourcing: How the power of the crowd is driving the future of business. Random House (2008)
7. Ipeirotis, P.G., Provost, F., Wang, J.: Quality management on amazon mechanical turk. In: Proceedings of the ACM SIGKDD Workshop on Human Computation, pp. 64–67. ACM (2010)
8. Karger, D.R., Oh, S., Shah, D.: Iterative learning for reliable crowdsourcing systems. In: Advances in Neural Information Processing Systems, pp. 1953–1961 (2011)
9. Liu, X., Lu, M., Ooi, B.C., Shen, Y., Wu, S., Zhang, M.: Cdas: A crowdsourcing data analytics system. Proceedings of the VLDB Endowment 5(10), 1040–1051 (2012)
10. Marcus, A., Wu, E., Karger, D.R., Madden, S., Miller, R.C.: Demonstration of qurk: a query processor for humanoperators. In: Proceedings of the 2011 ACM SIGMOD International Conference on Management of Data, pp. 1315–1318. ACM (2011)
11. Marcus, A., Wu, E., Karger, D.R., Madden, S.R., Miller, R.C.: Crowdsourced databases: Query processing with people. In: CIDR (2011)
12. Mason, W., Suri, S.: Conducting behavioral research on amazon mechanical turk. Behavior Research Methods 44(1), 1–23 (2012)
13. Park, H., Garcia-Molina, H., Pang, R., Polyzotis, N., Parameswaran, A., Widom, J.: Deco: A system for declarative crowdsourcing. Proceedings of the VLDB Endowment 5(12), 1990–1993 (2012)
14. Raykar, V.C., Yu, S., Zhao, L.H., Valadez, G.H., Florin, C., Bogoni, L., Moy, L.: Learning from crowds. The Journal of Machine Learning Research 99, 1297–1322 (2010)
15. Wang, J., Kraska, T., Franklin, M.J., Feng, J.: Crowder: Crowdsourcing entity resolution. Proceedings of the VLDB Endowment 5(11), 1483–1494 (2012)
16. Whitehill, J., Wu, T.-F., Bergsma, J., Movellan, J.R., Ruvolo, P.L.: Whose vote should count more: Optimal integration of labels from labelers of unknown expertise. In: Advances in Neural Information Processing Systems, pp. 2035–2043 (2009)
17. Yuen, M.-C., King, I., Leung, K.-S.: A survey of crowdsourcing systems. In: 2011 IEEE Third International Conference on Social Computing (socialcom), pp. 766–773. IEEE (2011)

Repair Diversification for Functional Dependency Violations

Chu He, Zijing Tan*, Qing Chen, Chaofeng Sha, Zhihui Wang, and Wei Wang

School of Computer Science,
Shanghai Key Laboratory of Data Science,
Fudan University, Shanghai, China
{12210240018,zjtan,13210240082,cfsha,zhhwang,weiwang1}@fudan.edu.cn

Abstract. In practice, data are often found to violate functional dependencies, and are hence inconsistent. To resolve such violations, data are to be restored to a consistent state, known as "repair", while the number of possible repairs may be exponential. Previous works either consider optimal repair computation, to find one single repair that is (nearly) optimal *w.r.t.* some cost model, or discuss repair sampling, to randomly generate a repair from the space of all possible repairs.

This paper makes a first effort to investigate repair diversification problem, which aims at generating a set of repairs by minimizing their costs and maximizing their diversity. There are several motivating scenarios where diversifying repairs is desirable. For example, in the recently proposed interactive repairing approach, repair diversification techniques can be employed to generate some representative repairs that are likely to occur (small cost), and at the same time, that are dissimilar to each other (high diversity). Repair diversification significantly differs from optimal repair computing and repair sampling in its framework and techniques. (1) Based on two natural diversification objectives, we formulate two versions of repair diversification problem, both modeled as bi-criteria optimization problem, and prove the complexity of their related decision problems. (2) We develop algorithms for diversification problems. These algorithms embed repair computation into the framework of diversification, and hence find desirable repairs without searching the whole repair space. (3) We conduct extensive performance studies, to verify the effectiveness and efficiency of our algorithms.

1 Introduction

Data consistency is one of the issues central to data quality. We say data is inconsistent when they violate predefined data dependencies, *e.g.*, functional dependencies, and repairing data means restoring the data to a consistent state, *i.e.*, a *repair* of the data. Violations of functional dependencies are commonly found in practice. There is generally no single deterministic way of rectifying inconsistencies; former works address this issue mainly based on two approaches.

---

* Corresponding Author.

S.S. Bhowmick et al. (Eds.): DASFAA 2014, Part II, LNCS 8422, pp. 468–482, 2014.
© Springer International Publishing Switzerland 2014

Input Instance I

	ID	Name	City	Zip
t_1	230012(0.6)	Michael(0.9)	Beijing(0.8)	100000(0.5)
t_2	230056(0.9)	Alex(0.2)	Shanghai(0.4)	100000(0.1)
t_3	230012(0.8)	Taylor(0.3)	Chengdu(0.9)	100000(0.3)

Functional Dependencies:
ID->Name, City, Zip
Zip->City

Repair r_1

ID	Name	City	Zip
230012	Michael	Beijing	100000
230056	Alex	Beijing	100000
230012	Michael	Beijing	100000

Repair r_2

ID	Name	City	Zip
230012	Michael	Beijing	100000
230056	Alex	Shanghai	$v^{(t_2,Zip)}$
230012	Michael	Beijing	100000

Repair r_3

ID	Name	City	Zip
230012	Michael	Beijing	100000
230056	Alex	Shanghai	$v^{(t_2,Zip)}$
$v^{(t_3,ID)}$	Taylor	Chengdu	$v^{(t_3,Zip)}$

Repair r_4

ID	Name	City	Zip
230012	Michael	Chengdu	100000
230056	Alex	Shanghai	$v^{(t_2,Zip)}$
230012	Michael	Chengdu	100000

Fig. 1. Instance, functional dependencies and repairs

The first approach, namely optimal repair computation [3,5,6,10,13], aims to find a repair with the minimum repair cost, while the second one, namely repair sampling [4], aims to randomly generate a repair from the space of all repairs.

This paper presents a novel approach, referred to as *repair diversification*. We leave the formal framework to Section 3, where repair diversification is formalized as a bi-criteria optimization problem *w.r.t.* objective functions in terms of repair cost and distance. Informally, the goal of repair diversification is to generate a set of repairs, aiming at minimizing costs of repairs and simultaneously maximizing their diversity: generating repairs that have small cost to be practical, and that are dissimilar to each other to avoid the redundancy.

Example 1: We give an illustrative example, and leave formal definitions to Sections 2 and 3. Fig. 1 presents an instance I of schema $(ID, Name, City, Zip)$ and some functional dependencies defined on this schema. A weight (number in bracket) is associated with every attribute value in I, reflecting the confidence the data owner places in that value [3,5,13]; a larger weight implies a more reliable value. Instance I violates functional dependencies, and hence is inconsistent.

We also give some (not all) repairs for instance I, denoted by r_i ($i \in [1,4]$). Each repair is a possible consistent state for this instance, obtained by modifying some attribute values. Note that some repairs have variables as attribute values, e.g., $v^{(t_2,Zip)}$ in r_2, where each variable indicates a new value not used in that attribute in I. As an example, in r_3, $v^{(t_2,Zip)} \neq v^{(t_3,Zip)} \neq 100000$. □

We highlight the properties of repair diversification.

(1) Cost. Intuitively, it costs more to modify a value with a large weight in a repair, and a repair with a small cost is more likely to occur in practice. Therefore, small-cost repairs are expected to carry more meaningful information, and are hence more instructive. Optimal repair computation employs the notion of cost to find a single optimal repair, while discarding all other repairs even if they have a similar (or even the same) cost as the found optimal repair. In contrast, repair sampling generates a random repair each time; some repairs may

be impractical in terms of repair cost. As will be seen later, repair diversification includes optimal repair computation as its special case, which is already proved to be NP-complete in various settings. Worse still, neither of the former work has considered finding a set of repairs when considering their costs.

(2) Diversity. Repairs each with a small cost, may be similar to each other, and hence may fail to show its own uniqueness. Therefore, it is generally insufficient to consider only costs, when generating a set of repairs; each repair is expected to carry some *novel* information, in order to avoid redundancy. Recall repairs in Fig. 1. When diversity is concerned, repair r_3 is preferable to r_2 in forming a set with r_1. This is because r_3 shows a totally different way to restore consistency compared to r_1; their sets of modified values are disjoint. On the contrary, r_1 and r_2 deal with tuples t_1 and t_3 in the same way. To our best knowledge, neither of the former work on data repairing has considered the notion of diversity.

Repair cost and diversity may compete with each other; we therefore need to take them both into consideration, and find a tradeoff between them.

Motivating scenarios. Intuitively, repair diversification results in a set of repairs that are likely to occur (small cost), and simultaneously, that are dissimilar to each other (high diversity). Complementary to existing methods, we contend that repair diversification is useful in various practical situations. For example, as remarked in [4], the notion of consistent query answering [1] can be generalized to *uncertain query answering*, when each possible repair is regarded as a possible world. In this setting, repair diversification technique can be employed to compute a set of repairs that effectively summarizes the space of all repairs, which may be sufficient to obtain meaningful answers. As another example, since fully automated repair computation may bring the risk of losing critical data, the guided data repair framework [15] is recently proposed to involve users (domain experts) in repair computation: several representative repairs are first presented to the users, and machine learning techniques are then applied to user feedback or comments on these repairs, for improving the quality of future repairs. Due to the large number of possible repairs, techniques for selecting good representative repairs become very important. As noted earlier, repair diversification technique lends itself as an effective approach to generating such repairs, while neither optimal repair computation nor repair sampling is suitable for this setting.

Contributions. We make a first effort to investigate repair diversification.

(1) We provide a formal framework for repair diversification problem (Section 3). We first study two functions to rank repairs: a *distance* function that measures the dissimilarity of repairs, and a *cost* function that measures the possibility of a repair. We then present two diversification objectives, both to minimize repair cost and to maximize repair distance. Based on them, we propose two versions of repair diversification problem, both modeled as bi-criteria optimization problem, and show that their decision problems are NP-complete.

(2) Despite the intractability, we develop algorithms for repair diversification problems, *i.e.*, to find diversified top-k repairs based on diversification objectives (Section 4). Our algorithms embed repair computation into the framework of diversification, by employing diversification objectives to guide repair generations.

Hence, these algorithms have the early termination property: they stop as soon as desirable repairs are found, without searching the whole repair space.

(3) Using both real-life and synthetic data, we conduct an extensive experimental study to verify the effectiveness and efficiency of our algorithms (Section 5).

Related Work. As remarked earlier, repair diversification aims to generate a set of repairs by considering both repair cost and diversity, and hence significantly differs from former works on data repairs. Specifically, (1) optimal repair computation [3,5,6,10,13] generates exactly one (nearly) optimal repair in its cost; and (2) repair sampling [4] is proposed to generate a random repair from the space of all possible repairs in each run, considering neither cost nor diversity.

There has been a host of work on data repairing (see [2,9] for a survey of more related works). Specifically, guided (interactive) data repairing [15] is proposed to incorporate user feedback in the cleaning process for improving automatic repairing results. Moreover, the dashboard discussed in [7] supports user interaction, which provides several summarization of data violations.

One may find there are some similarities between repair diversification and query result diversification [8,11], if regarding repair computation as a query finding repairs from the repair space, and regarding the cost of a repair as the opposite of its relevance to the query. However, repair diversification introduces new challenges that we do not encounter in query result diversification. Query result diversification is proposed to pick diversified results from a known set of relevant results. In contrast, repair diversification aims to compute and pick diversified repairs from those low-cost ones, when taking as input the exponential repair space with many costly repairs that are supposed to rarely occur in practice. These costly repairs should always be excluded from the result, even if they contribute large distances to other repairs. It is infeasible to compute all the repairs in advance, since the number of repairs is exponential and the computation of a single repair is also expensive, especially when repair cost is involved. Therefore, a tricky technique of repair diversification concerns employing diversification objectives to guide repair generation, hopefully avoiding as much as possible repair computations.

2   Preliminaries

We review some basic notations and the definition of repair [4].

An instance I of a relation schema $R(A_1, \ldots, A_m)$ is a set of tuples in $Dom(A_1) \times \cdots \times Dom(A_m)$, where $Dom(A_i)$ is the domain of attribute A_i. We denote by $Dom_I(A_i)$ the set of values that appear in attribute A_i in I. We assume that every tuple is associated with an identifier t that is not subject to updates, and use the terms tuple and tuple identifier interchangeably. We denote an attribute A_i of a tuple t in an instance I by $I(t, A_i)$, called a *cell*, and abbreviate it as $t[A_i]$ if I is clear from the context.

For an attribute set $X \subseteq \{A_1, \ldots, A_m\}$, an instance I satisfies a functional dependency (FD) $X \to A_i$, written as $I \models X \to A_i$, if for every two tuples t_1, t_2 in I such that $t_1[X] = t_2[X]$, we have $t_1[A_i] = t_2[A_i]$. We say I is *consistent w.r.t.* a set Σ of FDs if I satisfies every FD in Σ; otherwise I is *inconsistent w.r.t.* Σ. Similar to former works on repairs [4,5,13], we adopt attribute value modification as the only repair operation, which is sufficient to resolve FD violations. Specifically, for any two tuples t_1, t_2 in I that violate an FD $X \rightarrow A_i$, we fix this violation either (1) by modifying $I(t_1, A_i)$ to be equal to $I(t_2, A_i)$ (or vice versa), or (2) by introducing a new value in $Dom(A_k) \backslash Dom_I(A_k)$ to $I(t_1, A_k)$ or $I(t_2, A_k)$, where $A_k \in X$.

To distinguish introduced new values from existing constant values in I, we denote these new values by variables. Specifically, the new value introduced to $I(t_i, A)$ is denoted by $v^{(t_i, A)}$; $v^{(t_i, A_j)}$ and $v^{(t_{i'}, A_{j'})}$ denote the same value if and only if $i = i'$ and $j = j'$. Observe that $v^{(t_i, A)}$ introduced to $I(t_i, A)$ may also be used as the value of $I(t_j, A)$ $(i \neq j)$ in repairing I, due to the enforcement of an FD of the form $X \rightarrow A$. To simplify notation, in this case we replace $v^{(t_i, A)}$ as $v^{(t_k, A)}$ where k is the minimum tuple identifier such that $I(t_k, A) = v^{(t_i, A)}$. Note that this is just a symbol mapping between variable names.

Repair *w.r.t.* FD. Assume that instance I of a relation schema $R(A_1, \ldots, A_m)$ is a set of tuples $\{t_1, \ldots, t_n\}$, and that I is inconsistent *w.r.t.* a set Σ of FDs. We can apply attribute value modifications to I, for a new instance I_r that is consistent *w.r.t.* Σ, called a *repair* of I. As an auxiliary notion, we denote by $mod(I, I_r)$ cells $t_i[A_j]$ that have different values in I and I_r.

We now formally define the notion of a *repair*. A repair of I *w.r.t.* Σ is an instance I_r such that (1) I_r has the same set of tuples (with identifiers) as I, and is obtained from I by a list of attribute value modifications, in which some $I(t_i, A_j)$ $(i \in [1, n], j \in [1, m])$ is replace by either (a) a different constant in $Dom_I(A_j)$, or (b) a variable denoting a new value in $Dom(A_j) \backslash Dom_I(A_j)$; (2) I_r is consistent *w.r.t.* Σ; and (3) there is no $I_{r'}$ that satisfies (1) and (2), such that $mod(I, I_{r'}) \subset mod(I, I_r)$ and for each $t_i[A_j] \in mod(I, I_{r'})$, $I_{r'}(t_i, A_j) = I_r(t_i, A_j)$, where $I_{r'}(t_i, A_j)$ is either a constant or a variable.

Example 2: All repairs presented in Fig. 1 satisfy our notion of repair. $\square$

3   Framework of Repair Diversification

Distance & Cost. We are to define a *distance* function d to measure the dissimilarity between two instances carrying the same set of tuples (identifiers), with modified attribute values. Following former works [3,5,13], we use *weight* to reflect the confidence of the accuracy of data. These weights can be manually placed by the users, be collected through analysis of existing data, or be set to an equal value by default; our approach does not depend on a particular setting. Specifically, a weight $w(t, A) \in [0, 1]$ is attached to each attribute A of each tuple t, *i.e.*, $t[A]$. We assume that two instances carrying the same set of tuple identifiers, *e.g.*, instance I and its repair I_r, or two repairs $I_r, I_{r'}$ of I, have the same weight $w(t, A)$ for $t[A]$. Distance function d is defined as follows:

$$d(I_1, I_2) = \sum_{i \in [1,n], j \in [1,m]} w(t_i, A_j) \times \triangle(I_1(t_i, A_j), I_2(t_i, A_j))$$

I_1, I_2 are instances of schema $R(A_1, \ldots, A_m)$, both carrying a set of tuple identifiers $\{t_1, \ldots, t_n\}$. $\triangle(a, b) = 0$ if $a = b$, otherwise $\triangle(a, b) = 1$, where a (resp. b) is either a constant or a variable.

Example 3: Considering the weights presented in Fig. 1, the distance between r_3, r_4 is $0.8+0.8+0.3+0.3=2.2$. Note that $v^{(t_2, Zip)}$ in $t_2[Zip]$ of r_3, r_4 does not increase their distance: any new value that can be assigned to $v^{(t_2, Zip)}$ in r_3 can also be used in r_4, and vice versa. □

The function d, when applied to an instance I and its repair I_r, measures the cost of I_r. defined as follows:
$$c(I_r) = d(I, I_r).$$

Example 4: r_1, r_2, r_3, r_4 have a cost of 1.6, 1.3, 1.2, 1.2, respectively. □

Since I is clear from the context, below we write repair I_{r_k} of I as r_k. We use $U_{(I, \Sigma)}$ to denote the space of all possible repairs of I w.r.t. a set Σ of FDs. To formalize repair diversification problem, we introduce two natural diversification objective functions of the form $f(S, c(\cdot), d(\cdot, \cdot))$, where $S \subseteq U_{(I, \Sigma)}$, and $c(\cdot), d(\cdot, \cdot)$ is the above mentioned cost and distance function, respectively. $c(\cdot)$ and $d(\cdot, \cdot)$ are assumed to be fixed here; we hence abbreviate the function as $f(S)$.

(1) **Diversification function** $f_d(S)$. On a set $S = \{r_1, \ldots, r_k\}$ of k repairs, one natural bi-criteria objective concerns the difference between the cost sum and the distance sum of the repairs in S, which is encoded as follows:
$$f_d(S) = (1 - \lambda) \sum_{r_i \in S} c(r_i) - \frac{2 \cdot \lambda}{(k-1)} \sum_{r_i, r_j \in S, i < j} d(r_i, r_j).$$
The distance sum is scaled down with $\frac{2}{(k-1)}$, since there are $\frac{k \cdot (k-1)}{2}$ numbers for distance sum, while only k numbers for cost sum. Here λ is a parameter defining a trade-off between repair cost and distance; preference to distance increases as λ increases. Observe the following. (1) As λ increases, it tends to introduce large-cost repairs for the optimization of $f_d()$, since their overhead in cost is outweighed by their distances to small-cost repairs. However, repair diversification aims to pick diversified repairs from low-cost ones; large-cost repairs that rarely occur in practice should be banned. (2) It is generally infeasible to avoid large-cost repairs by setting an upper bound for repair cost; for a given number a, it is NP-complete to determine whether there exists a repair r_i such that $c(r_i) \leq a$. Putting these together, we find λ used in $f_d()$ is typically a small number, e.g., $\lambda \in [0, 0.3]$. We will further discuss this through experiments (Section 5).

(2) **Diversification function** $f_m(S)$. Our second bi-criteria objective concerns the maximum difference between the cost and the distance of the selected set :
$$f_m(S) = (1 - \lambda) \max_{r_i \in S} c(r_i) - \lambda \min_{r_i, r_j \in S, i \leq j} d(r_i, r_j),$$
As shown in experiments, $f_m()$ can better adapt to different settings of $\lambda \in [0, 1]$.

Example 5: When $\lambda = 0.3$, $(f_d(\{r_1, r_3, r_4\}) = 0.64) < (f_d(\{r_2, r_3, r_4\}) = 0.73)$, while $(f_m(\{r_1, r_3, r_4\}) = 0.46) > (f_m(\{r_2, r_3, r_4\}) = 0.4)$. □

Repair Diversification Problem. We next state the repair diversification problem. Given a positive integer k, a parameter λ, an inconsistent instance I w.r.t. a set Σ of FDs, a repair diversification objective function $f(S)$, it is to find a set S_k^* of k repairs, such that $S_k^* = \underset{\substack{S \subseteq U_{(I, \Sigma)} \\ |S| = k}}{argmin} \, f(S)$.

Specifically, we denote the proposed repair diversification problem by MindiffP (resp. MinmaxP) when $f(S) = f_d(S)$ (resp. $f_m(S)$).

Note that an optimal solution of MindiffP may be far from the optimum *w.r.t.* MinmaxP and vice versa. Also observe that neither $f_d(S)$ nor $f_m(S)$ satisfies the *stability* property [11]: $f_d(S)$ ($f_m(S)$) *cannot* guarantee that $S_k^* \subset S_{k+1}^*$. Therefore, the optimal solution cannot be constructed incrementally with the increase of k, which necessarily complicates the optimization of MindiffP (MinmaxP).

Not surprisingly, MindiffP (MinmaxP) is intractable; it includes optimal repair computation [13] as a special case.

Theorem 1: *The* MindiffP *(resp.* MinmaxP*) problem is NP-complete.* □

4   Finding Diversified Repairs

We present algorithms for diversification problems MindiffP and MinmaxP. In light of the intractability, our algorithms are necessarily heuristic. The first set of algorithms relate repair diversification problems to known techniques for facility dispersion problems (Section 4.1). These algorithms, however, may suffer from both effectiveness and efficiency. We then present the second set of algorithms (Section 4.2). These algorithms embed repair computation into the framework of repair diversification, and therefore have the early termination property.

4.1   Baseline Algorithms

Algorithm BMindiff. Given a positive integer k, a parameter λ and an inconsistent relation I *w.r.t.* a set Σ of FDs, BMindiff is designed for problem MindiffP, aiming at picking k repairs to minimize the objective function $f_d(S)$ (Section 3). In a nutshell, BMindiff works in two steps. (1) It randomly computes a set S of n ($n \geq k$) repairs. This can be achieved by employing existing techniques, *e.g.*, repair sampling. (2) It selects a set S_k of k repairs from S. To do so, BMindiff relates the diversification objective $f_d()$ to the objective of the well-known *Maximum Sum Dispersion problem* (MAXSUMDISP) [12]; this is a technique commonly used for optimization problems [11]. Recall that the objective of MAXSUMDISP problem is to select k points from a set of n points, such that the sum of all pairwise distances between these k points is maximized.

We show that an instance of MindiffP can be transformed to an instance of MAXSUMDISP. Given a set S of n repairs, we construct a set S' of n points, in which each point v_i represents a repair $r_i \in S$. The objective function of MAXSUMDISP is $f_d'(S') = \sum_{v_i, v_j \in S', i<j} d'(v_i, v_j)$, where $d'(v_i, v_j)$ is the distance between v_i, v_j. Herein, for two points $v_i, v_j \in S'$, we define the distance between them: $d'(v_i, v_j) = \frac{2 \cdot \lambda}{(k-1)} d(r_i, r_j) - \frac{1-\lambda}{(k-1)} (c(r_i) + c(r_j))$, where $d(\cdot, \cdot)$, $c(\cdot)$ is the distance and cost function defined on repairs (Section 3). For a set S_k' of k points, it can be verified that $f_d'(S_k') = \frac{2 \cdot \lambda}{(k-1)} \sum_{r_i, r_j \in S_k, i<j} d(r_i, r_j) - (k - 1) \cdot \frac{1-\lambda}{(k-1)} \sum_{r_i \in S_k} c(r_i) = - f_d(S_k)$; each $c(r_i)$ is counted exactly k -1 times in the sum. Therefore, S_k' is the set of points that maximizes the objective function of MAXSUMDISP if and only if S_k is the set of repairs that minimizes the objective

function of MindiffP. Given this reduction, we can employ known techniques for MAXSUMDISP in the second step of BMindiff. Although MAXSUMDISP is proved to be NP-complete, various heuristic and approximation algorithms have been proposed to solve MAXSUMDISP (see *e.g.*, [12,14]).

Remark. (1) The bottleneck of BMindiff in terms of effectiveness lies in its first step, in which n candidate repairs are generated by considering neither cost nor diversity. Although BMindiff may simulate the known best algorithms for MAXSUMDISP in its second step, it still suffers from this significant limitation. (2) To improve the effectiveness of overall algorithm, BMindiff has to pick a $n \gg k$ in its first step, however, at the cost of efficiency.

Algorithm BMinmax. Taking the same input as BMindiff, BMinmax deals with problem MinmaxP, aiming to select k repairs that minimize the objective function $f_m(S)$ (Section 3). Just as BMindiff, BMinmax first randomly generates n repairs from the repair space, and then select k repairs from these repairs. To do so, BMinmax links the objective of MinmaxP to *Maximum Min Dispersion problem* (MAXMINDISP) [14]. For space limitation, we omit the details here.

4.2 Early Termination Heuristics

We are about to present algorithms Mindiff and Minmax for problems MindiffP and MinmaxP, respectively. These algorithms combine repair computation into repair diversification: instead of random repair generation adopted in BMindiff (BMinmax), Mindiff (Minmax) employs diversification objective to guide repair generation. This helps find repairs that are beneficial to diversification, and simultaneously avoids unnecessary computation. To facilitate repair generation, we first present a repairing algorithm, referred to as Genrepair. Genrepair is based on the sampling algorithm presented in [4], with non-trivial improvement by incremental techniques. It can produce different repairs of an inconsistent instance I, by taking as input different sorting of cells in I. As will be seen later, a tricky technique concerns sorting cells in Mindiff (Minmax) to produce different inputs for Genrepair, such that desirable repairs can be generated.

Algorithm Genrepair. Genrepair is a common basis of Mindiff and Minmax. It takes as input a list of cells of an inconsistent instance I *w.r.t.* a set Σ of FDs, and produces a repair of I after processing all cells one by one in the list order.

Genrepair has to detect conflicts that are among cells processed so far, and that are deduced by FD reasoning. To this end, Genrepair employs the notion of *equivalence class* (EC) [3,4]. Each EC is associated with an attribute: an EC e^A on attribute A is a set of cells of the form $t_i[A]$; these cells have a same value in the repair generated by Genrepair. We use the following notations. (1) Any cell c belongs to exactly one EC at any time, denoted by $ec(c)$; cell c is initially in the EC $\{c\}$, *i.e.*, a singleton set containing itself. (2) Any EC e^A is associated with a value that is assigned to all cells in e^A in the generated repair, denoted by $val(e^A)$; $val(e^A)$ is initially NULL, which implies that the value is not yet determined. (3) We denote by ξ the set containing all equivalence classes (ECs).

After initializing the set ξ of ECs (line 1), Genrepair processes cells one by one. (1) when the current cell $t_i[B]$ has already been put into an EC containing other cells, and the value of that EC is determined, Genrepair modifies $t_i[B]$ if its value is different from the value chosen for that EC (lines 4-5). (2) Otherwise, Genrepair has to determine the value for the EC that $t_i[B]$ belongs to, and maintains ξ for further processing (lines 6-11). Genrepair first tries to use the value of $t_i[B]$ as the value of its EC (line 7), while this may cause its EC to be merged with other EC having the same value (line 8). To address this issue, Genrepair calls function Merge (line 9). If the merging fails (NULL is returned), Genrepair turns to introduce a new value as the value of $t_i[B]$ and its EC (line 10). Otherwise, Merge returns a modified set of ECs, accepted by Genrepair (line 11).

Algorithm 1. Genrepair

input : *a list L of cells of inconsistent instance I w.r.t. a set Σ of FDs.*
output: *a repair of I.*

1 initialize ξ: each cell c is in the EC $\{c\}$;
2 **while** L *is not empty* **do**
3   remove the first cell c from L; suppose $c = t_i[B]$ and $ec(t_i[B]) = e^B$;
4   **if** $|e^B| \neq 1$ **and** $val(e^B) \neq NULL$ **then**
5   **if** $t_i[B] \neq val(e^B)$ **then** $t_i[B] := val(e^B)$;
6   **if** $(|e^B| \neq 1$ **and** $val(e^B) = NULL)$ **or** $(|e^B| = 1)$ **then**
7          $val(e^B) := t_i[B]$;
8   **if** *there exists* $e'^B \in \xi$ *such that* $val(e'^B) = t_i[B]$ **then**
9              $\xi' := $ **merge**(e^B, e'^B, ξ, I) ;
10   **if** $\xi' = NULL$ **then** introduce a new value to $t_i[B]$ and $val(e^B)$ in ξ ;
11   **else** $\xi := \xi'$;

Function Merge

input : *ECs e^B, e'^B, set ξ of ECs and instance I*
output: *a modified ξ if merging of e^B, e'^B succeeds, or NULL*

1 put (e^B, e'^B) into an empty list K ;
2 **while** K *is not empty* **do**
3   remove the first element, say (e^C, e'^C), from K;
4   **if** $val(e'^C) \neq NULL$ **and** $val(e^C) \neq NULL$ **and** $val(e'^C) \neq val(e^C)$ **then** return NULL;
5   remove e^C, e'^C from ξ, and add to ξ a new EC $e''^C := e^C \cup e'^C$, where $val(e''^C) := val(e'^C)$ if $val(e^C) = $ NULL, otherwise $val(e''^C) := val(e^C)$;
6   **while** *there exists (1) $X \rightarrow A \in \Sigma$ and (2) tuples t_1, t_2 in I such that (1) $C \in X$, (2) $t_1[C] \in e^C$, $t_2[C] \in e'^C$, (3) $\forall D \in X$, $ec(t_1[D]) = ec(t_2[D])$ and (4) $ec(t_1[A]) \neq ec(t_2[A])$* **do**
7   put $(ec(t_1[A]), ec(t_2[A]))$ into K if $(ec(t_1[A]), ec(t_2[A])) \notin K$;
8 **return** ξ;

The difficulty of function Merge arises from the fact that merging of two ECs may cause mergings of other ECs iteratively. To this end, Merge maintains a list K of pairs of ECs to be merged, and continues until this list is empty. The merging of two ECs fails when the values of these ECs are not NULL, and are different (line 4). Otherwise, Merge merges two ECs by replacing them with a new EC that is equal to their union (line 5). Merge then collects in K all pairs of ECs to be merged (lines 6-7). To improve the efficiency, Genrepair provides an *incremental* version of the *chase* approach. Specifically, given two ECs on attribute C, Merge only checks tuples t_1, t_2 when $t_1[C], t_2[C]$ originate from the two ECs respectively, and applies FDs that have C as left-hand side attribute. Merge returns a modified ξ when all merging succeeds (line 8).

Example 6: Recall repairs in Fig 1. Genrepair produces repair r_1 when taking as input cells first sorted in tuples t_1, t_2, t_3 and then in attributes $ID, Name,$ $Zip, City$, while r_2 is generated when sorting cells first in attributes $ID, Name,$ $City, Zip$, and then in tuples t_1, t_2, t_3. $\square$

Complexity. The number of iterations in Genrepair is equal to the number of cells, *i.e.*, $m \cdot n$, where m is the number of attributes and n is the number of tuples. In each iteration, Genrepair may call Merge to merge ECs when necessary, and the worst case complexity of Merge is $O(m \cdot n \log n)$. Therefore, Genrepair takes $O(m^2 \cdot n^2 \log n)$ in the worst case. Note that (1) cells not involved in any FD can be safely skipped; and (2) ECs are incrementally maintained in Genrepair and chase is incrementally conducted in Merge. The actual complexity of Genrepair is hence much better than its worst case complexity.

Remark. Compared to [4], Genrepair improves equivalence class techniques by maintaining all ECs in an incremental way, while [4] requires to rebuild ECs from scratch when new cell is processed. Combining this with the incremental chase adopted in Merge, Genrepair outperforms [4] in the efficiency of producing a single repair, as will be verified by experiments (Section 5).

Algorithm Mindiff. We then present algorithm Mindiff for MindiffP. Combining diversification methods, this algorithm is based on the following observations. (1) Genrepair postpones resolution of FD violations until it has to do so. For example, suppose cells $t_1[A], t_1[B], t_2[A], t_2[B]$ violate an FD $A \rightarrow B$, Genrepair will resolve this violation by modifying the value of the last cell among this four cells in the input list. (2) Genrepair uses as the value of each EC the value of the first cell of that EC in the input list, when this does not cause merging of ECs to fail. (3) It is preferable *not* to modify cells that have been modified by other repairs, when producing a repair r for diversity. Intuitively, such modifications increase the cost of r, while may fail to simultaneously increase the distances between r and other repairs, and hence hinder repair diversification.

Taken together, these tell us the following. (1) When calling Genrepair with an input list L, it is better to put at the front of L those cells that we prefer not to modify, while put at the rear of L those cells that we prefer to modify. (2) When generating diversified repairs, cells modified in former repairs should be adjusted towards the front of the input list, to reduce the possibility of modifying these cells in future repairs.

Example 7: We adjust the list for producing r_1 in Example 6, by putting at its front the three cells that are modified in r_1, *i.e.*, $t_2[City]$, $t_3[Name]$, $t_3[City]$, Genrepair generates r_3 when taking this adjusted list as its input. $\square$

We present details of Mindiff. It employs Genrepair to compute S_k of k repairs by providing different list L as input (lines 1-8). Specifically, function $w'(c)$

measures the preference of modifying a cell c, say $t[A]$, in the current repair (line 3). In $w'(c)$, $cnt(S_k, t[A])$ denotes the number of repairs in S_k that have modified cell $t[A]$, and $vio(t[A])$ is computed as the number of tuples violating t on A. Intuitively, it is preferable to modify cell c with a small $w'(c)$ value, *i.e.*, cell with a small cost, modified by less repairs, and involved in more FD violations; this follows the semantics of repair diversification: finding diversified repairs among low-cost ones. In addition to this idea, Mindiff also uses parameter $\alpha \in [0, 1]$ to follow a greedy randomized approach, in producing input list L for Genrepair (lines 5-7). Note that max (resp. min) in line 6 may require recomputation after cells are removed from P. When $\alpha = 0$, cells are ordered in L by decreasing $w'()$ values; when $\alpha = 1$, L is purely constructed at random; otherwise, each time a random cell, among cells whose $w'()$ values are within a range, is added to list L. Complementary to greedy strategies, randomization is introduced to Mindiff by following this approach; this allows for trying different sorting of cells.

Algorithm 2. Mindiff

> **input** : k, λ and inconsistent relation I w.r.t. a set Σ of FDs.
> **output**: a set S_k of k repairs for diversification objective $f_d()$.

1 $S_k := \emptyset$;
2 **while** $|S_k| \leq k$ **do**
3   compute $w'(c)$ for each cell c, say $t[A]$, where $w'(c) = (w(t,A) \cdot (1-\lambda) + \frac{\lambda \cdot cnt(S_k, t[A]) \cdot w(t,A)}{(|S_k|-1)})/(vio(t[A]) + 1)$ if $|S_k| > 1$; otherwise $w'(c) = \frac{w(t,A)}{vio(t[A])+1}$;
4   suppose P is a set containing all cells in I; initialize an empty list L;
5   **for** $i:=1; i \leq |P|; i++$ **do**
6   randomly pick c from P such that $w'(c) \geq max - \alpha \cdot (max - min)$, where max (resp. min) is the maximum (resp. minimum) $w'()$ value in P ;
7   remove c from P, and add c to L;
8      $r :=$ Genrepair (L); add r to S_k if $r \notin S_k$;
9 **repeat**
10   pick repair r such that $f'_d(r)$ is maximum among all $f'_d()$ values of *unlabeled* repairs in S_k, where $f'_d(r) = (1-\lambda) \cdot c(r) - \frac{\lambda}{(k-1)} \sum_{r, r_i \in S_k, r_i \neq r} d(r, r_i)$;
11      $old := f_d(S_k)$; $S_k := S_k \backslash \{r\}$;
12   compute and add a repair r' to S_k, by following the same way as lines 2-8 ;
13   **if** $old \leq f_d(S_k)$ **then** replace r' by r in S_k and label r;
14   **else** label r' and remove all other labels in S_k.
15 **until** *termination condition is satisfied*;

Finally, Mindiff performs swaps between repairs in S_k and some newly generated repairs, in order to improve objective function $f_d()$ (lines 9-15). Specifically, Mindiff measures the contribution of repair r to $f_d(S_k)$ as $f'_d(r)$, and tests repairs in decreasing order of $f'_d()$ values (line 10). When trying to replace a repair r, Mindiff generates a new repair r' by running methods for producing S_k on the set $S_k \backslash \{r\}$, *i.e.*, employing lines 2-8 by setting the initial S_k as $S_k \backslash \{r\}$. Repair r' that improves diversification objective is kept in S_k; this will cause $f'_d(\cdot)$ to be recomputed for all repairs in line 10. For each repair whose testing does not change S_k, Mindiff attaches a label to it, such that it will not be tested again until S_k changes; such label is also attached to each new repair added to S_k. The termination condition (line 15) can be set based on the number of new repair computations, or based on the gain of swaps, *e.g.*, terminates when the average improvement in $f_d(\cdot)$ of the last n swaps is below some threshold.

Complexity. Computing $w'(\cdot)$ for all cells (line 3) and then sorting cells based on $w'(\cdot)$ (lines 5-7) take $O(m \cdot n)$ and $O(m \cdot n \log(m \cdot n))$, respectively, where m is the number of attributes and n is the number of tuples. The cost of a repair modifying p cells can be computed in $O(p)$, and the distance between two repairs modifying p and q cells respectively can be computed in $O(p+q)$. Then, suppose l is the maximum number of modified cells in any repair, computations of $f'_d(\cdot)$ take $O(k^2 \cdot l)$ (line 10). Typically, $l \ll n \cdot m$. Therefore, the overall complexity of Mindiff is governed by the running times of line 8 and line 12 that compute repairs, with the worst case complexity of $O(m^2 \cdot n^2 \log n)$.

Algorithm Minmax. Taking the same input as Mindiff, Minmax is designed for problem MinmaxP, aiming to optimize objective $f_m()$. Minmax follows the same framework as Mindiff, we therefore only highlight the differences between them. (1) Minmax computes the value of $\frac{w(t,A)}{vio(t[A])+1}$ for each cell $t[A]$, and fixes top $\beta\%$ cells at the front of the list L when calling Genrepair. Intuitively, this reduces the possibility of modifying these cells and hence avoids generating repairs with large costs. In our experiments, we set $\beta\% = 10\%$.
(2) Minmax uses function $f'_m(r)$ to measure the contribution of repair r to $f_m(S_k)$ in the swap stage, and tests repairs in decreasing order of $f'_m()$ values. $f'_m(r) = max((1 - \lambda) \cdot (c(r) - max_c(S_k \backslash \{r\})), \lambda \cdot (min_d(S_k \backslash \{r\}) - min_{r' \in S_k, r' \neq r} d(r, r')))$, where $max_c(S) = max_{r_i \in S} c(r_i)$, and $min_d(S) = min_{r_i, r_j \in S, i<j} d(r_i, r_j)$. Intuitively, repair r with the maximum $f'_m(r)$ value is the "worst" repair in terms of the optimization of $f_m(S_k)$.

5   Experimental Study

Experimental Setting. We use a PC with 3.3GHz Intel Duo CPU, 4GB memory and Windows 7. All experiments report the average over five runs.
Data. (1) DBLP data is extracted from dblp bibliography. (2) Synthetic Person data extends the relation in Fig. 1, and we populate the relation using a data generator. Each dataset in the experiments is produced by introducing noises, controlled by two parameters: (a) $|D|$: the number of tuples; and (b) $noi\%$: the noise rate, which is the ratio of the number of violating cells *w.r.t.* FDs to the total number of cells in the dataset. We randomly assign a weight to each cell.
Algorithms. We implement the following algorithms for problems MindiffP and MinmaxP, all in C++: (1) baseline algorithms BMindiff and BMinmax; and (2) Mindiff and Minmax. BMindiff (resp. BMinmax) randomly computes a set S of repairs, from which k repairs is then found. We implement repair sampling technique [4] to generate S, with a parameter $N=|S|$. To select k repairs from S in BMindiff (resp. BMinmax), we implement the facility dispersion technique presented in [12] (resp. [14]). In Mindiff and Minmax, we set the randomization parameter $\alpha=0.3$, and compute k new repairs in repair swapping phase.

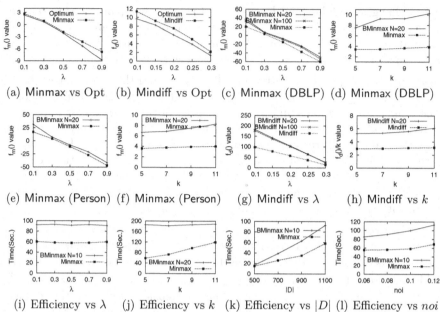

(a) Minmax vs Opt   (b) Mindiff vs Opt   (c) Minmax (DBLP)   (d) Minmax (DBLP)

(e) Minmax (Person) (f) Minmax (Person)   (g) Mindiff vs λ (h) Mindiff vs k

(i) Efficiency vs λ (j) Efficiency vs k (k) Efficiency vs $|D|$ (l) Efficiency vs noi

Fig. 2. Experimental Results

Exp-1. We verify the effectiveness of Minmax (resp. Mindiff) against the optimal result minimizing $f_m()$(resp. $f_d()$). Due to the intractability, we find the optimal result by enumerating all possible ones. Using Person data, fixing $|D| = 100$, $noi\% = 6.5\%$ and $k = 5$, it takes about 8 minutes to find the optimal result; this approach does not scale with data size. Fig. 2(a) shows results of $f_m()$, by varying λ from 0.1 to 0.9. Note that $f_m()$ values decrease with increasing λ; as λ increases, the minuend decreases while the subtrahend increases in $f_m(S)$ for any given set S. We find gaps between $f_m()$ values of Minmax and the optimal ones are about $[15\%, 35\%]$ of the optimal ones, and Minmax only takes 2 seconds.

In the same setting, Fig. 2(b) shows results of $f_d()$ by varying λ from 0.1 to 0.3. We see gaps between $f_d()$ values of Mindiff and optimal ones are about $[17\%, 32\%]$ of the optimal ones. Results on larger λ are omitted, because we find when $\lambda = 0.4$, the maximum repair cost in the optimal solution increases sharply, more than 150% of that when $\lambda = 0.3$. This shows that it favors large-cost repairs in minimizing $f_d()$; their overhead in cost sum is outweighed by their large distances to small-cost repairs in distance sum as λ increases. However, this significant rise in repair cost implies that an optimal solution for $f_d()$ deviates from a good result for repair diversification. To avoid such disturbance, λ has to be a small number when used in $f_d()$.

Exp-2. We verify the effectiveness of Minmax against BMinmax, using DBLP ($|D| = 1100$ and $noi\% = 10\%$). In Fig. 2(c), we fix $k = 5$, vary λ, and use $N=20$, 100 for BMinmax. We see the following. (1) Minmax significantly outperforms BMinmax even when $N = 100$, which requires more repair computations than Minmax by orders of magnitude. Specifically, Minmax reduces $f_m()$ values by 10%

to 55%, compared to BMinmax ($N = 100$). (2) As expected, the performance of BMinmax improves as N increases; however, the improvement is minor. The increased sample number is still too small compared to the exponential number of repairs, although it is already very costly to compute these repairs. (3) The gap between $f_m()$ values for Minmax and BMinmax decreases as λ increases. This is because preference to distance increases as λ increases, and it is relatively easy to pick a small number ($k=5$) of dissimilar repairs from an exponential repair space at random. Nevertheless, Minmax still outperforms BMinmax by about 10% when $\lambda=0.9$, and the cost sum of the k repairs in Minmax is about 75% of that in BMinmax (not shown). This demonstrates that Minmax is effective in finding diversified repairs from small-cost ones.

By fixing $\lambda = 0.3$ and varying k, Fig. 2(d) shows $f_m()$ values. Minmax consistently outperforms BMinmax. Indeed, the performance of Minmax is stable, while BMinmax degrades as k increases. The solution is incrementally constructed in BMinmax by its employed facility dispersion technique, $i.e.$, $S_k^* \subset S_{k+1}^*$, where S_k^* is the solution with k repairs. However, $f_m()$ does not have this property. Minmax addresses this issue by introducing randomization and by performing swaps between repairs, which is proved to be effective.

We then verify the effectiveness of Minmax using Person ($|D| = 1000$ and $noi\%$ = 6.5%). The results are reported in Fig. 2(e) and Fig. 2(f), in the same setting as Fig. 2(c) and Fig. 2(d), respectively. These results confirm our observations on DBLP data. For instance, Minmax outperforms BMinmax in reducing $f_m()$ value by 9% to 45%, as shown in Fig. 2(e).

Exp-3. Using DBLP ($|D| = 1100$ and $noi\% = 10\%$), the effectiveness of Mindiff is verified against BMindiff ($N=20$, 100), shown in Fig. 2(g) ($k=5$) and 2(h) ($\lambda=0.3$). Note that (1) λ is varied from 0.1 to 0.3 in Fig. 2(g). Mindiff still outperforms BMindiff for large λ; however, as noted earlier, those settings of λ deviate from the purpose of repair diversification; and (2) we show $f_d()/k$ values in Fig. 2(h), $i.e.$, the difference between the average cost and the average distance of repairs; these values are comparable for different k. The results show that Mindiff performs well. $f_d()$ values of Mindiff are only about [50%, 58%] of those of BMindiff; the performance of Mindiff is stable as k increases.

Exp-4. We evaluate the efficiency and scalability of Minmax against BMinmax ($N=10$, 20) on DBLP. We set $\lambda=0.3$, $k=5$, $|D|= 1100$ and $noi\%= 10\%$ by default, and vary one parameter in each of Fig. 2(i), 2(j), 2(k) and 2(l), respectively. For space limitation, we omit the results of Minmax on other dataset, and omit the results of Mindiff, since they are similar. We find Minmax outperforms BMinmax in terms of efficiency and scalability. (1) Fig. 2(i) shows that running times of all algorithms are not sensitive to λ, as expected. In this setting, numbers of repair computations conducted in Minmax (initial phase and swap phase) and BMinmax are the same. The results show that the time for Minmax is about 70% of that for BMinmax; this implies that Minmax is more efficient in computing a single repair, due to incremental computation techniques adopted in Genrepair. (2) Fig. 2(j) shows that running times of BMinmax are the same as k increases, since it always computes $N=20$ repairs and the time for selecting k repairs from these repairs is trivial. The running time of Minmax is almost linear in k; k controls the number of repair computations conducted in Minmax and as stated in the complexity analysis, time for repair computations governs the overall time. (3) Running times of all algorithms in Fig. 2(k) increase as $|D|$ increases, as expected. We see Minmax scales better with $|D|$ than BMinmax. (4) The increase of $noi\%$ also has a negative impact on the running times in Fig. 2(l), as expected. We find that Minmax scales well with $noi\%$.

6 Conclusions

We have presented a formal framework for repair diversification problems, established the complexity and developed algorithms for these problems, and experimentally verified our approach. We are currently experimenting with more real-life data sets to test the usefulness of the proposed algorithms, exploring different diversification objective functions, and studying optimization techniques to further improve our algorithms.

References

1. Arenas, M., Bertossi, L., Chomicki, J.: Consistent query answers in inconsistent databases. In: PODS (1999)
2. Bertossi, L.: Database repairing and consistent query answering. Morgan & Claypool Publishers (2011)
3. Bohannon, P., Fan, W., Flaster, M., Rastogi, R.: A cost based model and effective heuristic for repairing constraints by value modification. In: SIGMOD (2005)
4. Beskales, G., Ilyas, I., Golab, L.: Sampling the repairs of functional dependency violations under dard constraints. VLDB (2010)
5. Cong, G., Fan, W., Geerts, F., Jia, X., Ma, S.: Improving data quality: Consistency and accuracy. VLDB (2007)
6. Chu, X., Ilyas, I., Papotti, P.: Holistic data cleaning: Putting violations into context. ICDE (2013)
7. Dallachiesa, M., Ebaid, A., Eldawy, A., Elmagarmid, A., Ilyas, I., Ouzzani, M., Tang, N.: NADEEF: A commodity data cleaning system. SIGMOD (2013)
8. Drosou, M., Pitoura, E.: Search result diversification. SIGMOD Record 39(1), 41–47 (2010)
9. Fan, W., Geerts, F.: Foundations of data quality management. Morgan & Claypool Publishers (2012)
10. Fan, W., Li, J., Ma, S., Tang, N., Yu, W.: Towards certain fixes with editing rules and master data. VLDB (2010)
11. Gollapudi, S., Sharma, A.: An axiomatic approach for result diversification. WWW (2009)
12. Hassin, R., Rubinstein, S., Tamir, A.: Approximation algorithms for maximum dispersion. Operations Research Letters 21(3), 133–137 (1997)
13. Kolahi, S., Lakshmanan, L.: On approximating optimum repairs for functional dependency violations. ICDT (2009)
14. Ravi, S., Rosenkrantz, D., Tayi, G.: Heuristic and special case algorithms for dispersion problems. Operations Research 42(2), 299–310 (1994)
15. Yakout, M., Elmagarmid, A., Neville, J., Ouzzani, M., Ilyas, I.: Guided data repair. VLDB (2011)

BigOP: Generating Comprehensive Big Data Workloads as a Benchmarking Framework

Yuqing Zhu*, Jianfeng Zhan, Chuliang Weng, Raghunath Nambiar,
Jinchao Zhang, Xingzhen Chen, and Lei Wang

State Key Laboratory of Computer Architecture
(Institute of Computing Technology, Chinese Academy of Sciences), Huawei, Cisco
{zhuyuqing,zhanjianfeng,zhangjinchao,chenxingzhen,wanglei_2011}@ict.ac.cn,
chuliang.weng@huawei.com, RNambiar@cisco.com

Abstract. Big Data is considered proprietary asset of companies, organizations, and even nations. Turning big data into real treasure requires
the support of big data systems. A variety of commercial and open source
products have been unleashed for big data storage and processing. While
big data users are facing the choice of which system best suits their needs,
big data system developers are facing the question of how to evaluate their
systems with regard to general big data processing needs. System benchmarking is the classic way of meeting the above demands. However, existent big data benchmarks either fail to represent the variety of big data
processing requirements, or target only one specific platform, e.g. Hadoop.

In this paper, with our industrial partners, we present Big*OP*, an
end-to-end system benchmarking framework, featuring the abstraction of
representative *O*peration sets, workload *P*atterns, and prescribed tests.
BigOP is part of an open-source big data benchmarking project, *Big-
DataBench*[1]. BigOP's abstraction model not only guides the development of BigDataBench, but also enables automatic generation of tests
with comprehensive workloads.

We illustrate the feasibility of BigOP by implementing an automatic
test generation tool and benchmarking against three widely used big data
processing systems, i.e. Hadoop, Spark and MySQL Cluster. Three tests
targeting three different application scenarios are prescribed. The tests
involve relational data, text data and graph data, as well as all operations
and workload patterns. We report results following test specifications.

1 Introduction

Companies, organizations and countries are taking big data as their important
assets, as the era of big data has inevitably arrived. But drawing insights from
big data and turning big data into real treasure demand an in-depth extraction of
its values, which heavily relies upon and hence boosts the deployment of massive
big data systems.

Big data owners are facing the problem of how to choose the right system for
their big data processing requirements, while a variety of commercial and open

---

* Corresponding Author.

[1] BigDataBench is available at http://prof.ict.ac.cn/BigDataBench

S.S. Bhowmick et al. (Eds.): DASFAA 2014, Part II, LNCS 8422, pp. 483–492, 2014.

source products, e.g., NoSQL databases [9], Hadoop MapReduce [5], Spark [28], Impala [3], Hive [7] and Redshift [1], have been unleashed for big data storage and processing. On the other hand, big data system developers are in need of application-perspective evaluation methods for their systems. Benchmarking is the classic way to direct the evaluation and the comparison of systems.

Though many well-established benchmarks exist, e.g. TPC series benchmarks [13] and HPL benchmarks [8], no widely accepted benchmark exists for big data systems. Some benchmarks targeting *big data systems* appear in recent years [22,23,18,21], but they are either for a specific platform or covering limited workload patterns.

Together with our industrial partners, we present in this paper an end-to-end system benchmarking framework BigOP, which enables automatic generation of tests with comprehensive workloads for big data systems. We build BigOP for our urgent need to benchmark big data systems. BigOP is part of a comprehensive big data benchmarking suite *BigDataBench* [27], which is already used by our collaborators in testing architecture, network and energy efficiency of big data systems. The development of BigDataBench is guided by BigOP's abstraction model.

BigOP features an abstracted set of *O*perations and *P*atterns for big data processing. We work out the abstraction after considering the powerful representativeness of the five primitive relational operators [17] and the 13 computation patterns summarized in a report by a multidisciplinary group of well-known researchers [14]. The operations are extended from the five primitive relational operators, while the workload patterns are summarized based on general big data computation characteristics.

Figure 1 demonstrates an overview of BigOP. In BigOP, a benchmarking test is specified as a prescription for one application or a range of applications. A prescription includes a subset of operations and processing patterns, a data set, a workload generation method, and the metrics. The subset of operations and processing patterns are selected from BigOP's whole abstraction set. The data set can be obtained from real applications or generated through widely-obtainable tools. The workload generation method describes how operations are issued from clients, e.g, the number of client threads, the load, etc. The metrics can include the test duration, request latency metrics, and the throughput. With BigOP, a prescribed test can be implemented over different systems for comparison. System users can also prescribe a test targeting their specific applications.

BigOP leaves the choice of data set to users because the *variety* of big data makes a predefined data set for benchmarking irrelevant. Besides, data sets are usually related to the processing performance in big data scenarios, for example, highly isolated webpages vs. highly linked webpages for PageRank computations. The size of the chosen data set is required to be larger than the total memory size of the system under test (SUT) so that the volume characteristic of big data is covered. The velocity characteristic of big data can be represented in the workload generation specification. That is, BigOP design takes the three properties of Big Data into consideration.

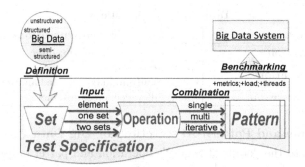

Fig. 1. Overview of BigOP

After presenting the design of BigOP (**Section 2**), we give three test prescriptions targeting three different application scenarios as an example. We benchmark against three widely used big data processing systems, i.e. MySQL cluster, Hadoop[5]+HBase[6] and Spark [28], using BigOP (**Section 3**). The tests involve relational data, text data and graph data, as well as all operations and workload patterns. We discuss workload representativeness by comparing YCSB [18], TPC-DS [26] and BigBench [22] to BigOP. We summarize related work (**Section 4**) and conclude (**Section 5**) in the end.

2   BigOP Design

2.1   Overview

BigOP is an end-to-end system benchmarking framework with a big data processing operation and pattern (*OP*) abstraction. Comprehensive workloads can be specified and thus constructed based on the *OP* abstraction.

An adequate level of abstraction and the end-to-end execution model leaves space for various system implementations and optimizations. Benchmarks with these two properties enable comparisons among different kinds of systems servicing the same goals. The success of TPC benchmarks [13] demonstrates this fact [16]. Therefore, BigOP takes an end-to-end benchmarking model. Benchmarking workloads are applied by clients issuing requests through interfaces. Requests are sent through a network to the system under test (SUT). Metrics are measured at the client side.

The success of TPC benchmarks also highlights the importance of benchmarking systems with functions of abstraction and the functional workload model [16]. Functions of abstraction are basic units of computation occurring frequently in applications, while the functional workload model includes functions of abstraction, the representative application load, and the data set. Functions of abstraction is the core. To generalize functions of abstraction for BigOP, we abstract a set of operations and patterns common to big data processing.

Furthermore, big data embodies great variety. So do big data applications. Therefore, BigOP allows users to flexibly specify data sets and workloads in test prescriptions. The test prescription allows for application-specific features, as well as comparisons across systems.

Table 1. Basic Operations

Categories	Typical Operations
Element Operation	put, get, delete; transform; filter
Single-Set Operation	project, order by aggregation(min,max,sum,median,average)
Double-Set Operation	union, difference, cross product

2.2 Operation and Pattern Abstraction for Big Data Processing

Before going into the details of BigOP's operation and pattern abstraction, we first review some facts about big data processing.

The concept of *set* in relational algebra [17] is still effective in big data processing scenarios, e.g. the MapReduce [19] model, which plays an important role in big data processing. In the model, a large piece of data is transformed into a set of elements denoted by key-value pairs in the *map* stage. The *reduce* stage does further processing over the mapped set. We thus adopt the general concept of *set* in BigOP. Elements in a set can be uniquely identified.

Data accesses to memory and disk, as well as across network, must be considered in big data benchmarking. Memory size plays an important role as for system performance. As technology improves, the starting data size of TPC-H increases along with the obtainable amount of memory. Thus, benchmarks must consider the whole system, including the system composition. Besides, big data systems can consist of not only nodes, but also datacenters, due to the huge volume of big data. Thus, communication must also be considered in benchmarks. Furthermore, the huge volume of big data and the resulting processing complexity demand distribution and parallelization of computation tasks. Otherwise, the time required for big data processing would be intolerably long. Hence, BigOP requires the data set size to exceed the total memory size of the SUT, but a single element in the data set must fit into a single node's memory to be processed; or, the element can at least be read serially and transformed into a set of smaller elements for processing.

Operation Abstraction. Considering the above facts, we first abstract big data processing operations into three categories, i.e., *element operation, single-set operation* and *double-set operation*. Element operation can be computed based on an individual element, which might require only local memory access. Single-set operations are computed based on elements of one set. Double-set operations require input from two sets of data. We did not include multi-set operations, which can be composed by combining multiple double-set operations. The more sets are involved in a processing task, the more demands for data accesses involving memory, disk and network communication are there. Operations from the three categories can be combined and permutated to meet complex processing requirements. Table 1 illustrates the three categories of operations.

BigOP adopts the five primitive operators from *relational algebra* [17], which also takes a *set*-based perspective. They are filter(select), project, cross product, set union, and set difference. The five primitive operators are fundamental in the sense that omitting any of them causes a loss of expressive power. Many

Table 2. Workload Patterns

Patterns	Example Workloads
Single-Operation Processing	any abstracted operation
Multi-Operation Processing	operation combinations, SQL queries
Iterative Processing	graph traversal, finite state machines

other set operations can be defined in terms of these five. The *filter* operation is an *element* operation because it can operate a set element on given conditions with only element-local information. The *project* operation is in the *single-set* operation category, requiring the set information. The *union*, *difference*, and *cross product* fall in the *double-set* operation category.

The basic data access operations of *put*, *get* and *delete* are in the element operation category. We also add a *transform* operation to this category because it is common to turn a big element into a set of elements or another element, as demonstrated by the Map usages of MapReduce. *transform* is user-defined. Most big data are unstructured data, therefore *transform* is important to define data-specific computations.

We also include the commonly used *order by* and *aggregation* operations in the single-set operation category. *order by* is equal to *sort*, a fundamental database operation noted by Jim Gray [14]. Quite a few benchmarks have been built based on *sort* [11]. *Aggregation* is included because it is widely recognized important operation to turn sets into numerals.

Pattern Abstraction. The abstracted operations can be combined into more complex processing tasks following some patterns. We summarize three patterns as demonstrated with examples in Table 2. The three patterns are *single-operation*, *multi-operation* and *iterative* processing patterns. The single-operation pattern contains only a single operation in a processing task, while the multi-operation and iterative patterns can have multiple operations in a task. The inclusion of multiple operations allows the big data system to make a whole optimization plan for all operations in the task. The difference between multi-operation processing and iterative processing patterns is whether the exact number of operations to be executed is known beforehand. Iterative processing patterns only provide stopping conditions (which can be specified as user-defined functions), thus the exact number of operations can only be figured out in running time.

Different from SQL queries, the processing patterns in Table 2 can result in more than one set. Furthermore, the element definition of a data set relies on the *transform* operation, instead of *schema*. For example, a text document can be *transformed* into a set with *word* elements or with *sentence* elements. While element operation can be processed locally, global optimization techniques can be employed for single-set and double-set operations, as well as for multi-operation and iterative processing patterns.

The choice of the operations in Table 1 is in no way complete, but it is representative enough to represent a broad range of processing workloads when used with the patterns of Table 2. Besides, we think the efforts in benchmarking should be incremental and evolving. That is, more basic operations can be added to Table 1 in the future, as well as more patterns to Table 2.

Table 3. Test Prescriptions without Workload Generation Method Specifications

	Fast Storage	Log Monitoring	PageRank Computation
Operations	put, get, delete union	put, get, filter, aggregation	get, transform, filter, order by
Patterns	single-operation	single- and multi-operation	all patterns
Data Set	randomly generated structured data	real server logs	randomly generated directed graph
Metrics	throughput	request latency statistics	test duration

2.3   Prescriptions and Prescribed Tests

Each benchmarking test is specified by a prescription. Thanks to the abstraction of processing operations and patterns, a prescribed test can be implemented over different systems for comparison. A prescription includes a subset of operations and processing patterns, a data set, a workload generation method, and the measured metrics. The subset of operation and processing patterns is selected from BigOP's whole abstraction set. The data set can be taken from real applications or generated through widely-obtainable tools. The size of the chosen data set is required to be larger than the total memory size of the SUT, so that the volume characteristic of big data is covered. The workload generation method describes how operations are issued from clients. The velocity characteristic of big data can be represented in the workload generation specification. The measured metrics can be the duration of the test, request latency statistics, and the throughput.

We instantiate three prescribed test examples in Table 3, from the simplest to the most complex. The first and the third examples[2] are taken from Big-DataBench, while the second example is constructed from a common application scenario. The workload generation methods are not included in the prescription for space consideration. Instead, we describe them respectively in the following, together with an introduction to the application corresponding to each example.

Example 1. **Fast Storage.** Applications make frequent requests of data storage. This is the most basic scenario of big data acquisition. The data set is generated using BDGS' data generation tool [24]. The workload generation combines YCSB's workloads. The throughput of operations is the key metric.

Example 2. **Log Monitoring.** An application monitors its services by logs. Applications like user and server activity monitoring can be represented by this prescription. The workload generation is specified as follows. *put* is applied to some random log entries at a speed of 5000 operations per second (ops) till the total data size exceeds twice of the total system memory. Simultaneously, *get+filter+aggregation* is applied to the data set based on some random filtering condition for a *sum* result continuously.

Example 3. **PageRank Computation.** This is a core computation in the widely deployed Internet service. It is also a representative computation of graph applications. In workload generation, *get+filter+ transform* is applied to the data set *iteratively* till a given *condition* is met. The *order by* operation, a.k.a. *sort*, is executed to get the final result.

[2] Referred as Cloud OLTP and PageRank workloads respectively in BigDataBench.

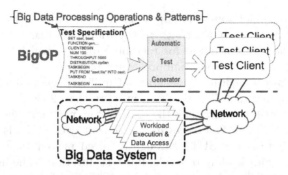

Fig. 2. Testing process and functional components of BigOP

In constructing test prescriptions, we can adjust factors in the prescription according to our needs. For example, to test for data scalability, we can increase the workload of *put*. Similarly, to test more complicated computations, we can increase the number of combinations for *multi-operation* and *iterative* processing patterns, as well as defining a sophisticated *transform* function.

3   Evaluation

Based on the BigOP framework, we implement an automatic test generator, which can generate tests from prescriptions. Figure 2 demonstrates the testing procedure of the evaluation.

We benchmark three widely used big data processing systems, i.e. MySQL cluster (SUT1), Hadoop+HBase (SUT2) and HDFS+Spark (SUT3). All the three SUTs are deployed over five physical nodes connected with 1Gbps network. Each node has 32 GB memory and a processor with six 2.40GHz cores. The operating system is Centos release 5.5 with Linux kernel 2.6.34. Among our three examples, we only choose the second and the third for the evaluation, since the first example has been extensively tested in other benchmarks.

Log Monitoring. Figure 3 demonstrates the resulting performances of SUT1 and SUT2. SUT2 excels under the frequent record insert task as expected. SUT1 performs much better in statistics computation tasks because of the long starting time of jobs in SUT2. For complex multi-operation tasks, SUT1 is very likely to excel the others as well. The reason is as follows. Even though MapReduce-like systems including Hadoop, Shark and Hive support complex user defined

Log Monitoring—Record Insert Task

Systems	Throughput (ops/sec)	Min Latency (us)	Max Latency (us)	Average Latency (us)
MySQL Cluster	3367.795612	0	46400516	295.3059112
Hadoop+HBase	16918.51187	3	16392237	56.32552309

Log Monitoring—Statistics Task

Systems	# of Runs	Max run time(sec.)	Min run time(sec.)
MySQL Cluster	2675	5204.483333	0.008
Hadoop+HBase	6	7312	614

Fig. 3. System performances under log monitoring workloads. (The zero min latency of SUT1 is due to the batch commit mode.)

functions, they have not considered optimizations on multi-operation and iterative processing patterns. On the contrary, traditional databases are strictly SQL compliant and heavily optimized for relational queries, which usually contain single-operation and multi-operation processing patterns.

Pagerank Computation. We generate 0.5 million pages and 3.7 million links using an open-source tool [24], resulting in more than 250GB data (including page contents), which is larger than the total system memory. The running time of the task over SUT2 is *363 seconds*, while that over SUT3 is *96 seconds*. We also run this test over SUT1 through a stored procedure, which contains costly large-scale joins leading to intolerable test durations. The indication here is *twofold*. First, distributed in-memory computation is effective for the *iterative* computation pattern, while frequent disk accesses and network communications can be costly. Second, there is still much optimization space for distributed computation in relational database, especially when the iterative pattern is involved.

We report a small fraction of evaluation results here due to the page limit. Further benchmarking results can be found on the BigDataBench webpage[3].

Discussion. Existent big data benchmarks only cover part of BigOP's abstraction of processing operations and patterns. We take YCSB [18], TPC-DS [26], and BigBench [22] for example. YCSB is mainly for NoSQL database benchmarking. Its workload consists of only *put* and *get* operations, though which can be combined by the *multi-operation* pattern to form *scan* and *read-modify-write* operations. TPC-DS covers all abstracted operations and the first two patterns, except for the *iterative* pattern. It targets a single application domain, i.e., decision support applications. The involved data is structured data. Thus, it is not as flexible in suiting different benchmarking requirements as BigOP. BigBench extends TPC-DS. It adds new data types of semi-structured data and unstructured data, but it still does not include the *iterative* pattern.

4   Related Work

BigBench [22] is the recent effort towards designing a general big data benchmark. BigBench focuses on big data analytics, thus adopting TPC-DS as the basis and adding atop new data types like semi-/un-structured data, as well as non-relational workloads like sentiment queries. Although BigBench has a complete coverage of data types, it targets only a specific big data application scenario, not covering the variety of big data processing workloads.

The AMPLab of UC Berkeley also proposes a big data benchmark [2] in recent years. It is the systems of Spark [28] and Shark [20] that inspire the design of the benchmark, which thus targets real-time analytic applications. The benchmark not only has a limited coverage of workloads, but also covers only relational data.

Industrial players also try to develop their benchmark suites. Yahoo! release their cloud benchmark specially for data storage systems, i.e. YCSB[18]. Having its root in cloud computing, YCSB is mainly for scenarios like that in the *Fast Storage* example. The characteristics of and the diverse application workloads

---

[3] http://prof.ict.ac.cn/BigDataBench/

for big data are not considered in YCSB. CALDA[25] effort represents a micro benchmark for Big Data analytics. It has compared Hadoop-based analytics to a row-based RDBMS one and a column-based RDBMS one.

There exist also benchmarks targeting a specific platform, e.g. Hadoop [5]. Hi-Bench [23] is a widely-known benchmark for Hadoop MapReduce. Hence, its four categories of workloads are limited to MapReduce based processing. It exploits stochastic methods to generate data for its workloads. However, its randomly generated data misses various features of real big data. Besides, its choice of workloads lacks of coverage, as well as solidly founded grounds. Gridmix [4] and Pigmix [10] are also two benchmarks specially designed for Hadoop MapReduce. They include a mix of workloads, including sort, grep and wordcount. They are also suffering from incomplete coverage of data and workloads.

Across different research fields, there are multiple famous and well established benchmarks. The TeraSort or GraySort Benchmark [11] considers the performance and the cost involved in sorting a large number of 100-byte records. The workload of this benchmark is too simple to cover the various needs of big data processing. SPEC [12] works well over standalone servers with a homogeneous architecture, but is not suitable for the emerging big data platforms with large-scale distributed and heterogeneous components. TPC [13] series of benchmarks are widely accepted for database testing, but only consider structured data. PARSEC [15] is a well-known benchmark for shared-memory computers, thus not suited for big data applications that mainly take a shared-nothing architecture.

5   Conclusion

BigOP is targeted at big data systems that support part of or all of its processing operation and pattern abstractions. Users of BigOP can flexibly construct tests through prescriptions for their application-specific or general benchmarking needs. Benchmarking tests can be constructed based on BigOP's abstracted operations and patterns. Big data systems that can implement the same prescribed test are comparable. System users can prescribe tests targeting at their specific application scenarios, while system developers can carry out general tests with all abstracted operations and patterns mixed and randomly generated in the workload. We design BigOP as a benchmarking framework in considering the variety of big data and its applications, which we believe is a good trade-off between benchmarking flexibility and conformity to real big data processing requirements.

Acknowledgments. This work is supported in part by the State Key Development Program for Basic Research of China (Grant No. 2014CB340402) and the National Natural Science Foundation of China (Grant No. 61303054).

References

1. Amazon redshift service, http://aws.amazon.com/cn/redshift/
2. Amplab big data benchmark, https://amplab.cs.berkeley.edu/benchmark/
3. Cloudera impala,
   http://www.cloudera.com/content/cloudera/en/
   products-and-services/cdh/impala.html
4. Gridmix, http://hadoop.apache.org/mapreduce/docs/current/gridmix.html
5. Hadoop project, http://hadoop.apache.org/

6. Hbase project, http://hbase.apache.org/
7. Hive project, http://hive.apache.org/
8. Hpl benchmark home page, http://www.netlib.org/benchmark/hpl/
9. Nosql databases, http://nosql-database.org/
10. Pigmix, https://cwiki.apache.org/confluence/display/PIG/PigMix
11. Sort benchmark home page, http://sortbenchmark.org/
12. Standard performance evaluation corporation (spec), http://www.spec.org/
13. Transaction processing performance council (tpc), http://www.tpc.org/
14. Asanovic, K., Bodik, R., Catanzaro, B.C., Gebis, J.J., Husbands, P., Keutzer, K., Patterson, D.A., Plishker, W.L., Shalf, J., Williams, S.W., et al.: The landscape of parallel computing research: A view from berkeley. Technical report, UCB/EECS-2006-183, EECS Department, University of California, Berkeley (2006)
15. Bienia, C., Kumar, S., Singh, J., Li, K.: The parsec benchmark suite: Characterization and architectural implications. In: Proc. of PACT 2008, pp. 72–81. ACM (2008)
16. Chen, Y., Raab, F., Katz, R.H.: From tpc-c to big data benchmarks: A functional workload model. Technical Report UCB/EECS-2012-174, EECS Department, University of California, Berkeley (July 2012)
17. Codd, E.F.: A relational model of data for large shared data banks. In: Pioneers and Their Contributions to Software Engineering, pp. 61–98. Springer (2001)
18. Cooper, B.F., Silberstein, A., Tam, E., Ramakrishnan, R., Sears, R.: Benchmarking cloud serving systems with ycsb. In: Proc. of SoCC (2010)
19. Dean, J., Ghemawat, S.: Mapreduce: Simplified data processing on large clusters. Communications of the ACM 51(1), 107–113 (2008)
20. Engle, C., Lupher, A., Xin, R., Zaharia, M., Franklin, M.J., Shenker, S., Stoica, I.: Shark: Fast data analysis using coarse-grained distributed memory. In: Proc. of SIGMOD 2012, pp. 689–692. ACM (2012)
21. Ferdman, M., Adileh, A., Kocberber, O., Volos, S., Alisafaee, M., Jevdjic, D., Kaynak, C., Popescu, A., Ailamaki, A., Falsafi, B.: Clearing the clouds: A study of emerging workloads on modern hardware. Architectural Support for Programming Languages and Operating Systems (2012)
22. Ghazal, A., Hu, M., Rabl, T., Raab, F., Poess, M., Crolotte, A., Jacobsen, H.-A.: Bigbench: Towards an industry standard benchmark for big data analytics. In: Proc. of SIGMOD 2013. ACM (2013)
23. Huang, S., Huang, J., Dai, J., Xie, T., Huang, B.: The hibench benchmark suite: Characterization of the mapreduce-based data analysis. In: Proc. of ICDEW 2010, pp. 41–51. IEEE (2010)
24. Ming, Z., Luo, C., Gao, W., Han, R., Yang, Q., Wang, L., Zhan, J.: Bdgs: A scalable big data generator suite in big data benchmarking. arXiv:1401.5465 (2014)
25. Pavlo, A., Paulson, E., Rasin, A., Abadi, D.J., De Witt, D.J., Madden, S., Stonebraker, M.: A comparison of approaches to large-scale data analysis. In: Proc. of SIGMOD 2009, pp. 165–178. ACM, New York (2009)
26. Poess, M., Nambiar, R.O., Walrath, D.: Why you should run tpc-ds: A workload analysis. In: Proc. of VLDB 2007, pp. 1138–1149. VLDB Endowment (2007)
27. Wang, L., Zhan, J., Luo, C., Zhu, Y., Yang, Q., He, Y., Gao, W., Jia, Z., Shi, Y., Zhang, S., Zhen, C., Lu, G., Zhan, K., Qiu, B.: Bigdatabench: A big data benchmark suite from internet services. Accepted by HPCA 2014 (2014)
28. Zaharia, M., Chowdhury, M., Das, T., Dave, A., Ma, J., McCauley, M., Franklin, M., Shenker, S., Stoica, I.: Resilient distributed datasets: A fault-tolerant abstraction for in-memory cluster computing. In: Proc. of NSDI 2012, p. 2. USENIX Association (2012)

A*DAX: A Platform for Cross-Domain Data Linking, Sharing and Analytics

Narayanan Amudha, Gim Guan Chua, Eric Siew Khuan Foo,
Shen Tat Goh, Shuqiao Guo, Paul Min Chim Lim,
Mun-Thye Mak, Muhammad Cassim Mahmud Munshi,
See-Kiong Ng, Wee Siong Ng, and Huayu Wu

Institute for Infocomm Research, A*STAR, Singapore
{naraa,ggchua,foosk,stgoh,guosq,limmc,mtmak,
mcmmunshi,skng,wsng,huwu}@i2r.a-star.edu.sg

Abstract. We introduce the A*STAR Data Analytics and Exchange Platform ("A*DAX"), which is the backbone data platform for different programs and projects under the Urban Systems Initiative launched by the Agency for Science, Technology and Research in Singapore. The A*DAX aims to provide a centralized system for public and private sectors to manage and share data; meanwhile, it also provides basic data analytics and visualization functions for authorized parties to consume data. A*DAX is also a channel for developers to develop innovative applications based on real data to improve urban services.

In this paper, we focus on presenting the platform components that address challenges in data integration and processing. In particular, the A*DAX platform needs to dynamically fuse heterogeneous data from unpredictable sources, which makes traditional data integration mechanisms hard to be applied. Also, the platform needs to process data with different dynamics (i.e., database vs. data stream) in large scale. In our design, we use a semantic approach to handle data fusion and integration problems, and propose a hybrid architecture to process static and dynamic data to answer queries at the same time. Other issues about the A*DAX platform, e.g., security and privacy, are not covered in this paper.

1 Introduction

1.1 Background

Cities around the world are growing at an extraordinary pace, with the current population of 1.7 billion urban dwellers expected to grow to 2.2 billion by 2020. This translates to an increase of 1 million people living in cities every week - particularly cities in emerging regions like Asia and Africa. The challenge of all cities, regardless of its stage of development, is to be able to grow sustainably, create job opportunities for its dwellers and ensure social wellness through efficient provision of infrastructure and services.

The Agency for Science, Technology and Research (A*STAR) of Singapore launched the Urban Systems Initiative in 2012, which is a five-year multi-program

S.S. Bhowmick et al. (Eds.): DASFAA 2014, Part II, LNCS 8422, pp. 493–502, 2014.

initiative to address the new technological needs of the rapidly urbanizing world. It aims to enable the development of solutions for complex urban challenges in collaboration with the relevant government agencies and industries to enhance the competitiveness of Singapore in the construction of "smart city".

Under the Urban Systems Initiative, four inter-related programs have been launched to address these challenges:

Integrated Urban Planning. To develop an integrated platform for quantitative and evidence-based urban planning.

Sense & Sense-abilities. To develop a unified platform to "sense" and "make-sense" of the living environment in real-time.

Complex Systems. To develop complex system theories and models to unravel the complexity of city dynamics.

City Logistics & Supply Chain. To develop city logistics platforms based on complex systems approach and data analytics for addressing freight traffic congestion.

The four programs identify research problems, design solutions and implement infrastructures in each domain. Meanwhile, they are also dependent on each other for resource reuse, research achievement sharing and decision making.

Data is essential to all the programs under the initiative. In each domain, both problem identification and solution validation are supported by data, which are contributed by the government agencies, industrial partners and social media. Furthermore, another objective of the Urban Systems Initiative is to encourage companies, research institutions and individual developers to fuse data from different domains and come out with new applications or services for better social good.

To achieve the goal, large amount of social, economic, geographic, business and educational data that are collected by the public and private sectors need to be effectively merged and stored, efficiently accessed for use, and securely shared among programs and third-party developers. Motivated by this, the A*STAR Data Analytics and Exchange Platform ("A*DAX") is built. The A*DAX is a scalable and open standards based platform for data management, sharing and analytics. It is the backbone for the various projects under the Urban Systems Initiative.

1.2   Challenges

Designing the A*DAX platform faces difficult challenges in system, networking, security, etc. In this paper, we focus on the challenges in data management. The first challenge we need to resolve is integrating unpredictable source data. Traditional data integration [3] assumes the deterministic set of input database schemas, and applies either Local-As-View (LAV) or Global-As-View (GAV) approach to logically link up the source data and provide a unique view to data users. However, in our scenario, ideally more and more parties will dynamically join the A*DAX platform for data sharing. It is not possible to determine a permanent unique view across source data. Thus, it is difficult to use the existing approach for schema mapping.

Another challenge is the combination of static data and dynamic data. Based on the current tenants of the A*DAX platform, some data are one-time loaded, while some data are continuously streaming into the system for storing. The platform needs an efficient solution to archive and to query two types of source data, so that the data are well consumed by different users.

1.3   Contribution and Organization

In this paper, we first depict the A*DAX platform with respect to different components for different functions. Then we focus on the several components that resolve challenging problems such as data fusion and static and dynamic data processing. We propose a metadata-level semantic graph to guide data fusion, and we show how the semantic graph is incorporated with new data source merging. For data processing across both static datasets and dynamic data streams, we adapt the λ-architecture and manage both view and materialized view for users to query the underlying data. This design benefits both the data owners and data users in access control and query issuing.

The rest of the paper is organized as below. In Section 2 we describe the A*DAX platform in a high level. In Section 3 we present how the platform adaptively manage and fuse source data. In Section 4, the data processing architecture is introduced. We introduce the access control mechanism in the A*DAX platform in Section 5. Finally, we conclude the paper in Section 6.

2   Platform Overview

Fig. 1 shows the logical architecture of the A*DAX platform. Basically, there are three layers for the platform, namely Storage Layer, Master Layer and Application Layer. The Storage Layer maintains several database management systems for different types of data. The relational database management system (RDBMS) is used for storing structured data. There is also an XML extension to the RDBMS to handle semi-structured data. The Geospatial DB is used for managing GIS data. To serve the urban planning, map-based GIS data processing and visualization is a main part of the platform. The Text DB archives text-based documents. It supports keyword-based search and retrieval. The platform also incorporates a Hadoop cluster to store large-scale datasets and provide parallel data processing. All the data storages are inter-connected and communicate with the View Manager in the Master Layer for data access.

The Master Layer contains all the major components of the platform. The Data Semantics Manager summarizes the semantics of the metadata of all data sources. It constructs a semantic graph, which can be dynamically extended as new data sources arriving, to link up all data sources and merge identical entities. It also guides the view generation to provide the interface for users to use the data from different sources holistically. The View Manager creates both materialized and non-materialized views on top of the data storage. We require that users can only query the data store via provided views. There are two advantages of this restriction. First, the data can be better protected. The data providers will authorize and supervise the creation of views on their data. Thus they can choose what data can be showed to users. Second, it eases the management of pools of data. On one hand, the users do not need to worry about different query formats to different source data, as all queries will be interpreted against views rather than raw datasets; on the other hand, access control policies can be enforced on top of views so that it is easy to check whether a user query is legitimate to access the information it is interested in. The Query Manager accepts user queries, validates them and then passes them to the View Manager. The View Manager will do query parsing and transformation and forward sub-queries to different data storages for processing. More details about the Data Semantics Manager and View Manager will be discussed in the next two sections.

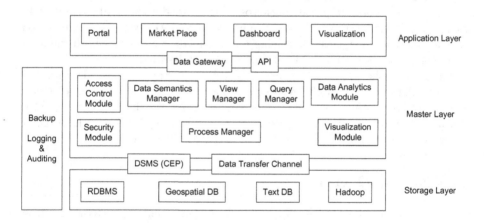

Fig. 1. Logical Architecture of the A*DAX platform

The Master Layer also contains the Data Analytics Module which provides classic data mining tools, e.g., for regression and classification, to users. The Visualization Module can visualize data and data analytics results to users. There are also Access Control Module and Security Module to ensure the security and privacy of the system. All the operations will be logged and audited.

Finally, the Application Layer provides user interfaces to upload data to the A*DAX platform, or to consume data. The platform also provides APIs for developers to use the data (by authorization) to develop innovative applications.

3   Data Fusion

The main purpose of the A*DAX is to provide a platform for different agencies to share their data. Despite the uniqueness of the data owned by each agency, there are still many overlapped attributes among different data sources. These overlapped attributes will link up different datasets and then give opportunities for users to mesh them and come out with new insight. We call the process of linking up different data sources by their common attributes as *data fusion*.

As mentioned in Section 1, traditional data integration techniques are not proper for the data fusion in the A*DAX platform. The reason is that the number of data sources keeps increasing as more sectors join the platform. We need a more dynamic approach to extend the data fusion by keeping the existing fused data framework unchanged. In our platform, we propose to use a metadata-level semantic graph to reflect the linking between data sources.

The semantic graph is similar to the ER (Entity-Relationship) diagram [1] which models a relational database at conceptual level. There are two types of nodes in the semantic graph, i.e., entity nodes and relationship nodes. An entity node models an entity class of a data source. If two datasets from different sources have any common attributes, the related entity classes in the two datasets will be linked by a relationship node in the semantic graph, and the relationship node contains the information of the common attributes from the two datasets. Once a new dataset is added to the data sharing platform, it will be linked to the existing nodes in the semantic graph by common attributes, and keep the existing graph unchanged. Similarly, if a dataset is removed from the platform, only the relevant entity nodes and corresponding relationship nodes in the semantic graph are removed, without affecting other nodes.

In fact, we do not really require the owner of each dataset to identify entity class and to decompose their data. When a data owner tries to upload data to the system, he/she only needs to go through the schema of the existing datasets through the semantic graph, and tell the system which attributes in his/her dataset overlap with the attributes in the existing data store. Then the new dataset can be linked to existing ones in the semantic graph. In other words, each dataset can be treated as a whole as an entity class in the semantic graph. Fig. 2 shows an example semantic graph for the data sources from different government agencies.

In this example, let us assume that the Immigration & Checkpoints Authority (ICA) shares citizen data, the Inland Revenue Authority (IRAS) shares personal income tax data, the Housing Development Board (HDB) shares residential data and the Urban Redevelopment Authority (URA) shares land planning data in the A*DAX platform. There are many common attributes between different datasets, and the semantic graph can be constructed as shown. There is an entity table and a relationship table describing the semantic graph. In the entity table, the name of each entity dataset and its database location, owner and upload time are recorded. As a result, given an entity in the semantic graph that is queried by a user, from the entity table we can locate the physical storage of the dataset and then route the relevant sub-query (composed by the View Manager, as described later) to the data storage for processing. The relationship table tells how two entities are linked. For example, for r1, the NRIC attribute from the ICA citizen data can be matched to the ID attribute in the IRAS data.

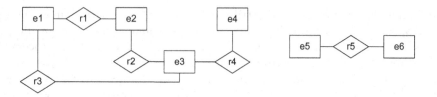

Entity						Relationship			
ID	name	location	onwer	timestamp		ID	E1	E2	Linkable Attributes
e1	citizen	RDB.citizen	ICA	2008-01-01		r1	e1	e2	E1.NRIC ~ E2.ID
e2	tax	RDB.tax	IRAS	2009-01-01		r2	e2	e3	E1.ID ~ E2.owner
e3	housing	RDB.hdb	HDB	2009-01-01		r3	e1	e3	E1.NRIC ~ E2.onwer
e4	landuse	RDB.land	URA	2007-01-01		r4	e3	e4	E1.postcode ~ E2.pc
...	...	...	...	...		...	...	...	...

Fig. 2. Logical Architecture of the A*DAX platform

The semantic graph and the description tables are managed by the Data Semantics Manager in the platform. It also communicates with the View Manager for creating views across different data sources, as well as the Query Manager for accepting new queries and finally translating the queries into sub-queries issued to different data sources.

4   Data Processing

As mentioned, the A*DAX platform can be used for sharing heterogenous data, e.g., data with different representation type and data with different dynamics. We need to resolve this data heterogeneity during data processing.

4.1   Mixed-type Data Processing

In the A*DAX platform, we support three types of data, i.e., structured data, semi-structured (XML) data and unstructured (text) data. As such, we support both structured query and unstructured query to the mixed-type data storage. Structured query can be issued to the structured and semi-structured datasets, via SQL and XQuery query languages (though we do not require users to master all query languages as discussed later). Unstructured query, i.e., keyword query can be used to either select relevant records in the structured and semi-structured databases, or retrieve relevant text documents.

4.1.1 Structured Query Processing

Although there are standard structured query languages for different types of data, i.e., SQL for relational data and XQuery/XPath for XML data, we cannot prompt the platform users to specify queries in different languages to search different parts of data store. This is because users may be blind to the underlying data structures, and do not know the type of the data they are interested in. To solve this problem, we leverage on the views, which are a set of type-unified virtual representation of underlying data. Views are driven by user queries, and created under the supervision of owners of involved datasets. Views are the only data interface presented to users to let them know what data they can query. The internal relationship between a view and involved datasets in either relational format or XML format will be handled by the system. A user only need to issue SQL queries to the view, and the system will translate the query into sub-queries, in either SQL or XQuery to query involved physical datasets.

Continuing with the example data in Fig. 2, if a user would like to know the details of particular house owners, he/she may raise a request to the A*DAX. After getting approvals from the ICA and the URA, a view with citizen's particulars and their housing information will be created. The user can just issue SQL queries to the view, and the system will internally join the two datasets to process queries.

There are also materialized views, which will be discussed later. All views are managed by the View Manager.

4.1.2 Keyword Query Processing

Keyword queries can be issued to structured/semi-structured datasets, or text document store. If the keyword query is to search structured or semi-structured data, the query engine will return data records that contain query keywords and are ranked based on interpreted query intention. Both structured data and semi-structured data will be semantically indexed and searched by our algorithms. The detailed semantics-based keyword search algorithms are omitted in this paper. They can be found in [2].

If the keyword query is issued to the text database, an inverted list based document retrieval will be executed. Relevant text documents will be returned. We follow the existing document retrieval techniques, and omit the details in this paper.

4.2 Mixed-Dynamics Data Processing

4.2.1 Data Storage

Data shared in the A*DAX platform can be either uploaded in batch to the system, or streamed into the system in real-time. For batch data, depending on the data size, data characteristics and the data owner's preference, the system will either store the data into databases, or the Hadoop Distributed File System. Basically, if the data size is manageable and the data will be used for OLTP-like query processing, the data will be stored in database. On the other hand, if the data is too large and only for BI-like analytical purpose, the data will be stored in the Hadoop system and processed by MapReduce.

For streaming data, a data stream management system (DSMS) will receive the data and perform complex event processing (CEP) over the data to answer continuous queries. Meanwhile, the data will also be archived in databases, unless the data owner denies this operation. The system will keep a size threshold for each data stream, and the database archive for each stream will be periodically transferred to the Hadoop system to release database space and ensure database query performance.

4.2.2 Query Processing

In this part, we talk about how the A*DAX processes queries over both static data and dynamic data. We adapt the λ-architecture [4] to design the query processing module for both static data and dynamic data in the A*DAX platform. Fig. 3 shows the architecture in the A*DAX for query processing.

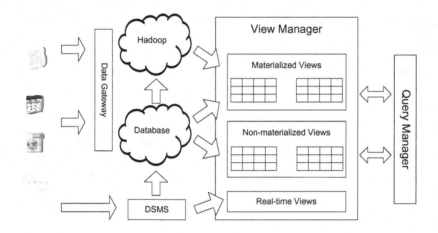

Fig. 3. Query processing over static and dynamic data in A*DAX

As shown in Fig. 3, we consider three layers of data storage. The DSMS system temporarily store the most recent set of streaming data in its memory, in order to process continuous queries registered in the system. Then it pushes all data to the database system. The database system is expected to execute SQL queries on-the-fly. Periodically, the database system will further archive old data in the Hadoop Distributed File System.

The DSMS offers a virtual real-time view to users to use the data streams. Users may issue continuous queries, in SQL format, to the real-time view, which will eventually register into the DSMS system for data filtering. There are also virtual views on top of the database system. As mentioned, such views may be constructed across different database instances, guided by the semantic graph in the Data Semantics Manager. Users may issue SQL query to search the datasets in the A*DAX platform via these views, and all queries will be eventually executed in the Storage Layer.

The materialized views are constructed based on approved analytical queries (such as aggregation queries) over the large-scale historical data (mainly) in the Hadoop system. In other words, materialized views store the result of analytical queries processed in the Hadoop system. Since data processing over large-scale data, e.g., using MapReduce in Hadoop is time consuming, we pre-cache processing results for certain frequently asked queries for users to perform efficient online analytical data processing and visualization. The data in the materialized views will be periodically updated as new data are archived in the Hadoop system. Note that the platform also allows users to program MapReduce jobs and process data on the Hadoop Distributed File System.

Let us assume the scenario that the HDB continuously sends house transaction data to the A*DAX, and another government agency would like to find out the highest transaction price for each month, after being approved to view such data. Assume the house transaction data are partially stored in both the Hadoop system and the database system, and a materialized view to summarize each month's transaction data stored in the Hadoop system was created. This query will be sent to all the three views. In the materialized view, historical summarized records can be easily retrieved. For the non-materialized view, the query will be executed against the database data, and return the result. During the data execution, the continuous query issued to the real-time view will monitor the most recent data, and compare the result with the database search result to find out the highest transaction price for the most recent month.

If the materialized view for the issued query does not exist, the query will be programmed as a MapReduce task and executed in the Hadoop system. In this case, the Hadoop execution will be long, and probably dominates the overall query processing time. Then the continuous query issued to the real-time view becomes significant, which ensures that no data is missing during the query processing.

Finally, the query results from all views will be merged and post-processed, if necessary, and returned to the user.

5   Access Control

The A*DAX platform involves a lot of sensitive data from public and private sectors in Singapore. The security and privacy of the system is crucial. The Security Module of the platform guarantees the system security using cryptographic techniques. In this section, we focus on privacy control.

Each data provider has a right to control the access to his/her data. Since the data in the platform are very sensitive and the platform users can be quite diverse with different levels of profiles, we do not follow the typical role-based access control model. Instead, we provide the channel and require each data user to get approval from the data owner before he/she can access the data. The data owners do not need to specify any access control policy on the data. They will evaluate each data access request in ad-hoc manner, and grant the access or partial access to their data based on data users' profiles and access purposes.

Because the A*DAX shares data among government agencies and big companies, rather than in individual level, the number of users is limited and the workload for ad-hoc access approval is not high.

The Access Control Module of the A*DAX platform maintains the access privilege of each user, and compiles it against the views to make decision that whether the user is allowed to access the queried attributes of each view.

For example, a user is querying the view contains citizen particulars and housing information, i.e., the view across the ICA and the HDB's data. Suppose the user is granted the full access to the citizen data by the ICA, and granted partial access to the housing data on which only the transaction price for each house is not accessible. If the user query does not involve the attribute of transaction price, the query will be processed. Otherwise, the query will be rejected.

6   Conclusion

In this paper, we introduced the A*DAX platform, a platform for cross-domain data linking, sharing and analytics for public and private sectors in Singapore, to improve urban planning and development. We described the general framework of the platform, and focused on the components that handle data storage, data fusion and query processing. We designed an adaptive semantics-based metadata to guide the system integrating data from unpredictable data sources. Furthermore, we designed a system architecture based on the λ-architecture to process heterogeneous data in the A*DAX data storage. In particular, our architecture provides virtual and materialized views between the data storage and the query engine. It offers convenience for users to issue queries in unique format, and it also provides a way for data owners to control the disclosure of their data to users.

Acknowledgement. This work was supported by the A*STAR SERC Grant No. 1224604057 and 1224200004.

References

1. Chen, P.P.: The ER model: Toward a unified view of data. ACM Trans. Database Syst. 1(1), 9–36 (1976)
2. Le, T.N., Wu, H., Ling, T.W., Li, L., Lu, J.: From Structure-based to Semantics-based: Towards Effective XML Keyword Search. In: Ng, W., Storey, V.C., Trujillo, J.C. (eds.) ER 2013. LNCS, vol. 8217, pp. 356–371. Springer, Heidelberg (2013)
3. Lenzerini, M.: Data Integration: A Theoretical Perspective. In: PODS, pp. 233–246 (2002)
4. Marz, N., Warren, J.: Big Data - Principles and best practices of scalable realtime data systems. MEAP Began (2012)

Optimizing Database Load and Extract for Big Data Era

K.T. Sridhar and M.A. Sakkeer

XtremeData Technologies, Bangalore, India
{sridhar,sakkeer}@xtremedata.com

Abstract. With growing and pervasive interest in Big Data, SQL relational databases need to compete with data management by Hadoop, NoSQL and NoDB. Database research has mainly focused on result generation by query processing. But SQL databases require data in-place before queries may be processed. The process of DB loading has been a bottleneck leading to external ETL/ELT techniques for loading large data sets. This paper focuses on DB engine level techniques for optimizing both data loads and extracts in an MPP, shared-nothing SQL database, *dbX*, available on in-house commodity hardware and cloud systems. The agile, data loading of dbX exploits parallelism at multiple levels to achieve TBs of data load per hour making it suitable for cloud and continuous actionable knowledge applications. Implementation techniques at DB engine level, extensions to load/extract syntax and performance results are presented. Load optimization techniques help to speed up data extract to flat files and CTAS type SQL queries too. We show linear scale up with cluster scale out for load/extract in public cloud and commodity hardware systems without recourse to database tuning or use of expensive database appliances.

Keywords: parallel DBMS, bulk load/extract, cloud, big data, dbX.

1 Introduction

The *Big Data Era* is upon us, ushering in data that has higher volume, velocity and variety compared to structured data processed by traditional SQL databases. In its recent report on the *Digital Universe* [6], IDC revises its 2010 size estimate to 1,227 exabytes and projects 40,000 exabytes for 2020, with analyzed volume for 2012 at 0.5%, growing to 33% in 2020. In today's industry, multi-TB databases running to hundreds of TBs are indeed common.

Big data arrives in a variety of formats. SQL database can do query processing only after data is *in-place* in its internal format. This very first step for data analytics and mining, *loading data*, is considered a performance bottleneck [1, 2, 4, 11]. Estimated ingest rate in 2006 for a multi-petabyte scientific applications database [7] was about 1 TB in ten hours! Data load performance expectations of today's applications have risen multi-fold, way beyond the 102 GB/hr of LSST. And, such performance is demanded on commodity hardware or cloud, not on expensive, special purpose database appliances.

S.S. Bhowmick et al. (Eds.): DASFAA 2014, Part II, LNCS 8422, pp. 503–512, 2014.
© Springer International Publishing Switzerland 2014

Traditional SQL databases have to urgently revisit the problem of data load, and the less frequently used *data extract*, for their very survival in the Big Data era. In this paper, we focus on issues of load/extract problem of SQL databases and propose solutions for mitigating the bottleneck. Our proposals have been incorporated at architectural and implementation levels of a new product, *dbX*, a parallel, SQL database, to provide load/extract rates of multi-TBs/hr.

dbX is a high performance, full featured ANSI SQL database designed as an MPP, shared-nothing system that can be deployed on commodity hardware systems and cloud [12]. Based on open source rdbms PostgreSQL [8], it is not a *federated* system but has been fully re-architected and re-engineered to support a *vector execution* model exploiting parallelism at multiple levels: IO, data partitioning, intra-operator, operator and intra-query. The architecture of dbX comprises *1 head and n data* nodes with *scale out* options and *multi-tenancy* (a physical node runs multiple *virtual data nodes*). Currently, dbX is available on two public, pay-for-use, cloud systems: Amazon (AWS) and Rackspace.

The rest of the paper is organized as follows. Section 2 discusses related work highlighting importance of data load performance. Section 3 summarizes load/extract statements of dbX SQL. Section 4 highlights technical issues that affect load/extract performance of DB engines with solutions proposed in sections 5 and 6. Section 7 presents and discusses dbX performance results for different use-case scenarios on commodity hardware and cloud; the final section concludes with directions for future work.

2   Why Is Load Performance Important?

Loading an external, ASCII file into a DB as *pages* adds to query cost and has led to several criticisms in related work that is summarized below.

Inherently Slow: Despite faster query processing of databases, the pre-requisite of loading is termed slow and complex. [11] presents a compelling argument against data loading: until entire flat file is loaded, queries cannot be run precluding the possibility of quick peeks into data; they explore hybrid techniques with flat files as part of DB. The *time-to-first-analysis* cost for structured data with steps of modeling, loading and tuning is termed *unjustifiable* [2] in the context of Hadoop systems, leading to a proposal that piggybacks on Map/Reduce to load conventional DBs. [1] shows that databases underperform for loads and over perform for querying vis a vis Hadoop.

Alternative Approaches: Hadoop's Map/Reduce paradigm deals with flat files without incurring cost of data loading, but has slower query performance [1] and this *NoSQL* approach has gained wide acceptance. The *NoDB* approach of [4] takes ideas of [11] further and modifies a traditional row based database to support querying over raw files and claims competitive query performance.

Continuous Actionable Knowledge: An emerging use-case is *near real-time data warehousing*, whose proponents argue that more recent data, from frequent loads, leads to better decision support almost on a continuing basis from analytics and mining [10] and propose external methods to achieve it.

Load-Analyze-Drop: Pay-for-use, public cloud systems offer several levels of storage at different costs and IO performance, e.g., Amazon's cheapest storage S3 is not the best for query processing. Consequent to prohibitive cost of high performance storage (idle servers are charged), an emerging trend is to setup need based clusters on cloud to load data residing in low cost storage, run analytic queries and drop the cluster within a limited period of a few days. To service such requirements, faster loads and linear scale out gains are essential on cloud.

Multi-file Bulk Data: Emerging big data applications in digital advertising, click streams, stock/equity trade, etc., generate a large number of structured data files from multiple sources, to be loaded into a single table of the DB. The number of small files, of a few 100 MBs, in a day may be 30,000 or more, totaling to several TBs. Multiple load commands for many small files underperform compared to a single command to load all files.

Load Metric: The importance of data load performance is now recognized by the Big Data community: unlike TPC benchmarks that focus only on query processing, recent proposal for a Big Data benchmark [3] includes data load as part of the benchmark metric.

3   Load and Extract in dbX SQL

dbX supports bulk loading and extraction of data through its *COPY* statement [13], which is an enhanced version of the COPY statement of PostgreSQL. We highlight only special features of dbX COPY statement. Data source for load may be a file in staging area, a Linux pipe or even the DB port. Multi-file bulk data requirements of section 2 are addressed: when source path is staging area directory or contains a regular expression pattern for filename.

An important COPY option is the use of reserved words NODE or NODEID in the statement; the latter specifies a *parallel mode* of bulk load/extract to take advantage of MPP, shared-nothing architecture of dbX to land/emanate multiple source/target pipes on/from data nodes for higher rates.

```
COPY all_trn FROM '/v1/2013/nov/14oct.dat' DELIMITER '|' NODE;
COPY all_trn FROM '/v1/2013/nov/14*.d?t' DELIMITER '|' NODEID;
```

When data is loaded in parallel, COPY may distribute data across nodes based on *scatter* method (round robin or hash) specified for target table, and range partition data stored on a node. In addition to standard load options for DELIMITER, NULL AS, ESCAPE and QUOTE, other special options include: *EOL* (non-LF/CR char as line terminator to handle multi-line text columns), *IGNORE TRAILING* (columns set as DB null for missing trailing values) and *INSTEAD NULL* (columns stored as DB null when value is all blanks).

As COPY is also a statement under transaction control, errors due to data type mismatches, invalid syntax, missing columns, integrity violations, bad escape sequences, etc., may rollback correctly loaded data leading to iterative cycles. To avoid such iterations, error handling methodology may be specified using *ON ERROR CONTINUE* (ignore error and continue with load; no rollback),

ON ERROR LOG (log error data to file and continue with load; no rollback), *LIMIT* (abort and rollback only if number of errors exceeds a specified limit) and *ON ERROR ABANDON* (default option to rollback on first error).

The syntax of extract is similar to COPY of load without error option; COPY also supports CSV and fixed format files. Another form of load/extract statement is CTAS query (CREATE TABLE AS, SELECT INTO, INSERT SELECT) that loads results of a SQL query into another table. More details of load/extract statements and CTAS queries may be found in [13].

4   Issues in Data Load/Extract

The load task parses all source data breaking them into lines, and converts them to rows in DB pages for persistent storage. Though without any algorithmic complexity, the task is fraught with issues at several levels: IO system, computational steps, SQL standard requirements, parallelism and error handling.

Synchronous IO: The general purpose file system of an OS like Linux supports POSIX calls that perform *synchronous* or *serial IO*, which imposes a serial order on (read, compute) or (compute, write) requirements of the application.

Kernel Buffering: Linux IO is generally buffered by the kernel with small buffers (4k or 8k) and is not conducive to big sized IO preferred by databases for bulk writes (data load) or even reads (sequential scan). Further, the look-ahead policies of the kernel for reading may be at variance with database requirements.

Disk Fragmentation: Disk systems perform better with contiguous block allocation. General purpose file system usage varies widely leading to disk fragmentation; attendant head seek costs lead to poorer IO performance.

IO Contention: During load or extract, concurrent read and write are performed on a system with limited IO bandwidth. Also, a DB may write temp files, or its own logs for transaction durability. Despite use of RAID, the IO contention reduces IO bandwidth available to the DB engine.

Compute Load: Though not so, bulk loading is generally believed to be IO bound without any serious compute load. The task of parsing a large file for syntax and semantic correctness, and generating DB tuples imposes a significant compute load; any potential parallelism must be exploited on multi-core systems.

Locking: Databases need to support concurrent clients; e.g., during a bulk load other concurrent clients may run queries, or multiple loads, on the same table and access pattern for table file becomes N-1 [5, 9] requiring locks. Locks affect performance and load rate is subject to workload variations of database.

ACID Compliance: Databases build a *write-ahead log (WAL)* to support durable transactions. Persistent table writes of database must be logged increasing IO cost of loads. Generally, WAL records are small in size but for bulk loads this could become doubling of system write volume for DB pages.

Integrity Constraints: May be simple predicates on column values, a primary key or duplicate constraint on distinctness of column values or more complex foreign key constraints. Integrity constraint checks add to compute demand of load and also IO bandwidth contention as other tables may be read.

Indexes: May be user defined for query performance or internally created to manage primary key, foreign key and duplicate constraints. User defined indexes may be deferred, but internal indexes must be built during load. Index building is costly and significantly reduces load performance.

Data Partitioning: Parallel databases distribute data across nodes either by hashing, on 1 or n columns, or in round robin mode. This entails data movement across nodes through the network layer, which has lower performance than IO hardware. Range partitioning distributes data across child tables based on column predicates: entails compute load and handling of multiple files.

Parallel Configuration: Data source for load may be a single pipe landing on the DB: flat file in staging area or a Linux pipe. The transfer rate of the single incoming pipe limits load rate; can be improved for parallel databases with multiple source pipes. Multi-tenancy configurations also add to contention.

Error Handling: Source data isn't always error free: type mismatches, syntax, integrity violations, etc. Due to transaction requirements, good data loaded until error may be lost by rollback; iterative load cycles are not acceptable; retaining good data with log of bad rows has more compute and IO demand.

Data extraction from a DB is affected by several of the data load issues: sync IO, kernel buffering, IO contention, disk fragmentation, compute load (to convert tuples to formatted strings), locking and parallel extract.

5   Optimizing Load Performance

Solutions for issues of section 4 are outlined below in terms of architectural changes, careful engineering decisions and implementation techniques for optimizing load performance of databases.

Parallel IO: An alternative to synchronous IO is *asynchronous IO* (*aio*) of Linux systems with a POSIX interface. Unlike sync IO, aio calls do not block the caller on an IO request and completion is notified through an interrupt; computations may be overlapped with IO. As a policy, we adopt async IO for all IO including flat file reads/writes. The use of parallel IO opens up opportunities for database specific read-aheads, and threaded parallel computation.

File System & Big IO: Kernel buffering of Linux file system may be bypassed and IO performed as *direct IO* (*O_DIRECT*) with big buffers that accommodate multiple DB pages. The database has a better control on its caching requirements as it no longer relies on kernel caching. XFS is a scalable, high performance file system that performs well for such requirements.

Parallel Task Scheduling: By careful analysis, the loading problem may be broken into tasks scheduled in parallel through a framework of master and worker threads on multi-core CPUs. Despite synchronization overheads, gains are significant. The parsing and tuple creation task of bulk loads is parallelizable with each worker thread dealing with different parts of input file; multiple DB pages may be created in two steps. For data intensive applications, a similar scheduling framework is reported at file system level in recent work [9].

Minimal Locking: A DB table is organized as pages of fixed size identified by a position index, used for non-sequential IO. By forcing every new load to begin at a page boundary, worker threads may use private buffers to create DB pages eliminating lock costs. As a N-1 access pattern to table file, loading cannot avoid locks. Consequent to table file growth, contention for pages must be resolved without degrading to a 1-1 access pattern. We use *bunched locking* to acquire the required, multiple pages with a single lock. The short duration, transient locks support concurrent clients on target table for querying or load.

Transaction Logging: A transaction layer using MVCC, snapshot isolation and two phase commits may be based on the positive assumption that *users want a transaction to commit, not rollback* and support minimal WAL logging to reduce IO overheads for bulk loading and CTAS queries. Log data written for such tasks is just a *minimal WAL record (minWAL)*, carrying adequate information for recovery to deal with aborted load transactions, and not the entire page or its rows. A consequence of this approach: bloat in table file size for aborted load; such bloat may be compacted using other DB commands.

Data Distribution: For hash distribution of data, parallel DB may shift data across data nodes through the network: node #3 may first process a row that must be stored on node #5. Data shipping across nodes is done using large buffers of 4 MB and managed by independent worker threads that avoid file writes; network cost is reduced by networks such as InfiniBand and 10G.

Parallel Loads: A parallel database must support loads that land source pipes directly at data nodes; with local disks on data nodes, IO bandwidth contention reduces and throughput increases with compute and IO parallelism. But the scale-up in load rate may be affected by contention between flat file reads and database writes within a node. Parallel load implementation must take advantage of cluster scale out for linear scale up in performance.

Constraints & Indexes: Simple check constraints on column values may be evaluated by worker threads. Local indexes and primary key index may also be processed before pages are written out. When constraints involve use of global indexes, network traffic across data nodes increases. To avoid resource lock-ups, such checks must be performed in buffered mode, and may be deferred after page writes. Index creation is costly and is often done sequentially.

6   dbX Load/Extract Implementation

dbX architecture and its implementation incorporate solutions outlined in section 5. We depict a simple scenario in Fig.1: a single, non-parallel data source in head staging area for load to n nodes that adopts a pipelined architecture. The master process (*mast-h*) on head reads flat file with big aio requests. Independent threads (*iohn*) handle aio callbacks and ensure only integral number of lines are part of raw buffers (*rb*) dispatched round robin to nodes by threads (*xferh*). The master process never blocks on IO or network, but monitors channels for completion or errors. On node side, a set of threads (*xfern*) manage channels from head and other data nodes. Raw buffers of max size 4 MB are queued up

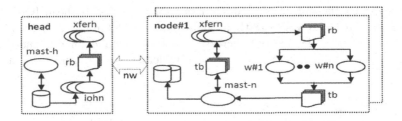

Fig. 1. dbX Load Architecture

for processing by a set of 4 to 8 worker threads (w#1 to w#n) to parse data buffers (*rb*) and convert to DB tuple buffers (*tb*). The node master (*mast-n*) may dispatch a tuple buffer (*tb*) to a peer node or process it locally to generate DB pages in a private page buffer of 2 MB. Relation is extended under bunched locking and in the post-lock phase page buffers are updated with page index values and pages written to relation file through aio; a minWAL record is written into WAL on first write to relation file. The page buffer filling and write method is akin to handling of N-1 strided access pattern in parallel file systems [5].

Though less common, extracting DB data as external, ASCII files suffers from same issues as data load. dbX architectural model for extract is similar to load and uses several load optimization techniques. For non-parallel extract to head, the process is driven from data node side, where a master process issues look-ahead, big aio read requests on table to fill its private buffers. 4 to 8 worker threads process pages for transaction visibility and tuple conversion to delimited, ASCII lines with JIT compilation. Filled buffers are handled by network threads on both node and head, with the master on head generating the output file.

7   Performance Results

Load and extract performance results are from dbX on Amazon cloud and commodity system in our lab with 1 head and 8 data nodes: 2x AMD Opteron 2431, 2.4 GHz, 6 core, 32 GB RAM, Infiniband, RAID 6, 7.2k rpm, 1 tb disks (16 on head, 12/node); Linux Centos with XFS. For bulk loading, we use TPC-H 1 TB lineitem (6.14 billion rows, 765 GB) and define *database rate* as size of uncompressed input or output file divided by time taken for load or extract statement expressed as a per sec (*mb/s*), or per hour (*tb/h*) value; on commodity system rate values are averaged over 3 runs. Configuration 8x4 denotes a system with 8 physical, data nodes running 4 virtual nodes (multi-tenancy) per data node.

7.1   Load and Extract

Commodity Hardware: Fig. 2 shows results of load/extract runs in parallel and non-parallel modes varying node multi-tenancy. Best results for parallel mode are on 8x1: load at 1059 mb/s (3.64 tb/h) and extract at 1306 mb/s

(4.48 tb/h); single source, non-parallel mode, bottlenecked by head IO, is best on 8x4: load at 927 mb/s (3.18 tb/h) and extract at 583 mb/s (2 tb/h). Multi-tenancy without real scale out increases resource demand on data nodes, as confirmed by *dstat* monitoring of CPU and IO demands, and drops rates at 8x4.

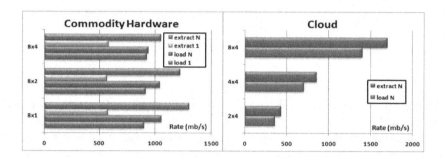

Fig. 2. Load and Extract: dbX on Commodity Hardware and Cloud

Cloud: On Amazon, we use cluster of *hs1.8xlarge* instance with 1 head and 8 data nodes of multi-tenancy 4 each. Fig. 2 shows results of 1 TB parallel load and extract on 2x4, 4x4 and 8x4. We observe a linear scale-up in performance with cluster scale out; 8x4 has max rates of 1402 mb/s (4.8 tb/h) for load and 1706 mb/s (5.9 tb/h) for extract. Without any special tuning for cloud, dbX delivers higher rates for parallel COPY on better hardware of cloud.

Data Partitioning & Errors: Single source, non-parallel load of 100 GB, 300 GB and 1 TB lineitem on lab system of 8x1 with the three data partitioning options of dbX showed no significant change in rates. Load rate of 1 TB file with 2.5% errors dropped to 2.72 tb/h generating a log of about 156.7 million rows; rate drop was found proportional to number of error rows and log file size.

7.2 Rate Scale up on Cluster Scale Out

A scale out test requires enhancement to cluster configuration; increasing multi-tenancy does not add computing power. On lab system of max 8 nodes, scale out test runs 2x1 to 8x1 configurations while on cloud we use 2, 4 and 8 node clusters with multi-tenancy 4. Results in Fig. 3 show linear scale up in performance for parallel load with scale out of cluster nodes on both lab and cloud systems. Scale out test results indicate that techniques adopted to implement load/extract in dbX can provide several TBs per hour rate dependent purely on cluster size.

7.3 Concurrent Clients

Multiple (2 to 6) concurrent loaders load 1 TB into same target table on 8x1 and 8x2 lab system (total 36.84 billion rows after 6 clients). As system resources are shared, load rate is calculated as a *throughput rate*: divide total file size across all concurrent loaders by slowest loader's time. Fig. 3 shows a 12% drop in non-parallel load throughput rate from 2 to 6 clients due to head IO bottleneck. Parallel loads saturate at 3 clients, but show a 10% gain from 2 to 6. Almost linear curve shows that N-1 access pattern of load is handled reasonably well.

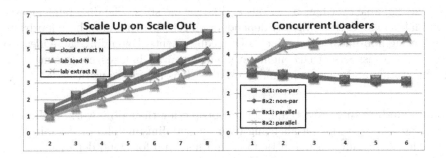

Fig. 3. dbX Scale Up on Scale Out and Concurrent Loaders

In the second test on 8x1 lab system, parallel loader runs concurrently with 10 query (count, join, group, union, etc.) clients in a loop accessing load table and other tables. Query and load clients share system resources and throughput rate drops by about 50% (single loader) or 20% (4 loaders) of non-query, single loader rate. When loaders and query clients compete for resources, load rates may be improved by controlling query scheduling with user set priority.

7.4 Others

To validate per hour rates got by extrapolating figures from a smaller sized file, we run a *rate sustainability* test. Parallel load of a 5.88 TB flat file (got by copying 1 TB lineitem multiple times) on 8x4 cloud was at 4.72 tb/h (extrapolated at 4.8); extract at 5.8 tb/h (5.9 extrapolated); both measured variations are under 1.7%. Similar sustainability test on 8x2 lab system: load at 3.23 tb/h (3.59 extrapolated); extract at 4.25 tb/h (4.22 extrapolated); lower load rate on lab system may be due to *disk fragmentation* for flat file with head seek cost.

A disk occupancy issue, *fragmentation*, is highly dependent on file system workload and affects load/extract rates of DBs; even if fragmentation is restricted only to flat file, rates can drop by 15 to 25%. All tests on lab system were run at about 45 to 55% free space on disks. When data load includes index creation or constraint checks for duplicates or foreign keys, load rate is significantly affected, and for such cases dbX performance is not optimized.

8   Conclusion

We have outlined several DB engine level techniques for optimizing performance of SQL load/extract statements. An optimal partitioning of load/extract problem has been proposed and implemented in dbX to deliver over 3.5 tb/h load rate on commodity hardware and 4.8 tb/h on Amazon cloud, both on small clusters. The careful engineering decisions, implementation techniques and use of parallelism at multiple levels help to mitigate issues in supporting DB specific requirements for load/extract. Without recourse to database tuning or use of expensive database appliances, we show linear performance scale up with cluster scale out on a DB running in public cloud and commodity systems, with rates sustainable over longer periods and larger file sizes. The dbX implementation makes load/extract linearly dependent on IO/CPU power of base hardware eliminating non-linear effects due to DB requirements. Extending these ideas to other related aspects of DB engine in management of indexes/constraints, disk fragmentation and query scheduling are directions for future work.

Acknowledgments. We thank several people for their dbX work: at Bangalore: Shiju Andrews for parallelizing IO layer and Jimson Johnson for minWAL handling; at Schaumburg: Jim Benbow and his team for cloud deployment.

References

1. Pavlo, A., et al.: A Comparison of Approaches to Large Scale Data Analysis. In: SIGMOD 2009, pp. 165–178. ACM (2009)
2. Abouzied, A., Abadi, D.J., Silberschatz, A.: Invisible Loading: Access-Driven Data Transfer from Raw Files into Database Systems. In: EDBT/ICDT 2013, pp. 1–10. ACM (2013)
3. Baru, C., Bhandarkar, M., Nambiar, R., Poess, M., Rabl, T.: Benchmarking Big Data Systems and the Big Data Top 100 List. BIG DATA 1, 60–64 (2013)
4. Alagiannis, I., Borovica, R., Branco, M., Idreos, S., Ailamaki, A.: NoDB: Efficient Query Execution on Raw Data Files. In: SIGMOD 2012, pp. 241–252. ACM (2012)
5. Bent, J., et al.: PLFS: A Checkpoint Filesystem for Parallel Applications. In: SCO 2009. ACM (2009)
6. Gantz, J., Reinsel, D.: The Digital Universe in 2020: Big Data, Bigger Digital Shadows and Biggest Growth in the Far East. In: IDC IVIEW, IDC (2012)
7. Becla, J., et al.: Designing a Multi-petabyte Database for LSST. In: SPIE Conference on Observatory Operations, Strategy, Processes and Systems, SLAC-PUB-12292 (2006)
8. PostgreSQL: http://www.postgresql.org
9. Xu, R., et al.: Filesystem Aware Scalable I/O Framework for Data Intensive Parallel Applications. In: IPDPSW 2013, pp. 2007–2014. IEEE (2013)
10. Santos, R.J., Bernardino, J.: Real-time Data Warehouse Loading Methodology. In: Desai, B.C. (ed.) IDEAS 2008, pp. 49–58. ACM (2008)
11. Idreos, S., et al.: Here are my Data Files. Here are my Queries. Where are my Results? In: 5th Biennial Conference on Innovative Data Systems Research, CIDR, pp. 57–68 (2011)
12. XtremeData: dbX, http://www.xtremedata.com
13. XtremeData: dbX SQL User Guide, Vol. II, Document X4631-02. XtremeData (2011)

Web Page Centered Communication System Based on a Physical Property

Yuhki Shiraishi[1], Yukiko Kawai[2], Jianwei Zhang[1], and Toyokazu Akiyama[2]

[1] Tsukuba University of Technology
[2] Kyoto Sangyo University
{yuhkis,zhangjw}@a.tsukuba-tech.ac.jp, kawai@cc.kyoto-su.ac.jp,
akiyama@cse.kyoto-su.ac.jp

Abstract. We have developed a novel communication system that enables *ALL* users to share information and emotion effectively through *ALL* Web pages based on a physical property: a user can perceive others browsing the same pages through avatars on the Web browser, like a face to face communication in real world. The system optimize the displayed and communicable users by the relationship among them based on the information searching and sharing activities. Furthermore, the system also enables users to search not only Web pages but appropriate users: we construct a new ranking model by combining the real space information (the user activities) with the Web space information (the hyperlinks). We also verify the effectiveness of the system by the user study experiments in the condition where we suppose the system works especially well, i.e., information retrieval, collaboration tasks, and sharing of feelings.

Keywords: Communication; Search; Ranking.

1 Introduction

People have spent time in the Web space as much as in real space now. In real world, people in the same place can efficiently share information, e.g., opinions, questions, impressions, etc. However, in the Web, users can not effectively and immediately share them because the Web space lacks a physical property: the perception of others browsing the same Web contents pointed by URI; in real world, people in the same place can recognize each other by involuntary physical conditions, whose property enables to construct *a loose relationship*, which is we believe a natural and important feature but lacked in nowadays Web experience.

On the other hand, the social networks, micro blogs, or Q&A sites are becoming popular for the information sharing services based on a semantic distance: these services are restricted within specific users having friendly relationship or specific Web sites, not applicable for all users and all Web contents.

Therefore, we aim to realize seamless Web page centered communication for all sites and users by adopting the physical property. Thus, the user can perceive other users browsing the same contents or Web pages pointed by URI as avatars on the Web browser, as if people can see the faces of others happened to be in

S.S. Bhowmick et al. (Eds.): DASFAA 2014, Part II, LNCS 8422, pp. 513–522, 2014.

(a) Communication while browsing   (b) Rankings of pages and users

Fig. 1. A physical approach to seamless communication in Web and real space

the same place in real. Moreover, since the communication system is developed as a extension of the existing browser, there is no need to construct an another virtual space, e.g., Second Life, in addition to existing Web space. The developed system[11][12] visualizes Web browsing users as avatars and enables them to communicate with each other while browsing the similar pages (Fig. 1(a)). As a result, the developed system enables real world-like communication as follows:

- Lost person asks the direction to a passer-by.
- The audiences happened to attend the lecture discuss the same document.
- People looking at a baseball game at the stadium, a painting at the museum, the sight of city, etc., talk with each other for sharing the feeling.

However, the distribution of visiting people is different between Web and real space. That is, the number of people visiting the famous sites in Web space is larger than famous place in real space. Then, in the Web space, it is difficult to communicate with each other. Therefore, the displayed and communicable users are optimized for the target user based on the relationship among them by utilizing the browsing or communication history, as stated afterwards.

The developed system also enables users to search not only desired Web pages but appropriate users by utilizing the information retrieval and sharing activities in the Web (Fig. 2). We construct a new ranking model, a page-user ranking model (PURank), by combining the real space information (user activity) with the Web space information (hyperlinks). Communication with pages or users in the model means recommendation like hyperlinks in PageRank[3]. The user activity is represented by a link because we only need to extend PageRank which would be best algorithm to search vast web pages and users immediately.

In this paper, we explain the developed system and also verify the effectiveness of the system by the user study experiments in the condition where we suppose the system works especially well, i.e., information retrieval tasks, collaboration, and usability.

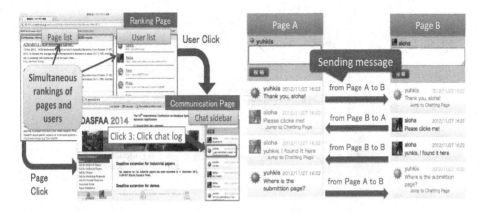

Fig. 2. Relationship between communication and ranking functions

Fig. 3. Communication over different pages

2 Page Centered Communication

If the user has a question about contents on the page, s/he can immediately ask others by typing a sentence into chat sidebar. The others can see the question in the chat sidebar and respond accordingly.

For some unpopular pages, the number of users accessing them may be small, even zero. Thus, the system provide two functions: 1) a chat log sidebar, and 2) a communication with other users over different pages.

The chat log sidebar enables asynchronous communication, making the past chat logs on Web pages available to the current users; they can search the answer to the previous similar question from the chat log.

We also extend the communication function by enabling real-time communication not only in the same page but similar pages as shown in Fig. 3. The details are as follows: a target user can communicate with not only users browsing the same page that the target user is browsing now, but also 1) users having browsed the page that the target user is browsing now, and 2) users browsing the page that the target user browsed before. To distinguish the difference, the former is displayed transparently and is not saved to the log (Fig. 3) — as if we heard a hum of far-off voices in the real space.

At all pages, the user can click a transparent user avatar and go to the page that the clicked avatar is browsing now. The user also click the transparent chat log to jump to the page in the same way.

The developed system also displays avatars related to the query, i.e., avatars related to the information the target user want to acquire. Of course, users can click the avatars to go to their browsing pages. To encourage users to talk more with each other, users can easily find the frequently talking avatar; if a user enters more sentences, his or her user avatar is enlarged.

The communication function is achieved by the collaboration between the server side and the client side. The system construction is shown in Fig. 4.

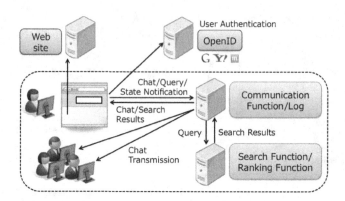

Fig. 4. System construction diagram

3   Page-User Ranking Model

The communication with pages or users in our model means recommendation like hyperlinks in PageRank. Thus, we regard all kinds of communication as positive recommendation even though these includes the negative one as well.

3.1   Page-User Graph and Adjacency Matrix

The nodes of proposed graph consist of four parts (Fig. 5).

1. Hyperlinks between pages (Part I)
   If a page p_i has a hyperlink to another page p_j, the element of matrix $M(p_j, p_i)$ is set to 1_1; otherwise, the value is set to 0_1.
2. Social page links from users to pages (Part II)
   If a user u_i has communication through a page p_i or had communication through a page p_j before, the elements matrix $M(p_i, u_i)$ and $M(p_j, u_i)$ are marked as $\bar{1}_2$ and $\underline{1}_2$, respectively; otherwise, the value is set to 0_2.
3. Social page links from pages to users (Part III)
   This part represents whether a page has communication or had communication by a user. Thus, the links of Part III are represented as the inverse matrix of those of Part II ($\bar{1}_3$, $\underline{1}_3$, or 0_3).
4. Social user links from users to users (Part IV)
   If two users have similar interests, a bidirectional social link is generated between them. Our implementation is based on comparing the overlap of pages that two users have browsed as follows:

$$M(u_i, u_j) = M(u_j, u_i) = \begin{cases} 1_4 \text{ if } \dfrac{\sum_k (p_k u_i \wedge p_k u_j)}{\sum_k (p_k u_i \vee p_k u_j)} > \tau, \\ 0_4 \text{ else,} \end{cases} \tag{1}$$

where $p_k u_i$ and $p_k u_j$ are determined by whether u_i and u_j have browsed p_k or not, respectively. If $M(u_i, u_j)$ is larger than a threshold τ, the corresponding element of Part IV is set to 1.

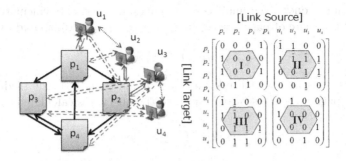

Fig. 5. Page-user graph and adjacency matrix

3.2   Probability Transition Matrix

First, $\rho\alpha/a$, $\sigma\beta/b$, and $\sigma\gamma/b$ are multiplied by the elements of the adjacency matrix depending on the kinds of elements: 1_1, $\bar{1}_3$, $\underline{1}_3$, respectively. Similarly, $\rho x/a$, $\rho y/a$ and $\sigma z/b$ are multiplied by the elements: $\bar{1}_2$, $\underline{1}_2$, and 1_4, respectively.

In this model, α, β, and γ are the weight parameters assigned to 1_1, $\bar{1}_3$, $\underline{1}_3$, respectively. x, y, and z are the weight parameters as well. Thus, the higher the parameter, the more the importance of the kind of the links.

On the other hand, a and b are scaling parameters: the number of pages and the number of users, respectively. Then, ρ and σ are the weight parameters assigned to the kinds of elements of the matrix: pages and users, respectively. The higher the parameter, the more the importance of the kind of the elements.

Then, the elements each row is normalized by linear transformation:

$$M(i,j) = \frac{M(i,j)}{\sum_i M(i,j)} \tag{2}$$

4   Ranking and Query Processing

In the preprocessing, the system crawls Web pages using Apache Nutch and indexes them using Apache Solr by search and ranking server shown in Fig. 4. Next, a page-user graph, its adjacency matrix, and probability transition matrix are constructed. Then, PURank scores of pages and users are calculated. In the implementation, the PURank scores can be obtained by calculating the eigenvector of the probability transition matrix corresponding to the largest eigenvalue:

$$\boldsymbol{r} = d\mathbf{M}\boldsymbol{r} + \frac{(1-d)}{n}\boldsymbol{e}, \tag{3}$$

where n is the total number of pages and users in the page-user graph, d is a damping factor given by a user, $\mathbf{M}$ denotes the probability transition matrix, and $\boldsymbol{e}$ denotes an n-dimension column vector with all elements set to 1. All elements of $\boldsymbol{r}$ are initialized to 1 and the calculation terminates when $\boldsymbol{r}$ converges.

Submitting a query, the final scores of pages and users are calculated immediately: 1) the score of page p according to the query q is calculated by

$$s_p(q) = r_p \text{tfidf}_p(q), \tag{4}$$

where r_p is the score of PURank with page p and $\text{tfidf}_p(q)$ is the tf-idf score of query q in page p, 2) the score of user u to the query q is calculated by

$$s_u(q) = r_u \sum_p s_u(q,p), \tag{5}$$

$$s_u(q,p) = s_p(q)f(t_u(p)), \tag{6}$$

where r_u is the score of PURank with user u, $s_u(q,p)$ is the score of query q with user u in page p, and $f(t_u(p))$ is a decay function depending on $t_u(p)$, the browsing time of page p with user u.

After that, the lists of pages and users are displayed as shown in Fig. 1(b) according to $s_p(q)$ and $s_u(q)$ in descending order. The lists of pages of recommendation of users also displayed according to $s_u(q,p)$ in descending order.

5   Evaluation

In this paper, we evaluate the communication function of the developed system[1]. To verify the effectiveness of the system in the condition where we suppose the system works especially well, i.e., information retrieval tasks, collaboration, and usability, the beginning evaluation of the system focuses on the communication function among users who use the system for the first time within one hour.

We have evaluated the calculation time of PURank using real crawling data under the condition displayed in the Table 1. First, for the 13,166 URIs of Web pages of a specific city and 5 users, the PURank matrix (having $13,171 \times 13,171$ elements) are constructed. Nest, PURank scores are calculated using SLPEc library. As a result, it is confirmed that the calculation time is 356.21[sec].

The calculation time is not enough fast for real-time search, but satisfies the requirements for background processing. Then, the search results according to a user's query is determined immediately by the query processing process.

It is also confirmed that the parameter of PURank affects the ranking results [11]. Therefore, in this experiment, the evaluated system provides both the Web page centered communication function and the ranking function based on the number of accessing Web pages, but provides neither user rankings results nor page rankings results based on hyperlinks and social links. This limitation is exposed because the experiment is carried out for users who use the system for the first time. In this case, it is difficult to evaluate the user ranking results because of the short period of the usage of the system. Another practical reason is that the cost of crawling all Web sites is too expensive. Thus, the evaluated system provides full communication function and the simple ranking function

---

[1] This system is now available in [10]. Firefox 13 or newer version is required.

Table 1. Search and ranking server environment

Server	HP ProLiant SL390s G7 2U ×4
CPU	Intel(R) Xeon(R) X5650 (2.67GHz, 12MB Cache, 6.40GT/s QPI) ×2 (number of cores: 6)
Memory	24GB (4GB×6/2R/1333MHz/ DR3 RDIMM)
Storage	500GB, 7,200RPM SATA
Network	NC362i dual-port Gigabit Ethernet
Network Switch	Cisco SG300-20-JP
OS	CentOS release 5.5 (Final)
Libraries	PETSc 3.2-p7, SLEPc 3.2-p5, FBLASLAPACK 3.1.1, OpenMPI 1.4.2
Crawling target	Web pages of a specific city
Matrix size	13,171 × 13,171
Number of URI	13,166
Number of users	5

based on the browsing history. This situation would also occurs in the case the PURank model works well and selects the appropriate users.

In this case, we verify especially the communication function of the system by analyzing the questionnaire results[4], after the realization of pseudo environment in which the system could improve the user experience of searching tasks. Thus, the followings three case studies are analyzed:

1. Task becoming enjoyable or useful by communicating with each other
2. Difficult task obtaining the useful information by nonprofessional user
3. Task becoming efficient by collaboration with each other

In these cases, we suppose that a few users are faced with the similar searching task simultaneously [2]. Since users can communicate with each other over different pages, this situation would appear for not too frequent queries (daily usage search words) such as 10,000 times a month. Then, these users can have a chance to communicate with each other by using our developed communication function.

5.1 Experimental Method

At first, all participants are explained how to use the system for ten minutes. Next, about five participants of each group are asked to solve each task within ten minutes. Almost all participants each group are unfamiliar each other.

In the first experiment for three groups consisting of five experienced users who are university students in computer science including at least one experienced users with searching experience more than ten years, two tasks for Case 1 (e.g., "Plan to go on a sightseeing."), one task for Case 2 (e.g., "Write down a programming code to sum up 1 to n in Haskel."), and one task for Case 3 ("List the special products of each prefecture in Japan.") are presented sequentially.

In the second experiments for other five group consisting of two to five beginners (totally eighteen participants) who are teenager or above 40 years old

---

[2] This situation would often occur after the developed system spreads to enough users.

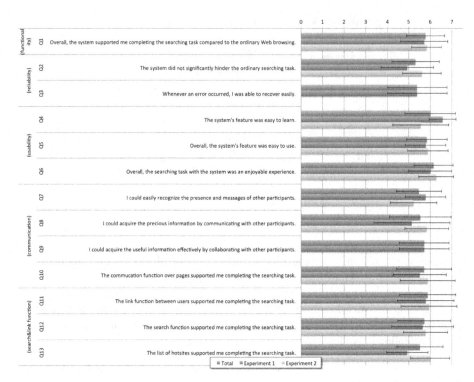

Fig. 6. Questionnaire and evaluation results: 1 (strongly disagree) $\sim$ 4 (neither agree nor disagree) $\sim$ 7 (strongly agree), N=32 (N=14 for experiment 1, N=18 for experiment 2), the mean μ and the standard deviation $\mu - \sigma \sim \mu + \sigma$ are shown

persons including at least one beginner with searching experience less than three years, one tasks for Case 1 and one tasks for Case 2 are presented sequentially.

The communication with each other is allowed only by using the system. The searching method is also restricted by using the toolbar (Fig. 1(b)), by which the system shows only the top 20 results[3] with the hot levels (which represent how may users access the pages according to eleven levels). The system also provides "hotsites" button, which shows the list of pages that have highest hot levels[4].

After the experiments, the questionnaire shown in Fig. 6 are completed by the participants with seven-level Likert scale[6] with the reason. The contents of questionnaire are determined based on the ISO/IEC TR 9126 standard for product quality[5] and the collaboration catalog[8]. In the experiment 2, Q3 (error recovery) is not questioned because of beginner participants.

5.2   Experimental Results

The fourteen effective questionnaire results are obtained for the first experiment; the eighteen effective results for the second experiment as shown in Fig. 6.

---

[3] This condition is set to raise the probability of meeting each other.

[4] This represent how may users access the page according to eleven levels.

Fig. 6 shows the evaluation results with the mean μ and the standard deviation σ for valid completions of questionnaire. These results show the effectiveness of the system because $\mu - \sigma >= 4$ for all questionnaire in total experiment.

In Q2, $\mu - \sigma >= 4$ for Experiment 2; for Experiment 1, $\mu - \sigma < 4$ but $\mu \approx 5$. This results indicate that the avatars sometimes seem to prevent from browsing Web only for experienced users who always watch around all around in browser. In Q4 and Q5, as $\mu - \sigma >= 4$ for both Experiment 1 and Experiment 2, it is confirmed that the way to learn the usage and the way to use the system is easy not only experienced users but also beginners.

As the evaluations of Q6 are positive, it is also confirmed that almost all users felt enjoyable while using our system n the case 1.

In Q8, $\mu - \sigma >= 4$ for Experiment 2; for Experiment 1, $\mu \approx 5$. These results show that many users acquired useful information from others; the remained a few users have different opinions as follows: 1) other participants do not give me any useful information, 2) only top 20 sites are too small.

As for 1), a few users only provided the useful information but could not acquire one by other participants. This means that they supported non-professional users. Thus, we consider that the system works effectively in the case study 2. 2) is caused by the experimental conditions, and this restriction would be removed.

In Q9, the results shows our system supported the collaboration tasks efficiently. Thus, the functionality of the system in the case study 3 is confirmed.

In addition, "hotsites" seem to support the beginner users' Web experience by referencing the other users browsing activities.

As written above, it is confirmed that our system supports information sharing even among unfamiliar users by Web page centered communication.

6 Related Work

Twitter, Facebook, etc. can have similar function by collaborating with specific sites showing tweets or posts there. These might be useful just to know others' comments but difficult to communicate with each other especially among unfamiliar users. Flicker, Instagram, etc., are also becoming popular now, and are called object centered social networks. However, the communication is also restricted to the users having similar interest, e.g., favorite photo. We provide Web page centered communication, i.e., place-centered not object-centered.

ComMentor[9] provides an annotation function to Web contents. Wakurawa[1] display the browsing paths of other users currently accessing the web page. However, the users have to find interesting and popular pages for connecting with other users, and only a communication service is provided.

Various approaches based on user interactions have been developed for ranking Web pages. For example, popular pages can be identified on the basis of the users' browsing histories[2,7]. In BrowseRank[7], to calculate the ranking, the huge log data are basically needed; in our model, as utilizing not only user activities but also page links, large data would not be needed.

7   Conclusion

We have developed a novel communication system that enables users to share information seamlessly in the Web and the real space based on a physical approach: users can perceive other users browsing the Web contents as avatars in Web browsers like faces in real world. The developed system also enables users to search not only desired Web pages but appropriate users; we construct a new ranking model by combining user activity with the hyperlinks. Experimental results showed the system has a potential to efficiently provide a novel physical approached, i.e., Web page centered communication experience to the users.

Acknowledgments. This work was partially supported by MIC SCOPE and JSPS KAKENHI (24780248).

References

1. Akatuka, D.: "Wakurawa" Communication Media Focusing on Weak Ties. In: Proc. the 14th Workshop on Interactive Systems and Software, pp. 139–140 (2006)
2. Balmin, A., Hristidis, V., Papakonstantinou, Y.: ObjectRank: Authority-Based Keyword Search in Databases. In: Proc. VLDB 2004, pp. 564–575 (2004)
3. Brin, S., Page, L.: The Anatomy of a Large-scale Hypertexual Web Search Engine. Computer Networks 30(1-7), 107–117 (1998)
4. Heinrich, M., Lehmann, F., Springer, T., Gaedke, M.: Exploiting Single-User Web Applications for Shared Editing - A Generic Transformation Approach. In: Proc. WWW 2012, pp. 517–526 (2012)
5. ISO/IEC: Software engineering - Product quality, ISO/IEC 9126-1 (2001)
6. Likert, R.: A technique for the measurement of attitudes. Archives of Psychology 22(140), 5–55 (1932)
7. Liu, Y., Gao, B., Liu, T.-Y., Zhang, Y., Ma, Z., He, S., Li, H.: BrowseRank: Letting Web Users Vote for Page Importance. In: Proc. SIGIR 2008, pp. 451–458 (2008)
8. Pinelle, D., Gutwin, C., Greenberg, S.: Task analysis for groupware usability evaluation: Modeling shared-workspace tasks with the mechanics of collaboration. ACM Trans. Comput.-Hum. Interact. 10(4), 281–311 (2003)
9. Röscheisen, M., Winograd, T., Paepcke, A.: Content Ratings and Other Third-Party Value-Added Information Defining an Enabling Platform. D-Lib Magazine (1995)
10. Seamless Communication and Search System with Ranking, http://klab.kyoto-su.ac.jp/~mito/index.html
11. Shiraishi, Y., Zhang, J., Kawai, Y., Akiyama, T.: Proposal of Combination System of Page-centric Communication and Search. In: Proc. 1st Int. Workshop on Online Social Systems (2012)
12. Shiraishi, Y., Zhang, J., Kawai, Y., Akiyama, T.: Simultaneous Realization of Page-centric Communication and Search. In: Proc. CIKM 2012 (Demo Paper), pp. 2719–2722 (2012)

TaxiHailer: A Situation-Specific Taxi Pick-Up Points Recommendation System

Leyi Song, Chengyu Wang, Xiaoyi Duan, Bing Xiao, Xiao Liu,
Rong Zhang, Xiaofeng He, and Xueqing Gong

Institute for Data Science and Engineering, Software Engineering Institute,
East China Normal University, Shanghai, China
{songleyi,chengyuwang,duanxiaoyi,bingxiao,xiaoliu}@ecnu.cn,
{rzhang,xfhe,xqgong}@sei.ecnu.edu.cn

Abstract. This demonstration presents TaxiHailer, a situation-specific recommendation system for passengers who are eager to find a taxi. Given a query with departure point, destination and time, it recommends pick-up points within a specified distance and ranked by potential waiting time. Unlike existing works, we consider three sets of features to build regression models, as well as Poisson process models for road segment clusters. We evaluate and choose the most proper models for each cluster under different situations. Also, TaxiHailer gives destination-aware recommendations for pick-up points with driving directions. We evaluate our recommendation results based on real GPS datasets.

Keywords: location-based service, taxi, recommendation system.

1 Introduction

The applications of location-based services (LBS) have sprung up in recent years. Based on historical trajectory data, a number of taxi-related recommendation systems and research issues have been proposed. Y. Ge et al.[1] present a mobile recommender system, which has the functionality to recommend a sequence of pick-up points for taxi drivers, in order to maximize the probability of business success. The non-homogeneous Poisson process (NHPP) model has been employed by X. Zheng[2] to describe the behavior of vacant taxis and then to estimate the waiting time of passengers. T-Finder[3] adopts the NHPP model to make road segment recommendations for passengers. It also recommends the top-k parking places and routes to these places for taxi drivers.

It is valuable to make recommendations for passengers using crowd sensing with GPS data of taxi. However, making an accurate recommendation is still a challenge due to many factors, such as the noisy GPS data, the fluctuated weather, the complexity associated with real-world traffic patterns, etc. For example, the NHPP model generally works well for road segments recommendations, but it is not suitable in all situations, such as unpopular roads[2]. Thus, it can achieve higher accuracy through adopting different models for road segments according to the specific situation and adding more features to train models.

S.S. Bhowmick et al. (Eds.): DASFAA 2014, Part II, LNCS 8422, pp. 523–526, 2014.

We develop a situation-specific pick-up points recommendation system for passengers, which takes many factors into account, such as departure point, destination, time, weather and traffic patterns. TaxiHailer distinguishes itself from existing taxi-hailing and recommendation systems in four major ways. First, it focuses on the efficiency of offline processing on large trajectory datasets by taking advantage of the MapReduce framework. Second, we use three sets of features that can influence the waiting time: (i) trajectory-related features, which are statistics calculated based on historical data for road segments, e.g. traffic volume, pick-up rate, etc, (ii) road segment features that describe road properties such as lanes, direction and so on and (iii) additional features including weather conditions. Third, We cluster road segments into groups. For each group, we build regression models and Poisson process model for prediction and deploy the most efficient ones for different situations. Finally, the route to the destination is used to prune the ranked pick-up point candidates, since it is unreasonable to hail a taxi driving in the opposite direction especially on main roads.

2   System Overview

Given a query with specified departure point, destination, time and weather, TaxiHailer gives a ranked list of pick-up points within a specified distance. The points are ranked by the potential waiting time with walking penalty and have been pruned by the direction of route to destination. The models used for prediction are built on historical trajectories collected from taxi GPS devices.

As Figure 1(a) illustrates, TaxiHailer consists of two major parts: (1) offline processing and (2) online computing. The offline modules re-build the waiting time prediction models using recent data periodically. Firstly, in the preprocessing module, we filter the raw GPS data with noises and errors and persist

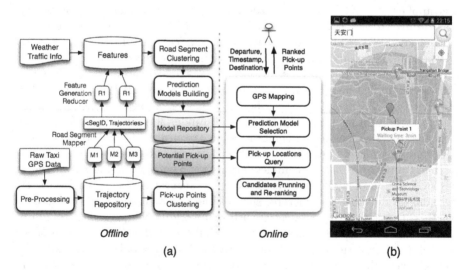

Fig. 1. (a)System Architecture (b)TaxiHailer Screenshot

trajectories into repository. Each taxi trajectory is a collection of GPS points with attributes such as taxi ID, timestamp, taxi state, speed and so on. In the next step, a MapReduce job starts to calculate the statistical features of road segments. We collect additional features, such as weather conditions through web services. Afterwards, road segments are clustered into hundreds of groups to reflect different traffic situations. Besides, to generate potential pick-up points, we clustered historical pick-up points on segments with frequency and distance rules. Then, one point in each cluster is chosen as potential pick-up point for recommendation. Finally, we build regression models and Poisson process models for each road segment cluster under different situations. We choose the best models for each cluster by evaluating using sampled test set.

In the online part, TaxiHailer processes queries and recommends pick-up points within the specified distance. A brief query processing flow is shown as follows. Firstly, the origin of user query is mapped to the road network. Then, we can select the prediction model according to the query context. Next, a collection of result candidates can be fetched by distance limitation. Afterwards, we use the route towards the destination to prune the candidate set, and finally, we re-rank the pruned candidates and return them with predicted waiting time.

3   Challenges and Approaches

In this section, we discuss our implementation details coping with some major challenges in TaxiHailer.

- **Error and Noise Filtering**
  Raw GPS data are never perfectly accurate and contain noise and outlier data points due to sensor noise and other factors. In TaxiHailer, we correct part of inaccurate or noise data from raw GPS data by path inference. For outliers and recording errors, we remove them by setting rules on GPS record attributes. For example, a GPS point should be removed if the distance to the previous point is longer than the possible driving distance in the time interval.
- **Efficient Offline Trajectory Processing**
  Tens of thousands of taxis drive on the roads and generate billions of GPS points per month in big cities, e.g. there are more than 60,000 taxis in Beijing. Huge volume of data accumulated in several months makes accurate predictions possible. Due to the large scale of taxi GPS datasets, we need an efficient mechanism to perform calculation. We implement Map-Reduce flows for the generation of trajectory statistical features. Briefly, the Map tasks map the taxi points to road segments in parallel. The Reduce tasks calculate all statistics we need for the road segments.
- **Situation-Specific Model Building**
  The waiting time for the same road segment may differ in different situations, such as weather, time period. Also, social activities can cause traffic jams, leading to a longer waiting time there. Thus, before model building, we separate anomalous situations from the statistics. Besides, we divide time

into hours and weekdays/weekends/holidays. In practice, we first train set of models for each road segment group. Then, the most proper waiting time prediction model for different situations can be selected by periodically evaluation. We generate sufficient test sets which are sampled on the roads and calculated from the historical trajectory data. In this way, TaxiHailer can yield different recommendations in different situations.

- **Destination-Aware Pick-up Points Recommendation**
  The driving direction is rarely considered in previous taxi recommendation services. However, the volumes of taxis driving on opposite side of the same road are quite different in most cases. It is undesirable to get recommended taxis that drive in the other side. As we generate potential pick-up points including their directions, destination awareness is included in our query algorithm. When the candidate pick-up points are retrieved, we firstly prune them by the direction of planned route to destination. The items in final list are ranked by waiting time plus their walking penalty.

With all the approaches mentioned above, TaxiHailer can effectively select pick-up points for passengers according to destinations and specific queries.

4   Demonstration Scenario

We will demonstrate TaxiHailer and showcase its pick-up points recommendation service. The demonstration is based on taxi GPS datasets in Beijing and Shanghai. More specifically, our demonstration includes two parts shown below:

- **TaxiHailer backend.** We will show how GPS data are processed and models are built in TaxiHailer. Trajectories, road segments, potential pick-up points will be visualized on the map. For a specific situation, different categories of models chosen for road segments will be labelled in colors. We will also report our evaluation results of prediction models in charts.
- **TaxiHailer application.** We will show a mobile application of TaxiHailer as the screenshot is shown in Figure 1(b). Users can generate two kinds of queries: (i) simulated queries based on historical data and (ii) real-time queries. The application will mark the recommended pick-up points on the map. If the destination is specified, an optimized ranked list will be returned.

Acknowledgements. This work is partially supported by Natural Science Foundation of China (Grant No. 61103039 and 61232002) and the Key Lab Foundation of Wuhan University (Grant No. SKLSE2012-09-16).

References

1. Ge, Y., Xiong, H., Tuzhilin, A., Xiao, K., Gruteser, M., Pazzani, M.J.: An energy-efficient mobile recommender system. In: KDD, pp. 899–908 (2010)
2. Zheng, X., Liang, X., Xu, K.: Where to wait for a taxi? In: UrbComp, pp. 149–156 (2012)
3. Yuan, N.J., Zheng, Y., Zhang, L., Xie, X.: T-finder: A recommender system for finding passengers and vacant taxis. IEEE Trans. Knowl. Data Eng. 25(10), 2390–2403 (2013)

Camel: A Journey Group T-Pattern Mining System Based on Instagram Trajectory Data

Yaxin Yu[1], Xudong Huang[2], Xinhua Zhu[3], and Guoren Wang[1]

[1] College of Information Science and Engineering, Northeastern University, China
{Yuyx,Wanggr}@mail.neu.edu.cn
[2] College of Software, Northeastern University, China
Xudong.huang664@gmail.com
[3] QCIS, University of Technology, Sydney, Australia
Xinhua.Zhu@uts.edu.au

Abstract. A Journey Group T-Pattern (JG T-Pattern) is a special kind of T-Pattern(Trajectory Pattern) in which a large number of users walked through a common trajectory; also, it allows users depart from the trajectory for several times. Travel route is an instance of Journey Group and hot travel route can then be mined under the help of Camel. Instagram is a popular photo-sharing smart phone application based on social network, it is widely used among tourists to record their journey. In this paper, we focus on data generated by Instagram to discover the JG T-pattern of travel routes. Previous researches on T-pattern mining focus on GPS-based data, which is different from the UGC-based(User Generated Content based) data. Data of the former is dense because it is often generated automatically in a certain pace, while the latter is sparse because it is UGC-based, which means the data is generated by the uploading of users. Therefore, a novel approach, called Journey Group T-pattern Mining strategy, is proposed to deal with the trajectory mining on sparse location data. The demo shows that Camel is an efficient and effective system to discover Journey Groups.

1 Introduction

This demo presents Camel: a Journey Group T-Pattern mining system built with Instagram data and Google Map API. Camel distinguishes itself from existing trajectory mining systems as it mines sparse location data generated by users, *i.e.*, UGC-based data, not dense data, *i.e.*, GPS-based data, generated automatically in a certain pace by GPS devices. Existing trajectory mining system almost includes temporal information with dense and regular time span and snapshot can then be applied. However, location data in Camel system is generated by the uploading of photos taken by Instagram users, which results in sparse and irregular time span. For this reason, existing T-Pattern mining methods under GPS-based data [1–5] cannot be used for UGC-based data directly. For example, a travel agency may place more emphasis on caring for common trajectories among travelling routes of all tourist to discover popular tourist areas. Here, the

S.S. Bhowmick et al. (Eds.): DASFAA 2014, Part II, LNCS 8422, pp. 527–530, 2014.

common trajectory, generated by linking clusters in which travellers walked together with departing from for appointed several times at most, is called Journey
Group T-Pattern in our Camel system, abbreviated to be JG T-Pattern.

In fact, JG T-Pattern is derived from the concept "Swarm" proposed in paper
[6]. The difference between "Swarm" and "Journey Group" is that the former
has temporal information while the latter uses the index number instead. Some
similar concepts in trajectory mining such as "Flock", "Convoy" and "Swarm"
are compared by researchers [7], and further, a new concept "Gathering" is put
forward. A key feature of these existing related concepts mentioned above is
that all of them need temporal information, which results in the availability
of snapshot in trajectory mining systems involving these concepts. In the JG
T-Pattern mining problem, however, the time span between adjacent location
points is irregular, in other words, the time span is not fixed, and the upper limit
is not known either. Due to this reason, the snapshot is unavailable. As a result,
a novel algorithm is proposed in this paper to solve the JG T-Pattern mining
problem, and implemented in the prototype Camel, which is based on Instagram
trajectory data and Google Map API.

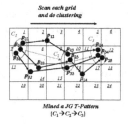

Fig. 1. Mining of JG
T-Patterns

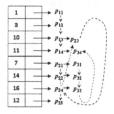

Fig. 2. Hash Table

To illustrate JG T-Pattern concept further, we firstly give two definitions in
the following. Here, let n_p represent the number of participators in a cluster, k_p
denote the lowest number of cluster one participator must appears in and m_p is
the number of points belonging to common participator in two linkable clusters.

Definition 1. (*Participator*) *A participator means the owner of a location
point, and the owner is a user generating the trajectory where the point in.*

Definition 2. (*Journey Group*) *The concept Journey Group derives from
Swarm [6]. In a Journey Group, each cluster must satisfy that $n_p \geq m_p$ and
each participator must appear in at least k_p clusters.*

Based on the definitions mentioned above, the basic meaning of JG T-Pattern
is shown in Fig.1. There are three directional trajectories $(P_{11} \rightarrow P_{12} \rightarrow P_{13} \rightarrow
P_{14})$, $(P_{21} \rightarrow P_{22} \rightarrow P_{23} \rightarrow P_{24} \rightarrow P_{25})$, $(P_{31} \rightarrow P_{32} \rightarrow P_{33} \rightarrow P_{34})$ where the
direction depends on the uploading time sequence, respectively belongs to users

O_1, O_2, and O_3. JG T-Pattern mining is to find a T-pattern in which the three users all walked through a certain number of location points in some common areas. As shown in Fig.1, there are three clusters denoted as dotted circle and a Journey Group denoted as dotted line with arrows. P_{12}, P_{22}, P_{32} and P_{24} are locations departed from their trunk trajectory. Each point can be denoted as a complex binary tuple, i.e., P_{ki}={(latitude, longitude), i}, where P_{ki} means the ith location point of kth trajectory, and parameter i means the index number of a point. As shown in Fig.1, let k_p=1 and m_p=3. Then, C_1, C_2, C_3 are clusters due to $n_p \geq m_p$, and C_1 is linked to C_2 because the number of points belonging to common participator in C_1 and C_2 is equal to m_p, so as C_2 and C_3, and the cluster sequence $C_1 \rightarrow C_2 \rightarrow C_3$ forms a Journey Group.

2   JG T-Pattern Mining Algorithm

Based on T-Patterns data in Fig.1, a hash table with the form ⟨key,value⟩ is generated, which is shown in Fig.2. The key is grid index number coming from divided map at an appointed size and the value is a pointer pointing to a list containing the trajectory points located in same grid. Every trajectory path belonging to an user is connected by dotted line. JG T-Pattern Mining Algorithm is implemented as follows.

1) Set Thresholds. There are 2 parameters need to be set. One is the lowest number of participators appearing in both two adjacent clusters, and the other is the size of radius when do density-based clustering. The former is represented by m_p and the latter is denoted as *Eps*.

2) Cluster Points. Aiming at hash table, the algorithm scans it from up to down, and finds an item whose pointer field is not empty. Once it finds a grid with location points, it does density-based clustering in that gird. For those satisfying the condition to be a cluster, i.e., the number of points in a grid is over or equal to the threshold, we add it to the set Mined Clusters, if the set is not empty, we must compare the current cluster with existing clusters in the set.

3) Generate JG T-Patterns. When at least m_p participators appearing in both the current cluster and other clusters in the Mined Clusters, we link the two clusters, that is to say, at least m_p users walked through from the current cluster to that cluster, which can form a segment of a Journey Group T-Pattern.

3   Demonstration

We collected image data from Instagram users of five cities (Sydney, Melbourne, Brisbane, Perth and Darwin) in Australia. For the purpose of displaying trajectories clearly in the demo GUI, we just only used a part of the whole data. We load 2000 user trajectories into the Camel to execute our JG T-Pattern mining algorithm. In fact, all data can be processed in Camel. The running results are displayed in Google Map, which is shown in Fig.3. Circular region generated by Google Map API represents marker cluster where each marker marks a location

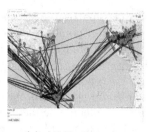
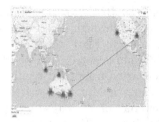

(a) All Trajectories

(b) Mined Clusters and JG
T-Patterns

Fig. 3. Demonstration of Camel System

point. The region will be scaled up as zooming in or out the map. Fig.3(a) shows all users trajectories, where the labels shaped like water drops represent locations where user taking photos and can be connected one by one by directional lines when these labels belong to the same user. Fig.3(b) shows the mined clusters and JG T-Patterns, here, two Journey Groups, denoted as red directional line, are discovered. Here, parameter *Eps* can be changed from 0.5 to 0.3, and m_p (denoted by *MinPts* in Camel GUI) from 100 to 200. From these figures mentioned above, we can conclude that lots of Australian travellers have been to the United States.

References

1. Geng, X., Arimura, H., Uno, T.: Pattern Mining from Trajectory GPS Data. In: Proceedings of International Conference on Advanced Applied Informatics, pp. 60–65 (2012)
2. Giannotti, F., Nanni, M., Pedreschi, D., et al.: Unveiling the complexity of human mobility by querying and mining massive trajectory data, vol. 3, pp. 377–383. Springer (2011)
3. Zheng, Y., Zhang, L., Xie, X., Ma, W.: Mining Interesting Locations and Travel Sequences from GPS Trajectories. In: Proceedings of the 18th International Conference on World Wide Web, pp. 791–800 (2009)
4. Ma, S., Zheng, Y., Wolfson, O.: T-Share: A Large-Scale Dynamic Taxi Ridesharing Service. In: Proceedings of the 29th IEEE International Conference on Data Engineering, pp. 410–421 (2013)
5. Liu, S., Wang, S., Jayarajah, K., Misra, A., Krishnan, R.: TODMIS: Mining Communities from Trajectories. In: Proceedings of the 22nd International Conference on Information and Knowledge Management, pp. 2109–2118 (2013)
6. Li, Z., Ding, B., Han, J., Kays, R.: Swarm: Mining Relaxed Temporal Moving Object Clusters. Proceedings of the VLDB Endowment 3(1), 723–734 (2010)
7. Zheng, K., Zheng, Y., Yuan, N., Shang, S.: On Discovery of Gathering Patterns from Trjaectories. In: Proceedings of the 29th IEEE International Conference on Data Engineering, pp. 242–253 (2013)

Harbinger:
An Analyzing and Predicting System for Online Social Network Users' Behavior

Rui Guo, Hongzhi Wang, Lucheng Zhong, Jianzhong Li, and Hong Gao

Harbin Institute of Technology
Harbin, Heilongjiang, China
{ruiguo,wangzh,zlc,lijzh,honggao}@hit.edu.cn

Abstract. Online Social Network (OSN) is one of the hottest innovations in the past years. For OSN, users' behavior is one of the important factors to study. This demonstration proposal presents *Harbinger*, an analyzing and predicting system for OSN users' behavior. In *Harbinger*, we focus on tweets' timestamps (when users post or share messages), visualize users' post behavior as well as message retweet number and build adjustable models to predict users' behavior. Predictions of users' behavior can be performed with the established behavior models and the results can be applied to many applications such as tweet crawlers and advertisements.

Keywords: Social Network, User Behavior, Message Timestamp.

1 Introduction

Online social networks have exploded incredibly in the past years. For instance, Twitter has 200 million active users who post an average of 400 million tweets altogether every day [4]. Since the large group of users make OSNs valuable for both commercial and academical applications, the understanding of users' behavior could help these applications to improve efficiency and effectiveness.

The understanding of users' behavior brings challenges. The crucial one is that they keep on challenging computing resources. Another difficulty is that users are influenced by many factors, most of which are invisible through OSNs.

Existing works study OSN users' behavior in several different ways. [5] characterizes behavior by clickstream data. [7] downloaded user profile pages, and modeled users' online time with Weibull distributions.

To analyze and predict users' behavior, we present *Harbinger* system. *Harbinger* has two major functions: users' post behavior analyzing and single message retweet number analyzing. Statistics methods in [6] are applied to avoid the effect of invisible factors. We observe both a group of users and single tweets, collect message timestamps and message retweet numbers, visualize users' post behavior and the variation of message retweet number, and describe them by Gaussian Mixture Model as well as Logarithm Model. Unlike previous works,

S.S. Bhowmick et al. (Eds.): DASFAA 2014, Part II, LNCS 8422, pp. 531–534, 2014.

Harbinger analyzes users' post behavior through tweets' timestamps rather than users' clickstream, online time or friendship.

2 System Overview

Our System contains two major functions: the users' post behavior analyzing function and the single message retweet number analyzing function.

For the OSN users' post behavior, the user of *Harbinger* is expected to choose an analyzing function, a target OSN user and an analyzing pattern (daily, weekly or monthly pattern). Then corresponding data are selected and statistics are preformed. We set the time span of the statistics to be one hour (from 00:00 to 24:00) to the daily pattern, and one day from the beginning to the end of the week or month to the weekly or monthly pattern. Finally, the figure of message number and time, and the analysis results such as Figure 2(a) will be plotted.

For the retweet number function, a user chooses a tweet function and target tweet rather. The system analyzes the data and plots the relationship between time and the retweets of the selected tweet such as Figure 2(b).

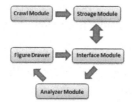

Fig. 1. Modules of *Harbinger*

As shown in Figure 1, *Harbinger* has five major modules: crawl, storage, interface, analyzer and figure drawer modules.

In the crawl module, we develop an OSN crawler. The crawler collects the message information, including content and timestamp, and then stores it in the storage module, where all data are stored in a database. The User Interface (UI) of the Interface Module connects visitors to *Harbinger* and other modules. The user of *Harbinger* can select target OSN user or tweet, and the analyzing pattern in the UI module. The selection is sent to the analyzer. After statistics, analysis and calculation in the analyzer are based on the models in Section 3, and the results are sent to the figure drawer, which draws figures in the UI according to the analysis results.

3 Models and Algorithms

In this section, we describe the models and algorithms used in our system. They are the major parts of the analyzer. Based on the technology in [6], we develop the

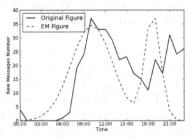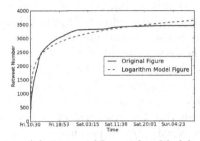

(a) Figure of Gaussian Mixture Model   (b) Figure of Logarithm Model

Fig. 2. Figure of retweet number and time

Gaussian Mixture Model (GMM) to describe OSN user behavior and Logarithm Model to illustrate the relationship between retweet number and time. Figure 2 shows the results of the GMM and Logarithm Model.

Gaussian Mixture Model

In the daily pattern of the users' post behavior, we find that the relationship between new messages' number and the time in a day follows the mixture of two Gaussian Distributions (or Normal Distributions). OSN users mostly work during the day, while rest at the noon and dusk. Thus there are two peaks of fresh OSN messages. The curve around each peak is similar to a Gaussian Distribution. As the result, the figure can be treated as a mixture of two Gaussian Distributions.

Thus we develop Gaussian Mixture Model [1] (GMM) to compute unknown parameters of the figure. Assume the daily time is t, the number of new messages during t is $f(t)$, the two Gaussian Distributions are $N_1(\mu_1, \sigma_1^2)$ and $N_2(\mu_2, \sigma_2^2)$, and there is

$$f(t) = \frac{1}{\sqrt{2\pi}\sigma_1} e^{-\frac{(x-\mu_1)^2}{2\sigma_1^2}} + \frac{1}{\sqrt{2\pi}\sigma_2} e^{-\frac{(x-\mu_2)^2}{2\sigma_2^2}}$$

To figure out the exact parameters $(\mu_1, \sigma_1, \mu_2, \sigma_2)$ in GMM, we apply Expectation-Maximization (EM) algorithm [2], which is the computing process of GMM. Figure 2(a) shows the results of GMM. In Figure 2(a), the solid line means the original figure of time and retweet number (the user post frequency in a day), and the dotted line means the results of EM algorithm. The results show that the post frequency is indeed similar to the sum of two Gaussian distributions.

Logarithm Model

We find that the relationship between retweet number and posted time of a specific message follows the logarithm function. After stretching and shifting, a basic logarithm function can describe retweet number properly. In this curve, x-axis is the posted time (e.g. how long the tweet is posted) of the message and y-axis is the retweet number.

The retweet number grows very fast after the tweet is posted, and with x increasing, the growth becomes more and more unchanged. Thus we compute the retweet number RN in the posted time t (e.g. how long the message is posted) as $RN = k_1 log_{base}(k_0 x + k_2) + k_3$, where $base$ is the base of logarithm function representing the steepness of the curve. k_0 is the x-axis stretch parameter, k_1 is the y-axis stretch parameter, k_2 is the x-axis shift parameter, and k_3 is the y-axis shift parameter.

To compute the exact parameters of Logarithm Model, we apply Least Squares Algorithm [3]. The basic idea of Least Squares algorithm is to approximate the model by a linear function and to refine the parameters with iterations.

Figure 2(b) describes the result of Logarithm Model. The solid line means the original figure of time and retweet number growth (how the retweet number changes over time after being posted), and the dotted line means the result of Logarithm Model. The result shows that the original figure is indeed similar to a logarithm function figure.

4   Demonstration Scenario

Harbinger is well encapsulated with a friendly interface. Though the system has a large database and specialized analyzer module, what the user faces is only a simple UI. The interface is shown in the video (http://www.youtube.com/watch?v=xgXcsbNYqqQ&feature=youtu.be).

Acknowledgments. This paper was partially supported by NGFR 973 grant 2012CB316200, NSFC grant 61003046, 61111130189 and NGFR 863 grant 2012AA011004. the Fundamental Research Funds for the Central Universities(No. HIT. NSRIF. 2013064).

References

1. http://en.wikipedia.org/wiki/Mixture_model
2. http://en.wikipedia.org/wiki/Expectation-Maximization
3. http://en.wikipedia.org/wiki/Least_squares
4. (2013), https://business.twitter.com/whos-twitter
5. Benevenuto, F., Rodrigues, T., Cha, M., Almeida, V.: Characterizing user behavior in online social networks. In: Proceedings of the 9th ACM SIGCOMM Conference on Internet Measurement Conference, pp. 49–62. ACM (2009)
6. Guo, R., Wang, H., Li, K., Li, J., Gao, H.: Cuvim: Extracting fresh information from social network. In: Wang, J., Xiong, H., Ishikawa, Y., Xu, J., Zhou, J. (eds.) WAIM 2013. LNCS, vol. 7923, pp. 351–362. Springer, Heidelberg (2013)
7. Gyarmati, L., Trinh, T.A.: Measuring user behavior in online social networks. IEEE Network 24(5), 26–31 (2010)

Cloud-Scale Transaction Processing with ParaDB System: A Demonstration

Xiaoyan Guo, Yu Cao, Baoyao Zhou, Dong Xiang, and Liyuan Zhao

EMC Labs China
{xiaoyan.guo,yu.cao,baoyao.zhou,dong.xiang,liyuan.zhao}@emc.com

Abstract. Scalability, flexibility, fault-tolerance and self-manageability are desirable features for data management in the cloud. This paper demonstrates ParaDB, a cloud-scale parallel relational database system optimized for intensive transaction processing. ParaDB satisfies the aforementioned four features without sacrificing the ACID transactional requirements. ParaDB is designed to break the petabyte or exabyte barrier and scale out to many thousands of servers while providing transactional support with strong consistency.

1 Introduction

Relational database management systems (RDBMS) have been extremely successful in traditional enterprise environments for more than three decades. However, RDBMSs are no longer competitive choices for cloud-scale applications, since data management in the cloud desires scalability, flexibility, fault-tolerance and self-manageability, which cannot be completely well supported by existing RDBMSs.

Recently, numerous NoSQL systems have been proposed for scalable data management in the cloud, such as Amazon Dynamo, Google BigTable, Yahoo PNUTS, Facebook Cassandra. These systems do not (fully) support SQL and usually build atop key-value data storage, where data are partitioned, replicated and then distributed over multiple nodes to achieve high performance, scalability and availability. However, all these systems guarantee only eventual consistency or other weak consistency variants.

Although a small number of large-scale Web applications can tolerate weak consistency, almost all enterprise applications demand ACID-compliant transaction processing, so as to guarantee the application correctness and simplify the application logic design. As such, Google developed Megastore and Percolator on top of BigTable for more general support of transaction processing. Other recent research works towards the similar direction include ElasTraS [1] and CloudTPS [2]. However, these systems still cannot perfectly support cloud-scale enterprise transactional applications.

We thereby present ParaDB, a scalable, flexible, fault-tolerant and self-manageable parallel database system supporting ACID transaction consistency. Unlike the aforementioned systems, ParaDB facilitates the data management in the cloud by scaling out a traditional centralized RDBMS. Moreover, since current NoSQL databases usually expose a subset of functionalities of RDBMS, a straightforward encapsulation of ParaDB can easily enable existing NoSQL applications to run with ParaDB. ParaDB is designed to break the petabyte or exabyte barrier and scale out to many thousands of servers while providing transactional support with strong consistency.

S.S. Bhowmick et al. (Eds.): DASFAA 2014, Part II, LNCS 8422, pp. 535–538, 2014.

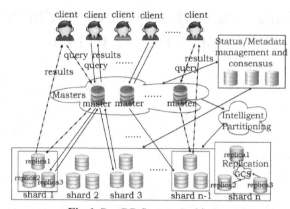

Fig. 1. ParaDB System Architecture

The major contributions of ParaDB are summarized as follows:

(1) ParaDB employs an intelligent iterative hyper-graph based database partitioning engine, which minimizes the number of distributed transactions, the major performance bottleneck. ParaDB also conducts intelligent data re-partitioning and live database migration without losing load balance and transparency.

(2) ParaDB implements an efficient eager active-active replication mechanism, which ensures strong ACID consistency in the multi-master configuration, and improves system availability and fault-tolerance at the cost of only small performance reduction.

(3) ParaDB deploys multiple identical masters to avoid the single node bottleneck and improve system scalability, with the help of an efficient and elastic multi-master metadata synchronization mechanism.

2 System Overview

Figure 1 depicts the high-level system architecture of ParaDB. There are two types of system nodes: *shard server* node and *master server* node. The shard server handles table storage and query processing. The tables in the database are partitioned, replicated and then stored at different shard server nodes. The master server serves as a query router and a distributed transaction coordinator. It only stores the metadata and system status information. There could be multiple master servers, which can concurrently accept and process clients' query requests. ParaDB supports concurrency control, transaction recovery and consistency management. ParaDB utilizes two-phase commit protocol to guarantee the ACID properties.

As shown in Figure 1, ParaDB consists of three major functional components,i.e. *intelligent data partitioning*, *eager and active-active data replication* and *metadata management and synchronization*, which are described as follows. Due to the space limitation, we ignore some technical details and the comprehensive performance study, both of which can be found in the technical report [5].

Intelligent Partitioning. ParaDB considers both the database schema and workloads to derive an intelligent scheme for data partitioning and query routing, so as to minimize the number of distributed transactions and thus optimize the system performance.

In addition, in case of dramatic system configuration changes, database upgrades or workload shifts, ParaDB also applies intelligent database re-partitioning and live data migration without losing load balance and transparency. We realize the above functionalities by constructing an intelligent iterative hyper-graph based database partitioning engine, which first analyzes the database and workload and constructs a weighted hyper-graph, then conducts iterative hyper-graph partitioning to obtain the optimal partitioning scheme. More details about the partitioning engine can be found in the paper [3].

Eager and Active-Active Replication. In ParaDB, data are replicated within the boundaries of *shard groups*. Each shard group consists of three shard server nodes, which store identical data. Eager replication means that modifications by a transaction to one replica of a data item will be applied to other replicas before the transaction commits, which guarantees strong consistency at all times. Active-active replication means that every data replica can accept read/write requests of transactions, as well as coordinate the data synchronization with other replicas. We implement an active-active eager replication protocol based on Postgres-R(SI) [4] with the help of an open-source group communication system called Spread[1], which guarantees that messages will be sent to all members of a group following a strict and user-specified order. With the atomicity and total order guaranteed by Spread, our replication protocol can ensure the transactions to be executed (or the changes to be applied) at different nodes in the same order, therefore enforcing the transaction consistency across different shard server nodes.

Metadata Management and Synchronization. ParaDB deploys multiple master servers to avoid single node bottleneck. Efficient metadata synchronization mechanisms are required among these masters. The metadata management component in ParaDB is responsible for synchronizing the system metadata and the snapshots (e.g. name and status) for both master nodes and shard nodes. All these distributed coordination functions are realized with the aid of Zookeeper[2], an open-source coordination service for distributed applications, which simplifies the consensus protocol design, and are elastic when facing new or unpredictable synchronization and management expectations.

3   Demonstration Scenarios

Our demonstration will illustrate both the general properties of ParaDB as a parallel database system, as well as its unique and novel features like intelligent partitioning and eager replication. For the purpose of system demonstration, we will install and run ParaDB on three physical machines, which are virtualized into two master server nodes and nine shard server nodes, managing sample databases generated by the TPC-C[3] benchmark. In this demonstration, we design four demonstration scenarios.

System Introduction. In this scenario, we will first introduce the motivation of ParaDB and explain its overall system design and novel features. Then we will dive deeply into the technical details of the data partitioning, active-active eager replication and metadata synchronization in ParaDB.

---

[1] http://www.spread.org/
[2] http://zookeeper.apache.org/
[3] http://www.tpc.org/tpcc/

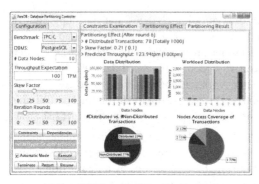

(a) ParaDB Data Partitioning Controller   (b) ParaDB Log Viewer

Fig. 2. ParaDB Demonstration System

Intelligent Partitioning. In this scenario, we will demonstrate how the data partitioning controller of ParaDB (shown in Figure 2a) semi-automatically and intelligently partitions and replicates database tables. We will show the audience the visualized partitioning results for the TPC-C database. Finally, we will illustrate how ParaDB conducts data re-partitioning and live migration by removing one shard server node.

Eager and Active-Active Replication. In this scenario, we will demonstrate how the eager active-active replication protocol works, by analyzing the query execution logs output by the log viewer of ParaDB (shown in Figure 2b). We will verify the correctness of our protocol by conducting a set of conflicting transactions to be executed concurrently at two shard server nodes. We will also demonstrate that shard servers of a shard group can simultaneously accept and process read/write requests of transactions.

Performance Study. In this scenario, we will demonstrate the performance and scalability of ParaDB, with the log viewer in Figure 2b. We will first conduct experiments to study the performance impact of our replication protocol. After that, we will sequentially run the TPC-C benchmark queries against three TPC-C databases containing 1000 warehouses, 2000 warehouses and 3000 warehouses respectively, in order to prove that ParaDB can scale linearly along with the database size.

References

1. Das, S., Agrawal, D., El Abbadi, A.: ElasTraS: An Elastic Transactional Data Store in the Cloud. In: USENIX HotCloud (2009)
2. Wei, Z., Pierre, G., Chi, C.-H.: CloudTPS: Scalable Transactions for Web Applications in the Cloud. In: IEEE Transactions on Services Computing (2011)
3. Cao, Y., Guo, X., Zhou, B., Todd, S.: HOPE: Iterative and Interactive Database Partitioning for OLTP Workloads. In: ICDE (2014)
4. Wu, S., Bettina, K.: Postgres-R(SI): Combining Replica Control with Concurrency Control Based on Snapshot Isolation. In: ICDE (2005)
5. Guo, X., Cao, Y., Zhou, B., Xiang, D., Zhao, L.: ParaDB: A Cloud-Scale Parallel Database System for Intensive Transaction Processing. Technical Report (2013), https://tinyurl.com/paraDB-techreport

BSMA-GEN: A Parallel Synthetic Data Generator for Social Media Timeline Structures

Chengcheng Yu, Fan Xia, Qunyan Zhang, Haixin Ma,
Weining Qian, Minqi Zhou, Cheqing Jin, and Aoying Zhou

Center for Cloud Computing and Big Data, Software Engineering Institute,
East China Normal University, Shanghai, China
{52111500011,52101500012,51121500043,51111500010}@ecnu.cn,
{wnqian,mqzhou,cqjin,ayzhou}@sei.ecnu.edu.cn

Abstract. A synthetic social media data generator, namely BSMA-GEN is introduced in this demonstration. It can parallelly generate timeline structures of social media. The data generator is part of BSMA, a benchmark for analytical queries over social media data. Both its internal process and generated data are to be shown in the demonstration.

1 Introduction

Social media has become an important kind of source of many applications. Besides systems based on general purpose big data management or processing tools, specifically designed systems, such as the Little Engine[1] and Feed Frenzy[2], also exist. Since these systems diverge in design principles, benchmarking their performance and understanding their advantages become a problem. Efforts on designing such benchmarks include LinkBench[3], LDBC[4], and BSMA[5]. In this demonstration, we will show BSMA-GEN, the data generator of BSMA.

BSMA[1] is a benchmark for social media analytical queries, which contains a formal description of social media data, a set of social media analytics workloads, and a tool for measuring query processing performance. BSMA-GEN is the component for generating synthetic timeline structures when users need dataset with different volume and/or distribution compared with the real-life data.

Intuitively, a synthetic data generator should be *flexible* to generate *realistic* data *efficiently*. BSMA-GEN simulates the process of users' behavior on tweeting and retweeting. Operations of other social media services can be mapped to these terms. It provides configurable parameters to generate synthetic social media timelines with various distributions. It also has a parallel version for generating high-throughput tweet streams.

2 System Overview

The architecture of BSMA-GEN is shown in Fig. 1 (a). It contains a master that schedules workloads. On each slave, after recieving the workload, it outputs

---

[1] Available at: https://github.com/c3bd/BSMA

S.S. Bhowmick et al. (Eds.): DASFAA 2014, Part II, LNCS 8422, pp. 539–542, 2014.

and output items in the timeline. An item is a tuple $< t, c, u, f >$, in which t is the timestamp when the tweet is published, c is the content of the tweet, u is the author, and f is the *father* of the tweet, which can be *nil* for denoting the message is original, or a pointer to another tweet n, denoted by $n \leftarrow m$, meaning that m is a repost of n. Apparently, m's timestamp is always later than that of n, i.e. $t_m > t_n$.

BSMA-GEN determines the value of timestamp t, author u, and the pointer f online, based on the information from the social network and history. It utilizes two in-memory data structures, namely, the *followship network* (Fig. 1 (a)) and a *buffer* (Fig. 1 (b)) for recent tweets. The part of social network on each slave is assigned by the master, while the buffer is dynamically maintained.

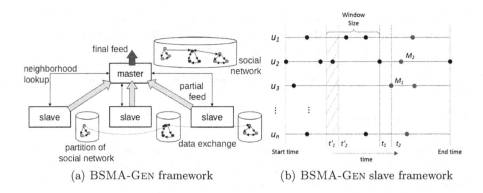

(a) BSMA-GEN framework   (b) BSMA-GEN slave framework

Fig. 1. The architecture of BSMA-GEN

There are two components in each slave for generating tweets.

Tweet Generation. This component determines values of timestamp t and author u. A nonhomogeneous Poisson process with changing intensity function is used to model each user.

Retweet Generation. This component determines values of the pointer f. The distribution of number of retweets, which is often a power-law distribution, is used, in combination of a time decay function, to choose tweets from the buffer as candidate values of f.

The parameters of the above models can be determined based on statistics over real-life data, or specified by users directly.

2.1   Parallel Execution

The performance bottleneck on a BSMA-GEN node is the querying of the social network. Thus, BSMA-GEN uses a master, that partitions the social network and assigns partitions to working nodes, i.e. slaves. Each slave is responsible for generating tweets posed by users in the partition. BSMA-GEN may use any graph partitioning algorithm. In current implementation, the fast unfolding algorithm[6] is used.

Most real-life social networks cannot be partitioned into relatively equal-size disconnected components. Thus, interactions between nodes with connected partitions is inevitable. In BSMA-GEN, a simple protocol is used. A slave looks up neighborhood information in master, and then retrieves history tweets from neighboring slave's buffer directly.

An asynchronized model and the *delayed update* strategy are used. While retrieving information from other slaves, the node continues to generate new feeds that do not need interactions with others. While the required information is received, the tweet depending on it is generated with a new delayed timestamp. Thus, the items in the feed are still ordered. Experiments show that this strategy does not affect the distribution of generated data.

3 Evaluation of the Data Generator

Part of the experimental results are shown in Fig. 2. The distributions of generated data are compared with those from S3G2, the data generator of LDBC[4], and from real-life Sina Weibo data, in Fig 2 (a) and (b). It is shown that BSMA-GEN may generate more realistic data compared with S3G2.

The scaling experiment is shown in Fig. 2 (c). It is shown that along with the increasing of slaves, the performance of BSMA-GEN increases almost linearly.

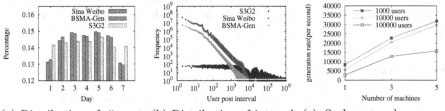

(a) Distribution of #tweets over time. (b) Distribution of intervals between tweets. (c) Scale-ups when more slaves are provided.

Fig. 2. Experimental results

4 Demonstration Outline

The demonstration will be conducted on a laptop, with threads running on different CPU cores generating tweet feeds parallelly. Our demonstration includes the following two parts:

BSMA-GEN **internals** We will show how BSMA-GEN generates timelines:
- We will show how the buffer is organized. The buffer contains recent tweets. The information and statistics of these tweets and their authors, as well as how the buffer is updated, will be shown.
- The social network, which is an important source for determination of retweeting, is to be shown in the demonstration. We will show the active part of the social network that are queried by BSMA-GEN. Furthermore, we will show how opinion leaders, who are users with many followers, generate hot tweets, that are retweeted many times, and affect users that are not following them.
- We will demonstrate the parallel execution of the threads. The partitioned social network, as well as the workloads of the threads, and the communication among them, are to be shown.

BSMA-GEN **monitor** We will show the output of BSMA-GEN monitor, which is the tool visualizes the performance and the synthetic data:
- The throughput of BSMA-GEN will be shown in the demonstration.
- The dynamics of retweet graphs of top retweeted tweets will be monitored and visualizing online, while the data are being generated.
- The distributions of different aspects of the generated data, including distributions over time, users, and tweets, are to be shown online, while BSMA-GEN is running. The baseline, which is the expected distribution based on parameter setting, will also be shown.

Acknowledgment. This work is partially supported by National Science Foundation of China under the grant number 61170086, National High-tech R&D Program (863 Program) under grant number 2012AA011003, and Innovation Program of Shanghai Municipal Education Commission under grant number 14ZZ045.

References

1. Pujol, J.M., Erramilli, V., Siganos, G., Yang, X., Laoutaris, N., Chhabra, P., Rodriguez, P.: The little engine(s) that could: Scaling online social networks. IEEE/ACM Trans. Netw. 20, 1162–1175 (2012)
2. Silberstein, A., Terrace, J., Cooper, B.F., Ramakrishnan, R.: Feeding frenzy: Selectively materializing users' event feeds. In: SIGMOD Conference, pp. 831–842 (2010)
3. Armstrong, T.G., Ponnekanti, V., Borthakur, D., Callaghan, M.: Linkbench: A database benchmark based on the facebook social graph. In: Ross, K.A., Srivastava, D., Papadias, D. (eds.) SIGMOD Conference, pp. 1185–1196. ACM (2013)
4. Boncz, P.A., Fundulaki, I., Gubichev, A., Larriba-Pey, J.L., Neumann, T.: The linked data benchmark council project. Datenbank-Spektrum 13, 121–129 (2013)
5. Ma, H., Wei, J., Qian, W., Yu, C., Xia, F., Zhou, A.: On benchmarking online social media analytical queries. In: GRADES, p. 10 (2013)
6. Blondel, V.D., Guillaume, J.L., Lambiotte, R., Lefebvre, E.: Fast unfolding of communities in large networks. Journal of Statistical Mechanics: Theory and Experiment 2008, P10008 (2008)

A Mobile Log Data Analysis System Based on Multidimensional Data Visualization[*]

Ting Liang[1], Yu Cao[2], Min Zhu[1,**], Baoyao Zhou[2], Mingzhao Li[1], and Qihong Gan[1]

[1] College of Computer Science, Sichuan University, Chengdu, China
[2] EMC Labs, Beijing, China

Abstract. The log data collected from mobile telecom services contains plenty of valuable information. The critical technical challenges to mobile log analysis include how to extract information from unformatted raw data, as well as how to visually represent and interact with the analysis results. In this paper, we demonstrate MobiLogViz, which seamlessly combines multidimensional data visualization and web usage mining techniques. By automatically processing the log data and providing coordinated views with various interactions, MobiLogViz effectively aids users in analyzing and exploring mobile log data.

Keywords: Mobile log, visualization, multidimensional.

1 Introduction

As smart phones connecting to Internet and running various applications become more and more pervasive, a vast amount of mobile log data is continuously generated. These mobile logs record comprehensive and concrete user behavioral information, and thus are of great value. Analyzing mobile logs, discovering valuable knowledge and discerning patterns hidden within the massive and multidimensional logs may help mobile service providers spot subtle trends, investigate phone usage and aggregated statistics, and then predict network traffic flow or refine their services. However, the log data are often semi-structured, dynamically generated, disordered, of high redundancy, noisy, and thus hard to process. Furthermore, their huge volume and strong temporal correlation also impose challenges to their processing.

It has two limitations to apply web usage mining techniques to extract information from mobile log data. On one hand, the mining process is usually not dynamically adjustable at runtime. On the other hand, the presentation of mining results is not intuitive and provides no user-friendly interface.

In this paper, we demonstrate MobiLogViz, an interactive visual mobile log analysis system, which seamlessly couples existing web usage mining techniques and multi-dimensional data visualization techniques. MobiLogViz enables analysts to investigate the mobile log data in a more exploratory and intuitionistic manner. In MobiLogViz, multidimensional attributes of logs, such as time, phone brand, website

---

[*] This work is a part of the joint research project with EMC Labs China (the Office of CTO, EMC Corporate), and is funded by EMC China Center of Excellence.
[**] Corresponding author.

S.S. Bhowmick et al. (Eds.): DASFAA 2014, Part II, LNCS 8422, pp. 543–546, 2014.

visited, browser used and operate system embedded, are encoded into visual components (e.g. color hue). As a result, the variation trends of dimensions and the relationships among them can be easily configured and fully understood.

2   System Overview

Following the visualization process referred in [1], we divide the overall mobile log analysis (shown in Figure 1) procedure into two consecutive phases.

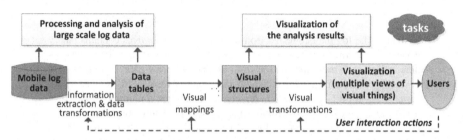

Fig. 1. The mobile log analysis overview

2.1   Phase 1: Processing and Analysis of Large Scale Log Data

In the processing phase, information is extracted from the raw data and structured into tables. The raw data exists in a collection of text files. Each file consists of series of records, each of which comes from the HTTP header and is separated by a special separator. A record usually starts with a timestamp and four IP addresses. In addition, it contains various header fields as "Get", "Host", "Accept", "User-Agent", "Accept-Encoding" and "Accept-Language", etc. In this demonstration, we focus on the fields "timestamp", "Host" and "User-Agent" and conduct the analysis tasks to be further discussed in section 3.The processing phase is carried out in the following steps.

STEP1: Detect if there is a newly arrived log file. If true, go to STEP2.

STEP2: Extract the header fields "Time", "Host" and "User-Agent".

STEP3: Extract time and website information from field "Time" and "Host".

STEP4: Extract mobile phone information (e.g. phone brand, operating system and browser used) from field "User-Agent".

STEP5: Insert the extracted information into a formatted table.

STEP6: Apply statistics functions to get summarizations along with time, including the summary of website hits for each website, the summary of the use of different browsers, operating system or phone brand, as well as the statistics of website hits on a specific operating system.

At last, we generate several statistical tables based on different attributes and their correlations. For an instance, table "Time-Host" contains the mainstream websites traffic every each hour. Table "Time-Host-OS" contains the correlations between mainstream websites and operating systems over the time. In the following part, we will take these two tables for example.

2.2   Phase 2: Visualization of the Analysis Results

Based on the output table generated in the previous phase and following the design guidelines of visualization, we designed MobiLogViz which consists of three visualization metaphors. ThemeRiver has been developed to visualize the time-varying information in log data. HeatMap has been enhanced to mainly convey the relationship between pairs of attributes and temporal distribution of individual attributes. And HistoMatrix has been used for correlation visualization.

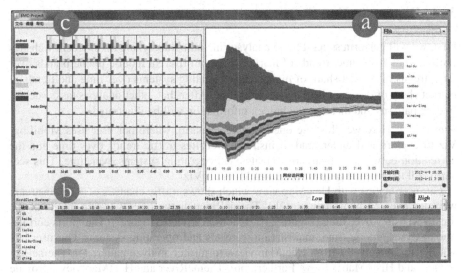

Fig. 2. User interface of MobiLogViz

The ThemeRiver view mainly represents the changes of website hits, the use of browser, operating system or phone brand along with time. Take website hits analysis as example, the website in table "Time-Host" is mapped to different colored horizontal layer, as illustrated in Figure 2a. Blue layer represents website qq.com while orange layer represents website sina.com. The vertical width of each layer at a particular time point indicates the value of website hits. A layer narrows (or widens) along with the time, means that the hits of corresponding website decreases (or increases).

In HeatMap view (Figure 2b), color is used to visually represent different value. We enhance HeatMap visualization by changing the mapping scheme from linear to exponential for each range of value. As shown in Figure 2b, each square represents the hits of a website in a particular time interval. The closer to the red of a square is, the larger the value is; and a blue square has the smallest value. When the mouse hovers over the squares, the concrete value will be displayed. Furthermore, user can select an item in drop-down box to switch HeatMap view.

HistoMatrix View is used to display correlations among attributes. Figure 2c shows its application on correlations among website, operating system and time. Each raw of the graph represents the relationship between a particular website and several operating systems. The histogram in each grid, which is discriminated by different colors, represents different types of operating systems. The height of a bar chart represents the value of website hits on a specific operating system in a time interval. When user selects some interested websites to explore their temporal hit distributions in HeatMap view, the view will update to represent the circumstances of the relevant operating systems and the website hits along with time.

In addition to three visualization views, our system also provides convenient interactions that enable users to further explore log data. The interactions include attributes selection, view switches, time range selection and color choices, etc.

3   Demonstration Scenarios

First, we will perform several trend analyses including analyzing mainstream websites traffic trends over time, trends of market share of different mobile phone brands over time, trends of market share of mainstream operating systems over time and trends of market share of different user agents over time. After that, other analysis tasks such as finding patterns and discovering hidden information will be illustrated.

Scenario 1. Here we illustrate one example scenario, where our user uses MobiLog-Viz to explore and understand mainstream websites traffic trends over time and the correlation between mainstream websites and operating systems over time. This scenario will touch upon major features of MobiLogViz.

Our user begins with HeatMap view which shows traffic changes of 50 mainstream websites over 24 hours (Figure 2b). ThemeRiver view presents traffic trends of the top 10 websites over time (Figure 2a). HistoMatrix view (Figure 2c) supports correlation analyses among time, mainstream websites and operating systems. Our user can select 1 to 10 websites from HeatMap view, to further analyze patterns in ThemeRiver view and HistoMatrix view. Furthermore, ThemeRiver and HeatMap views provide the analyses of historical data.

Scenario 2. Here we demonstrate the capability of MobiLogViz in real-time flow data analysis. After the new arrival of log files at every each hour, we extract useful information and structure them into tables. Then the HeatMap view, ThemeRiver view and HistoMatrix view in our system will update to show the latest 24 hours' data. Furthermore, the audience can choose one or more visualization views to focus. Both static and dynamic analyses of flow data are available.

Reference

1. Card, S.K., Mackinlay, J.D., et al. (eds.): Readings in information visualization: Using vision to think. Morgan Kaufmann Publishers Inc., San Francisco (1999)

Tutorials

Similarity-Based Analytics for Trajectory Data: Theory, Algorithms and Applications

Kai Zheng

The University of Queensland

The prevalence of GPS sensors and mobile devices has enabled tracking the movements of almost any kind of moving objects such as vehicles, humans and animals. As a result, in the past decade we have witnessed unprecedented increase of trajectory data both in volume and variety. With some attributes such as variable lengths, uncontrolled quality, high redundancy and uncertainty and so on, trajectory data challenge the traditional methodologies and practices in many research areas including data storage and indexing, data mining and analytics, information retrieve, etc. Trajectory data management has been attracting numerous research interests from both academia and industry due to its tremendous value and benefits in a variety of critical applications like traffic analysis, fleet management, trip planning, location-based recommendation, etc. In this tutorial, we will talk about the challenges, techniques and open problems with the focus on similarity-based analytics, the foundation of trajectory management, and covering a range of topics from fundamental theory, algorithms to advanced applications.

Outline

This tutorial consists of the following parts:

- Basic concept and background information
- Trajectory similarity measures
- Trajectory index
- Trajectory pattern mining

Some related work that will be covered in this tutorial are listed as follows:

References

1. Chen, Z., Shen, H.T., Zhou, X., Zheng, Y., Xie, X.: Searching trajectories by locations – An efficiency study. In: SIGMOD 2010 (2010)
2. Pfoster, D., Jensen, C.S., Yannis, T.: Novel approaches to the indexing of moving object trajectories. In: VLDB (2000)
3. Vlachos, M., Gunopulos, D., Kollios, G.: Discovering similar multidimensional trajectories. In: ICDE 2002 (2002)
4. Chen, L., Ozsu, M.T., Oria, V.: Robust and Fast Similarity Search for Moving Object Trajectories. In: SIGMOD 2005 (2005)
5. Lin, B., Su, J.: One Way Distance: For Shape Based Similarity Search of Moving Object Trajectories. Geoinformatica (2008)
6. Potamias, M., Patroumpas, K., Sellis, T.K.: Sampling trajectory streams with spatiotemporal criteria. In: SSDBM 2006 (2006)

S.S. Bhowmick et al. (Eds.): DASFAA 2014, Part II, LNCS 8422, pp. 549–550, 2014.

 7. Chakka, V.P., Everspaugh, A.C., Patel, J.M.: Indexing Large Trajectory Data Sets
    With SETI. In: CIDR 2003 (2003)
 8. Giannotti, F., Nanni, M., Pinelli, F., Pedreschi, D.: Trajectory pattern mining. In:
    SIGKDD, pp. 330–339 (2007)
 9. Mamoulis, N., Cao, H., Kollios, G., Hadjieleftheriou, M., Tao, Y., Cheung, D.W.:
    Mining, indexing, and querying historical spatiotemporal data. In: SIGKDD,
    pp. 236–245 (2004)
10. Zheng, Y., Zhang, L., Xie, X., Ma, W.-Y.: Mining interesting locations and travel
    sequences from gps trajectories. In: WWW, pp. 791–800 (2009)
11. Chen, Z., Shen, H.T., Zhou, X.: Discovering popular routes from trajectories. In:
    ICDE, pp. 900–911 (2011)
12. Zheng, K., Zheng, Y., Xie, X., Zhou, X.: Reducing Uncertainty of Low-Sampling-
    Rate Trajectories. In: ICDE 2012 (2012)
13. Wei, L.-Y., Zheng, Y., Peng, W.-C.: Constructing popular routes from uncertain
    trajectories. In: ACM SIGKDD, pp. 195–203 (2012)
14. Luo, W., Tan, H., Chen, L., Ni, L.M.: Finding Time Period-Based Most Frequent
    Path in Big Trajectory Data. In: SIGMOD 2013 (2013)

Graph Mining Approaches: From Main Memory to Map/Reduce

Sharma Chakravarthy

Information Technology Laboratory and CSE Department
University of Texas at Arlington, Texas
sharma@cse.uta.edu

1 Audience

Practitioners and professionals requiring up-to-date information on latest trends in newer forms of mining paradigms and how to apply these techniques for various applications, such as dealing with very large graph sizes, partitioning techniques, graph query answering etc. will benefit from this tutorial. The presenter has been working for over a decade on graph mining, scalability issues of graph mining, and its applications. Although graph mining itself has been around for a long while, it has come to the forefront due to its ability to make a difference in such domains as fraud monitoring and more recently analyzing very large social networks. Conventional mining techniques do not lend themselves to some of these applications as they cannot represent inherent structural relationships and exploit them during mining. We will present several graph mining approaches that have been proposed in the literature and new ones that are being developed. Practitioners will benefit from the practical nature of the topics and find the solutions presented applicable to problems they have encountered. Researchers will benefit from the issues that need to be addressed in one of the hot areas currently being revolutionized by increasing amounts of information available using large computing farms.

2 Tutorial Description

In this tutorial, we argue that graph mining techniques are extremely important and it is getting its share of attention recently as the sizes of graphs have exploded due to social networks and the need for their analusis. There is also renewed interest in answering graph queries in both exact ad approximate way due to the presence of large graph based such as Freebase and entity-based very large graphs. Most of the earlier mining approaches assume transactional and other forms of data. However, there are a large number of applications where the relationships among data objects are extremely important. For those applications, use of conventional approaches results in loss of information that will critically affect the knowledge that is discovered/inferred using traditional mining approaches. Mining techniques that preserve and exploit the domain characteristics is extremely important and graph mining is one such general purpose technique as it uses arbitrary graph representation using which complex relationships can be represented.

S.S. Bhowmick et al. (Eds.): DASFAA 2014, Part II, LNCS 8422, pp. 551–552, 2014.

Graph mining, as opposed to transaction mining (association rules, decision trees and others) is suitable for mining data with structural relationships. Graph mining is certainly appropriate for mining web graphs, social networks, and other graphs created by large data sets (e.g., author-citation, query/answer graphs from Question/Answer sites). The complex relationships that exist between entities that comprise these structures (for example, friendship relations, and question/answer relationships) can be faithfully represented using graph format. A graph representation comes across as a natural choice for representing complex relationships because it is relatively simple to visualize as well. The various associations between objects in a complex structure are easy to understand and represent graphically. Most importantly, the representation in graph format preserves the structural information of the original application which may be lost when the problem is translated/mapped to other representation schemes.

Graph mining aims at discovering interesting and repetitive patterns/ substructures within structural data. Relevant work in graph mining starts the Subdue substructure discovery system. Subdue discovers interesting and repetitive substructures in graph representations of data. It employs a beam search to discover interesting subgraphs and compresses the original graph with instances of the discovered substructures. The frequent subgraphs (FSG) approaches use *a priori* algorithms to structural data represented in the form of a labeled graph and finds frequent itemsets that correspond to recurring subgraphs in the input. Scalability has always an issue which has given rise to disk-based graph mining and database-oriented graph mining. More recently, Map/reduce, Pregal and partitioning of graphs have become new paradigms for further scaling graph mining to unprecedented sizes.

In this tutorial, we start with main memory graph mining algorithms, need for scalability and how database-oriented algorithms served that purpose. The tutorial will continue with some of the more recent approaches using map/reduce and other paradigms due to the explosion in the sizes of graphs coming from social networks. We compare these approaches with respect to the problems addressed, scalability, and how graph querying can use some of the techniques developed earlier. We present a few applications that have used graph mining techniques beneficially.

3   Significance

Mining has become an important technique for extracting higher forms of abstraction and knowledge from large amounts of connected data and as a result has become an enabling technology for information management. Information management is a challenging task as the amount of information is increasing dramatically. Graph mining techniques offer a unique advantage that conventional mining techniques do not offer. This is extremely significant as relationships among data are becoming more and more complex. This is an area that is witnessing renewed emphasis and we believe that the area has lots of potential for both research and practical applications.

This tutorial brings various aspects of mining, information management, applications, and recent trends in meeting the needs of knowledge discovery.

Crowdsourced Algorithms in Data Management

Dongwon Lee

College of Information Sciences and Technology
The Pennsylvania State University
University Park, PA 16802, USA
dongwon@psu.edu

Abstract. As a novel computation paradigm, crowdsourcing is being actively pursued in diverse academic disciplines. Within computer science, many sub-fields have embraced the concept of crowdsourcing with open arms and applied the concept to solve diverse challenging problems. Database community is no exception to this phenomenon and there have been many exciting new results using crowdsourcing appearing in recent database literature. This tutorial in particular seeks to cover state-of-the-art crowdsourced algorithms in data management. After gentle introduction on the concept of crowdsourcing, this tutorial provides the overall landscape of crowdsourced database research, with the focus on the latest crowdsourced algorithms that extend conventional database algorithms (e.g., count, sort, match, and search).

1 Introduction

As the notion of "crowdsourcing" emerges as a novel computing paradigm, the database community has actively adopted it. By utilizing the human-based computation within DBMS, researchers have proposed novel database systems (*e.g.,* Deco [1], CrowdDB [2], Qurk [3], CDAS [4]) that are particularly effective in addressing the computations hard for machines but easier for humans. At a more micro-level, in addition, people have developed novel human-powered database operations by extending existing techniques. Often, human-powered algorithms attempt to optimize (one of) three objectives–i.e., *monetary cost, latency,* and *quality of answers* [5]. Among many recent advancements in such crowdsourced algorithms, in this tutorial, we focused on representative ones, categorized with respect to the four database operations as follows:

1. *Count.* Being able to count the number of items in a collection with the help of humans has been studied in [6,7,8].
2. *Sort.* In sorting items using human powers, one can use conventional techniques such as tournament sort [9] or novel sorting [10]. Beyond sorting, more advanced sort-related problems such as crowd-enabled max queries [11,5], top-k queries [12,13], and skyline queries [14] are considered.
3. *Match.* This covers the crowdsourced database join [10,15] as well as human-powered entity resolution [16,17] techniques.

S.S. Bhowmick et al. (Eds.): DASFAA 2014, Part II, LNCS 8422, pp. 553–554, 2014.

4. *Search.* How to filter, search, and retrieve a set of constrained items fast and accurately with the help of human judges is an interesting problem. Recent advancements such as [18,19,20,21] are covered in this tutorial.

References

1. Parameswaran, A.G., Park, H., Garcia-Molina, H., Polyzotis, N., Widom, J.: Deco: Declarative crowdsourcing. In: CIKM, pp. 1203–1212 (2012)
2. Franklin, M.J., Kossmann, D., Kraska, T., Ramesh, S., Xin, R.: CrowdDB: Answering queries with crowdsourcing. In: SIGMOD, pp. 61–72 (2011)
3. Marcus, A., Wu, E., Karger, D.R., Madden, S., Miller, R.C.: Demonstration of Qurk: A query processor for humanoperators. In: SIGMOD, pp. 1315–1318 (2011)
4. Liu, X., Lu, M., Ooi, B.C., Shen, Y., Wu, S., Zhang, M.: CDAS: A crowdsourcing data analytics system. PVLDB 5(10), 1040–1051 (2012)
5. Venetis, P., Garcia-Molina, H., Huang, K., Polyzotis, N.: Max algorithms in crowdsourcing environments. In: WWW, pp. 989–998 (2012)
6. Marcus, A., Karger, D., Madden, S., Miller, R., Oh, S.: Counting with the crowd. PVLDB 6(2), 109–120 (2012)
7. Trushkowsky, B., Kraska, T., Franklin, M.J., Sarkar, P.: Crowdsourced enumeration queries. In: ICDE, pp. 673–684 (2013)
8. Gao, J., Liu, X., Ooi, B.C., Wang, H., Chen, G.: An online cost sensitive decision-making method in crowdsourcing systems. In: SIGMOD, pp. 217–228 (2013)
9. Cormen, T.H., Leiserson, C.E., Rivest, R.L., Stein, C.: Introduction to Algorithms, 3rd edn. MIT Press (2009)
10. Marcus, A., Wu, E., Karger, D.R., Madden, S., Miller, R.C.: Human-powered sorts and joins. PVLDB 5(1), 13–24 (2011)
11. Guo, S., Parameswaran, A.G., Garcia-Molina, H.: So who won?: Dynamic max discovery with the crowd. In: SIGMOD, pp. 385–396 (2012)
12. Davidson, S.B., Khanna, S., Milo, T., Roy, S.: Using the crowd for top-k and group-by queries. In: ICDT (2013)
13. Polychronopoulos, V., de Alfaro, L., Davis, J., Garcia-Molina, H., Polyzotis, N.: Human-powered top-k lists. In: WebDB (2013)
14. Lofi, C., Maarry, K.E., Balke, W.T.: Skyline queries in crowd-enabled databases. In: EDBT (2013)
15. Wang, J., Li, G., Kraska, T., Franklin, M.J., Feng, J.: Leveraging transitive relations for crowdsourced joins. In: SIGMOD, pp. 229–240 (2013)
16. Wang, J., Kraska, T., Franklin, M.J., Feng, J.: CrowdER: Crowdsourcing entity resolution. PVLDB 5(11), 1483–1494 (2012)
17. Bellare, K., Iyengar, S., Parameswaran, A.G., Rastogi, V.: Active sampling for entity matching. In: KDD, pp. 1131–1139 (2012)
18. Parameswaran, A.G., Garcia-Molina, H., Park, H., Polyzotis, N., Ramesh, A., Widom, J.: Crowdscreen: Algorithms for filtering data with humans. In: SIGMOD, pp. 361–372 (2012)
19. Yan, T., Kumar, V., Ganesan, D.: CrowdSearch: Exploiting crowds for accurate real-time image search on mobile phones. In: MobiSys, pp. 77–90 (2010)
20. Bozzon, A., Brambilla, M., Ceri, S.: Answering search queries with crowdsearcher. In: WWW, pp. 1009–1018 (2012)
21. Amsterdamer, Y., Grossman, Y., Milo, T., Senellart, P.: Crowd mining. In: SIGMOD, pp. 241–252 (2013)

Author Index